Mental Health and Illness Worldwide

Series Editors
Norman Sartorius
Mental Health Programmes (AMH)
Association for the Improvement
Geneva, Switzerland

Ee Heok Kua
National University of Singapore
Singapore, Singapore

Most books on mental health and illness are published for readers in North America and Europe, and not much is known about psychiatric practice, services and research in Asia, Africa, and South America. This series will include contributions of clinicians and researchers worldwide. Each volume will cover broad issues including epidemiology, cross-cultural comparison, clinical research, stigma of mental illness, cultural issues in mental healthcare, health economics, innovative services, preventive programs and health service outcome research. The volumes will find a wide readership among psychiatrists, psychologists, sociologists, health policy makers, social workers, health economists, anthropologists and philosophers. It will provide the readers a broader perspective of mental health and illness worldwide and also future research initiatives.

More information about this series at http://www.springer.com/series/14178

Niels Okkels • Christina Blanner Kristiansen
Povl Munk-Jørgensen
Editors

Mental Health and Illness in the City

With a special contribution by D. Goldberg and G. Thornicroft

With 20 Figures and 16 Tables

Springer

Editors
Niels Okkels
Department of Affective Disorders
Aarhus University Hospital Risskov
Risskov, Denmark

Christina Blanner Kristiansen
Department of Psychiatry
Odense University Hospital
Odense, Denmark

Povl Munk-Jørgensen
Department of Psychiatry
Odense University Hospital
Odense, Denmark

ISBN 978-981-10-2325-5 ISBN 978-981-10-2327-9 (eBook)
ISBN 978-981-10-2326-2 (print and electronic bundle)
https://doi.org/10.1007/978-981-10-2327-9

Library of Congress Control Number: 2017945746

Printed on acid-free paper

This Springer imprint is published by Springer Nature
The registered company is Springer Nature Singapore Pte Ltd.
The registered company address is: 152 Beach Road, #21-01/04 Gateway East, Singapore 189721, Singapore

Series Preface

Psychiatry lives exciting and challenging times. Advances of knowledge stemming from basic sciences and epidemiological and clinical research have provided a better understanding of the etiopathogenesis, psychopathology, and natural history of mental disorders. Improved methods of treatment have changed clinical practice and prolonged the life of people with mental illness. Economic consideration and the emphasis on human rights of people with mental illness made a profound impact on the way in which psychiatry is to be practiced.

Regrettably, however, psychiatry is not practiced in the same manner around the world. Undergraduate and postgraduate education in psychiatry varies in content and duration from country to country. Psychiatrists use different doses of medication for the same disorders. The systems of care for people with mental illnesses differ in the organization and content of their interventions. Support to scientific investigations of matters related to psychiatry fluctuates and in many countries amounts to very little. Information about the function of psychiatric services varies in quantity and quality.

This series of seven books on *Mental Health and Illnesses Worldwide* aims to help in reducing these differences and facilitate international collaboration in psychiatry. We have invited top experts from different countries to edit the volumes, and they have in turn selected authors from different parts of the world. We have also decided to approach the body of psychiatry from a public health and epidemiological perspective rather than have books dealing with different groups of diseases. The series includes books examining and presenting knowledge assembled according to social and public health variables – gender, urbanity, migratory status, age, and education. Each of the volume has adopted a wide perspective and included chapters based on knowledge stemming from epidemiology, on results of the investigation of cultural issues, on the best of psychopathology, on the results of the investigation of biological factors, mental health care and its innovations, health economics, and experience gained in preventive programs. The volume editors have agreed to aim at producing volumes marked by the balance of information and knowledge from basic social and behavioral sciences and from clinical practice.

The seven volumes of this opus are:

1. Mental Health and Illness in the City
   Editors (Denmark): Niels Okkels, Christina Blanner Kristiansen, Povl Munk-Jørgensen

2. Mental Health and Illness of the Elderly
   Editors: Helen Chiu (Hong Kong) and Ken Shulman (Canada)
3. Mental Health and Illness of Women
   Editors: Prabha Chandra (India) and Helen Herrman and Jane Fisher (Australia)
4. Mental Health and Illness in Rural Communities
   Editor: S. Chaturvedi (India)
5. Mental Health and Illness in Migrant Populations
   Editors: Driss Moussaoui (Africa), Dinesh Bhugra (United Kingdom), and Antonio Ventriglio (Italy)
6. Mental Health and Illness of Children and Adolescents
   Editors: Eric Taylor (United Kingdom), John Wong (Singapore), and Frank Verhulst (Netherlands)
7. Education About Mental Health and Illness
   Editors: Marc Herman (Belgium), Tan Chay Hoon (Singapore), and Edmond Pi (USA)

We were delighted to see that the volume editors have succeeded in recruiting outstandingly knowledgeable authors for the chapters of their books. Most of them have received worldwide recognition for their contributions in their fields of specialization, and all of them have written their texts with authority and excellent judgment concerning the materials to be included.

We believe that this series of books demonstrates the importance and value of interdisciplinary and international collaboration and that it will provide readers a global perspective of mental health and mental illness. We also hope that it will help to make our discipline more homogenous and bring its practitioners worldwide closer together in the pursuit of helping people with mental illness worldwide.

We wish to express our profound gratitude to Niels Okkels, Christina Blanner Kristiansen, and Povl Munk-Jorgensen, the editors of this volume on *Mental Health and Illness in the City* – the first of the series – for their commitment and dedication to this excellent work.

Norman Sartorius
Kua Ee Heok

Volume Preface

In this book, we give you different perspectives on urban mental health from all around the world. Authors from different kinds of cities, both regarding size, geography, culture, and income, all present important issues in mental health and illness. However – and this is most important – they do not merely give you the usual summary of problems that researchers often tend to. They also give you their perspectives and suggestions for solutions.

The biggest city in Denmark is Copenhagen with about one million inhabitants. To many people in Asia, where several megacities have more than ten millions of inhabitants – more inhabitants than many smaller countries – this would probably be considered a town. Despite the differences in sizes and geography, however, the cities of the world have much in common. From the person living in the smallest European provincial town to his counterpart in an Asian megacity, city living has its pros and cons. The endless possibilities of the city come with a prize: Long opening hours in shops and stores offer nonstop shopping, but also long working hours for the employees. Crowded multicultural neighborhoods make the city an inspiring cultural melting pot, but carry the risk of marginalizing minority groups. Larger cities offer impressive public transportation networks, but at the same time a massive intruding and destructive traffic, resulting in both air and noise pollution as well as causing a stressful environment for both pedestrians and those behind the wheels. Though big reports and bold statements from major political actors and international organizations are obviously necessary steps for actions to be taken it seems that these are too general – they are often elusive and lack "what-can-actually-be-done" opportunities. As such, the motivation for the book is both to explore how city living affects mental health, but at the same provide real-world solutions to the problems we may find.

Today, more than half of the world's population live in cities. The ongoing urbanization occurs in parallel with an increase in the prevalence of mental illness. The question is then, if urban living may be associated with mental illness? In the book, we first present the historical evidence for an urban-rural difference in mental illness. We then explore the city environment and investigate a sample of challenges that are both important to mental health and frequently related to city living, such as single person households, traffic, noise, high rates of drug use and crime, homelessness, and limited access to green areas. These are all existing and perhaps even increasing challenges despite many politicians in power – and in opposition – are aware of the harmful effects, also on mental health.

On the positive side, the city may offer opportunities for the prevention and treatment of mental illness that rural areas cannot. One such advantage lies within the organization of mental health services. In the middle section of the book, we therefore look at the mental health services of major cities and how these have developed over the past decades. As it turns out, the recent deinstitutionalization and growth of community mental health services has changed psychiatric services considerably – both for better and for worse. We highlight a few interesting initiatives that accommodate the current challenges, such as new model to find housing for homeless people. The final section of the book is dedicated to chapters we find to be particularly visionary, regarding real-life solutions aiming to improve mental health in the city.

In 1996, professors David Goldberg and Graham Thornicroft published the book *Mental Health in Our Future Cities*. In the book, authors from megacities all around the world were invited to describe the organization of their local urban mental health services. The cities were chosen because they had published interesting work on the past and recent challenge of deinstitutionalization.

Now, 20 years later, Goldberg and Thornicroft have invited all the same authors to do a follow-up on the changes and developments in the past two decades. In their special contribution "Mental Health Care in 11 Major Cities Over Two Decades," chapter 9 in this book, they give an overview of the current status and challenges of mental health services in a series of large cities from all the world's continents. Their conclusion is interesting and important to all with interest in mental health care.

Throughout the book you will find opinions of what could – and should – be done toward the challenges presented in each of the chapters. Thus, it is our hope that by the end of reading the book, you will not only be more aware of the major problems of mental health and city living – you will also be encouraged that actions *can* be taken, that there *are* actual possibilities and solutions, and that we *can* each participate in improving mental health in the city.

The editors
Niels Okkels
Christina Blanner Kristiansen
Povl Munk-Jørgensen

Contents

Series Editors Biography

Professor Norman Sartorius, M.D., M.A., D.P.M., Ph.D., FRC. Psych, obtained his M.D. in Zagreb (Croatia). He specialized in neurology and psychiatry and subsequently obtained a Master's Degree and a Doctorate in Psychology. He carried out clinical work and research and taught at graduate and postgraduate levels at the University of Zagreb, at the Institute of Psychiatry in London, at the University of Geneva, and elsewhere. Professor Sartorius joined the World Health Organization (WHO) in 1967 and soon assumed charge of the program of epidemiology and social psychiatry. In 1977, Professor Sartorius was appointed Director of the Division of Mental Health of WHO. He was the Principal Investigator of several major international studies on schizophrenia, depression, and of mental and neurological disorders. In 1993, Professor Sartorius was elected President of the World Psychiatric Association (WPA) and served as President-elect and then President until August 1999, after which he was elected President of the European Psychiatric Association. Professor Sartorius is currently the President of the Association for the Improvement of Mental Health Programmes and he is a member of the Geneva Prize Foundation, having been its President from 2004 to 2008. Professor Sartorius holds professorial appointments at universities in different countries including China, UK, and USA.

Professor Sartorius has published more than 400 articles in scientific journals, authored or coauthored 12 books, and edited more than 80 others. He is the coeditor of three scientific journals and is a member of editorial and advisory boards of many scientific journals. Professor Sartorius is also a corresponding member and fellow of a large number of international organizations and advisory boards. He has several honorary doctorates and is a member of Academies of Science and of Medicine in different countries. He speaks Croatian, English, French, German, Russian, and Spanish.

Dr. Kua Ee Heok is the Tan Geok Yin Professor of Psychiatry and Neuroscience at the National University of Singapore (NUS) and Senior Consultant Psychiatrist at the National University Hospital, Singapore.

He was trained as a doctor at the University of Malaya and received postgraduate training in psychiatry at Oxford University and geriatric psychiatry at Harvard University.

A member of the World Health Organization team for the global study of dementia, he is the previous Head of the Department of Psychological Medicine and Vice-Dean, Faculty of Medicine, at NUS, and the Chief Executive Officer and Medical Director at the Institute of Mental Health, Singapore.

His research interest includes depression, dementia, and alcoholism, and he has written 23 books on psychiatry, aging, and addiction. A novel he wrote, *Listening to Letter from America*, is used in a module on anthropology at Harvard University.

The former President of the Pacific-Rim College of Psychiatrists and President of the Gerontological Society of Singapore, he was Editor of the *Singapore Medical Journal* and *Asia-Pacific Psychiatry* journal.

About the Editors

The three editors, **Niels Okkels**, **Christina Blanner Kristiansen**, and **Povl Munk-Jørgensen**, all M.D.'s, have since 2011 established themselves as a well-functioning team in psychiatric research, education, and publishing.

Povl Munk-Jørgensen represents a long-lasting international experience in academic psychiatry. He has been the Editor of the *Acta Psychiatrica Scandinavica* for 18 years and President of the European Psychiatric Association Section Epidemiology and Social Psychiatry for 6 years. Furthermore, he has both been the author and editor on a number of international books and book chapters on both clinical and academic psychiatry. His main interest is recruiting and education on an eliterian level. Niels Okkels and Christina Blanner Kristiansen represent the paradigmatic shift in psychiatry toward an early entrance into academic psychiatry seen these years. Both started doing research very early during medical school, both focusing on clinical relevant topics. Both of them already have solid experiences within academic psychiatry from both peer-reviewed articles and debate papers as well as authoring textbook chapters. They have already both developed strong international networks: Christina Blanner Kristiansens's Ph. D. study involves three European university clinics, and Niels Okkels has gained international editorial experiences from writing both a Scandinavian and an English Psychiatric e-Textbook, as well as from editorial board membership with the *Acta Psychiatrica Scandinavica*.

In 2012, Niels Okkels and Povl Munk-Jørgensen established the Psychiatric Research Academy, inviting the most gifted and interested medical students to be trained in practical clinical research on an eliterian basis. Christina Blanner Kristiansen is a member of the academy's steering committee. Niels Okkels is our team leader in editing the present book.

Contributors

Maryam Abbasinejad Department for Mental Health and Substance Abuse, Roozbeh Hospital, Tehran, University of Medical Sciences, Tehran, Iran

Lúcia Abelha Institute of Studies in Collective Health, Federal University of Rio de Janeiro, Rio de Janeiro, Brazil

Francesco Amaddeo Department of Neurosciences, Biomedicine and Movement Sciences, University of Verona, Verona, Italy

Gavin Andrews CRUfAD, UNSW at St Vincent's Hospital, Darlinghurst, NSW, Australia

Dinarte Ballester Psychiatry Service, Hospital Mãe de Deus, Porto Alegre, Brazil

Stephen T. Baron Behavioral System Baltimore, Inc. (BHSB), Baltimore, MD, USA

Stuart Bell Oxford Health NHS Foundation Trust, Northampton, UK

Dinesh Bhugra Institute of Psychiatry, Psychology and Neuroscience, King's College London, London, UK

Ingrid Bremer Department of International Health, Georgetown University, Washington, DC, USA

Ellis Busnello Universidade Federal do Rio Grande do Sul, Porto Alegre, Brazil

Marcel Buster Department of Epidemiology and Health Promotion, Public Health Service, Amsterdam, The Netherlands

Juan Chen Department of Applied Social Sciences, The Hong Kong Polytechnic University, Hung Hom, Kowloon, Hong Kong SAR, China

Jorge Mauricio Cuartas Arias Research Group in Psychiatry, Department of Psychiatry, Faculty of Medicine, Universidad de Antioquia, Medellin, Colombia

Faculty of Psychology, Universidad de San Buenaventura, Medellin, Colombia

Ana M. Díaz Zuluaga Research Group in Psychiatry, Department of Psychiatry, Faculty of Medicine, Universidad de Antioquia, Medellin, Colombia

Rubens Dantas Fundação do ABC, Santo André, SP, Brazil

Joana Corrêa de Magalhães Narvaez Laboratório de Psiquiatria Molecular, National Institute for Translational Medicine, Hospital de Clínicas de Porto Alegre, Federal University of Rio Grande do Sul, Porto Alegre, RS, Brazil

Center for Drug and Alcohol Research, Hospital de Clínicas de Porto Alegre, Federal University of Rio Grande do Sul, Porto Alegre, RS, Brazil

Jack J. M. Dekker Arkin Mental Health Care, Research Department, Amsterdam, The Netherlands

Jennifer Dykxhoorn PsyLife Research Group, Division of Psychiatry, Faculty of Brain Sciences, UCL, London, UK

Klement Dymi Psychiatric Centre Ballerup, Copenhagen University Hospital, Copenhagen, Denmark

Tarik Endale Department of International Health, Georgetown University, Washington, DC, USA

Caroline Galli Moreira Center for Drug and Alcohol Research, Hospital de Clínicas de Porto Alegre, Federal University of Rio Grande do Sul, Porto Alegre, RS, Brazil

Brasilian Psychiatric Association, Florianópolis, SC, Brazil

David Goldberg Institute of Psychiatry, King's College London, London, UK

Alfredo Gutierrez Carrizo Artech, Tucumán, Argentina

Ahmad Hajebi Department of Psychiatry, Tehran Psychiatry Institute, University of Medical Sciences, Tehran, Iran

Denisse Jaen-Varas Department of Psychiatry, Universidade Federal de São Paulo, São Paulo, Brazil

Mathew James Mental Health Service, St Vincent's Hospital, Sydney, Australia

Marjia Jannati Department of International Health, Georgetown University, Washington, DC, USA

Sean Z. Kaliski Department of Psychiatry and Mental Health, University of Cape Town, Cape Town, South Africa

Shigenobu Kanba Department of Neuropsychiatry, Graduate School of Medical Sciences, Kyushu University, Fukuoka, Japan

Marianne Kastrup Psychiatric Centre Ballerup, Copenhagen University Hospital, Copenhagen, Denmark

Takahiro A. Kato Department of Neuropsychiatry, Graduate School of Medical Sciences, Kyushu University, Fukuoka, Japan

James B. Kirkbride PsyLife Research Group, Division of Psychiatry, Faculty of Brain Sciences, UCL, London, UK

Christina Blanner Kristiansen Department of Psychiatry, Odense University Hospital, Odense, Denmark

Ee Heok Kua Department of Psychological Medicine, National University Hospital, National University of Singapore, Singapore, Singapore

Carlos López-Jaramillo Research Group in Psychiatry, Department of Psychiatry, Faculty of Medicine, Universidad de Antioquia, Medellin, Colombia

David LeCount Field Faculty Associate School of Social Work, University of Wisconsin, Madison, USA

Loiva dos Santos Leite Municipal Health Authority, Porto Alegre, Brazil

Giovanni Marcos Lovisi Institute of Studies in Collective Health, Federal University of Rio de Janeiro, Rio de Janeiro, Brazil

Eric Macnaughton Department of Psychology, Wilfrid Laurier University, Waterloo, ON, Canada

Jair de Jesus Mari Department of Psychiatry, Universidade Federal de São Paulo, São Paulo, Brazil

Layla McCay Centre for Urban Design and Mental Health, London, UK

Department of Global Health Entrepreneurship, Tokyo Medical and Dental University, Tokyo, Japan

Department of International Health, Georgetown University, Washington, DC, USA

Ahmad Mohit Tehran University of Medical Sciences and Health Services, Tehran, Iran

Rubén Alvarado Muñoz Mental Health Unit, School of Public Health, Faculty of Medicine, University of Chile, Santiago, Chile

Povl Munk-Jørgensen Department of Psychiatry, Odense University Hospital, Odense, Denmark

R. Srinivasa Murthy National Institute of Mental Health and Neurosciences (NIMHANS), Bangalore, India

Shinfuku Naotaka International Center for Medical Research and Treatment, Kobe University, Kobe, Japan

Tarek A. Okasha Institute of Psychiatry, Faculty of Medicine, Ain Shams University, Cairo, Egypt

Niels Okkels Department of Affective Disorders, Aarhus University Hospital Risskov, Risskov, Denmark

Abbas Ostadtaghizaddeh Department of Disaster and Public Health, School of Public Health, Tehran University of Medical Sciences, Tehran, Iran

Soumitra Pathare Centre for Mental Health Law and Policy, Indian Law Society, Pune, India

Lesley Jane Robertson Department of Psychiatry, School of Clinical Medicine, Faculty of Health Sciences, University of the Witwatersrand, Johannesburg, Gauteng, South Africa

Sedibeng District Health Services, Johannesburg, Gauteng, South Africa

Fernando Rosa da Rosa Social Service for the Industry (SESI-RS), Porto Alegre, RS, Brazil

Sara Jane Escouto dos Santos Municipal Health Authority, Porto Alegre, Brazil

Norman Sartorius Association for the Improvement of Mental Health Programmes (AMH), Geneva, Switzerland

Sara Schilling Mental Health Unit, School of Public Health, Faculty of Medicine, University of Chile, Santiago, Chile

Naotaka Shinfuku Department of Social Welfare, School of Human Sciences, Seinan Gakuin University, Fukuoka, Japan

Edward Shorter History of Medicine Program, Faculty of Medicine, University of Toronto, Toronto, ON, Canada

Angelia Sia National Parks Board, Ministry of National Development Centre for Urban Greenery and Ecology Research, Singapore, Singapore

Francesca Solmi PsyLife Research Group, Division of Psychiatry, Faculty of Brain Sciences, UCL, London, UK

Carolyn Stephens UCL Institute of Health Equity, London, UK

London School of Hygiene and Tropical Medicine, London, UK

Universidad Nacional de Tucumán, Tucumán, Argentina

Christopher Paul Szabo Department of Psychiatry, School of Clinical Medicine, Faculty of Health Sciences, University of the Witwatersrand, Johannesburg, Gauteng, South Africa

Charlotte Maxeke Johannesburg Academic Hospital, Johannesburg, Gauteng, South Africa

Crista Taylor Behavioral System Baltimore, Inc. (BHSB), Baltimore, MD, USA

Jan Theunissen GGZ in Geest Research Department, Amsterdam, The Netherlands

Graham Thornicroft Institute of Psychiatry Psychology and Neuroscience of King's College, London, UK

Ana Cristina Tietzmann Medical Residency Program in Psychiatry – Universidade Federal de Ciências da Saùde de Porto Alegre, Porto Alegre, Brazil

Sam Tsemberis Pathways Housing First Institute and Department of Psychiatry, Columbia University Medical Center, New York, NY, USA

Wilco Tuinbreijer Department of Public Mental Health Care, Municipal Health Service, Amsterdam, The Netherlands

Antonio Ventriglio Department of Mental Health, Department of Clinical and Experimental Medicine, University of Foggia, Foggia, Italy

Kathleen Westcoat Behavioral System Baltimore, Inc. (BHSB), Baltimore, MD, USA

Matty A. S. de Witt Department of Epidemiology and Health Promotion, Public Health Service, Amsterdam, The Netherlands

Jihyun Yi Department of International Health, Georgetown University, Washington, DC, USA

Carolina Ziebold Jorquera Department of Psychiatry, Universidade Federal de São Paulo, São Paulo, Brazil

Review of Urban Mental Health

1

Niels Okkels, Christina Blanner Kristiansen, and
Povl Munk-Jørgensen

Abstract

Today, more than half the world's population lives in cities. We know very little, however, about how city living affects mental health. There are aspects of the urban environment that may be harmful. One such example is traffic. A recent review found that living in close proximity to a major road increases the risk of developing dementia. Another example is artificial light at night that seem to alter the sleep-wake cycle and negatively affect sleep quality and mood. Lack of access to green areas may also contribute to the current high rates of mental illness.

After presenting the potential problems of city living, we turn to look at the solutions. There is great potential in city planning, that is, how we incorporate our knowledge of green areas, noise, and aesthetics in the construction of neighborhoods and streets. Such initiatives, however, require political vision, and it is of outmost importance that those in power are aware of urban mental health if we wish to improve mental health in our future cities.

Keywords

Urban health • Urban population • Urban planning • Cities • Rural health • Mental health • Mental illness • Mental disorders

Contents

N. Okkels (✉)
Department of Affective Disorders, Aarhus University Hospital Risskov, Risskov, Denmark
e-mail: niels.okkels@clin.au.dk

C.B. Kristiansen • P. Munk-Jørgensen
Department of Psychiatry, Odense University Hospital, Odense, Denmark
e-mail: christinablanner@gmail.com; d115975@dadlnet.dk

N. Okkels et al. (eds.), *Mental Health and Illness in the City*, Mental Health and Illness Worldwide, https://doi.org/10.1007/978-981-10-2327-9_1

> "There is a need to identify the psychologically harmful elements of urban life in the hope that cities can be made happier and healthier places in which to live." Glyn Lewis, 1994 (Lewis and Booth 1994)

Introduction

Urbanization is continuously increasing throughout the world. Today more than half of the world's population lives in cities, and conservative projections say that by 2030, this number will double (UN-Habitat 2016) (Fig. 1). The most rapid inclines are seen in the developing parts of the world (Box 1). In Africa, the rate of urbanization is currently 11 times faster compared to Europe (UN-Habitat 2016). On the positive side, urbanization is associated to growth, socioeconomic transformation, and increased wealth. Unfortunately, urbanization in its current form is

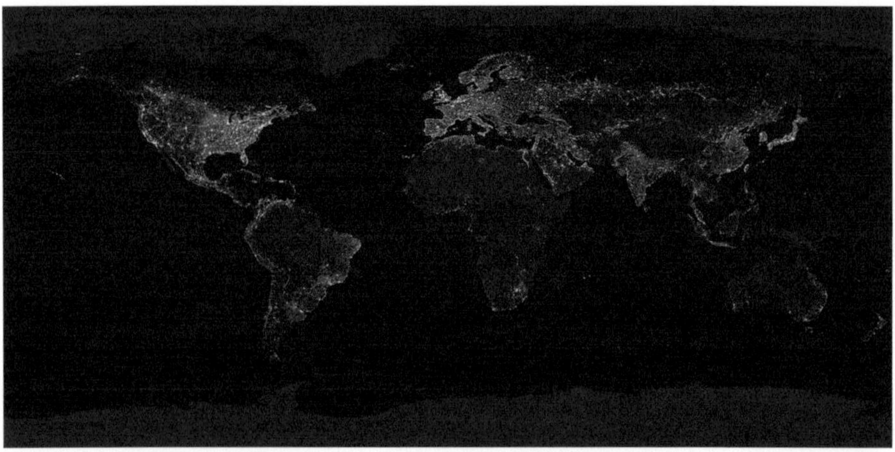

Fig. 1 World lights by night (Source: https://en.wikipedia.org/wiki/File:Earthlights_dmsp_1994–1995.jpg)

highly unsustainable, and many of the world's major problems are closely related to urbanization (UN-Habitat 2016).

Box 1 Highlights on Urbanization

- All regions of the world are expected to become more urbanized, with the largest increase in Asia and Africa.
- Globally 54% of the world population live in cities today. By 2050 it is expected that two-thirds of the world's population will live in cities.
- Tokyo (Japan) is the world's largest city with 38 million inhabitants, followed by Delhi (India) with 25 million, Shanghai (China) with 23 million and then Mexico City (Mexico), Mumbai (India), and Sao Paulo (Brazil) with each approximately 21 million.
- By 2050 there will be 2.5 billion more citizens. India is expected to add about 400 million, China 300 million, and Nigeria 200 million.
- The rural population of the world is today 3.4 billion and expected to peak within a few years and thereafter decline gradually.

Source: https://esa.un.org/unpd/wup/publications/files/wup2014-highlights.Pdf World Urbanization trends 2014

The challenges of urbanization are global and include rising inequalities, insecurity, and housing problems. The number of people living in informal settlements, such as slums, is increasing (UN-Habitat 2016). These areas present major challenges in terms of sustainable urbanization and access to basic urban services such as sanitation. In terms of mental health, informal settlements are problematic because of their high levels of social problems including violence, drug trafficking and extreme poverty (▶ Chap. 6, "Crime, Illegal Drugs, and Mental Health" by Narvaez J., in this book). An example of an urban area afflicted by these issues is the *favelas* of Rio de Jainero in Brazil (▶ Chap. 14, "Prospects and Challenges of Mental Health Services in a Major City in Latin America: Rio de Janeiro" by Lovisi G., in this book).

In China, urban inequality is even further exacerbated by their local household registration, *the hukou system* (UN-Habitat 2016). In this system a person's hukou status defines the right of access to urban public services such as employment, education, and healthcare. The mental health of many citizens is affected since they live in the urban areas without having urban hukou status and consequently have no rights to the urban public services (▶ Chap. 10, "Migration and Urban Mental Health in China" by Chen J., in this book).

For as long as psychiatry has existed as a medical specialty, scholars have speculated that city environments may be "toxic" to mental health and that more people with mental illness live in cities compared to the countryside (▶ Chap. 2, "History of Urban Mental Illness" by Shorter, E., in this book). The common notion is that urban environments are harmful because they are noisy and burdened by pollution, traffic, prostitution, narcotics, and homelessness, whereas the rural areas offer access to nature, peace, and rest.

Following this logic, urban environments should be improved by making them more "rural" and build more parks, make car-free zones, and reduce noise and pollution. But is it really that simple?

First, we must take a step back and ask which mental illnesses that are more – or less – common in the city compared to the countryside. The idea that mental illnesses are more common in the cities may be changing. In a survey from the United States, Lara Robinson concludes that rural children are in fact experiencing financial difficulties and feel unsafe and isolated, and overall have more mental, behavioral, and developmental problems (Robinson et al. 2017). Maybe modern cities (compared to the old days) actually have more protective factors. Is it meaningful then, to ask if cities favor mental illness? "Likely no," historian Edward Shorter writes. "Some cities do, and other don't. It all depends on the balance of risk and protective factors that prevails in the particular place at that particular time, for cities and countrysides alike."

A central question is if – and how – city living affects our mental health. To answer this question, we will take a critical standpoint and investigate the classical notion that there are more people with mental illnesses in cities and that urban environments have a negative impact on our mental health. The questions are explored using both historical sources and modern surveys. Finally we look at challenges that are characteristic for city settings, such as homelessness and lack of access to green areas, and present two visionary suggestions with the potential to improve mental health and illness in cities worldwide.

In parallel with the rising urbanization, prevalences of mental illnesses have increased dramatically, and for several years, mental illness has been among the leading causes of disability worldwide. Therefore, in this chapter, we will try to answer the following questions:

- Could there be a connection between increased urbanization and increased rates of mental illness?
- Is mental illness more common in cities, as compared to the countryside? And if so, does this mean that the city environment is harmful to our mental health, or do certain environments attract people with dispositions to develop mental illness?
- What implications may city living have to mental health and illness?
- Which aspects of the urban environment are unhealthy to its citizens?

What Is a City?

According to the Oxford English Dictionary a city is "a municipality of larger size or population than a town" (Murray et al. 1970). The Oxford English Dictionary definitions of town, city, metropolis, and megapolis/megalopolis are presented in Box 2. The reader may note that the urban definitions are ordered in a hierarchy, but say nothing specifically about the number of inhabitants. As such, there exist no precise way to define a city or any other urban settlement. Richard Sennett, in *The Fall of Public Man* from 1977, put it this way: "*There are probably as many different ways of conceiving what a city is as there are cities. A simple definition therefore has its attractions. The simplest is that a city is a human settlement where strangers are likely to meet*" (Sennett 1992).

Box 2 Urban Definitions
Town: "An inhabited place which is larger than a village, contains more businesses and amenities, and typically has more complete and independent local government; now esp. one smaller than, or not officially designated, a city."

City: "A municipality of larger size or population than a town."

Metropolis: "The chief town or city of a country (occasionally of a province or district), esp. the one which is the seat of government; a capital. In extended use: any large, bustling city."

Megalopolis/Megapolis: "A very large or principal city, a metropolis (occasionally as a proper name); (now sometimes) spec. a continuous urban sprawl formed from the expansion of adjoining cities.'

Urban: "Relating to, situated or occurring in, or characteristic of, a town or city, especially as opposed to the countryside."
Source: Oxford English Dictionary

What History Can Teach Us

There are several historical texts that describe higher occurrences of mental illness in cities, as compared to the countryside (▶ Chap. 2, "History of Urban Mental Illness" by Shorter, E., in this book). In 1839 the Danish medical doctor Jens Rasmussen Hübertz wrote a letter to all pastors in his local community and asked them to report on the number of "Seelenkranke" ("soul sick") in their parishes. He found that in the countryside 1 in 466 were ill, whereas in Copenhagen (the capital of Denmark) 1 in 331 were ill. Another historical survey took place in the Kingdom of Saxony, today part of Germany and Poland. The survey found that the prevalence of "Irrsinnige" (mentally ill) in cities were 103 out of 100.000, against 49 per 100.000 in the countryside.

Most historical sources from the nineteenth and early twentieth century seem to find higher occurrences of mental illness in cities, compared to the countryside. In 1838 the French psychiatrists Jean-Étienne Dominique Esquirol noted that there were a high number of "insané" (mentally ill) in Paris. He therefore postulated that the mentally ill from the provinces moved to Paris. He speculated that certain aspects of the city environment attracted this group of people. Other scholars of the time thought, however, that it was the urban environment that made the citizens ill and not the other way around. In 1942 the German psychiatrist Karl Jaspers wrote that living in city milieus under difficult conditions contributed to higher rates of mental illness (Jaspers 2013). In line with the views of Karl Jasper, professor of history Edward Shorter describes the differences between the city and the countryside as follows:

"The big bustling metropoles like Munich and Breslau, where people drank hard and brought syphilis home from prostitutes, served as matrixes of pathology in contrast to the more rustic Heidelberg" (▶ Chap. 2, "History of Urban Mental Illness" by Shorter, E., in this book).

Recent Evidence

In 2010 a meta-analysis was published in the scientific journal *Acta Psychiatric Scandinavica*. The analysis included all studies published on the differences between mental illness in city setting and rural areas since 1985 (Peen et al. 2010). It concluded that the occurrence of mental illness is 38% higher in cities. Similarly in 2001, the influential Dutch psychiatrist Jim van Os described that city living was associated with high risk of psychosis (van Os et al. 2001). Other recent studies, however, are less clear in their findings. They indicate that the findings vary considerably depending on the city or countries examined. In fact, several recent studies report higher occurrences of certain types of mental illness in rural areas, compared to urban areas (Breslau et al. 2014).

What could explain these contradictory findings? We must consider if our notion of city environments as harmful is overly simplified or at least premature. In the following, we will take a detailed look at the evidence for a geographical distribution of mental illness across rural and urban environments.

Psychosis

In 1930 two Chicago sociologists, Robert Farris and Warren Dunham, showed that schizophrenia was most common in densely populated urban environments (Faris and Dunham 1939). Since 1930 these results have been supported by a large number of studies (▶ Chap. 3, "Urban-Rural Differences in Major Mental Health Conditions" by Solmi F., in this book). A recent meta-analysis look at the literature from high-income countries (where most studies have been conducted) and find that people born and raised in cities have twice the risk of developing schizophrenia compared to people from the countryside (Vassos et al. 2012).

Only nine studies have been conducted in low-income countries, and here the results are less clear: In China and Chile, the occurrence of schizophrenia in cities seems higher, whereas in India schizophrenia seems more prevalent in the countryside. Studies from Tibet and Uganda find comparable prevalences in urban and rural areas. So far there is no satisfactory explanation for why the difference between city settings and countryside are less conclusive in studies from low-income countries.

Several studies discuss if the geographical variation in occurrence of mental illness could be mediated by deprivation (Bhavsar et al. 2014). Deprivation describes the collective effects of low income, high unemployment, poor living conditions, and many people per household. Furthermore, researchers have gained interest in the significance of relative deprivation, or inequality, in other words. A hypothesis says that exposure to high inequality in deprivation may lead to a feeling of social exclusion, mistrust, and insecurity, which then increases stress and the risk for transition to psychosis. The idea is that it is not the total amount of deprivation that puts the individual at risk, as much as it is the *differences* in deprivation between two areas. An example is ethnicity, which can be a risk factor if a person is part of a minority which is marginalized or a protective factor if one lives in close proximity to other people of similar sociocultural background (Shaw et al. 2012).

Another explanation to the geographical differences in occurrence of schizophrenia and psychoses may be that people who experience prodromal symptoms, or already have transitioned into psychosis, are more likely to move to certain deprived urban environments. This theory is called "social drift" and implies that it is not the environments that are unhealthy and cause mental illness, but rather that people with mental illness are attracted to certain environments.

Another, perhaps more simple, explanation could be access to mental health services. The simple point here is that cities compared to rural areas have more psychiatrists and easier access to psychiatric hospitals, which increase the likelihood that people will receive a diagnosis. Or, on the other hand, people from the countrysides with mental illness are more likely to remain undiagnosed and untreated.

Suicide

In his book from 1897, the French sociologist Émilie Durkheim described that suicides have a tendency to cluster in certain geographical areas (Durkheim 1897). Several studies have since showed that suicide has a lower prevalence in urban settings, compared to rural areas. As with studies on rural-urban differences in psychosis and schizophrenia, most studies have been conducted in high-income countries. Recently, a handful of studies have been published from low-income countries, among these Taiwan, Belarus, India, China, and South Korea. The majority of these studies point to higher prevalences of suicide in rural areas.

There is also a difference in rates of suicide between neighborhoods. Suicide is associated with deprivation, inequality, and lack of trust. A systematic review of 54 studies found that risk factors, such as unemployment, single-person households, and being part of an ethnic minority, are of less importance if one lives in an area where these risk factors are common (Crawford et al. 2009). Again, it seems that *relative* deprivation is more important than total deprivation.

A possible contributing factor for the higher suicide rates in rural areas may be that the use of violent suicidal methods such as firearms are more common in the countryside. A study found that suicide by shooting happened in 75% of suicides in rural areas, compared to about 50% of people living in the city. Comparably, in low-income countries, there are high rates of poisoning by pesticides in suicides in the countryside. As such, part of the difference in prevalence of suicides between rural and urban areas may simply be explained from the availability of methods.

A more complex theory is that emigration of young people from the countryside and into the cities leaves a rural population with more risk factors, including high unemployment, social isolation, and low income. Studies suggest that certain issues are more common in rural areas, including mental health illiteracy, less service utilization, stigma and discrimination, and lack of anonymity as a barrier to seek help.

Urban Characteristics

We will now take a closer look at the urban environment and explore how the characteristics of the city environment may have an impact on mental health and illness.

Traffic, Pollution, and Noise

Increasing urbanization places more people close to traffic. A recent study from the scientific journal *the Lancet* concludes that living in close proximity to a major road increases the risk of developing dementia (Chen et al. 2017, article). The study is based on a large cohort from Ontario, the most populous province in Canada. Laboratory research has found that particles from car exhaust can induce inflammation in the nervous system, activate microglia, and stimulate antibodies against neurons. Also, constant noise affects cognitive function in rats. The study from the Lancet however found no association between major roads and risk of neurological disorders such as Parkinson's disease or multiple sclerosis. Studies as these suggest that proper public health strategies and city planning, with the aim of decreasing the exposure to heavy traffic, may prevent certain mental illnesses.

Artificial Light at Night

For thousands of years, humans have adopted their activity to a predictable 24 h light cycle (Bedrosian and Nelson 2013). But since the invention of the light bulb, almost 130 years ago, our lighting environments have changed considerably. Instead of adjusting our activity to the daylight, humans can now adjust their lighting environments to accommodate their needs for activity. Today, television, tablets, and computer screens expose humans to light at times of the day that we are not evolutionary adapted to. Recently, the World Health Organization has denoted night shift work as inductive of cancer, and Denmark has become the first country in the world to compensate women with shift work if they develop breast cancer.

About 20% of people in western economies do shift work, such as healthcare staff, flight personnel, and factory workers. Shift work is associated with increased risk of depression, especially in women. An explanation can be that exposure to artificial light at night may disrupt the circadian rhythm. It is difficult, however, to establish a direct causation between night work, light, and mood, since night work influence a number of variables with importance to mood, such as the possibility of social interaction in the day.

Another example is urban light pollution (Fig. 1). More than 99% of all people in North America or Europe are exposed to artificial light at night. In humans, as in other mammals, light is registered in the retina, a lining of cells in the eye. Cells in the retina then confers information to areas in the brain with importance to circadian rhythm, secretion of sleep hormone, wake hormone, and mood. The prevalence of mood disorders and sleep disorders have increased considerably in parallel with the increased urbanization. It is plausible that the unnatural exposure to light at night may explain part of this development.

The use of the light bulb spread before we understood the significance of light to physiological processes, including mood. Now that we know more about the impact of light on our brains, it is of outmost importance that we design our lighting environments so that they do not alter our circadian rhythm, disturb our sleep, and increase our risk of depression.

Green Areas

Compared with citizens with good access to green areas, people in densely populated cities without access to green areas have a higher risk of developing mental illness (▶ Chap. 18, "Green Environment and Mental Health in the City" by Kua EH., in this book). A considerable amount of research now indicates that access to green areas such as parks may have a positive influence on psychological and physiological factors, for example, by increasing physical activity. The aesthetical impact might also be significant. A study of admitted patients found that recovery was faster for patients in rooms with a view to green areas. Other studies show how the memory and attention of elderly citizens improve with access to green areas, perhaps mediated by increased physical activity and more social interaction. A recent survey from Ontario, Canada, found that people in neighborhoods with many trees had a higher self-reported health, compared with citizens in areas without trees. The association was persistent after controlling for sociodemographic variables such as income.

Once again city planning and political will seem crucial if we wish to improve the mental health of people in a world that is increasingly urbanized.

Mental Health Services in Major Cities

In the twenty-first century, there is evidence that deinstitutionalization of psychiatric patients has continued along with decrease of inpatient beds and closure of large mental health hospitals (▶ Chap. 9, "Mental Health Care in 11 Major Cities Over Two Decades" by Goldberg D., in this book). In parallel, community services have generally increased, and resources have shifted from mental health hospitals to community centers. Perhaps as a consequence of this, there are now more people in contact with psychiatric services in Copenhagen, Denmark, despite the lowest number of beds since the first asylums were built almost 200 years ago. In Porto Alegre, Brazil, community centers for mental health have now become the main focus of care for psychiatric patients; and the cities whose community services were most advanced 20 years ago, such as Verona, Italien, and Baltimore and Madison, USA, have all expanded their services further (Goldberg and Graham 2013).

A few cities report new interesting initiatives. In Bangalore, India, volunteers offer basic mental healthcare in many parts of the city. Mental health centers in Sydney, Australia, now offer internet-based cognitive-behavioral therapy. The costs are equivalent to 2 months of medication and show similar efficacy and accept among the Australians. In London, UK, general practitioners have invited psychology students to collaborate. This results in better access to psychological treatment for depression and anxiety for more people. A similar initiative has been implemented in Tehran, Iran, to accommodate the increased urban population.

Improving Mental Health in the Cities

The challenges of organizing mental health services seem to increase in parallel with the city size. In megacities like London, there will often be issues with homelessness and high complexity in patients because of comorbid mental illness and drug abuse. These challenges are accentuated in megacities and underline the importance of a firm political planning. Many governments have not developed clear strategies for the provision of urban mental healthcare despite the increasing urbanization (Sartorius 2002). We will now examine at two promising initiatives that may improve mental health in the cities of the world.

Housing First

The Pathways Housing First model offers housing to a group of people that are notoriously difficult to help; those with comorbid mental illness and drug abuse (▶ Chap. 16, “Homelessness and Challenges of Community Care” by Tsemberis S., in this book). The model is unique because it considers housing a basic right, and not something people have to earn through, for example, abstinence or regular visits to an outpatient mental health clinic. Apart from housing the model includes a mobile team of therapists, support persons, and social workers. The participants in the program are not excluded if they relapse or are put on the street by their landlord; in this case the mobile team will quickly help the person to be rehoused. The Pathways Housing First model presents a U-turn compared to the previous attempts to find housing to homeless people with comorbid mental illness and drug abuse. In more common “treatment first” models, there will typically be prerequisites in terms of abstinence or compliance with treatment. “Housing first” has shown to outperform on housing stability, quality of life, and cost-effectiveness. Several cities in the USA have implemented the Housing First model and declared they have ended homelessness, and recently Norway declared they have ended homelessness in the western part of their country.

Urban Design

The untapped potential for improving mental health through urban design extends beyond green spaces (▶ Chap. 17, "Urban Design and Mental Health" by McCay L., in this book). Other examples are areas that promote physical activity, such as good bicycle paths, outdoor training facilities, and sidewalks. And other examples are spaces where people feel safe and confident, facilitate encounters with nature or other people, promote social interactions, or reduce the stress associated with commuting.

As such, the questions and responsibilities of public mental health now expand beyond psychiatrists and psychologists. City planners, architects, engineers, transport specialists, and politicians, all have the power to affect mental health of city people. Urban designers are beginning to consider physical health when planning city spaces; now it is time to consider mental health also.

Summary and Conclusion

Is mental illness more prevalent in cities? The answer is both yes and no. Some mental illnesses are more prevalent in certain cities and at certain points in history; other mental illnesses are, on the other hand, more prevalent in rural areas. Mental illnesses are diverse and so are cities. A better question may be which harmful and protective factors are characteristic of urban environments.

There are challenges that are particularly common in urban environments that may be associated to mental illness. These include limited access to green areas, homelessness, violence, high rates of crime, easy access to drugs, traffic and noise, and high variance in social and economical status. Recently, however, there are new and promising initiatives that deal with these challenges.

In a historical perspective the term "big cities" is relatively new. Researchers are only beginning to explore how the characteristics of the city; pollution, noise, light, traffic, and lack of green environments may contribute to the development of mental illness. The speed of urbanization is faster than we could have ever imagined, and we are behind in our understanding of how the urban environment affects mental health. We believe that city planning is a crucial area with enormous potential to improve mental health and illness in our cities.

Cross-References

▶ Crime, Illegal Drugs, and Mental Health
▶ Green Environment and Mental Health in the City

▶ History of Urban Mental Illness
▶ Migration and Urban Mental Health in China
▶ Prospects and Challenges of Mental Health Services in a Major City in Latin America: Rio de Janeiro
▶ Urban Design and Mental Health
▶ Urban-Rural Differences in Major Mental Health Conditions

References

Bedrosian TA, Nelson RJ (2013) Influence of the modern light environment on mood. Mol Psychiatry 18(7):751–757

Bhavsar V, Boydell J, Murray R, Power P (2014) Identifying aspects of neighbourhood deprivation associated with increased incidence of schizophrenia. Schizophr Res 156(1):115–121

Breslau J, Marshall GN, Pincus HA, Brown RA (2014) Are mental disorders more common in urban than rural areas of the United States? J Psychiatr Res 56:50–55

Chen H, Kwong JC, Copes R, Tu K, Villeneuve PJ, van Donkelaar A, Hystad P, Martin RV, Murray BJ, Jessiman B, Wilton AS (2017) Living near major roads and the incidence of dementia, Parkinson's disease, and multiple sclerosis: a population-based cohort study. Lancet 389:718–726

Crawford MJ, Kuforiji B, Ghosh P (2010) The impact of social context on socio-demographic risk factors for suicide: a synthesis of data from case-control studies. J Epidemiol Community Health 64(6):530–534

Durkheim E (1897) Le suicide: étude de sociologie. F. Alcan, Paris

Faris REL, Dunham HW (1939) Mental disorders in urban areas. Chicago University Press, Chicago

Goldberg D, Graham T (2013) Mental health in our future cities. Psychology Press, East Sussex, UK

Jaspers K (2013) Allgemeine psychopathologie. Springer-Verlag, Berlin

Lewis G, Booth M (1994) Are cities bad for your mental health? Psychol Med 24(04):913–915

Murray JA, Bradley H, Craigie WA, Onions CT (1970) The Oxford English dictionary. Clarendon, Oxford

Peen J, Schoevers RA, Beekman AT, Dekker J (2010) The current status of urban-rural differences in psychiatric disorders. Acta Psychiatr Scand 121(2):84–93

Robinson L, Holbrook J, Bitsko R et al (2017) Differences in health care, family, and community factors associated with mental, behavioral, and developmental disorders among children aged 2–8 years in rural and urban settings – United States, 2011–2012. MMWR Surveill Summ 66(SS-8):1–11

Sartorius N (2002) Fighting for mental health: a personal view. Cambridge University Press, Cambridge

Sennett R (1992) The fall of public man. WW Norton & Company, New York, p 39

Shaw RJ, Atkin K, Bécares L, Albor CB, Stafford M, Kiernan KE, Nazroo JY, Wilkinson RG, Pickett KE (2012) Impact of ethnic density on adult mental disorders: narrative review. Br J Psychiatry 201(1):11–19

United Nations Human Settlements Programme (UN-Habitat) (2016) Urbanization and development: emerging futures. World cities report 2016. United Nations Human Settlements Programme (UN-Habitat), Nairobi. http://wcr.unhabitat.org/wp-content/uploads/2017/02/WCR-2016-Full-Report.pdf. Downloaded 10 Mar 2017 van Os J, Hanssen M, Bijl RV, Vollebergh W (2001) Prevalence of psychotic disorder and community level of psychotic symptoms: an urban-rural comparison. Arch Gen Psychiatry 58(7):663–668

Vassos E, Pedersen CB, Murray RM, Collier DA, Lewis CM (2012) Meta-analysis of the association of urbanicity with schizophrenia. Schizophr Bull 38(6):1118–1123

Part I

Primarius

History of Urban Mental Illness

2

Edward Shorter

Abstract

Observers have always considered cities to have higher rates of mental illness than the "purer" countryside. Historical statistics have confirmed higher urban than rural illness rates. Recently, however, this relationship has reversed itself, and rates of psychiatric illness are higher in the countryside than the city. Of interest here are the risk factors and the protective factors that regulate urban rates of mental illness compared to rural. Traditional urban-rural differences might be reversing. Yet the balance of risk and protection may well vary from city to city, and from nation to nation, making the circumstances that control this balance the most intriguing subject of investigation.

Keywords

History of urban mental illness • History of rural mental illness • Risk factors in the city • Protective factors in the city • Urban-rural differences in mental illness • Attitudes towards urban mental illness

Contents

E. Shorter (✉)
History of Medicine Program, Faculty of Medicine, University of Toronto, Toronto, ON, Canada
e-mail: edwardshorter@gmail.com; history.medicine@utoronto.ca

N. Okkels et al. (eds.), *Mental Health and Illness in the City*, Mental Health and Illness Worldwide, https://doi.org/10.1007/978-981-10-2327-9_18

Epigram: "Like all great cities, [Vienna] consisted of irregular things, of change, of slipping ahead, falling behind, things and affairs bumping together, with bottomless moments of silence in between, of pathways and uncharted ways, of a great rhythmic beat and the eternal cacophony and disharmony of all the rhythms in conflict" (Musil 1930/1978). Robert Musil, novelist, *The Man Without Qualities* (*Der Mann ohne Eigenschaften*), 1930.

Introduction

Do the cities produce a particular *form* of mental illness? There are some that are stamped by the character of urban life. Yet from the beginning, psychiatrists have reflected that the major illnesses do not bear any particular urban logo. What Johann Christian Reil, the founder of German psychiatry, wrote in 1803, doubtless continues to apply. "Insanity," he said, "comes from a weakness that is rooted in our very humanity, to which we are all subject . . . and that we cannot hold off, even through rank and riches." An attack of illness, a blow of fate "that might shake individual families or entire states, may prepare for us a permanent place in the asylum" (Reil 1803/1968). Thus, psychiatric illness could strike anyone, anywhere, any time.

But is there something about urban life that increases or inhibits the *frequency* of mental illness? That is a different question, one that has always been of lively interest. In 1839, JR Hübertz, a physician in Copenhagen, wrote to all the local pastors in Denmark, asking about the number of people "of diseased mind" (Seelenkranken) in their parishes. (The procedure was not as absurd as it would sound today, given that pastors in this closely knit, largely rural society, usually knew the parishioners very well.) It turned out that in the countryside one person in every 466 was mentally ill, in Copenhagen one in 331 (Hübertz 1844): There was more mental illness in the metropolis.

The "Toxic City" Tradition Builds

There has always been a tendency to believe that the cities, vast wens of iniquity, are more productive of mental illness than the sturdy countryside. In 1857, the French poet Charles Baudelaire half-adored half-deplored "the chaos of the living cities" ("le chaos des vivantes cités") (Baudelaire 1857/2014). And as the urbanization of Western society increased during the nineteenth century, the numbers seemed to confirm the prejudices: mental health in urban environments did appear steadily worse compared to rural.

The systematic epidemiological surveys that began around the 1860s took note of mental illness. In the Kingdom of Saxony, the census of 1861 asked about the "insane" (Irrsinnige) at a household level. In the cities 103 per 100,000 were marked as insane, in the countryside 49 (Oettingen 1882).

In a census in 1880 for the Grand Duchy of Oldenburg, by then part of the German Empire, the enumerators noted with a simple "yes" or "no" whether the person had a "mental illness." In the cities, 2.5 per 1000 population did so, in the countryside 3.0 (Kollmann 1883/84). For once, the cities seemed healthier.

But Oldenburg was exceptional. In the United States, the census of 1880 inquired about "insanity." The rate of insanity per 100,000 population in cities over 50,000 was 232, in the whole country 183 (White 1903).

Various inquiries elsewhere in the earlier twentieth century found mental illness to be commoner in cities, but not by a great deal. A follow-up study into adulthood of 5395 Icelanders born in 1895–97 found that by 1957, 13.4% of the urban males had developed "neuroses," 8.8% of the rural males (21.4% of the urban females, 17.6% of the rural) (Helgason 1964). The differences were statistically significant. (Significance was not calculated for the earlier studies.)

This slight urban surplus continued in further investigations. A survey of 106 general practices in England and Wales in 1955–56 learned that, of patients in practices in conurbations, 51.3% had "psychoneuroses," of patients in rural districts, 37.3%. Of patients in conurbations, 3.4% had "depression," rural 2.3%. As for "debility and undue fatigue," the respective figures were 7.1% and 5.3% (Logan and Cushion 1958). Thus once again, the cities had outpaced the countryside in terms of mental pathology.

A raft of studies showed that the villages were not idylls of innocence (Fenayrou 1894). Yet in the city vs. country discussion, one would wish to untangle the question: Was it the stress of urban life that made healthy people ill? Or did the villages seem more tranquil because all the restless spirits had left for the city? Paris psychiatrist Etienne Esquirol noted in 1838 that the high number of "insane" (insensés) in Paris was not typical of France as a whole, because, "It is the insane from all the provinces that come to Paris" (Esquirol 1838). The big question is thus whether people become mentally ill because of urban life, or whether people with pre-existing illnesses move to the city?

For contemporary observers, the evidence seemed to point to "urban toxicity." In 1912, New York sociologist Howard Woolston described "the urban habit of mind": "Men are assailed at every sense by the presence of their neighbors. The sound of footsteps and hoof-beats, the rattle of wagons and rush of cars, the clang of bells and hoot of whistles . . . 'The consequences for the nervous system, he said, were dire': The natural result of city life is increased nervousness . . . the highly-strung society woman, who repairs to a sanitarium to nurse hysteria after an unusually gay season; the man who steadies his nerves with a cocktail before business . . . the broken wretch who flings himself into the river or gibbers in the psychopathic ward. For these the pace has been too hot. Their nerves are burning out" (Woolston 1912).

Woolston joined early a vast sociological stream that insisted that the city, with its "gibbering psychotics," equaled pathology. And who could think otherwise! US cities in those years were receiving vast number of immigrants, the "wretched refuse of your teeming shore," as Emma Lazarus's famous poem had it; these unwashed hordes were seen as deep vats of illness (Stella 1908). And the noise and confusion! It would surely drive a normal person to distraction. Charles Henderson at the University of Chicago asked if "modern industry and city life [are] unfavorable to the family" (Henderson 1909)? And of course the answer was yes. The school of urban sociology that flourished at Chicago led by Robert E Park composed the main bill of indictment (Park et al. 1925).

Urban Toxicity?

In 1906, unmindful of Reil's 1803 dictum that insanity is common everywhere, Robert Gaupp, at the time a staff psychiatrist at the Munich University Psychiatric Clinic, said that in general the organic psychoses were commoner in cities than in the countryside. "Alcoholics, psychopaths, epileptics, paralytics [neurosyphilis], hysteria [pseudo-epilepsy is meant], organic brain patients, and symptomatic delirium cases – those are the chief psychic disorders the frequency and form of which differ from that seen in rural asylums." He added that traumatic neurosis (later called post-traumatic stress disorder) was more often seen in cities. He noted no urban-rural differences in what people were then calling "dementia praecox," later "schizophrenia." And he thought that manic-depressive illness was commoner in the countryside, especially among women (who perhaps because of the illness did not have the resolve to migrate to the city). The big bustling metropoles like Munich and Breslau, where people drank hard and brought syphilis home from prostitutes, served as matrixes of pathology in contrast to the more rustic Heidelberg (Gaupp 1906).

For psychiatrist and psychopathologist Karl Jaspers there was no doubt of the toxic influence of the cities. He wrote in 1942, "The difficult [urban] living conditions with their harmful mental impact eventuate in a much greater frequency of psychopathies (hysteria and so forth)" (Jaspers 1942/1973).

Many later studies since seemed to confirm this wisdom. For example, a large meta-analysis of all urban-rural studies since 1985 concluded in 2010 that psychiatric disorders in general were 38% higher in the city than in the countryside, that mood disorders were greater by 39%, and anxiety disorders by 21% (Peen et al. 2010). Thus, the former "excitement" of urban life seemed to have turned into the "stress" of urban life. When in 1994 two researchers at the Institute of Psychiatry and the London School of Tropical Medicine asked, "Are cities bad for your mental health?" the answer was yes. Based on 6572 respondents to the General Health Questionnaire, the authors concluded, "There is a need to identify the psychologically harmful elements of urban life in the hope that cities can be made happier and healthier places in which to live" (Lewis and Booth 1994).

Data from the first "wave" of the Epidemiologic Catchment Area study of mental illness launched in the early 1980s by the National Institute of Mental Health of the US government further reinforced the case that the metropolis encouraged depressive illness: An analysis in 1998 showed that 3.83% of the high-status women in cities had experienced a "recent major depression," 1.55% of women in "nonmetro" areas. (For comparable men, 3.2% and 0.89%) (Goldsmith et al. 1998).

Research by Jim van Os and team in 2001 strengthened this notion of urban toxicity, finding psychosis strongly associated with increasing urbanicity. The investigators suggested "that the increased prevalence of psychotic disorders in urban environments should be interpreted in light of increased levels of 'psychosis proneness' in urban populations" (Van Os et al. 2001). In other words, urbanites themselves were prone.

Are these true findings of urban "toxicity"? Or is there a compositional effect that comes from cities having more single people, a demographic segment at higher risk of psychiatric illness than the married, who stayed behind on the farm? More careful national studies have been unable to show big urban-rural differences. An analysis in 2005 of urban-rural differences among common mental disorders in Britain, concluded, "In general, the effects of place on rates of the common mental disorders appear modest." Recognizing that they were flying in the face of a century of sociological tradition, the authors conceded that their finding was "counter-intuitive" (Weich et al. 2005).

A careful European Union study in 2004 turned on further warning lights, at least for depression. Reviewing a number of large studies of urban-rural differences in depression across the EU, the study noted that depression was indeed often higher in urban areas, yet there were several exceptions, such as Norway. Moreover, "most of the urban risk disappears when marital status is controlled for." The main socio-demographic variables, such as marital status, may be very different from city to countryside, from city to city, and from country to country (European Commission, Health & Consumer Protection, Directorate-General 2004). It is therefore difficult to pinpoint as the source of the pathology some urban atmospheric "feeling" – not unlike the "praecox feeling" said by some observers to be a diagnostic sign of schizophrenia (Shorter 2015): hard to specify but you knew it when you saw it.

Risk Factors and Protective Factors

The great Austrian novelist Robert Musil, in *The Man Without Qualities* (1930), captured in the above epigraph the uncertainty of the metropolis in the 1920s, triumph and despair on the Ringstrasse of Vienna!

Is such an atmosphere a hatchery of psychiatric illness or not? The question is too simple, because there is no single effect of urban life. The psychic impact depends on such circumstances as the recency of the immigrants, the friendliness of the setting to alcohol and drugs, or the presence of neighborhoods where disorganization may flourish. Rather, we should ask what *risk* factors in cities seem to propagate psychiatric illness, and what *protective* factors seem to inhibit it.

Among the risks of urban life, contemporary authorities point almost in unison to "stress," although stress itself is not a psychiatric illness; it does not appear independently in the *DSM* aside from "post-traumatic stress disorder," and it has an uncertain medical status. Textbooks of pathology and physiology do not typically include "stress" in their indexes. To be sure, stressful events activate the hypothalamic-pituitary-adrenal (HPA) axis, but in psychiatry it is not HPA activation that is associated with such illnesses as serious depression but HPA *dysfunction*, a different concept (Shorter and Fink 2010). Nonetheless, stress has become a kind of mantra and is chanted whenever the city is involved. A group of German scholars indict as stressful "a lack of green zones," meaning access to park land (Lederbogen et al. 2013).

Yet stress is a recent concept, and possibly an unscientific one. A long tradition, beginning in 1903 with the writing of German sociologist Georg Simmel, indicts the corrosive effect on human relationships and community solidarity of big-city money-hunger and anonymity (Simmel 1903). This analysis was grafted onto a socially conservative strain of thought that considered the city as a solvent of the entire society. Social statistician and moralist Alexander von Oettingen spoke of "rationalistic atomism," and cited as evidence of the city's toxic radiation: the closer one lived to Paris, the greater the number of local young women who scampered to the metropolis to become prostitutes (Oettingen 1882).

Indeed, the cities threw up a huge risk factor in the "social isolation" that they made possible. In 1939, Robert EL Faris and H Warren Dunham, close to though not part of the Chicago School as led by Robert Park (see above), suggested that "seclusiveness" was a key trait of schizophrenia, and that schizophrenics would be drawn to those areas of the city, in this case Chicago, that fostered the "disorganization" in which isolation might flourish (Faris and Dunham 1939). Thus, isolation was a two-edged sword: the isolation that made possible the secret pleasures of the boudoir of Emile Zola's novel *Nana* (1880), also made possible schizophrenia.

While a century of sociology has emphasized the chaos of the living city, little has been said of possible urban protective factors against mental illness. As one protective factor, the city offered relief of mind-benumbing sameness. Said one physician in 1898 of the population that toiled in the endless cornfields of the state of Iowa, "Aspiration and hope may become lost in a morass of drudgery, and finally degenerate into unconscious despair, to terminate in a despondent and melancholic phase" (Rawson 1898). Much urban appeal lay in what sociologist Ernest Groves called in 1916 the relief of "rural restlessness." Groves, later author of a classic work on rural sociology (Groves 1922), described "the barrenness of country life to the girl growing into womanhood, hungry for amusement" (Groves 1916). Migration to the city might protect her from gloom.

A consortium of the NORC Walsh Center for Rural Health Analysis and the Rural Health Policy Research Center of the University of North Dakota has recently analyzed two US government databases – the Urban and Rural Health Chartbook data and the National Surveys on Drug Use and Health – for 2010–2011. This consortium analysis suggests that there are indeed protective factors (Rural Health Reform Policy Research Center 2014). By almost any measure of health or mental pathology, the large central metropolitan areas emerge as far safer and healthier than the "non-core" nonmetropolitan counties. The study divided counties into the "large central (inner cities)," the suburban, the "small metro," the "large rural," and the "small rural." Here are the findings:

It turns out that in the big cities, people keep their teeth, in rural areas less so: edentulism for males rose steadily across size of place from big city to rural; there was a smooth relationship between teenage cigarette smoking by size of place: teens smoke least in big cities, most in rural areas; the same applied for obesity and physical activity (almost a smooth relationship). Suicide rates rose steadily as size of place declined, as did death rates. The supposedly sturdy yeomanry of the village turn out to be an epidemiologic nightmare, the small town and farm being more pathological by almost every measure. (For a thorough review of urban-rural differences in mental illness, please see ▶ Chap. 3, "Urban-Rural Differences in Major Mental Health Conditions," by Francesca Solm.)

This pathology included mental health: "Serious mental illness" for males increased steadily as size of place declined (a less smooth relationship for females, but still, in the big city lower than rural); the same pattern applied for major depression; and female adolescents in the smaller towns had considerably more "major depression" than in the big cities.

One group of scholars who analyzed this database concluded, "Contrary to expectations, the prevalence of mental disorders was not higher in the most urban compared with the most rural areas, suggesting that the move to identify mechanistic explanations for risk associated with the urban environment is premature" (Breslau et al. 2014). Premature, to say the least, after 100 years of orthodoxy.

Conclusion

These new findings demand interpretation. Is it that the basic social subtext of the cities has changed in the twenty-first century, making them repositories of benign rather than malignant behavior? One might call this a result of the renaissance of the central city in US life. Or is it that life in the small towns and countryside has deteriorated from the halcyon days of yore, making them foci of drug abuse, especially crack cocaine, and of the pathologies of the disaffiliated? Or is it the power of the consortium's database itself, and the differentiation of the analysis into five subcategories rather than simply "urban-rural"? These are big questions that cannot be answered in this chapter.

Rather, the main point is that it is meaningless to ask whether cities "favor" mental illness. The findings are highly contradictory. Some cities do, and some do not. It all depends on the balance of risk and protective factors that prevails in that particular place at that particular time, for city and countryside alike.

In sum, cities have had a bad rap in psychiatry. The thoroughfares seem to have become marshaling yards for homeless schizophrenics; the malls, assembly grounds for adolescent drug-dealing. Yet there is a deeper psychic reality: In the conditions of privacy and self-realization that the cities make possible, the human spirit may take wing.

Cross-References

- ▶ Green Environment and Mental Health in the City
- ▶ Loneliness and Single-Person Households: Issues of *Kodokushi* and *Hikikomori* in Japan
- ▶ Mental Health Care in 11 Major Cities Over Two Decades
- ▶ Urban-Rural Differences in Major Mental Health Conditions

References

Baudelaire C (1857/2014) Les Petites Vieilles. In: Les Fleurs du Mal, new ed. Arvensa, Paris, pp 235
Breslau J, Marshall GN, Pincus HA et al (2014) Are mental disorders more common in urban than rural areas of the United States? J Psychiatr Res 56:50–55

Esquirol E (1838) Des maladies mentales, vol 2. Baillière, Paris, pp 741
European Commission, Health & Consumer Protection, Directorate-General (2004) The state of mental health in the European Union. Brussels, European Commission, p 48–51
Faris REL, Dunham HW (1939) Mental disorders in urban areas: an ecological study of schizophrenia and other psychoses. University of Chicago Press, Chicago, pp 175–177
Fenayrou A (1894) Contribution à l'étude des folies rurales: La folie dans l'Aveyron. Duclos, Toulouse
Gaupp R (1906) Die klinischen Besonderheiten der Seelenstörungen unserer Grossstadtbevölkerung. Münch Med Wochenschr 53:1250–1252,1310–1313
Goldsmith HF. Holzer CE. Manderscheid RW (1998) Depression in metropolitan and non-metropolitan neighborhoods. In: Manderscheid RW, Henderson MJ (eds) Mental Health, United States, Center for Mental Health Services, Washington, DC, pp 188–203. DHHS Pub. No, (SMA)99–3285; Supt. Of Docs. Print Off, 1998, Tab 4, 196
Groves ER (1916) Psychic causes of rural migration. Am J Sociol 21:623–627
Groves ER (1922) The rural mind and social welfare. University of Chicago Press, Chicago
Helgason T (1964) Epidemiology of mental disorders in Iceland. Munksgaard, Copenhagen, p 103
Henderson CR (1909) Are modern industry and city life unfavorable to the family? Am J Sociol 14:668–680
Hübertz JH (1844) Statistik des Irrenwesens in Dänemark. Allg Z Psychiatr 1:457–481
Jaspers K (1942) Allgemeine Psychopathologie, 9th edn. Springer, Berlin, p 605
Kollmann P (1883/84) Die geisteskranke Bevölkerung im Grossherzogthum Oldenburg nach den Ergebnissen der Volkszählung vom 1. Dezember 1880. Allg Z Psychiatr 40:481–542, 489
Lederbogen F, Haddad L, Meyer-Lindenberg A (2013) Urban social stress – Risk factor for mental disorders. The case of schizophrenia. Environ Pollut 183:2–6
Lewis G, Booth M (1994) Are cities bad for your mental health? Psychol Med 24:913–915
Logan WPD, Cushion AA (1958) Morbidity statistics from general practice, vol 1 (general). Her Majesty's Stationery Office, London, p 131, 137
Musil R (1930) Der Mann ohne Eigenschaften, vol 1. Rowohlt, Hamburg, p 10
Oettingen A (1882) Die Moralstatistik in ihrer Bedeutung für eine Socialethik, 3rd edn. Deichert, Erlangen, p 367, 674
Park RE, Burgess ES, McKenzie RD (1925) The City. University of Chicago Press, Chicago
Peen J, Schoevers R, Beekman AT (2010) The current status of urban-rural differences in psychiatric disorders. Acta Psychiatr Scand 121:84–93
Rawson AA (1898) Rural insanity. Iowa Stat Med Soc J 16:73–78
Reil JC (1803/1968) Rhapsodien über die Anwendung der psychischen Curmethode auf Geisteszerrüttungen, Reprint edn. Bonset, Amsterdam, p 11
Rural Health Reform Policy Research Center (2014) The 2014 Update of the Rural-Urban Chartbook, University of North Dakota Center for Rural Health. https://ruralhealth.und.edu/projects/health-reform-policy-research-center/pdf/2014-rural-urban-chartbook-update.pdf. downloaded 14 March 2016
Shorter E (2015) What psychiatry left out of the DSM-5: historical mental disorders today. Routledge, New York, pp 119–123
Shorter E, Fink M (2010) Endocrine psychiatry: solving the riddle of melancholia. Oxford University Press, New York
Simmel G (1903) Die Grossstädte und das Geistesleben. Jahrb Gehe-Stiftung Dresden 9:185–206
Stella A (1908) The effects of urban congestion on Italian women and children [in the US]. Med Rec 73:722–732
Van Os J, Hanssen M, Bijil RV et al (2001) Prevalence of psychotic disorder and community level of psychotic symptoms: an urban-rural comparison. JAMA Psychiat 58:663–668
Weich S, Twigg L, Glyn L (2005) Rural/non-rural differences in rates of common mental disorders in Britain. Br J Psychiatry 188:51–57
White WA (1903) The geographical distribution of insanity in the United States. J Nerv Ment Dis 30:257–279
Woolston HB (1912) The urban habit of mind. Am J Sociol 18:602–614

Part II

Urban Mental Health: The Facts

Urban-Rural Differences in Major Mental Health Conditions

3

Francesca Solmi, Jennifer Dykxhoorn, and James B. Kirkbride

Abstract

In this chapter, we review the evidence for rural-urban and spatial variation of three major sets of mental health disorders: schizophrenia and related psychotic disorders, common mental disorders including depression and anxiety, and suicide. For each, we review the recent literature which has addressed these issues and report the main strengths and limitations of the available evidence. The most consistent evidence for rural-urban gradients in mental health risk are found to exist for schizophrenia and suicide, with more mixed evidence in relation to common mental disorders. For schizophrenia and suicide, we go on to review the major hypotheses that have been put forward to account for rural-urban variation in risk.

Keywords

Schizophrenia • Psychotic disorders • Common mental disorders • Depression • Anxiety • Suicide • Rural-urban • Epidemiology

Contents

F. Solmi • J. Dykxhoorn • J.B. Kirkbride (✉)
PsyLife Research Group, Division of Psychiatry, Faculty of Brain Sciences, UCL, London, UK
e-mail: francesca.solmi@ucl.ac.uk; jennifer.dykxhoorn.15@ucl.ac.uk; j.kirkbride@ucl.ac.uk

N. Okkels et al. (eds.), *Mental Health and Illness in the City*, Mental Health and Illness Worldwide, https://doi.org/10.1007/978-981-10-2327-9_7

Introduction

How the social and physical environment may affect our mental health – both positively and negatively – is a topic of increasing global importance. By 2014, more than half of the world's population lived in urban areas, and this is set to rise to a staggering 66% by 2050. In this chapter, we examine the potential mental health implications that this global shift toward living in urbanized environments may have at the population level. We have restricted our analysis to three major psychiatric conditions, where most research has been conducted: psychotic disorders, including schizophrenia; common mental disorders, i.e., depression and anxiety; and suicide. We have organized the chapter into four main sections. First, we review the current evidence linking psychotic disorders to our urban environments. We identify the main hypotheses which have been proposed to explain such associations and review their plausibility, given the available evidence. We repeat this analysis for common mental disorders in section "Common Mental Disorders (Depression and Anxiety)" and for suicidal outcomes in section "Suicide." In the final section, our focus turns to consider the extent to which the evidence presented in this chapter informs possible preventive medicine strategies with respect to our urban environments.

For each mental health outcome, we have distinguished between studies which examine urban-rural differences in risk and those which have provided more detailed examination of small area variation in the rate at which psychiatric conditions occur. These second set of studies have often sought to move beyond rural-urban dichotomies to investigate how differences in the environmental milieu of small areas – or "neighborhoods" – within a single setting may affect the risk of mental health disorders. Such effects are typically studied using a variety of "multilevel" statistical approaches which are designed to examine whether risk factors which occur at several levels of causation, including both the individual and neighborhood level, are associated with a given outcome of interest. Typically, such small areas have been defined by administrative units to investigate "neighborhood-level" effects. While the extent to which these represent meaningful neighborhoods as perceived by their occupants has been debated (Weich 2005), small area units typically provide a more homogeneous and precise basis for analytical investigation than simpler rural-urban distinctions. Although primarily used for administrative purposes, such as census enumeration or elections, area-level units such as postcode tracts, census wards, or postal areas are often demarcated by physical barriers such as major roads, rivers, or other topographical features, providing some ecological validity to their "neighborhood" status.

The best data we have on the role cities play in shaping mental health presently come from high-income country (HIC) settings. The paucity of data on mental health and urbanization from low- and middle-income country (LAMIC) settings, where economic development, basic services, and physical health care may present more fundamental topics for service providers and urban planners, reflects an important gap in the literature. It also presents an opportunity to carefully tailor the best available evidence on mental health and city life from other settings into policy and planning recommendations for the design of mentally healthy cities. Since it has been suggested that population dynamics and urban development trajectories will differ in LAMIC settings (Menezes 2014), this calls on researchers and funding bodies to invest time, energy, and capital into research programs which identify the burden and correlates of psychiatric morbidity in rapidly developing populations. For this reason, in this chapter, we endeavor to make special reference to any studies conducted in LAMIC settings, where pertinent and robust evidence is available.

Schizophrenia and Other Non-affective Psychoses

Schizophrenia affects between 4.6 (range, 1.9–10.0), 3.3 (range, 1.3–8.2), and 4.0 (range, 1.6–12.1) people per 1,000, depending on whether measured at a single point in time (point prevalence), over the course of a year (annual or period prevalence), or over the entire lifetime (lifetime prevalence) is considered (Saha et al. 2005). Incidence, which describes the number of *new* cases within a defined population in 1 year, is estimated to be approximately 15.2 new cases per 100,000 people per year (range, 7.7–43.0) (McGrath et al. 2004). However, incidence is heterogeneously distributed throughout the population according to several sociodemographic indicators (Kirkbride et al. 2006), including age, sex, ethnicity, and place. Schizophrenia and related disorders are more common among men, young adults (before 35 years old), and people from disadvantaged socioeconomic backgrounds (Kirkbride et al. 2012). Consistent evidence suggests that higher rates are experienced by people from minority ethnic backgrounds, most notably those of black African and black Caribbean origin living in the UK, the Netherlands, and elsewhere (Bresnahan et al. 2007; Schofield et al. 2011; Boydell et al. 2001; Veling et al. 2006). A long history of research, beginning with the pioneering work of two Chicago sociologists Robert E. L. Faris and Warren H. Dunham in the 1930s, have shown that the distribution of the incidence of schizophrenia and other non-affective psychoses is not random, but occurs with greater-than-expected frequency in more densely populated urban settings (Faris and Dunham 1939). Such geographical patterning exists between countries (Saha et al. 2006), within countries along a urban-rural gradient (March et al. 2008; Heinz et al. 2013), and between neighborhoods within urban areas (March et al. 2008). Over the last 70 years, spatial variation in the incidence of non-affective psychotic disorders such as schizophrenia has been replicated in a number of early (Hare and Family 1956; Löffler and Häfner 1999; Giggs and Cooper 1987) and more recent studies (Tables 1 and 2 summarize those studies published since 1990).

Table 1 Summary of literature on rural-urban differences in the distribution of schizophrenia (From 1990 onward, chronologically ordered)

Study ID	Diagnostic criteria and outcome	Case ascertainment	Sample size	Definition of urban	Timing of urban	Findings	Confounders	Comments
(Lewis et al. 1992) Sweden	ICD-8 (schizophrenia)	Register	Sample = 49,191 (male conscripts) Cases = 268	*Degree of urbanization*	Upbringing	Higher odds of schizophrenia in men who grew up in the city	Family finances, cannabis use, psychiatric diagnosis at conscription, parental divorce, number of friends, "nervous" feelings, family history of schizophrenia	Fully adjusted model showed weaker statistical significance between urban and rural risk
(Thornicroft et al. 1993) Italy	ICD-9 (schizophrenia and other psychoses, including affective)	Registers	Not reported	*Urban versus rural*	At diagnosis	Higher prevalence and incidence rates in the rural compared with urban areas	Household composition, employment, population density, number of dependents, number of family members economically inactive, age, number of rooms per person	Living alone, being unemployed, and % total population who are unmarried, separated, or divorced were associated with schizophrenia in urban but not in rural area
(Takei et al. 1995) UK	ICD-9 (schizophrenia)	Registers	Sample = 24,858 Cases = 6,553	*Urban versus rural*	Birth	Risk of schizophrenia higher for people born in urban areas and autumn/winter months	Age, sex, place of birth	Poor control group (psychiatric controls), cross-sectional data
(Marcelis et al. 1998) The Netherlands	ICD-9 (narrow schizophrenia, broad schizophrenia)	Registers	$N = 47,721$ Narrow schizophrenia (NS) ($N = 5,606$) Broad schizophrenia (BS) ($N = 16,716$)	*Urban versus rural (based on population density)*	Birth and upbringing	People with urban birth and urban exposure had higher risk of narrow and broad schizophrenia	Birth cohort, season of birth, age of onset	High correlation between urban birth and urban exposure. Unable to test for incidence among people born in urban areas who moved to rural areas Similar pattern for affective psychoses

(Marcelis et al. 1999) The Netherlands)	ICD-9 (narrow schizophrenia, broad schizophrenia)	Registers	$N = 1{,}351{,}637$ ($N = 338$, narrow schizophrenia; $N = 766$, broad schizophrenia)	*Urban versus rural (based on population density)*	Birth and admission	People born in urban areas had highest risk of schizophrenia. People born in rural environment and admitted in urban ones were at no greater risk cf. people born and admitted in rural areas		Findings validate the hypothesis that being born and raised in urban environment affects risk of schizophrenia
(Mortensen et al. 1999) Denmark	ICD-8 (schizophrenia)	Registers	$N = 1{,}750{,}000$ ($N = 2{,}699$ with schizophrenia)	*Degree of urbanization*	Birth	Highest risk in those born in the capital cf. rural area, with gradient in risk by degree of urbanization of birthplace	Age, sex, calendar month at diagnosis, maternal and paternal age at child's birth, family history of schizophrenia	
(Eaton et al. 2000) Denmark	ICD (schizophrenia, non-affective psychoses, affective psychosis, manic-depressive psychoses)	Registers	33,320 live births at 1 year occurred between 1973 and 1977 (10% random sample of all live births)	*Degree of urbanization*	Birth	Highest odds of schizophrenia and non-affective psychoses with urban birth (capital cf. rural areas), with a marked gradient by degree of urbanization. No differences by urban birth status for affective psychoses	Weight for age, gestational age, mother's age, parity, sex, birth year	Sample restricted to those under 21 years; only about 20% of people will have received diagnosis of a non-affective psychosis by then
(Schelin et al. 2000) Denmark	ICD-8 schizophrenia	Register	2,44 first admissions 1978–1982	*Degree of urbanization*	At diagnosis	Higher incidence (direct standardization) in the capital and suburban areas cf. rural areas in males and females. [NB: not formally tested, but 95% CI do not overlap]	Age and sex	People in the capital had higher odds of being diagnosed at the first admissions in the capital, but not in suburban areas cf. rural areas

(*continued*)

Table 1 (continued)

Study ID	Diagnostic criteria and outcome	Case ascertainment	Sample size	Definition of urban	Timing of urban	Findings	Confounders	Comments
(Suvisaari et al. 2000) Finland	ICD-8, ICD-9, schizophrenia	Register (hospital + pension)	Sample = all people born in 1950–1969 Cases = 15,892 patients hospitalized (1970–1991)	*Urban versus rural*	Birth	Lower risk of schizophrenia in urban versus rural areas	Age, sex, month of birth, birth cohort	Only 15–20% of population lived in urban areas over the study period
(Agerbo et al. 2001) Denmark	ICD-8, schizophrenia	Register	17,604 people (191 cases admitted in 1981–1993 and 17,413 controls matched by gender and age)	*Degree of urbanization*	Birth	Highest schizophrenia risk in people born in the capital cf. people born in rural areas, but no evidence of a gradient by degree of urbanization	Crowding, family history of schizophrenia (parents and siblings), season of birth	Crowding not associated with schizophrenia, although there might be limitations in choice of measurement
(Allardyce 2001) Ireland	ICD-9/ICD-10, schizophrenia	Register	439 cases diagnosed in 1979–1984 and 1992–1997 (177 in Galloway, 262 in Camberwell)	*Dumfries and Galloway, rural areas. Camberwell, inner city area*	At diagnosis	Higher incidence rate ratio in Camberwell cf. Dumfries and Galloway	Age, sex, time period	Timing of exposure at diagnosis does not rule out reverse causation
(Haukka et al. 2001) Finland	ICD-8/ICD-9, schizophrenia	Register	Sample = all people born 1950–1969 Cases = 14,828 patients hospitalized before 1992	*Density of postal addresses*	At birth	Higher schizophrenia risk in rural areas in the 1950s, higher risk in urban areas in the 1960s	Age, sex, birth year	
(Pedersen and Mortensen 2001a) Denmark	ICD-10 (schizophrenia)	Register	Sample = 2.66 million Danish citizens born 1950–1993 Cases = 10,264 people with schizophrenia	*Degree of urbanization*	Birth	Schizophrenia risk highest in the capital area cf. rural areas, with a gradient by degree of urbanization	Age, calendar year of diagnosis, history of mental illness in siblings, history of mental illness in parents	Same results found for different subgroupings of population, by place and time of birth of mother

(van Os et al. 2001) The Netherlands	DSM-III (any psychotic disorder, narrowly and broadly defined psychotic symptoms)	Survey, diagnostic interview (CIDI)	Sample = 7,076 individuals aged 18–64 years Cases = 107, any psychotic disorders. Psychotic symptoms = 295, narrow definition; 1,237 broad definition	*Degree of urbanization*	At diagnosis	Odds of reporting any outcome were highest in the most urbanized areas, with evidence of a gradient across levels of urbanization	Age, sex, education, country of birth of participant and parents	People born outside Finland excluded
(Pedersen and Mortensen 2001b) Denmark	ICD-8, ICD-10 (schizophrenia)	Register	Sample = 1.89 m Danish citizens born in 1956–1983 (807,000 born after 1971 for analyses on urbanicity during upbringing)	*Degree of urbanization*	Birth and upbringing	People born in the capital had highest schizophrenia risk, with evidence of gradient by urbanization	Age, sex, calendar year, history of mental health in parents and siblings	When adjusting for urbanicity at birth and upbringing, the former was no longer significantly associated with schizophrenia risk, although the latter was
			Cases = 8,253 people with schizophrenia (1,533 born after 1971)			Living in a higher degree of urbanization at age 5 than at birth was associated with higher schizophrenia risk cf. having always lived in rural areas. People living in a higher or lower degree of urbanization at age 10 cf. age 5 had higher or lower risks		
(Harrison et al. 2003) Sweden	ICD-9, ICD-10, schizophrenia, other non-affective psychoses	Register	Sample = 696,025 people born 1973–1980 Cases = 363 with schizophrenia, 590 with other non-affective psychoses	*Degree of urbanization*	Birth	Risk of non-affective psychoses, but not of schizophrenia, was higher for people born in the main cities and their suburbs	Birth weight, ponderal index, birth length, gestational age, season of birth, age of mother, APGAR score, parity, caesarean section, gender, mother's education	Short follow-up means results are generalizable only to early-onset cases

(*continued*)

Table 1 (continued)

Study ID	Diagnostic criteria and outcome	Case ascertainment	Sample size	Definition of urban	Timing of urban	Findings	Confounders	Comments
(van Os et al. 2003) The Netherlands	DSM-III-R (psychotic disorder, broad and narrow psychosis)	Survey, diagnostic interview (CIDI)	Sample = 7,076 people. Cases = 915 with subclinical symptoms, 295 clinical symptoms	*Degree of urbanization*	At diagnosis	Urbanicity was associated with higher risk of psychotic disorder in the proband and a broadly or narrowly defined psychosis in the family	Age, sex, level of education and country of birth of proband, proband's mother, and proband's father and for parental history of delusions or hallucinations or diagnosis	The study found a significant interaction between urbanicity and family history in their effects on psychotic disorder in the proband
(Byrne et al. 2004) Denmark	ICD-8, ICD-10, schizophrenia	Registers	Sample = 200,294 Cases =7,704	*Degree of urbanization*	At birth	Higher incidence rate ratios of schizophrenia among people born in the capital, capital suburbs, and provinces after adjusting for all covariates	Occupation, education, marital status, father's age at birth, death of a parent prior to admission, number of siblings, reference to father at birth, history of family psychiatric illness	Socioeconomic variables measured at time of diagnosis and not at birth
(Sundquist et al. 2004) Sweden	ICD-9, ICD-10 (and DSM-IV), psychosis	Registers	Sample = 4.4 m Swedish people aged 25–64 years on December 31, 1999 (35,727 people excluded as they had a previous admission) Cases = 6,160	*Quintiles of population density*	At diagnosis	Psychosis risk highest for both men and women in most urbanized quintile, with some evidence of a dose-response relationship	Marital status, education, immigrant status, age (stratified by gender)	Similar association found for depression, but not as strong
(Allardyce et al. 2005) Scotland, UK	ICD-9 (schizophrenia)	Registers	Sample/cases = 5,838 with hospital admission in 1989–1993	*Degree of urbanization*	At diagnosis	No differences in schizophrenia risk by degree of urbanicity	Social fragmentation, material deprivation	Higher levels of social fragmentation were associated with higher risk of schizophrenia

(Kirkbride et al. 2006) UK	DSM-IV	Population-based case finding, 1997–1999	Sample and cases = 568 (209 cases of schizophrenia)	*Southeast London (most urban), Nottinghamshire, and Bristol (least urban)*	At diagnosis	Higher incidence of schizophrenia in London cf. Nottingham and Bristol	Age, sex, ethnicity	Similar distribution found for affective psychoses
(Pedersen and Mortensen 2006a) Denmark	ICD-8, ICD-10 (schizophrenia)	Register	Sample = 711,897 people born in 1956–1986 alive at age 15 Cases = 2,720 with schizophrenia between 1970 and 2000	*Degree of urbanization*	Birth, birth of sibling, age 15	People born in capital areas had higher schizophrenia risk cf. those born in rural areas, regardless of older sibling place of birth. Among people born in rural areas, some evidence that having an older sibling born in the capital increased schizophrenia	Age, sex, calendar year, parental age, history of mental health in parents or siblings	Study suggests that some of the causes of schizophrenia due to the environment are rooted in families
(Pedersen and Mortensen 2006b) Denmark	ICD-8, ICD-10 (schizophrenia)	Register	Sample = 1.89 m people born 1956–1983; cases = 10,755 people with schizophrenia	*Degree of urbanization*	Upbringing (15th birthday)	People living in the capital city had greatest risk of developing schizophrenia, with some evidence of a gradient across levels of urbanization	Sex, calendar year, distance from main road, and mental illness in parent or sibling	Distance from main road was no longer significant when model adjusted for degree of urbanization
(Pedersen 2006) Denmark	ICD-8, ICD-10 (schizophrenia)	Register	Sample = 5.05 m people born between 1910 and 1986; cases = 23,051 diagnosed in 1970–2001	*Degree of urbanization*	Birth	Urbanization associated with greater schizophrenia risk in the youngest, but not oldest, age cohort	Age and sex	
(Laursen et al. 2007) Denmark	ICD-8, ICD-10 (schizophrenia)	Register	Sample = 2 million people born in 1955–1978; cases =13,297 between 1973 and 2001	*Degree of urbanization*	Birth	People born in urban areas had twice the schizophrenia risk than those born in rural areas		

(*continued*)

Table 1 (continued)

Study ID	Diagnostic criteria and outcome	Case ascertainment	Sample size	Definition of urban	Timing of urban	Findings	Confounders	Comments
(Weiser et al. 2007) Israel	ICD-10 (schizophrenia)	Register	Sample = 371,603 conscripts aged 16–17; cases = 1,174	*Degree of urbanization*	At military conscription	Increasing population density corresponded to increasing schizophrenia risk (linear trend)	Age, vulnerability, and SES	Male only sample. Effect more pronounced in "vulnerable" sample
(Kelly et al. 2010) Ireland	DSM-III-R (schizophrenia, other psychotic illness)	Diagnostic interview (SCID)	Sample/cases = 324 schizophrenia	*Urban versus rural*	At diagnosis	People living in Dublin had twice the risk of developing schizophrenia cf. people living in rural countries	Age, sex, SES, ethnicity	
(Sørensen et al. 2014) Denmark	ICD-8, ICD-10 (schizophrenia)	Register	Sample = 2.49 m people born 1955–1993; cases = 17,389 between 1970 and 2005	*Degree of urbanization*	Birth	People living in the capital area had greatest schizophrenia risk, with some evidence of a gradient across levels of urbanization	Age, sex, calendar year, parental history of mental illness, parental loss, immigration, parental age	
(Szöke et al. 2014) France	DSM-IV (non-affective psychoses, affective psychoses)	Reported by public and private services	Sample = 246,773 (aged 18–64 residing in catchment area and diagnosed with schizophrenia)	*Three areas according to town size (smallest towns, medium towns, largest towns)*	At diagnosis	People living in the largest towns had higher risk for both affective and non-affective psychoses	Age and sex	Risk for non-affective psychoses in largest towns was greater than that for affective psychoses
(Vassos et al. 2015) Denmark	ICD-10 (schizophrenia)	Register	Sample = 2,894,640 born in 1995–2012 Cases = 13,702	*Degree of urbanization*	Birth	Higher incidence of schizophrenia among people born in the capital cf. rural areas. Evidence of a dose response	Calendar period, age and sex, parental age at birth, family history of mental illness	Higher incidence of affective psychoses among people born in the capital versus rural areas, but no dose response

(Moreno-Küstner et al. 2014) Spain	ICD-10 (schizophrenia)	Register	Sample = 1,663 Cases = 1,052	*Urban versus rural*	At diagnosis	Higher schizophrenia rates in urban areas	None	No confounders included in analyses
LAMIC								
(Ganguli 2000) India	ICD and DSM (unspecified) schizophrenia	Multiple[a]	Review of 15 studies across Indian regions	*Urban versus rural*	At diagnosis	Higher prevalence of schizophrenia in rural cf. urban areas	Multiple[a]	Differences between urban and rural prevalence not tested statistically
(Vicente et al. 2006) Chile	DSM-III-R non-affective psychosis	Survey, diagnostic interview (CIDI)	Sample = 2,978 people	*Urban versus rural*	At diagnosis	Prevalence of non-affective psychoses lowest in area with greatest % rural population	Age, sex, education, marital status, income	Urban-rural areas not directly compared
(Wei et al. 2008) Tibet	DSM-IV schizophrenia	SCID-I	Sample = 5,145 Cases = N/A	*Urban versus rural*	At diagnosis	No significant differences between urban and rural areas	–	No multivariable analysis
(Xiang et al. 2008) China	Schizophrenia	CIDI	Sample = 5,926 Cases = 0.49% of the sample	*Urban versus rural*	At diagnosis	Higher odds of schizophrenia in urban areas after adjustment for confounders	Age, sex, marital status, education, employment, income, smoking status, family history of schizophrenia	
(Lundberg et al. 2009) Uganda	Psychotic symptoms	Interview (PDI-21)	Sample = 646	*Urban, semi-urban, rural*	Birth and upbringing	People born in urban areas had higher odds of having psychotic symptoms	Age, sex, education, marital status, SES, family history of psychiatric disorders	Small sample, findings could be due to chance
(Phillips et al. 2009) China	DSM-IV schizophrenia	Screening with GHQ followed up by SCID interview	Sample = 63,004 Cases = 2%	*Urban versus rural*	At diagnosis	No differences in odds of schizophrenia in urban and rural areas	Data were adjusted for design factors and clustering. Post-stratified to the sampling frame	No confounders included in analyses

(continued)

Table 1 (continued)

Study ID	Diagnostic criteria and outcome	Case ascertainment	Sample size	Definition of urban	Timing of urban	Findings	Confounders	Comments
(Song et al. 2009) China	CCMD schizophrenia	Screening (unspecified) followed by interview	Sample = 294,356 Age = 22–29	*Urban versus rural*	At diagnosis	Lower odds of schizophrenia in urban cf. rural populations	Sex, birth cohort (pre-during, post-famine)	Study refers to period of the famine of 1959–1961. Generalizability to other contexts might be limited
(Long et al. 2014) China	Schizophrenia	Multiple[a]	Meta-analysis of 52 studies	*Urban versus rural*	At diagnosis	Higher prevalence in urban cf. rural dwellers	Heterogeneous across studies	
(Chan et al. 2015a) China	Schizophrenia	Multiple[a]	Review of 42 prevalence studies	*Urban versus rural*	At diagnosis	Higher prevalence in urban cf. rural dwellers	Heterogeneous across studies	Differences between urban and rural prevalence not tested statistically

CIDI Composite International Diagnostic Interview, *CCMD* Chinese Classification of Mental Disorder, *DSM* Diagnostic and Statistical Manual of Mental Disorders, *GHQ* General Health Questionnaire, *ICD* International Classification of Disease, *SCAN* Schedules for Clinical Assessment in Neuropsychiatry, *SCID* Structured Clinical Interview for DSM-IV Axis I Disorders (**SCID**-I), *cf. Compared with, SES* socioeconomic status

[a]The study is a literature review/meta-analysis; therefore, it is not possible to retrieve specific information for each study included

Table 2 Summary of literature on neighborhood differences in the distribution of schizophrenia (From 1990 onward, chronologically ordered)

Study ID and setting	Diagnostic criteria and outcome	Case finding, age range, years	Sample size and cases	Definition of neighborhood	Timing of exposure	Area-levels measures	Individual-level measures	Findings	Comments
High-income countries									
(Dauncey et al. 1993) UK	Schizophrenia	Psychiatric registers (age 15–54 years, 1978–1980)	Cases = 67 schizophrenia	5 enumeration districts (Nottingham)	Birth, upbringing, diagnosis, follow up	Deprivation	Age, sex, ethnicity	Of 67 cases, 27 were born in the inner city. 23 out of these 27 were born in council estates/deprived areas. Over half of the Nottingham-born were in deprived areas 5 years before and after diagnosis	Small sample limits validity of findings
(Löffler and Häfner 1999) Germany	ICD-9 (schizophrenia)	Psychiatric registers (age 12–59 years, 1987–1989)	Cases = 276 incidents of schizophrenia	23 districts in Mannheim, 14 in Heidelberg	At diagnosis	Measure of segregation and population density	–	Highest admission rates in inner city areas with high population density and segregation	Ecological study could not separate the role of area and individual
(Croudace et al. 2000) UK	ICD-10 (admissions and first-episode schizophrenia)	Hospital registers (ages 16–64 years, all admissions 1992–1993, first episode 1992–1994)	Cases = 1,208 admissions, 168 first episode	104 electoral wards (Nottingham)	At admissions	Social deprivation	Age and sex	Higher incidence and admissions for psychosis in areas with above-average social deprivation	Nonlinear association

(*continued*)

Table 2 (continued)

Study ID and setting	Diagnostic criteria and outcome	Case finding, age range, years	Sample size and cases	Definition of neighborhood	Timing of exposure	Area-levels measures	Individual-level measures	Findings	Comments
(Van Os 2000) The Netherlands	ICD-9 (Schizophrenia)	Psychiatric registers (all ages, 1986–1997)	Sample = ~120,000 inhabitants, cases = 220	35 neighborhoods in Maastricht (120,000 inhabitants). Traditional and sociological entities	At diagnosis	% men, % single/married, % under age 25. % of population on rental support, nonvoters, welfare dependent, foreign born, unemployed, mobility, new housing	Age, sex, marital status	12% of variance in schizophrenia incidence at neighborhood level. % single/divorced men associated with higher schizophrenia risk. Higher risk of schizophrenia for single men living in areas with fewer single men	
(Boydell et al. 2001) UK	ICD-9, ICD-10 (schizophrenia)	Psychiatric registers (all ages, 1988–1997)	Sample = ~120,000 inhabitants Cases = 222	15 electoral wards in Camberwell, London	At diagnosis	% non-white ethnic minority, deprivation	Age, sex, ethnicity	Significant differences in incidence between neighborhoods. Higher rates in non-white ethnic minorities as % non-white ethnic inhabitants decreased	

(Silver et al. 2002) USA	DSM-III (schizophrenia)	Survey, diagnostic interview (DIS) (age 18–96 years)	Sample = 11,686 survey respondents	261 census tracts across New Haven, CT; Baltimore, MD; St. Louis, MO; Durham, NC; and Los Angeles, CA	At diagnosis	Index consisting of several measures of poverty, social isolation, and residential turnover	Age, sex, ethnicity, household income, years of education, marital status	Higher schizophrenia risk in women and with less education, lower income, single marital status, and neighborhoods with higher turnover and disadvantage; association disappeared after adjustment for SES	
(Boydell et al. 2004) UK	Unspecified (schizophrenia)	Psychiatric registers (1988–1997)	222 cases	15 electoral wards in Camberwell, London	At diagnosis	% non-white ethnic minorities, deprivation, inequality	Age, sex, individual ethnicity	Inequality only associated with higher rates of schizophrenia in most deprived areas	
(Drukker et al. 2006) The Netherlands	DSM-IV (schizophrenia)	Psychiatric registers (all ages, 1993–2002)	93 cases/ 3369 controls	36 districts in Maastricht	At diagnosis	Informal social control, social cohesion and trust, socioeconomic deprivation	Age, sex, education, SES, marital status, education, employment, and living conditions	Low social cohesion and trust, high levels of residential instability associated with higher treated incidence of schizophrenia. Associations no longer significant after adjustment for SES	Social capital obtained from questionnaire sent to sample of residents in each district (48% response)

(continued)

Table 2 (continued)

Study ID and setting	Diagnostic criteria and outcome	Case finding, age range, years	Sample size and cases	Definition of neighborhood	Timing of exposure	Area-levels measures	Individual-level measures	Findings	Comments
(Lofors and Sundquist 2007) Sweden	ICD-9/ICD-10 (psychosis)	Psychiatric registers (incident cases) (age 25–64 years, 1997–1999)	Sample = 4.5 million Swedish citizens. Cases = 10,930	Clusters of small area units (~2,000 residents each in Stockholm, 1,000 in the rest of the country)	At diagnosis	Linking social capital (voting participation), neighborhood deprivation	Sex, age, housing tenure, education, marital status, employment, country of birth	Low social capital and neighborhood deprivation associated with high rates of psychoses, even after adjustment for individual-level characteristics	Individual characteristics explain the whole association for depression
(Kirkbride et al. 2007b) UK	ICD-10 (schizophrenia, non-affective psychoses)	Clinical interview for cases identified in hospital registers [SCAN] (age 16–64 years, 1997–1999)	Sample = 2001 census ward population, cases = 218 non-affective psychosis	33 wards in Southeast London (Lambeth and Southwark, ~6,000 residents in each area)	At diagnosis	Population density, ethnic density, ethnic fragmentation, deprivation, voter turnout	Age, sex, ethnicity	Low social capital (measured by voting turnout) associated with higher rates of psychosis, even after adjusting for individual-level variables. Lower ethnic fragmentation associated with lower rates	

(Kirkbride et al. 2007a) UK	ICD-10 (schizophrenia, non-affective psychoses)	Clinical interview for cases identified in hospital registers [SCAN] (age 16–64 years, 1997–1999)	Sample = 2001 census ward population Cases = 294 (218 non-affective, 76 affective psychosis)	33 wards in Southeast London (Lambeth and Southwark, ~6,000 residents in each area)	At diagnosis		Age, sex, ethnicity	Incidence of broadly defined psychosis and non-affective psychosis followed nonrandom geographical distribution after accounting for individual-level variables. 12% of variance explained at neighborhood level (Kirkbride et al. 2011)	Pattern not observed for affective psychosis, but small sample
(Kirkbride et al. 2008) UK	ICD-10 (schizophrenia)	Clinical interview for cases identified in hospital registers [SCAN] (1997–1999, age 16–64)	Sample = 2001 census ward population, cases = 148 schizophrenia	33 wards in Southeast London (Lambeth and Southwark, ~6,000 residents in each area)	At diagnosis	Social cohesion and trust, social disorganization, ethnic density, ethnic fragmentation, socioeconomic deprivation	Age, sex, ethnicity	Evidence of nonlinear (U-shaped) association between social cohesion and trust and incidence of schizophrenia. No evidence of role for social disorganization. Lower ethnic fragmentation associated with lower rates	Social capital measured 5–9 years after admissions.

(*continued*)

Table 2 (continued)

Study ID and setting	Diagnostic criteria and outcome	Case finding, age range, years	Sample size and cases	Definition of neighborhood	Timing of exposure	Area-levels measures	Individual-level measures	Findings	Comments
(Werner et al. 2006) Israel	ICD-9 (schizophrenia)	Psychiatric registers (incident cases) (individuals born 1964–76; followed until 1997)	Sample = 68,794 people born and living in Jerusalem Cases = 520	24 areas (~2,900 residents in each area)	At diagnosis	Area-level SES index from multiple indicators	Age, sex, father's age and ethnicity, occupational prestige status, parental education	Lower area-level SES associated with higher schizophrenia rates after adjustment for individual-level SES	
(Veling et al. 2008) The Netherlands	DSM-IV (all psychotic disorders)	Clinical interview for cases identified in hospital registers (CASH) (age 15–54 years, 1997–1999, 2000–2005)	Sample = whole population, cases = 466	44 neighborhoods in The Hague (max 38,000 inhabitants per neighborhood)	At diagnosis	Socioeconomic deprivation ethnic density	Ethnicity age, sex, marital status, ethnicity	Higher incidence of psychotic disorders in immigrants in low ethnic density neighborhoods. No differences between native Dutch residents and immigrants in high ethnic density neighborhoods	Outcome used might conceal differences between affective and non-affective psychoses

(Zammit et al. 2010) Sweden	ICD-8, ICD-9, ICD-10 (schizophrenia, non-affective psychoses, affective psychoses)	Psychiatric registers (individuals born in Sweden in 1972–1977 at age 16, followed up until 2003)	Sample = [illegible]69,910	1,264 schools, 284 municipalities, 24 counties in Sweden	Upbringing	School level = average: foreign born, social fragmentation, low grade	Sex, country of birth, parental history of psychosis, change of residence between 8 and 16 years, parental SES, education, family income, marital status, school grade	1. Higher risk of non-affective psychoses in urban cf. rural areas explained by school-level social fragmentation and population density	2.2% of variance explained at neighborhood level
			Cases = 881 any non-affective psychoses; 355 affective psychoses			Municipality = urbanicity, population density, index of social fragmentation (residential mobility, voting, % married/single households)		2. Only social fragmentation remained associated with non-affective psychosis risk after control for individual variables	
(Schofield et al. 2011) UK	Nonorganic psychosis (excluding drug-induced ones)	Primary care database (age 16–74 years)	Sample = 135,827 patients	Postcode areas	At diagnosis	Area deprivation, ethnic density	Age, sex, ethnicity	FEP rates in people of black ethnicity increased in neighborhoods with lower own-group ethnic density. Deprivation associated with greater FEP rates	Some evidence of a dose-response effect between ethnic density and FEP rates
			Cases = 277 patients with FEP						

(*continued*)

Table 2 (continued)

Study ID and setting	Diagnostic criteria and outcome	Case finding, age range, years	Sample size and cases	Definition of neighborhood	Timing of exposure	Area-levels measures	Individual-level measures	Findings	Comments
(Bhavsar et al. 2014) UK	OPCRIT analysis to obtain RDC first-episode schizophrenia	Hospital records (age 16–35 years, 2000–2007)	Sample = 267,000 Lambeth residents, cases = 405	177 super-output areas (~1,100–1,700 residents)	At diagnosis	Deprivation, ethnic group density, population density	Age, sex, ethnicity, SES	Higher area-level deprivation associated with increase in schizophrenia incidence, after adjusting for all other factors	
(Kirkbride et al. 2014) UK	DSM-IV (first-episode non-affective and affective psychoses)	Clinical interview for cases identified in hospital registers [SCAN] (18–64 years, 1996–1998 in City and Hackney; 1998–2000 in Newham and Tower Hamlet)	Sample = 2001 census population; cases = 484 first-episode psychosis	56 neighborhoods in East London boroughs of City and Hackney, Newham, and Tower Hamlets	At diagnosis	Deprivation, inequality, population density, own-group ethnic density, own-group ethnic separation, social fragmentation	Age at contact, ethnicity, social class	Deprivation, inequality, and population density were independently associated with increased incidence of non-affective psychosis after adjustment for individual-level variables. No interaction between inequality and deprivation	Evidence of higher rates of schizophrenia in neighborhoods with low ethnic density

(Lasalvia et al. 2014), Italy	ICD-10 (all psychoses, affective and non-affective)	SCAN to individuals age 15–54 screened positive for a possible FEP	Cases = 558 (441 non-affective, 117 affective psychosis)	198 municipalities in Veneto region	At diagnosis	Population density, neighborhood deprivation	Age, sex, immigration status	Non-affective incidence rates were twice as high in the most deprived areas versus other areas; no differences by population density	No differences for affective psychoses
(Omer et al. 2014) Ireland	DSM-IV (first-episode psychosis)	Clinical interview for cases identified in hospital registers (age 16+ years, 1995–2007)	Sample = 199,139 residents in 2002 Cases = 132 non-affective; 123 affective psychoses	155 electoral divisions in Cavan and Monaghan counties (Ireland)	At diagnosis	Material deprivation, social fragmentation, urbanicity	Age and sex	Association between deprivation and higher incidence of psychoses	Effect may have been in women only First study in an entirely rural setting
(Sariaslan et al. 2015) Sweden	ICD-9, ICD-10 (schizophrenia)	Register (people born 1967–1989 followed up from age 15 to 2009)	Sample = 2,361,585 in Sweden	Small area neighborhood units, excluding <50 people (~1,000 residents each)	Upbringing	Population density, neighborhood deprivation	Sex, birth year, birth order	Higher schizophrenia rates in areas with greater population density and deprivation. Partially mediated by unobserved familial characteristics; disappeared when investigating association within familial nuclei	

(*continued*)

Table 2 (continued)

Study ID and setting	Diagnostic criteria and outcome	Case finding, age range, years	Sample size and cases	Definition of neighborhood	Timing of exposure	Area-levels measures	Individual-level measures	Findings	Comments
(Veling et al. 2015) The Netherlands	DSM-IV (FEP)	Clinical interview for cases identified in hospital registers (CASH) (age 15–54 years, 1997–1999, 2000–2005)	Sample = 277,008 residents in 2005 Cases = 611	42 of 44 neighborhoods in The Hague	At diagnosis	Social disorganization domains = socioeconomic level, residential mobility, ethnic diversity, % single-person households, % voter turnout at local elections, population density, crime	Age, sex, ethnicity	All area-level indicators showed association with schizophrenia incidence initially, but after mutual adjustment only association with crime level remained significant	
(O'Donoghue et al. 2016) Ireland	DSM-IV (FEP)	SCID-I for DSM-IV in early intervention settings (between 2006 and 2010, aged 16–65 years)	Cases = 292 FEP participants	139 electoral divisions (median 2,386 inhabitants per neighborhood)	At diagnosis	Social deprivation (using Haase-Pratschke index), unemployment, social fragmentation, social capital, population density	Age, sex, marital status, DUP, comorbid substance use, functioning, family history of psychosis	FEP rate higher in neighborhoods with greater social fragmentation and deprivation and lower social capital DUP greater in neighborhoods with greater social fragmentation	

LAMIC									
(Burns and Esterhuizen 2008) South Africa	DSM-IV (first-episode psychosis)	Psychiatric registers, (incident cases) (age 15–49, 2005)	Sample = 508,275 from 2001 census population; cases = 160	7 municipalities, in District Umgungundlovu	At diagnosis	Poverty and inequality indices, urbanicity (% urban population)	Age, sex, ethnicity, employment status	Positive correlation between inequality and FEP incidence after adjustment for individual-level characteristics. No significant correlation between poverty and FEP incidence	

CAS Census Area Statistics, *CASH* Comprehensive Assessment of Symptoms and History, *DIS* Diagnostic Interview Schedule, *DSM* Diagnostic and Statistical Manual of Mental Disorder, *FEP* first episode of psychosis, *ICD* International Classification of Disease, *LSOA* Lower Super Output Area, *OPCRIT* operational criteria, *QMPA* Questionnaire for Adult Psychiatric Morbidity, *SCAN* Schedules for Clinical Assessment in Neuropsychiatry, *SCID* Structured Clinical Interview for DSM Disorders, *cf.* Compared with, *SES* socioeconomic status, *LAMIC* low- and middle-income country, *DUP* duration of untreated psychosis

Urban Versus Rural Settings

Perhaps owing to the presence of comprehensive population and hospital registers, Western Europe has provided the setting for the vast majority of the 40 studies which have investigated urban-rural differences in the distribution of psychosis since 1990 (Table 1; Kirkbride et al. 2006; Lewis et al. 1992; Thornicroft et al. 1993; Takei et al. 1995; Marcelis et al. 1998, 1999; Mortensen et al. 1999; Schelin et al. 2000; Eaton et al. 2000; Suvisaari et al. 2000; Agerbo et al. 2001; Allardyce 2001; Haukka et al. 2001; Pedersen and Mortensen 2001a, b, 2006a, b; van Os et al. 2001, 2003; Harrison et al. 2003; Sundquist et al. 2004; Allardyce et al. 2005; Pedersen 2006; Laursen et al. 2007; Kelly et al. 2010; Sørensen et al. 2014; Szöke et al. 2014; Vassos et al. 2015; Moreno-Küstner et al. 2014; Byrne et al. 2004), with 13 studies in Denmark (Mortensen et al. 1999; Schelin et al. 2000; Eaton et al. 2000; Agerbo et al. 2001; Pedersen and Mortensen 2001a, b, 2006a, b; Pedersen 2006; Laursen et al. 2007; Sørensen et al. 2014; Vassos et al. 2015; Byrne et al. 2004), four in the Netherlands (Marcelis et al. 1998, 1999; van Os et al. 2001, 2003), three in Sweden (Lewis et al. 1992; Harrison et al. 2003; Sundquist et al. 2004), three in the UK (Kirkbride et al. 2006; Takei et al. 1995; Allardyce et al. 2005), two in Finland, (Suvisaari et al. 2000; Haukka et al. 2001) two in Ireland (Allardyce 2001; Kelly et al. 2010), one in Italy (Thornicroft et al. 1993), one in Spain (Moreno-Küstner et al. 2014), and one in France (Szöke et al. 2014); one study was conducted in Israel (Weiser et al. 2007). In comparison, research from developing countries has been more sparse (Xiang et al. 2008; Ganguli 2000; Vicente et al. 2006; Wei et al. 2008; Lundberg et al. 2009; Phillips et al. 2009; Song et al. 2009).

Urbanicity has commonly been defined by grouping geographical areas from the most rural to the most urban on the basis of population density (Marcelis et al. 1998, 1999; Sundquist et al. 2004; Weiser et al. 2007), degree of urbanization (Kirkbride et al. 2006; Lewis et al. 1992; Mortensen et al. 1999; Schelin et al. 2000; Eaton et al. 2000; Agerbo et al. 2001; Pedersen and Mortensen 2001a, b, 2006a, b; Harrison et al. 2003; Allardyce et al. 2005; Pedersen 2006; Laursen et al. 2007; Sørensen et al. 2014; Szöke et al. 2014; Vassos et al. 2015; Byrne et al. 2004; Vicente et al. 2006; Lundberg et al. 2009), density of postal addresses (Haukka et al. 2001; van Os et al. 2001, 2003), or a dichotomized urban-rural classification (Thornicroft et al. 1993; Takei et al. 1995; Suvisaari et al. 2000; Allardyce 2001; Kelly et al. 2010; Moreno-Küstner et al. 2014; Xiang et al. 2008; Ganguli 2000; Wei et al. 2008; Phillips et al. 2009; Song et al. 2009). With the exception of an Israeli study, which measured diagnosis at military conscription (Weiser et al. 2007), urbanicity has been measured at the time of diagnosis (Kirkbride et al. 2006; Thornicroft et al. 1993; Schelin et al. 2000; Allardyce 2001; van Os et al. 2001, 2003; Sundquist et al. 2004; Allardyce et al. 2005; Kelly et al. 2010; Szöke et al. 2014; Moreno-Küstner et al. 2014; Xiang et al. 2008; Ganguli 2000; Vicente et al. 2006; Wei et al. 2008; Phillips et al. 2009; Song et al. 2009), birth (Takei et al. 1995; Mortensen et al. 1999; Eaton et al. 2000; Suvisaari et al. 2000; Agerbo et al. 2001; Haukka et al. 2001; Pedersen and Mortensen 2001a, 2006a; Harrison et al. 2003; Pedersen 2006; Laursen et al. 2007; Sørensen et al. 2014; Vassos et al. 2015; Byrne et al. 2004), birth and upbringing (Marcelis et al. 1998; Pedersen and Mortensen 2001b; Lundberg et al. 2009), birth and diagnosis (Marcelis et al. 1999), or upbringing only (Lewis et al. 1992; Pedersen and Mortensen 2006b). While methods used to diagnose participants have varied (see Table 1), many studies used standardized clinical interviews to derive a diagnosis of schizophrenia (Kirkbride et al. 2006; van Os et al. 2001, 2003; Kelly et al. 2010; Xiang et al. 2008; Vicente et al. 2006; Wei et al. 2008; Lundberg et al. 2009; Phillips et al. 2009; Song et al. 2009) rather than reliance solely on case records. Many studies simultaneously investigated urbanization in relation to depression (see also section "Common Mental Disorders Depression and Anxiety") (Sundquist et al. 2004; Vicente et al. 2006; Wei et al. 2008; Lundberg et al. 2009), affective psychotic disorders (see section "A Note on the Affective Psychoses") (Takei et al. 1995; Marcelis et al. 1998; Eaton et al. 2000; Allardyce 2001; Laursen et al. 2007; Szöke et al. 2014; Vicente et al. 2006; Phillips et al. 2009), and other non-affective psychoses (Marcelis et al. 1998, 1999; Suvisaari et al. 2000; Harrison et al. 2003; Szöke et al. 2014; Vicente et al. 2006). Finally, an increasing number of studies have employed longitudinal cohorts (Kirkbride et al. 2006; Lewis et al. 1992; Thornicroft et al. 1993; Marcelis et al. 1998, 1999; Mortensen et al. 1999; Schelin et al. 2000; Suvisaari et al. 2000; Allardyce 2001; Haukka et al. 2001; Pedersen and Mortensen 2001a, 2006a, b; Harrison et al. 2003; Sundquist et al. 2004; Allardyce et al. 2005; Pedersen 2006; Laursen et al. 2007; Sørensen et al. 2014; Vassos et al. 2015; Byrne et al. 2004) (which usually provide stronger, prospective evidence of an association) over other types of study design, including cross-sectional (van Os et al. 2001; Moreno-Küstner et al. 2014; Xiang et al. 2008; Vicente et al. 2006; Wei et al. 2008; Lundberg et al. 2009; Phillips et al. 2009; Song et al. 2009) (more frequent in LAMIC) or case-control (Takei et al. 1995; Eaton et al. 2000; Agerbo et al. 2001) designs. Despite heterogeneity in design, populations, and definitions of urbanicity and diagnosis, all but eight studies (Thornicroft et al. 1993; Suvisaari et al. 2000; Haukka et al. 2001; Allardyce et al. 2005; Ganguli 2000; Wei et al. 2008; Lundberg et al. 2009; Phillips et al. 2009; Song et al. 2009) found higher schizophrenia risk in urban compared with rural areas (Kirkbride et al. 2006; Lewis et al. 1992; Takei et al. 1995; Marcelis et al. 1998, 1999; Mortensen et al. 1999; Schelin et al. 2000; Eaton et al. 2000; Agerbo et al. 2001; Allardyce 2001; Pedersen and Mortensen 2001a, b; van Os et al. 2001, 2003; Harrison et al. 2003; Sundquist et al. 2004; Pedersen and Mortensen 2006a, b; Pedersen 2006; Laursen et al. 2007; Kelly et al. 2010; Sørensen et al. 2014; Szöke et al. 2014; Vassos et al. 2015; Moreno-Küstner et al. 2014; Byrne et al. 2004; Weiser et al. 2007; Xiang et al. 2008; Vicente et al. 2006; Chan et al. 2015a; Long et al. 2014). Of those which did not, three reported no urban-rural differences (Allardyce et al. 2005; Wei et al. 2008; Phillips et al. 2009), while five observed higher rates in rural areas (Thornicroft et al. 1993; Suvisaari et al. 2000; Haukka et al. 2001; Ganguli 2000; Lundberg et al. 2009; Song et al. 2009). Summarizing much of the literature from high-income countries, a recent meta-analysis estimated that people living in urban areas at birth and upbringing had over twice the odds of developing schizophrenia compared with people living in rural areas (odds ratio [OR], 2.37; 95% confidence interval [95% CI], 2.01–2.81) (Vassos et al. 2012).

Findings from nine studies conducted in LAMIC were less consistent than those reported from high-income countries (see Table 1). Lifetime and 12-month prevalence of schizophrenia seems to be higher in urban areas in Chile (men) (Vicente et al. 2006) and China (Xiang et al. 2008; Chan et al. 2015a; Long et al. 2014), while the reverse has been observed in India (Ganguli 2000), and the Chinese famine study (Song et al. 2009), where rural populations experienced greatest levels of malnutrition. Other studies, including those from Tibet (Wei et al. 2008), Uganda (Lundberg et al. 2009), and China (Phillips et al. 2009), have found equivocal prevalence between urban and rural populations. Of the eight studies which did not find and observe higher schizophrenia rates in more urban areas (see previous paragraph), five were conducted in LAMIC settings (Xiang et al. 2008; Ganguli 2000; Lundberg et al. 2009; Phillips et al. 2009; Song et al. 2009), meaning only four of nine LAMIC studies (44.4%) observed this phenomenon (Xiang et al. 2008; Vicente et al. 2006; Chan et al. 2015a; Long et al. 2014), compared with 28 out of 31 (90.3%) studies in HIC settings; this difference was strongly statistically significant (chi^2 [χ^2] test; $p = 0.002$; Table 1). At present, it is not possible to determine the reason for this difference; on the one hand, the composition and risk profiles of people living in rural and urban settings in LAMIC settings may be very different to those in HICs, while on the other hand, methodological limitations of some studies from LAMIC settings (see below) may explain this difference. In general, these issues include the validity of definitions of urban exposure used (Vicente et al. 2006), small sample sizes (Lundberg et al. 2009), absences in statistical testing (Ganguli 2000; Chan et al. 2015b), or failure to use appropriate regression models (Wei et al. 2008). This may limit the validity and generalizability of these results, despite exceptions which have employed large sample sizes and robust statistical models to investigate urban-rural differences (Xiang et al. 2008; Song et al. 2009).

Neighborhoods

The majority of the studies which have investigated the spatial distribution of schizophrenia at finer-grained, small area "neighborhood levels" have been conducted in Europe (Schofield et al. 2011; Boydell et al. 2001; Löffler and Häfner 1999; Dauncey et al. 1993; Croudace et al. 2000; Van Os 2000; Boydell et al. 2004; Drukker et al. 2006; Lofors and Sundquist 2007; Kirkbride et al. 2007a, b, 2008, 2014; Veling et al. 2008; Zammit et al. 2010; Bhavsar et al. 2014; Lasalvia et al. 2014; Omer et al. 2014; O'Donoghue et al. 2016; Veling et al. 2015; Sariaslan et al. 2015): nine in the UK (Schofield et al. 2011; Boydell et al. 2001; Dauncey et al. 1993; Croudace et al. 2000; Boydell et al. 2004; Kirkbride et al. 2007a, b, 2008, 2014; Bhavsar et al. 2014), four in the Netherlands (Van Os 2000; Drukker et al. 2006; Veling et al. 2008, 2015), three in Sweden (Lofors and Sundquist 2007; Zammit et al. 2010; Sariaslan et al. 2015), two in Ireland (Omer et al. 2014; O'Donoghue et al. 2016), one in Italy (Lasalvia et al. 2014), and one in Germany (Löffler and Häfner 1999) with a minority undertaken in Israel (Werner et al. 2006), South Africa (Burns and Esterhuizen 2008), and the USA (Silver et al. 2002). Most studies employed hospital or early intervention service registers to identify cases of schizophrenia (Schofield et al. 2011; Boydell et al. 2001, 2004; Löffler and Häfner 1999; Dauncey et al. 1993; Croudace et al. 2000; Van Os 2000; Drukker et al. 2006; Lofors and Sundquist 2007; Zammit et al. 2010; Bhavsar et al. 2014; O'Donoghue et al. 2016; Sariaslan et al. 2015; Werner et al. 2006; Burns and Esterhuizen 2008), or to identify a "screened" population to further interview with clinical instruments (Kirkbride et al. 2007a, b, 2008, 2014; Veling et al. 2008, 2015; Lasalvia et al. 2014; Omer et al. 2014), and only one study employed a survey design (Silver et al. 2002). Neighborhood of residence was generally defined based upon administrative units at the time of diagnosis (Schofield et al. 2011; Boydell et al. 2001, 2004; Löffler and Häfner 1999; Dauncey et al. 1993; Croudace et al. 2000; Van Os 2000; Drukker et al. 2006; Lofors and Sundquist 2007; Kirkbride et al. 2007a, b, 2008, 2014; Veling et al. 2008, 2015; Bhavsar et al. 2014; Lasalvia et al. 2014; Omer et al. 2014; O'Donoghue et al. 2016; Werner et al. 2006; Burns and Esterhuizen 2008; Silver et al. 2002) or, more rarely, upbringing (Table 2; Zammit et al. 2010; Sariaslan et al. 2015).

These ecological and multilevel studies have suggested that up to 12% of variance in the distribution of schizophrenia could be explained by neighborhood characteristics (Van Os 2000; Lofors and Sundquist 2007; Kirkbride et al. 2007b; Zammit et al. 2010), although further work is required here; this pattern may vary by degree of urbanicity (higher in more urban areas (Kirkbride et al. 2006; Lewis et al. 1992; Takei et al. 1995; Marcelis et al. 1998, 1999; Mortensen et al. 1999; Schelin et al. 2000; Eaton et al. 2000; Agerbo et al. 2001; Allardyce 2001; Pedersen and Mortensen 2001a, b, 2006a, b; van Os et al. 2001, 2003; Harrison et al. 2003; Sundquist et al. 2004; Pedersen 2006; Laursen et al. 2007; Kelly et al. 2010; Sørensen et al. 2014; Szöke et al. 2014; Vassos et al. 2015; Moreno-Küstner et al. 2014; Byrne et al. 2004; Weiser et al. 2007; Xiang et al. 2008; Vicente et al. 2006; Chan et al. 2015a; Long et al. 2014)), and most studies, despite limited exceptions (Zammit et al. 2010; Sariaslan et al. 2015), have not examined other potential causal levels such as the role of the family or school environment.

Indicators of area socioeconomic deprivation, such as overcrowding (Boydell et al. 2001), housing tenure (Van Os 2000; Omer et al. 2014; Silver et al. 2002), unemployment (Boydell et al. 2001; Croudace et al. 2000; Van Os 2000; Drukker et al. 2006; Lofors and Sundquist 2007; Kirkbride et al. 2007b, 2008, 2014; Veling et al. 2008; Zammit et al. 2010; Bhavsar et al. 2014; Lasalvia et al. 2014; Omer et al. 2014; O'Donoghue et al. 2016; Burns and Esterhuizen 2008; Silver et al. 2002), and social/welfare benefits (Van Os 2000; Zammit et al. 2010; Silver et al. 2002), have extensively been used in the literature, either individually (Van Os 2000) or in combined indices (Schofield et al. 2011; Boydell et al. 2001, 2004; Löffler and Häfner 1999; Dauncey et al. 1993; Croudace et al. 2000; Drukker et al. 2006; Lofors and Sundquist 2007; Kirkbride et al. 2007b, 2008, 2014; Veling et al. 2008, 2015; Zammit et al. 2010; Bhavsar et al. 2014; Lasalvia et al. 2014; Omer et al. 2014; O'Donoghue et al. 2016; Sariaslan et al. 2015; Werner et al. 2006; Burns and Esterhuizen 2008; Silver et al. 2002). Although the plethora of measures used, and inclusion of different individual- and area-level variables, makes direct comparison of findings difficult, evidence of an association between measures of *absolute* deprivation at the neighborhood level and higher risk of schizophrenia has been consistently observed (Dauncey et al. 1993; Croudace et al. 2000; Boydell et al. 2004; Veling et al. 2008; Zammit et al. 2010; Bhavsar et al. 2014; Lasalvia et al. 2014; Omer et al. 2014; O'Donoghue et al. 2016; Werner et al. 2006; Burns and Esterhuizen 2008; Silver et al. 2002). Researchers have recently become increasingly interested in the role of *relative* deprivation, with higher schizophrenia rates reported in neighborhoods with greater inequality (Boydell et al. 2004; Kirkbride et al. 2014; Burns and Esterhuizen 2008). As further discussed in section "Main Hypotheses," it has been hypothesized that exposure to deprivation and inequality may lead to psychosocial stresses arising from perceptions of exclusion, threat, and mistrust, which in turn lead to the onset of psychotic symptoms (Bhavsar et al. 2014).

Subsequently, this has led researchers to investigate how neighborhood-level social capital affects psychosis risk. Social capital refers to the amount of "resources [...] embedded within an individual's social network" (Kawachi 2006) (p. 911) (i.e., structural social capital), as well as social cohesion (i.e., cognitive social capital Kawachi 2006; Mckenzie 2008), which taps into the presence of shared social norms, values, and belonging within a community. Although developing routine measures to operationalize these potentially nebulous constructs within the field of psychiatric epidemiology remains a challenge, studies which have measured social cohesion have shown that higher levels are associated with positive mental health outcomes (De Silva et al. 2005). Studies which have investigated this directly in relation to psychotic disorders have used various proxy markers of social cohesion, including voter turnout (Van Os 2000; Lofors and Sundquist 2007; Kirkbride et al. 2007b; Zammit et al. 2010; Veling et al. 2015), the proportion of the population engaging in voluntary work (O'Donoghue et al. 2016), and direct measures of social cohesion and trust via population surveys (Drukker et al. 2006; Kirkbride et al. 2008). Evidence to support an association between social cohesion and schizophrenia is mixed, with positive findings from some studies (Lofors and Sundquist 2007; Kirkbride et al. 2007b; O'Donoghue et al. 2016), but not others (Van Os 2000; Drukker et al. 2006; Zammit et al. 2010; Veling et al. 2015), following adjustment for other individual- and area-level risk factors. One study reported a nonlinear association between social cohesion and trust and schizophrenia rates (Kirkbride et al. 2008), with higher rates in areas with low or high levels of social cohesion (compared with areas with medium levels). Thus, although neighborhood-level social capital may be associated with schizophrenia, further research is required to investigate this in more detail.

A related concept of social fragmentation, first tested by Faris and Dunham (1939), has also been considered in relation to the incidence of psychotic disorders in urban areas. Social fragmentation has been heterogeneously operationalized in the literature, but can perhaps be thought of as indexing the absence of the social building blocks which would lead to the formation of social capital. Several neighborhood-level markers of this broad construct have been investigated in relation to psychosis incidence, including the proportion of single persons (Van Os 2000; Zammit et al. 2010; Kirkbride et al. 2014; Veling et al. 2015), proportion of rented households (as an indicator of housing turnover and therefore transience) (Kirkbride et al. 2014; Silver et al. 2002), and residential mobility (Van Os 2000; Zammit et al.

2010; Kirkbride et al. 2014; Veling et al. 2015; Silver et al. 2002). These measures have been associated with greater schizophrenia rates, in some (Van Os 2000; Zammit et al. 2010), but not all studies (Kirkbride et al. 2014; Veling et al. 2015; Silver et al. 2002), with some evidence they may be driven by individual-level social isolation rather than its neighborhood corollary (Veling et al. 2015).

Ethnic density – the proportion of someone's own ethnic group in their neighborhood – may represent a further ethnicity-specific construct of social fragmentation. Ethnic density may underpin a range of social processes, including the extent to which there are opportunities in your residential neighborhood to develop ties and bonds with other people who may share similar migration histories, cultures, religious beliefs, values, norms, customs, conventions, and modes of behavior. The association between ethnic density and schizophrenia risk was first observed by Faris and Dunham (1939), and has subsequently been observed in various settings, where the relative risk for ethnic minority groups (compared with the majority population) is reduced as one's own-group ethnic density becomes greater at the neighborhood level (Kirkbride et al. 2007b, 2008, 2014; Veling et al. 2008). Further research from Kirkbride et al. (2014) has extended this construct to consider the specific spatial patterning of ethnic groups within a community, which he has termed *ethnic fragmentation*. This research has revealed independent effects of both ethnic density and fragmentation on subsequent rates of non-affective psychotic disorders (Kirkbride et al. 2007b, 2008, 2014). A recent systematic review concluded that ethnic density may be protective against several adverse mental health outcomes, not limited to schizophrenia (Shaw et al. 2012). The authors suggest ethnic density could act as a buffer toward negative social experiences (such as racism and discrimination) by promoting resilience. Living in sociodemographically and socioeconomically homogeneous areas could also promote a stronger sense of identity and self-esteem in turn reducing experiences of conflict in social interactions (Shaw et al. 2012).

Strengths and Limitations

Findings of higher rates of schizophrenia in urban areas and in some neighborhoods compared with others are consistent in the schizophrenia literature across a number of study designs (e.g., longitudinal vs. cross-sectional or register-based studies vs. surveys), populations (e.g., service users, general population), settings (including various European countries and the USA), and sample sizes, suggesting that these results are unlikely to have arisen by chance. The increasing number of studies measuring urbanicity at birth also provides evidence that urban exposures early in life could be a risk factor for schizophrenia.

Despite these considerable advantages, there are a number of limitations which research have yet to fully address. For instance, while cohort studies which use large – often national – population registers can provide powerful, longitudinal information on a large number of individuals, the breadth of data they can include is typically narrower, often restricted to routine administrative sources. While these studies can account for several important sociodemographic indicators (e.g., age, gender, parental age, country of birth, income, education), measures of area deprivation, obstetric complications, and history of family mental illness, data on other individual-level characteristics may not always be available, including ethnicity (Veling et al. 2010; Tortelli et al. 2015; Morgan et al. 2010), traumatic life events (Myin-Germeys et al. 2003; Read et al. 2005; Norman and Malla 1993), substance use (Løberg et al. 2014; D'Souza et al. 2009), or markers of genetic vulnerability, although novel approaches are being developed (Sariaslan et al. 2015). Efforts to disentangle the possibility that associations between urban living and psychosis risk are due to other unmeasured factors, including the role of genes are still required.

A Note on the Affective Psychoses

Curiously, the relationship between psychotic disorders and the urban environment appears to be specific to non-affective psychotic disorder, with no robust evidence that the affective psychoses, including bipolar disorder with psychotic features and psychotic depression, share such environmental correlates, despite a shared genetic predisposition (Craddock et al. 2005; Blackwood et al. 2007). The balance of evidence with respect to affective psychoses, including analyses of urban birth using national register data, suggests that there are few differences in incidence rates between urban and rural areas (Eaton et al. 2000; Kirkbride et al. 2007a, 2014; Mortensen et al. 2003; Pedersen and Mortensen 2006c). Only three studies have observed higher rates of affective psychoses in urban areas compared with rural ones (Kirkbride et al. 2006; Marcelis et al. 1998; Szöke et al. 2014), and the magnitude of this association was smaller than observed for non-affective psychoses, while these studies only presented basic control for possible confounding factors (e.g., age, sex, ethnicity). Further research here is warranted, particularly as there is some evidence that nonpsychotic affective disorders (such as depression) may vary by urban-rural environments (see section "Common Mental Disorders (Depression and Anxiety)"). Finally, it is possible that the urban environment acts at the symptom, rather than disorder level; in a study of people with ICD-10-confirmed first-episode psychosis (including affective psychoses), Oher et al. (2014) observed that the urban environment was most strongly associated with positive symptoms (specifically hallucinations) and depressive symptoms.

Main Hypotheses

In this section, we review the evidence in support of the main hypotheses concerning the association between urban residency and later risk of schizophrenia and other non-affective psychoses.

Individual (Intragenerational) Social Drift

People who experience psychotic disorder usually have a lower socioeconomic status compared with their parents or peers who do not experience disorder. As proposed by Goldberg and Morrison (1963), this may result in the drift of people both socially and spatially into more deprived, urban or less socially desirable neighborhoods, particularly into more deprived and fragmented parts of inner cities, for which there is some evidence (Ngamini Ngui et al. 2013). Furthermore, studies which have investigated urban living close to the time of onset cannot exclude this possibility (Boydell et al. 2001; Faris and Dunham 1939; Hare and Family 1956; Löffler and Häfner 1999; Giggs and Cooper 1987; Thornicroft et al. 1993; Schelin et al. 2000; Allardyce 2001; van Os et al. 2001, 2003; Sundquist et al. 2004; Allardyce et al. 2005; Kelly et al. 2010; Szöke et al. 2014; Vicente et al. 2006; Croudace et al. 2000; Van Os 2000; Drukker et al. 2006; Lofors and Sundquist 2007; Kirkbride et al. 2007a, b, 2008, 2014; Veling et al. 2008, 2015; Bhavsar et al. 2014; Omer et al. 2014; Werner et al. 2006; Burns and Esterhuizen 2008; Silver et al. 2002). Nonetheless, the individual social drift hypothesis cannot explain the association between urban birth or upbringing and later psychosis risk (Lewis et al. 1992; Takei et al. 1995; Marcelis et al. 1998, 1999; Mortensen et al. 1999; Eaton et al. 2000; Suvisaari et al. 2000; Agerbo et al. 2001; Haukka et al. 2001; Pedersen and Mortensen 2001a, b, 2006a; van Os et al. 2001; Harrison et al. 2003; Pedersen 2006; Laursen et al. 2007; Sørensen et al. 2014). Further, a recent study by Kirkbride et al. (2015) found that the spatial patterning of people classified as "at-risk mental states" for psychosis was more similar to that of first episode of psychosis (FEP) participants than of healthy controls, suggesting that those who go on to develop frank psychosis may be exposed to adverse social environments prior to a full psychotic episode. Confirmation of the existence of social and spatial drift would not preclude the possibility that aspects of the urban environment remain causally relevant to psychosis onset; indeed the two may coagulate to increase risk still further in the lead-up to disorder.

Intergenerational Social Drift

An interesting hypothesis to emerge more recently posits that drift could occur intergenerationally (Pedersen and Mortensen 2006b; van Os et al. 2005). According to this idea, which encompasses the more formal notion of gene-environment correlation, families with an underlying vulnerability to psychosis may drift into lower socioeconomic positions and environments over time, even though frank psychosis may not reveal itself for several generations. Central to this hypothesis is that underlying genetic or environmental insults result in subclinical endophenotypes associated with non-affective psychoses, which compromise typical neurodevelopment or cognitive processes which subsequently inhibit people's upward social mobility relative to their peers across a range of domains including education, employment, and social domains. While these traits may never lead to frank psychosis in a given generation, accumulated genetic and environmental insults may eventually trigger psychosis several generations later. Thus, the apparent association between urban exposure and psychosis in the index case may be more readily explained by intergenerational drift, occurring via either gene-environment correlation or environment-environment correlation. It may be both an *active* and *passive* process. In the *active form*, successive generations gradually drift into lower socioeconomic positions or more marginal neighborhoods, whereas in the *passive form*, such groups remain in approximately the same social position, but unaffected families around them, over generations, exhibit a trend for upward social mobility. In a recent Swedish study, associations between population density, neighborhood deprivation, and the incidence of schizophrenia were partially mediated by unobserved familiar risk factors and disappeared within nuclear families (Sariaslan et al. 2015), suggesting intergenerational drift may explain some of the association between risk and urban living. The current evidence, however, is limited and intergenerational social drift will be difficult to detect.

Antenatal and Perinatal Risk Factors

These hypotheses propose that higher rate schizophrenia for people born in urban areas could arise from a set of biological influences confounded by urban birth, including exposure to infection in pregnancy and the perinatal period (Bleuler 1911) or obstetric complications (Mednick 1970; Lane and Albee 1966; Lewis and Murray 1987). While these hypotheses are intuitively appealing, given that risk of infections, poorer prenatal nutrition, or obstetric complications could more commonly occur in urban populations, evidence to support this is limited. Season of birth has been used as a proxy for infection during pregnancy (particularly with regard to influenza), but there is no evidence that this interacts with place of birth (Marcelis et al. 1998; Suvisaari et al. 2000). Further studies which have directly tested whether exposure to prenatal infections confound the association between urban birth and schizophrenia are required. Only two studies, to our knowledge, have been investigated whether obstetric complications confound or mediate the association between urban birth and schizophrenia, with no apparent relationship (Eaton et al. 2000; Harrison et al. 2003). The overall paucity of studies investigating this potential risk pathway marks this as an important topic for future research.

Social Stressors

The possibility that urban environments create stressful environments which in turn disrupts key neurobiological pathways relevant to psychosis has received considerable lip service (Mizrahi 2016; Akdeniz et al. 2014). Broadly, findings of higher rates of schizophrenia in neighborhoods with greater levels of deprivation, population density, and crime are consistent with the possibility that stress provides a mechanism to increase later psychosis risk (Lewis et al. 1992; Takei et al. 1995; Marcelis et al. 1998, 1999; Mortensen et al. 1999; Schelin et al. 2000; Eaton et al. 2000; Agerbo et al. 2001; Allardyce 2001; Pedersen and Mortensen 2001a, b, 2006a, b; van Os et al. 2001, 2003; Harrison et al. 2003; Sundquist et al. 2004; Pedersen 2006; Laursen et al. 2007; Kelly et al. 2010; Sørensen et al. 2014; Szöke et al. 2014; Moreno-Küstner et al. 2014; Weiser et al. 2007; Xiang et al. 2008; Vicente et al. 2006; Lundberg et al. 2009; Chan et al. 2015a; Vassos et al. 2012). Furthermore, neighborhoods with low levels of protection from such risk factors, indexed by higher social fragmentation, lower levels of social cohesion (Van Os 2000; Drukker et al. 2006; Lofors and Sundquist 2007; Kirkbride et al. 2007a, b, 2008; Omer et al. 2014; O'Donoghue et al. 2016), or a lack of ethnic density (Kirkbride et al. 2008, 2014), also appear to have increased rates. Nonetheless, studies which directly link exposure to urban environments with social stress, disrupted neurobiological process, and psychosis are lacking, although new experimental data show that stressful social environments can increase paranoid ideation in healthy controls and people with existing psychosis (Veling et al. 2016). Furthermore, another experimental study (limited to healthy subjects) found that urban residence and upbringing were associated with increased amygdala activity and affected the perigenual anterior cingulate cortex – an area of the brain which regulates negative affect and stress (Tost et al. 2015). Specific hypotheses have also been advanced suggesting that experiences of social defeat or social exclusion could mediate the association between urban exposure and neurobiological processes at the basis of vulnerability to psychosis (Selten and Cantor-Graae 2005, 2007; Selten et al. 2013), though empirical evidence is still required. Alternatively, prolonged activation of the human stress response might suppress the hypothalamic-pituitary-adrenocortical axis (Akdeniz et al. 2014), which, in turn, could result in dysfunctions of the immune system, predisposing some individuals to psychiatric illnesses.

Gene-Environment Interactions

A hypothesis that has increasingly gained attention is that environmental factors could combine with preexisting genetic vulnerability to increase schizophrenia risk (van Os and McGuffin 2003). This possibility is supported by findings of independent associations between older sibling and individual's urban birth and the risk of schizophrenia, suggesting that risk might reside both in familial and environmental factors (Pedersen and Mortensen 2006c). The lack of evidence to suggest that genetic risk differs between populations arranged along geographical or ethnic gradients argues against a purely genetic explanation of higher rates of schizophrenia in urban populations and ethnic minorities, respectively (Kirkbride and Jones 2011). Instead, underlying genetic vulnerabilities to psychosis may mean that subsequent exposure to stressful life events, psychosocial adversities, and hostile social environments has a more deleterious effect on key neurobiological pathways that affect psychosis risk. Unfortunately, at present direct evidence for gene-environment interactions implicating the urban environment are largely absent. Results from large, multisite gene-environment interaction studies in schizophrenia may reveal new directions for research including whether increased polygenic risk for psychosis exacerbates later risk of psychosis following exposure to adverse social environments (van Os et al. 2014).

Differences in Health-Care Provision and Sociodemographic Characteristics

Two further hypotheses have been suggested to explain observed differences. First, it has been suggested that differences in the provision of health care between areas could account for the observed spatial heterogeneity in the distribution of schizophrenia. While differential access to mental health care between urban and rural settings and possible variation in duration of untreated psychosis remain important issues in their own right, there is little evidence to suggest such issues will explain urban-rural differences in risk (Kirkbride et al. 2010), given the use of population-based case ascertainment in studies which do not solely rely on routine case registers (i.e., see Kirkbride et al. 2006, 2007a, 2008, 2014).

Second, urban-rural differences in risk could be explained by differences in sociodemographic characteristics of individuals living in different areas (compositional effects). However, there is little evidence for this, although one study found that area-level differences disappeared after adjustment for individual-level factors (Drukker et al. 2006) (though most have not (Boydell et al. 2001; Van Os 2000; Lofors and Sundquist 2007; Kirkbride et al. 2007a, b, 2008, 2014; Zammit et al. 2010; Bhavsar et al. 2014; Omer et al. 2014; Veling et al. 2015; Werner et al. 2006; Burns and Esterhuizen 2008)). Advances in multilevel modeling techniques (Kirkbride et al. 2007a, 2014) have helped to partition variation in incidence rates between individual- and neighborhood-level factors. Most studies have investigated this variance in relation to the neighborhood, but school environments may be highly relevant at certain ages (Zammit et al. 2010), and the role of the family environment has largely been ignored, until very recently (Sariaslan et al. 2015).

Common Mental Disorders (Depression and Anxiety)

The annual prevalence of common mental disorders (CMD), defined as mood and anxiety disorders, has been recently estimated as 15.4% (95%CI, 12.8%–18.6%) (Steel et al. 2014), with higher rates in women compared with men for both mood (women, 7.3%, 95% CI, 6.5%–8.1%; men, 4.0%, 95% CI, 3.5%–4.6%) and anxiety (women, 8.7%, 95% CI, 7.7%–9.8%; men, 4.3%, 95% CI, 3.7%–4.9%). Compared with schizophrenia (section "Schizophrenia and Other Non-affective Psychoses"), evidence for nonrandom spatial variation of CMD is more mixed. Although this may reflect real, context-specific differences, a number of methodological limitations could also account for these more heterogeneous findings. As opposed to the literature on schizophrenia, for instance, fewer epidemiological studies of CMD have used longitudinal, population-based study designs (i.e., cohort studies) or have measured incidence (cf. prevalence). Such studies are much harder to conduct for CMDs, given that a high proportion of people experiencing CMD may never present to mental health services. For this reason, cross-sectional surveys of the prevalence of CMDs in the general population or other settings have been more commonly used. Unfortunately, reliance on cross-sectional surveys and estimates of prevalence (which include both new and existing cases) makes it more difficult to separate cause from effect in respect to studying the role of the urban environment on risk of developing CMDs. Moreover, while ample literature exists, it has often been restricted to population subgroups (e.g., adolescents, the elderly, ethnic minorities, or individuals with chronic disease), further limiting generalizability. Here, we review the strengths, weaknesses, and overall level of evidence for spatial variation in CMDs.

Urban Versus Rural Settings

The majority of studies (Table 3) investigating the distribution of CMD between urban and rural settings have been conducted in Europe (three in the UK (Lewis and

Table 3 Summary of literature on rural-urban differences in the distribution of common mental disorders (From 1990 onward, chronologically ordered)

Study ID	Outcome	Instrument	Sample size, age	Definition of urban	Timing of urban	Findings	Confounders	Comments
High-income countries								
(Lewis and Booth 1994) UK	Psychiatric morbidity	GHQ	$N = 6{,}456$, Age 18 or over	*Urban (no open access), urban (open access), rural*	At time of survey	Participants living in urban areas had higher risk (odds) of experiencing psychiatric morbidity cf. those in rural areas, with access to open space associated with exact risk	Age, social class, marital status, employment, chronic illness, and region of residence	Partial dose-response relationship observed
(Kessler et al. 1994) USA	Any anxiety disorder/any mood disorder DSM-III-R (lifetime and 12-month prevalence)	CIDI	$N = 8.098$ Age 15–54	*Degree of urbanization*	At time of survey	No difference in the distribution of anxiety and mood disorders between urban and nonurban areas	Age, sex, ethnicity, education, marital status, region	
(Blazer et al. 1994) USA	Major depression DSM-III-R (1-month prevalence)	CIDI	$N = 8.098$ Age 15–54	*Degree of urbanization*	At time of survey	No difference in distribution of major depression between urban and nonurban areas	Age, sex, ethnicity, education, marital status, region, income, employment, household composition, religion	
(Parikh et al. 1996) Canada	Affective disorders DSM-III [1-month prevalence]	UM-CIDI	$N = 9{,}953$, Age 15–64	*Urban versus rural*	At time of survey	No differences between urban and rural areas	–	No adjustment for confounders

(continued)

Table 3 (continued)

Study ID	Outcome	Instrument	Sample size, age	Definition of urban	Timing of urban	Findings	Confounders	Comments
(Paykel et al. 2000) UK	Mood and anxiety disorders DSM-III [1-week prevalence]	CIS-R	$N = 9{,}748$ Age 16–64	*Self-reported rating of urban, semirural, and rural residence*	At time of survey	Higher risk (odds) of psychiatric morbidity in urban cf. rural areas; no differences between semirural and rural	Age, sex, social class, ethnicity, marital status, education, employment, housing type and tenure, life events in past year, social network, social support	
(Andrews 2001) Australia	Mood and anxiety disorders ICD-10, DSM-IV [12-month, 1-month prevalence]	MMSE	$N = 10{,}641$ Age: 16–64	*Urban versus rural*	At time of survey	No differences in risk (odds) of mood and anxiety disorders between urban and rural areas in univariable models		No further testing in multivariable models
(Ayuso-Mateos 2001) Finland, Ireland, Norway, UK, Spain	Depressive disorder ICD-10 [1-month prevalence]	BDI + SCAN	$N = 8{,}862$ Age 18–64	*Urban versus rural*	At time of survey	Higher prevalence of mood disorders in urban cf. rural areas. However, the difference was only statistically significant in the UK and Ireland	Age and sex standardization	Low response rates in the UK and Ireland may have biased results Differences expressed as means and 95% CI, no adjusted regression analyses

(de Graaf et al. 2002) The Netherlands	Mood and anxiety disorders DSMIII incidence	CIDI	$N = 5{,}618$ Age = 18–64	*Urban versus rural*	At time of survey	No difference between urban and rural areas	Age, sex, education, cohabitation status	
(Andrade et al. 2003) Canada, Czech Republic, Germany, the Netherlands, Turkey, USA	Major depressive episode DSM-III-R in all countries (exc. Germany DSM-IV) [lifetime prevalence]	CIDI	Age varying across studies, overall from 14 years or over	*Urban versus rural*	At time of survey	No difference between urban and rural areas, except in the Netherlands where risk (odds) were lower in rural areas	Other sociodemographic variables investigated, but not included in regression models	Response rate between 56 and 88% No multivariable adjustment for confounders
(Kessler et al. 2003) USA	Major depression DSM-IV [lifetime and 12-month prevalence]	CIDI	$N = 9{,}090$ Age 18 or over	*Degree of urbanization*	At time of survey	No difference between urban and rural areas	None	
(Sundquist et al. 2004) Sweden	Major depression ICD-9, ICD-10, and DSM-IV Incidence	Registers	Sample = 4.4 m Swedish people aged 25–64 years on December 12, 1999 (35,727 people excluded as they had a previous admission), cases = 7,751	*Degree of urbanization*	At diagnosis	Risk of depression highest for men and women in most urbanized quintile, but little evidence of dose response	Marital status, education, immigrant status, age (stratified by sex)	

(*continued*)

Table 3 (continued)

Study ID	Outcome	Instrument	Sample size, age	Definition of urban	Timing of urban	Findings	Confounders	Comments
(Wang 2004) Canada	Major depressive episode DSM-III-R [12-month prevalence]	CIDI	$N = 17{,}244$ Age 12 or over	*Urban versus rural*	At diagnosis	Higher risk (odds) of major depressive episode in urban areas after multivariable adjustment	Employment, marital status, ethnicity, immigration status	
(Kovess-Masfety et al. 2005) France	Mood and anxiety disorders DSM-IV [lifetime, 6-month, 1-year prevalence]	CIDI-SF	$N = 2{,}638$ Age 18–102 years	*Urban versus rural*	At time of survey	No differences between urban and rural settings after multivariable adjustment	Age, sex, education, employment status, marital status, country of birth, housing tenure and type	
(Patten et al. 2006) Canada	Major depressive disorder/ episode classification not specified [1-year/point prevalence]	CIDI	$N = 36{,}984$ Age 15 or over	*Unspecified*	At time of survey	No differences between urban and rural settings	Age, sex, marital status, income, education, employment, chronic conditions	

(Probst et al. 2006) USA	Depression Classification not specified (unspecified prevalence)	CIDI-SF	$N = 30{,}801$, Age 18–65	*Urban versus rural*	At time of survey	Higher risk (odds) of depression in rural populations, but no differences once other health indicators were included	Age, sex, ethnicity, self-reported health, education, income, marital status, employment, limitation in daily activities, asthma, hypertension, diabetes, health status change	
(Weich et al. 2006) UK	Common mental disorder (onset and maintenance) [point prevalence]	GHQ	$N = 9{,}518$ wave 1 (onset) $N = 7{,}659$ wave 2 (maintenance) Age 17–64	*Population density and non-rural versus rural*	At time of survey	Population density associated with maintenance of CMD, but not onset Higher risk (odds) of CMD in non-rural areas in univariable, but not multivariable models	Age, sex, marital status, education, employment, financial difficulties, physical health, income, housing tenure, type and problems, car access, overcrowding, social class	Non-rural areas do not necessarily correspond to urban areas Study is based on prevalence not incidence so “onset” is a partial misnomer
(Grant et al. 2009) USA	Generalized anxiety disorder (GAD) DSM-IV (incidence and prevalence)	AUDADIS-IV	$N = 34{,}653$	*Urban versus rural*	At time of survey	No differences in risk (odds) of reporting GAD between urban and rural areas	Age, sex, ethnicity, income, marital status, education, region	Only included GAD. Study also investigated major depressive disorder (MDD), not included in this review; no difference in urban-rural distribution of MDD

(*continued*)

Table 3 (continued)

Study ID	Outcome	Instrument	Sample size, age	Definition of urban	Timing of urban	Findings	Confounders	Comments
(Romans et al. 2011) Canada	Mood and anxiety disorders ICD-10, DSM-IV [12-month prevalence]	CIDI	$N = 31,321$ Age 15 or over	*Degree of urbanization*	At time of survey	Weak evidence of difference in prevalence across the four geographical areas, with lower risk (odds) of depression in most rural area	Age, sex, marital status, income, income adequacy, employed, housing tenure, country of origin, health, ethnicity, education, community support	Anxiety as outcome was not investigated in multivariable models
(de Graaf et al. 2012) The Netherlands	Mood and anxiety disorders ICD-10, DSM-IV [12-month prevalence]	CIDI	$N = 6,646$ Age 18–64	*Urban versus rural and degree of urbanization*	At time of survey	Lower risk (odds) of mood disorders in more urbanized areas. Highest risk of anxiety disorders in areas with very high degrees of urbanization cf. the most rural areas	Age and sex	
(de Graaf et al. 2013) The Netherlands	Mood and anxiety disorders ICD-10, DSM-IV [incidence]	CIDI	$N = 5,303$ (of the 6,646 eligible and invited)	*Urban versus rural*	At time of survey	No difference in incidence rate ratio of mood and anxiety disorders between urban and rural areas	Age and sex	

(Beyer et al. 2014) USA	Common mental disorders [point prevalence]	DASS	$N = 2,479$ Age 21–74	*Degree of urbanization*	At time of survey	Higher levels of depression and anxiety (though smaller coefficient and weaker association in the latter) in less urbanized areas but with no evidence of a linear relationship. Population density not associated with CMD	Age, sex, ethnicity, marital status, education, household income, occupational status, health insurance, green spaces, % tree canopy, urbanicity, population density, median household income below poverty, residential instability, % owner occupied households and unemployed, ethnic segregation	
LAMIC								
(Mohammadi et al. 2005) Iran	Mood and anxiety disorders (DSM-IV) [lifetime prevalence]	SADS	$N = 25,180$ Age 18 or over	*Urban versus rural*	At time of survey	Higher risk (odds) of mood and anxiety disorders in urban cf. rural area	Unspecified	Unclear if odds ratios were from univariable or multivariable analyses
(Vicente et al. 2006) Chile	Mood and anxiety disorders DSM-III-R [lifetime, 12-month prevalence]	(CIDI)	$N = 2,978$ Age 15 or over	Analyses grouped by four geographical regions, with different % of rural population	At time of survey	No differences across four regions in prevalence of disorders	Age, sex, education, marital status, income	Urban-rural areas not directly compared

(*continued*)

Table 3 (continued)

Study ID	Outcome	Instrument	Sample size, age	Definition of urban	Timing of urban	Findings	Confounders	Comments
(Amoran et al. 2007) Nigeria	Common mental disorders [point prevalence]	GHQ	*N* = 1,105 Age 15 or over	*Urban versus rural*	At time of survey	Higher prevalence of depression in rural areas	–	No control for confounders, statistical tests only conducted for depression, not anxiety
(Wei et al. 2008) Tibet	Depression DSM-IV point prevalence	SCID-I	Sample = 5,145 Cases = N/A	*City versus rural areas*	At diagnosis	No significant differences between urban and rural areas	–	No multivariable analysis
(Lundberg et al. 2009) Uganda	Depression and anxiety (classification not specified) [point prevalence]	HSCL–25	Sample = 646	*Number of inhabitants* Urban = Kampala Semi-urban = >25,000 inhabitants Rural ≤25,000 inhabitants	Birth and upbringing	People born in urban areas had higher risk (odds) of having depression and anxiety	Age, gender, education, marital status, SES, family history of psychiatric disorders	Small sample, type II error cannot be ruled out

AUDADIS-IV Alcohol Use Disorder and Associated Disabilities Interview Schedule DSM-IV Version, *BDI* Beck Depression Inventory, *CIDI* Composite International Diagnostic Interview, *CIDI-SF* Composite International Diagnostic Interview Short Form, *CIS-R* Clinical Interview Schedule-Revised, *DASS* Depression Anxiety Stress Scale, *GHQ* General Health Questionnaire, *DSM* Diagnostic and Statistical Manual of Mental Disorders, *HSCL* Hopkins Symptoms Checklist, *ICD* International Classification of Disease, *MMSE* Mini-Mental State Examination, *SADS* Somatic Anxiety Depressive Symptoms scale, *SCAN* Schedules for Clinical Assessment in Neuropsychiatry. *Cf.* Compared with

Booth 1994; Paykel et al. 2000; Weich et al. 2006), three in the Netherlands (de Graaf et al. 2002, 2012, 2013), one in Sweden (Sundquist et al. 2004), one in France (Kovess-Masfety et al. 2005), and one multi-country study across the UK, Spain, Finland, Ireland, and Norway (Ayuso-Mateos 2001). A number of studies have also been conducted in North America (four in Canada Parikh et al. 1996; Romans et al. 2011; Patten et al. 2006; Wang 2004) and six in the USA (Kessler et al. 1994, 2003; Blazer et al. 1994; Probst et al. 2006; Grant et al. 2009; Beyer et al. 2014), one across European and North American settings (Canada, USA, Turkey, Germany, the Netherlands, and Czech Republic) (Andrade et al. 2003), one in South America (Chile Vicente et al. 2006), one in Australia (Andrews 2001), two in Africa (Nigeria Amoran et al. 2007) and Uganda (Lundberg et al. 2009)), and two in Central Asia (Iran Mohammadi et al. 2005) and Tibet (Wei et al. 2008).

All studies employed a cross-sectional design, with the exception of one longitudinal cohort (Sundquist et al. 2004) and two incidence-based studies (de Graaf et al. 2002, 2013). Studies investigated CMD across a variety of ages, beginning in mid- to late adolescence. Even though all investigations recorded place of residence concomitantly to the time of the survey, definitions of urbanicity varied considerably; while the majority of studies defined urbanicity according to the degree of urbanization (Lundberg et al. 2009; de Graaf et al. 2012, 2013; Kovess-Masfety et al. 2005; Parikh et al. 1996; Romans et al. 2011; Kessler et al. 1994, 2003; Blazer et al. 1994; Probst et al. 2006; Beyer et al. 2014; Andrews 2001) or population density (Sundquist et al. 2004; Weich et al. 2006; de Graaf et al. 2002; Wang 2004; Grant et al. 2009), others were based on combined metrics (population density and employment in rural occupations such as fishing, agriculture, etc.) (Ayuso-Mateos 2001), housing type (Lewis and Booth 1994; Kovess-Masfety et al. 2005), self-report (Paykel et al. 2000), or the percentage of survey respondents classified as from a rural district (Vicente et al. 2006). Six studies provided no criteria to define urbanicity (Wei et al. 2008; Patten et al. 2006; Grant et al. 2009; Andrade et al. 2003; Mohammadi et al. 2005; Amoran et al. 2007). All studies, with the exception of one (at birth) (Lundberg et al. 2009), measured urbanicity at time of diagnosis.

Most studies investigated urban-rural variation of mood and anxiety disorders together (Lewis and Booth 1994; Paykel et al. 2000; Weich et al. 2006; de Graaf et al. 2002, 2012, 2013; Kovess-Masfety et al. 2005; Romans et al. 2011; Kessler et al. 1994; Beyer et al. 2014; Andrews 2001; Mohammadi et al. 2005; Amoran et al. 2007), with eight studies restricted to mood (depressive) disorders only (Ayuso-Mateos 2001; Parikh et al. 1996; Patten et al. 2006; Wang 2004; Blazer et al. 1994; Kessler et al. 2003; Probst et al. 2006; Andrade et al. 2003) and one restricted to anxiety disorders (Grant et al. 2009). A handful of other studies considered depression and anxiety alongside other psychiatric conditions (Sundquist et al. 2004; Vicente et al. 2006; Wei et al. 2008; Lundberg et al. 2009). Diagnostic criteria, however, varied greatly across studies; eight studies used DSM-III or DSM-III-R criteria (Vicente et al. 2006; Paykel et al. 2000; de Graaf et al. 2002; Parikh et al. 1996; Wang 2004; Kessler et al. 1994; Blazer et al. 1994; Andrade et al. 2003), four used DSM-IV (Wei et al. 2008; Kovess-Masfety et al. 2005; Kessler et al. 2003; Grant et al. 2009; Mohammadi et al. 2005), one used ICD-10 (Ayuso-Mateos 2001), five used a combination of ICD-10 and DSM-IV (Sundquist et al. 2004, 2012, de Graaf et al. 2013; Romans et al. 2011; Andrews 2001), and the remainder reported general psychopathology scores without reference to specific diagnostic classifications (Lewis and Booth 1994; Weich et al. 2006; Patten et al. 2006; Probst et al. 2006; Beyer et al. 2014; Amoran et al. 2007). A range of different instruments (CIDI (Vicente et al. 2006; de Graaf et al. 2002, 2012, 2013; Kovess-Masfety et al. 2005; Parikh et al. 1996; Romans et al. 2011; Patten et al. 2006; Wang 2004; Kessler et al. 1994, 2003; Blazer et al. 1994; Probst et al. 2006; Andrade et al. 2003), DASS (Beyer et al. 2014), AUDADIS-IV (Grant et al. 2009), GHQ (Lewis and Booth 1994; Weich et al. 2006; Amoran et al. 2007), SCID-I (Wei et al. 2008), CIS-R (Paykel et al. 2000), BDI (Ayuso-Mateos 2001), MMSE (Andrews 2001), HSCL-25 (Lundberg et al. 2009), and SADS (Mohammadi et al. 2005)) were employed to diagnose incidence (de Graaf et al. 2002; Grant et al. 2009) and point (Wei et al. 2008; Lundberg et al. 2009; Patten et al. 2006; Beyer et al. 2014; Amoran et al. 2007), week (Paykel et al. 2000), month (Ayuso-Mateos 2001; Parikh et al. 1996; Blazer et al. 1994; Andrews 2001), annual (Vicente et al. 2006; Kovess-Masfety et al. 2005; Romans et al. 2011; Patten et al. 2006; Kessler et al. 1994, 2003; Andrews 2001), 3-year (de Graaf et al. 2013), and lifetime CMD prevalence (Vicente et al. 2006; Kovess-Masfety et al. 2005; Kessler et al. 1994, 2003; Mohammadi et al. 2005). One study employed register data to measure incidence of depression (Sundquist et al. 2004).

Perhaps unsurprisingly, results from these diverse study designs were highly heterogeneous. Eight studies found higher prevalence or risk (expressed as odds ratios) of CMD in urban areas (Sundquist et al. 2004; Lundberg et al. 2009; Lewis and Booth 1994; Paykel et al. 2000; Ayuso-Mateos 2001; Romans et al. 2011; Wang 2004; Mohammadi et al. 2005), two in rural settings (Beyer et al. 2014; Amoran et al. 2007), while 14 observed no differences (Vicente et al. 2006; Wei et al. 2008; Weich et al. 2006, 2002; de Graaf et al. 2013; Kovess-Masfety et al. 2005; Parikh et al. 1996; Patten et al. 2006; Kessler et al. 1994, 2003; Blazer et al. 1994; Probst et al. 2006; Grant et al. 2009; Andrews 2001). One study reported a higher risk of depression in rural areas, but the opposite for anxiety disorders (de Graaf et al. 2012), whereas another study found no rural-urban differences in five out of six countries (Canada, USA, Turkey, Germany, and Czech Republic) (Andrade et al. 2003). Moreover, studies which used categorical (i.e., nonbinary) indicators of urbanization found little or no evidence supporting the presence of a dose-response relationship between the degree of urbanization and prevalence or risk of CMD (Sundquist et al. 2004; Paykel et al. 2000; de Graaf et al. 2013; Parikh et al. 1996; Romans et al. 2011; Kessler et al. 1994, 2003; Blazer et al. 1994; Beyer et al. 2014; Andrews 2001).

Neighborhoods

A large body of literature has also investigated whether CMDs vary between small area neighborhoods (Table 4), beyond variation which might exist between rural and urban populations (section "Urban Versus Rural Setting"). Given the breadth of the

Table 4 Summary of literature on neighborhood differences in the distribution of CMD (From 1990 onward, chronologically ordered)

Study ID and setting	Outcome and instrument	Case finding, age range, years	Sample size	Definition of neighborhood	Timing of exposure	Area-level measures	Individual-level measures	Findings	Comments
(Tweed et al. 1990) USA	Depressed mood (DIS) [6-month prevalence]	Survey Age: 18+	$N = 3{,}481$	Neighborhoods in Baltimore	At time of survey	Ethnic congruence	Age, sex, SES, ethnicity	Six-month prevalence was lower in neighborhoods with highest ethnic congruence. Effect was greater for participants of black ethnicity	
(Reijneveld and Schene 1998) The Netherlands	Common mental disorders GHQ-12 [point prevalence]	Survey Age: 16+	$N = 4{,}892$	Boroughs in Amsterdam	At time of survey	Area-level deprivation: registered income, household income below minimum, and unemployment rate	Age, sex, income, occupational status, and educational level	Higher risk (odds) in more deprived neighborhoods, but disappeared after adjustment	Similar results obtained when restricted to Dutch-born participants
(Yen and Kaplan 1999) USA	Depressive symptoms 18-item depressive symptom scale	Survey Age: 20+ Years: 1965–1970	$N = 1{,}296$	Census tracts	At time of survey	Neighborhood SES: (% population with low income, living in substandard housing, low education, unskilled male labors, children in single-parent homes)	Age, sex, ethnicity, income, chronic conditions, smoking, alcohol, and BMI	No differences in depressive symptoms between low and high SES neighborhoods	

(*continued*)

Table 4 (continued)

Study ID and setting	Outcome and instrument	Case finding, age range, years	Sample size	Definition of neighborhood	Timing of exposure	Area-level measures	Individual-level measures	Findings	Comments
(Halpern and Nazroo 2000) UK	Common mental disorders CIS-R [prevalent cases]	Survey Age: N/A	$N = 8{,}063$	UK census enumeration district	At time of survey	Group density	Age, sex, ethnicity, economic hardship, fluency in English, age at migration, attacks on person and home	Lower group density associated with more depressive symptoms. After adjustment for individual-level variables, association was attenuated for white participants, strengthened for black ethnic minorities	
(Ross et al. 2000) USA	Depression and anxiety (modified CES-D)	Survey Age: 18–92 Year: 1995	$N = 2{,}482$	US census/ postcode tracts	At time of survey	Neighborhood SES; perceived neighborhood disorder (physical and social)	Age, gender, ethnicity, income, education, marital, and parental status, employment, and urban residence	Neighborhood disorder associated with more symptoms of depression and anxiety in adjusted models, but not neighborhood SES	

(McCulloch 2001) UK	Common mental disorders	Survey	$N = 10{,}264$	N/A	At time of survey	Area-level deprivation (Townsend index)	Region of residence, year, education, ethnicity, housing tenure, access to car, employment	No association between deprivation and CMD after adjustment for other confounders	Household-level characteristics not included in multilevel model
	GHQ-12 [point prevalence]	Age: 16–75 (years 1991–1998)							
(Weich et al. 2002) UK	Depressive symptoms	Survey	$N = 1{,}887$	86 housing areas (geographically bounded areas in which the majority of the housing was homogeneous in form and character)	At time of survey	Built environment site survey checklist (BESSC)	Age, sex, marital status, employment, ethnicity, education. Household: tenure, level, structural problems	After adjusting for other characteristics, only living in areas with majority of buildings built after the 1970s (versus pre-1940s) and with less than 25% of dwellings with private garden was associated with higher risk (odds) of depressive symptoms	
	CES-D [prevalence]	Age: 16–75		Two wards in North London					
(Stafford and Marmot 2003) UK	Depression	Cohort study (wave 5)	$N = 5{,}539$	Electoral wards	At time of survey	Neighborhood deprivation (Townsend score, 1991 census); self-reported measure of neighborhood safety	Age, sex, SES, financial problems, satisfaction with standard of living	Deprivation was associated with higher risk (odds) of depression. Some weak evidence that this was worse for people of low SES in such deprived areas	
	GHQ-28 [point prevalence]								

(*continued*)

Table 4 (continued)

Study ID and setting	Outcome and instrument	Case finding, age range, years	Sample size	Definition of neighborhood	Timing of exposure	Area-level measures	Individual-level measures	Findings	Comments
(Weich 2003a) UK	Common mental disorders GHQ-12 [point prevalence]	Survey Age: 16–74	$N = 8{,}979$	Electoral wards	At time of survey	Area-level deprivation (Carstairs index of socioeconomic deprivation), population density	Age, sex, marital status, ethnicity, employment status, number of current physical health problems. Household level: head of the household SES, structural housing problems, low income, access to a car, and overcrowding	Negligible variance in CMD explained at area level , with no main effect for deprivation after adjusting for other characteristics. Possible subgroup effect of deprivation on CMD risk in the unemployed and economically inactive	

(Weich 2003b) UK	Common mental disorders	British Household Panel Survey (BHPS)	$N = 8,979$	Electoral wards (~2,400 addresses)	At time of survey	None (neighborhood employed as unit in multilevel modeling)	Age, sex, marital status, ethnicity, education, employment status, financial strain, and current physical health problems	CMD as binary outcome: No significant area-level variance. In adjusted models, 94% of variance was explained at the individual level, 6% at area level	
	GHQ-12 [point prevalence]						Household level: structural housing problems, household income, access to a car, tenure, SES, overcrowding, household type	GHQ score (continuous): 0.9% of variance at area level (significant) no longer significant in adjusted models, where 89% of variance was explained at individual level, 10% at household level	
(Leventhal and Brooks-Gunn 2003) USA	Depressive symptoms	RCT [treatment arm, moved to low-poverty neighborhoods + received voucher; comparison arm, voucher only; control arm, did not change residence]	$N = 550$	Not defined	As assigned by RCT	Neighborhood physical and social disorder measured by parental ratings of the size of problems (parent and interviewer rated)	Neighborhood satisfaction	Parents in treatment arm were less likely to report depressive mood	69% response rate at follow-up
	DMI (depressive mood inventory)						Family randomized based on age, ethnicity, sex, education, marital status, employment		

(*continued*)

Table 4 (continued)

Study ID and setting	Outcome and instrument	Case finding, age range, years	Sample size	Definition of neighborhood	Timing of exposure	Area-level measures	Individual-level measures	Findings	Comments
(Wainwright and Surtees 2004) UK	Mood disorder	Survey	$N = 19{,}687$	Electoral ward	At time of survey	Index of multiple deprivation	Age, sex, social class, marital status, employment status, and educational level	Higher risk (odds) of reporting mood disorder for participants living in more deprived areas after accounting for individual-level characteristics	
	Health and Life Experiences Questionnaire (HLEQ) [current and lifetime prevalence]	Age: 41–80 years						Nonsignificant 0.4% variation at area level once all individual-level variables were included	
(Greiner et al. 2004) USA	Depressive symptoms	Survey	$N = 4{,}601$	Community as perceived by respondent	At time of survey	Community rating, community involvement, population density	Age, sex, race, education, and identification of a personal health-care provider	Higher community ratings were associated with lower risk (odds) of reporting depressive symptoms	Self-rated measures of area could be affected by depressive symptom levels
	One (unspecified) option question [prevalence]	Age: 18 or over							

(Propper et al. 2005) UK	Common mental disorders GHQ-12 [point prevalence]	Survey (1991–2000)	$N = 8,184$ (for prevalence) $N = 7,047$ (5-year analyses)	Enumeration district (~500 people)	At time of survey	% unemployed, long-term sick age 16–64, tenure type, central heating, no car, crowding, lone parent, one person (pensioner/non-pensioner), black, Indian, Pakistani, Bangladeshi, migrant, working in agriculture, children, managerial	Individual: age, sex, ethnicity, and education Household level: net household income, household size, tenure, employment	No statistically significant neighborhood-level variance in CMD (<1%), 3% at household level	
(Weich et al. 2005) UK	Common mental disorders GHQ-12 [point prevalence]	Survey (1991–1992) Age: 16–74 years	$N = 5,809$ (onset analyses) $N = 1,850$ (for episode maintenance analyses)	Electoral wards (~2,000 people)	At time of survey	Socioeconomic deprivation (Carstairs index)	Age, sex, marital status, ethnicity, education, employment status, financial strain, and current physical health. Household level: structural housing problems, household income, car access, tenure, social class, overcrowding, and household type	No significant variance at area level on onset or maintenance of CMD. No association with deprivation after adjusting for HH and individual-level variables	

(*continued*)

Table 4 (continued)

Study ID and setting	Outcome and instrument	Case finding, age range, years	Sample size	Definition of neighborhood	Timing of exposure	Area-level measures	Individual-level measures	Findings	Comments
(Skapinakis et al. 2005) UK	Common mental disorders (SF-36) [point prevalence]	Survey Age: 18 years or over Years: 1998	N/A (circa 27,000)	Electoral division	At time of survey	Area-level deprivation	Age, sex, marital status, employment status, social class	1.5% variance explained at area level in null model reduced to 0.6% in full model accounting for individual characteristics and regional deprivation. Both significant. Deprivation associated with greater risk of CMD after accounting for individual-level variables	
(Veenstra 2005) Canada	Depressive symptoms 11-item questionnaire (Cronbach's alpha: 0.8)	Survey Age: 18+ Years: 2002	$N = 1,355$	Census metropolitan areas	At time of survey	Internal and external built environment, social capital, and neighborhood SES	Age, sex, ethnicity, income, education, community trust, political trust, participation in voluntary association	More public spaces per capita was associated with higher scores on the depressive symptoms scale. 2.1% of variance (significant) explained at area level in null model, 1.9% in multivariable model (significance not reported)	Community and political trust were associated with fewer symptoms

(Galea et al. 2005) USA	Major depression NWS depression module (DSM-IV) [lifetime/6-month prevalence]	Survey Age: 18 or over Years: 2002	$N = 1{,}355$	59 community districts in New York City	At time of survey	Internal built environment, external built environment, neighborhood income	Age, sex, ethnicity, income	In adjusted models, individuals living in neighborhoods with poorer built environment were more likely to report 6-month and lifetime prevalence of depression	Survey conducted soon after 9/11 World Trade Center terrorist attacks
(Lofors et al. 2006) Sweden	Anxiety (self-reported) [point prevalence]	Survey Age: 25–64 Years: 1995–2002	$N = 30{,}884$	SAMS (small area market statistics; ~2,000 people in Stockholm, ~1,000 people elsewhere)	At time of survey	% of individuals with income lower than national average, neighborhood quartiles	Age, sex, marital status, immigrant status, employment, social network, housing tenure	Higher risk (odds) of anxiety when moving from least to most deprived neighborhood, but not significant when adjusting for individual measures of SES	
(Fone and Dunstan 2006) UK	Common mental disorders (SF-36) [point prevalence]	Survey Age: 14–74 years Years: 1998	$N = 24{,}975$	Electoral division	At time of survey	Townsend deprivation score	Age, sex, social class, economic inactivity, occupational status, marital status, whether the person is a carer, housing tenure	1.3% of CMD variance explained at area-level in the null model. Adding compositional variables reduced variance to 0.5%, which remained significant. Both area-level and individual deprivation	

(continued)

Table 4 (continued)

Study ID and setting	Outcome and instrument	Case finding, age range, years	Sample size	Definition of neighborhood	Timing of exposure	Area-level measures	Individual-level measures	Findings	Comments
								associated with poorer mental health, but greater effect in economically inactive individuals	
(Matheson et al. 2006) Canada	Depression CES-D (short form) [point prevalence]	Survey Age: 18–74 Years: 2001–2001 and 2003–2004	$N = 56,428$	Census tracts (range 2,500–8,000 people)	At time of survey	18 neighborhood characteristics grouped into four factors: residential instability, material deprivation, dependency, ethnic diversity	Age, sex, marital status, education, visible ethnic minority	Material deprivation and residential stability associated with depression after adjustment for other characteristics	No control for household-level characteristics
(Guite et al. 2006) UK	Common mental disorders SF-36	Survey Age: 18 + Years: N/A	$N = 1,012$	LSOA (Lower Super Output Area) census tracts	At time of survey	Quality of internal (heat, light, noise, damp, and drafts) and external built environment, population density, long-term limiting illness, general health, crime level	Age, sex, ethnicity, residential stability, household size, housing type and tenure, housing benefit, rent arrears, requesting rehousing	Damp; not liking residential area; noise; overcrowding; dissatisfaction with green spaces, community facilities, and social facilities; feeling unsafe; presence of needles/syringes;	

								lack of spaces where people can interact were all associated with higher CMD risk (odds) when entered individually in multivariable models	
(Dupéré and Perkins 2007) USA	Depression (CES-D) [point prevalence]	Survey	$N = 412$	Empirically derived based on respondents' definitions	At time of survey	Fear of crime, civic participation, informal ties, and % of people with characteristics measured at the individual level in neighborhoods grouped in five categories: advantaged to disadvantaged	Age, sex, ethnicity, household income, education, unemployment, single parenthood, residential stability, personal stress, perceived social support	Depression: no associations found	Multilevel models for depression not reported in text/table
	Anxiety (STAI) [point prevalence]	Years: 1987						Anxiety: lower in middle/middle-high advantage neighborhoods and those with higher residential stability after multivariable adjustment	
(Galea et al. 2007) USA	Depression	Telephone survey in 2002 + follow-up at 6 and 18 months	$N = 1,570/1,22($ at FU	59 community districts in NYC	At diagnosis	Area-level SES based on 2000 census data	Age, sex, ethnicity, past 6-month depression, traumatic event exposure, income, education, marital status, social support, stressors	18-month cumulative incidence of depression was 19.4 versus 10.5 per 100 persons in more cf. less disadvantaged neighborhoods. Depression risk was doubled in disadvantaged	Response rate 56%
	SCID-III incidence	Age ≥18 years							(Over-sampled residents close to the World Trade Center)

(*continued*)

Table 4 (continued)

Study ID and setting	Outcome and instrument	Case finding, age range, years	Sample size	Definition of neighborhood	Timing of exposure	Area-level measures	Individual-level measures	Findings	Comments
								neighborhoods, after adjusting for individual-level confounders	
(Lofors and Sundquist 2007) Sweden	Depression Hospital diagnosis ICD-9/ICD-10 Incidence	Psychiatric registers Age: 25–64 Years: 1997–1999	Sample = 4.5 m Swedish citizens. Cases = 10,930	Clusters of small area neighborhood units	At diagnosis	Linking social capital (voting participation), neighborhood deprivation	Age, sex, housing tenure, education, employment, country of birth, marital status	Between 1.7% (women) and 2.5% (men) of total variance explained at neighborhood level. In multivariable models, deprivation, but not low levels of social capital associated with risk of depression	
(Stockdale et al. 2007) USA	Common mental disorders Age: 48 years Years:	Survey Age: mean (CIDI short form) [point prevalence]	N = 12,716	Census tract	At time of survey	Neighborhood SES, density of outlets selling alcohol, density of mental health/ alcohol/drug facilities,	Age, sex, ethnicity, income, education, experience of witnessing violence in past	No evidence of area-level effect on risk (odds) of depressive symptoms	

						neighborhood violent crime arrest rate, church density, average household occupancy	year, residential stability		
(Stafford et al. 2008) UK	Common mental disorder	Survey	$N = 9{,}082$	Census ward (England)	At time of survey	Social capital (family ties, friendship ties, associational membership, integration in society, trust, attachment to neighborhood, tolerance of others, reciprocity) and deprivation (Carstairs index)	Individual levels: age, gender, social class	No association between social capital measures and CMD in adjusted models	
		Age:		Postcode ward (Scotland)			Household level: deprivation score	For people living in deprived areas, lower social capital was associated with higher CMD risk (odds)	
	GHQ-12 [point prevalence]	Years: 1994–1995 and 1997–1999							
(Ahern and Galea 2011) USA	Major depression (unspecified prevalence) (PHQ-9)	Phone survey	$N = 3{,}946$	59 community districts in New York City	At time of survey	Neighborhood collective efficacy, social cohesion, informal social control	Age, sex, marital status, ethnicity, birthplace, main language, income, education, length of time in neighborhood, unemployment, illness, financial problems	Low neighborhood collective efficacy associated with higher risk of depression after adjustment for other factors	49% response rate
		Age: 18 or over							
		Years: 2005							

(*continued*)

Table 4 (continued)

Study ID and setting	Outcome and instrument	Case finding, age range, years	Sample size	Definition of neighborhood	Timing of exposure	Area-level measures	Individual-level measures	Findings	Comments
(Vallée et al. 2011) France	Depression Current prevalence (MINI)	Years: 2005	$N = 3{,}023$	50 census blocks in Paris and suburbs and perceived neighborhood	At time of survey	Activity space in neighborhood, neighborhood deprivation (self-reported, aggregate measure from respondents, actual deprivation), neighborhood location (inner city, suburbs)	Age, sex, nationality, education, occupational and employment status, marital status, functional limitation	People in deprived neighborhoods were significantly more depressed than those in advantaged neighborhoods, especially when activity spaces were limited Initial spatial variance in depression accounted for by neighborhood factors	
(Beyer et al. 2014)USA	Common mental disorders (unspecified prevalence) (DASS)	Survey Age: 21–74 Years: 2008–2011	$N = 2{,}479$	Wisconsin census block groups [600–3,000 people]	At time of survey	Green spaces, % tree canopy, urbanicity, population density, median household income below poverty, residential instability, % owner occupied households, % unemployed, ethnic segregation	Age, sex, ethnicity, marital status, education, household income, occupational status, health insurance	Higher % green spaces associated with better mental health outcomes after adjustment for all other variables Other social and economic aspects of neighborhood also associated with depression, anxiety, and stress	

(Jokela 2014) Australia	Mental health (unspecified prevalence) (mental health short form, SF-36)	Longitudinal survey over 10 years Age: Years:	$N = 20{,}012$	Statistical local areas (median population $n = 5{,}908$)	Repeated measurements at baselines and follow ups.	Index of neighborhood disadvantage; neighborhood remoteness	Age, sex, country of birth, smoking, alcohol, physical activity, neighborhood dissatisfaction and problems	Association between neighborhood disadvantage and poorer mental health explained by individual-level characteristics. Neighborhood remoteness was associated with better mental health	Participants moving to more disadvantaged areas had poorer mental health, supporting social drift hypothesis
(Jones-Rounds et al. 2014) Multiple European cities	Psychological wellbeing Point prevalence (6 questions capturing symptoms of depression and anxiety)	Survey Age: N/A Years: 2002–2003	$N = 5{,}605$	Not specified	At time of survey	Neighborhood quality	Individual level: SES, employment status, sex and marital status, housing quality Household level: size, composition, highest educational level of any member, number of people working full time, people over 60	Housing and neighborhood quality predicted psychological well-being in adjusted models. Effect of neighborhood quality less pronounced at higher levels of housing quality	

(continued)

Table 4 (continued)

Study ID and setting	Outcome and instrument	Case finding, age range, years	Sample size	Definition of neighborhood	Timing of exposure	Area-level measures	Individual-level measures	Findings	Comments
(Erdem et al. 2015) The Netherlands	Psychological distress	Survey	$N = 18{,}173$	City district (average 4,000 people)	At time of survey	Neighborhood SES (income, work, and level of education), neighborhood green space, urbanicity and home maintenance, social cohesion	Age, sex, ethnic background, marital status and years in current city, education, occupation, and financial difficulties	In univariable model, 2.87% of variance explained at neighborhood level, but reduced to 0.25% when individual variables included	
	Point prevalence (Kessler-10)	Age: 16 or over						Higher neighborhood social cohesion associated with lower psychological distress after adjustment	
		Years: 2008							
(Gariepy et al. 2015) Canada	Episodes of depression	Survey	$N = 7{,}114$	500 m radius buffer around center of postcode of each participant	At time of survey	Density of businesses, parks, and recreational facilities	Age, sex, marital status, education, income adequacy, family history of depression, chronic condition, childhood life events	Presence of parks, healthy food stores, fast-food restaurants, and health services associated with lower probability of depression episode among people in low depression trajectory. Parks were associated	
	Point prevalence	Age: N/A (average 42 at baseline)							
	(CIDI short form)	Years: 2000–2011							

								with lower probability of depression episode in people in the moderate depression trajectory	
LAMIC									
(Chiavegatto Filho et al. 2013) Brazil	Depression and anxiety (CIDI) [point prevalence]	Survey Age: 18 + Years: 2004–2005	$N = 3,542$	São Paulo boroughs	At the time of survey	Income inequality (Gini coefficient)	Age, sex, education, income, marital status	In multivariable analyses, living in areas with medium- to high-income inequality associated with higher risk (odds) of depression, but not anxiety	
(Tomita and Burns 2013) South Africa	Depression (CES-D) [point prevalence]	Survey Age: 15+ Years: 2008	$N = 13,469$	Neighborhoods (non-specified)	At the time of survey	Social capital (from four questions on (1) support network and reciprocity, (Menezes 2014) association activity, (Saha et al. 2005) collective norm and values, (McGrath et al. 2004), and safety	Civic participation, social trust, age, sex, ethnicity, marital status, employment, education, self-reported health status, social class, neighborhood attachment	In multivariable models, lower social capital was associated with higher levels of depressive symptoms; 2.61% of variance was attributable to neighborhood level	

CES-D Center for Epidemiological Studies Depression Scale, *CIDI* Composite International Diagnostic Interview, *CIS-R* Clinical Interview Schedule-Revised, *DASS* Depression Anxiety Stress Scales, *DIS* Diagnostic Interview Schedule, *DMI* Depressive Mood Inventory, *GHQ* General Health Questionnaire, *HLEQ* Health and Life Experiences Questionnaire, *NWS* National Women's Study, *PHQ-9* Patient Health Questionnaire, *SF-36* Short Form 36, *STAI* State-Trait Anxiety Inventory available research, we have restricted our primary focus on general adult population studies, consistent with other sections of this chapter. As before, most studies have been conducted in Western Europe (12 in the UK Guite et al. 2006; Weich et al. 2002, 2005; Weich 2003a, b; Stafford and Marmot 2003; Stafford et al. 2008; Halpern and Nazroo 2000; McCulloch 2001; Wainwright and Surtees 2004; Skapinakis et al. 2005; Fone and Dunstan 2006), two in the Netherlands (Reijneveld and Schene 1998; Erdem et al. 2015), two in Sweden (Lofors and Sundquist 2007; Lofors et al. 2006), one in France (Vallée et al. 2011), and one across several cities (Jones-Rounds et al. 2014)) and North America (12 in the USA Beyer et al. 2014; Tweed et al. 1990; Yen and Kaplan 1999; Ross et al. 2000; Stockdale et al. 2007; Leventhal and Brooks-Gunn 2003; Greiner et al. 2004; Galea et al. 2005, 2007; Dupéré and Perkins 2007; Ahern and Galea 2011) and three in Canada (Matheson et al. 2006; Gariepy et al. 2015; Veenstra 2005), with others undertaken in Australia (Jokela 2014), Brazil (Chiavegatto Filho et al. 2013), and South Africa (Tomita and Burns 2013).

As for the urban-rural literature, the evidence for neighborhood variation in CMD risk is more mixed than for schizophrenia. While up to 12% of variance in schizophrenia might have been explained at the area level, for CMD this proportion was much smaller, ranging from none to 4.4% (Lofors and Sundquist 2007; Weich 2003a, b; Weich et al. 2005; Skapinakis et al. 2005; Fone and Dunstan 2006; Erdem et al. 2015; Vallée et al. 2011; Veenstra 2005; Propper et al. 2005). Moreover, while many studies reported significant associations between CMD and at least one neighborhood factor (Beyer et al. 2014; Guite et al. 2006; Weich et al. 2002; Stafford and Marmot 2003; Halpern and Nazroo 2000; Wainwright and Surtees 2004; Skapinakis et al. 2005; Fone and Dunstan 2006; Erdem et al. 2015; Vallée et al. 2011; Jones-Rounds et al. 2014; Tweed et al. 1990; Greiner et al. 2004; Ahern and Galea 2011; Matheson et al. 2006; Gariepy et al. 2015; Jokela 2014; Chiavegatto Filho et al. 2013; Tomita and Burns 2013; Galea et al. 2005; Das-Munshi et al. 2010), a considerable number found no associations (Weich et al. 2002, 2005; Weich 2003a, b; Stafford et al. 2008; McCulloch 2001; Reijneveld and Schene 1998; Tweed et al. 1990; Yen and Kaplan 1999; Stockdale et al. 2007; Dupéré and Perkins 2007). Interestingly, while many studies did not find an overall main effect of neighborhood factors on CMD risk, many studies reported subgroup effects (Weich 2003a; Stafford and Marmot 2003; Halpern and Nazroo 2000; Fone and Dunstan 2006; Jones-Rounds et al. 2014; Tweed et al. 1990), which suggested that the role of neighborhood factors such as deprivation was associated with CMD risk in particularly disadvantaged individuals. While more research is needed, this suggests that social adversities at the neighborhood level may be particularly detrimental to the mental health of people who already face individual-level social disadvantage. Deprivation was most commonly investigated (Lofors and Sundquist 2007; Beyer et al. 2014; Weich 2003a; Weich et al. 2005; Stafford and Marmot 2003; Stafford et al. 2008; McCulloch 2001; Wainwright and Surtees 2004; Skapinakis et al. 2005; Fone and Dunstan 2006; Erdem et al. 2015; Lofors et al. 2006; Vallée et al. 2011; Ross et al. 2000; Stockdale et al. 2007; Dupéré and Perkins 2007; Galea et al. 2005, 2007; Matheson et al. 2006; Veenstra 2005; Jokela 2014), followed by ethnic density (Halpern and Nazroo 2000; Tweed et al. 1990; Matheson et al. 2006;

Das-Munshi et al. 2010), social capital (Lofors and Sundquist 2007; Stafford et al. 2008; Erdem et al. 2015; Greiner et al. 2004; Ahern and Galea 2011), and features of the built environment (virtually absent in the schizophrenia literature), such as housing (Guite et al. 2006; Weich et al. 2002), neighborhood quality (Guite et al. 2006; Jones-Rounds et al. 2014), or green space (Beyer et al. 2014; Guite et al. 2006; Gariepy et al. 2015).

With the exception of three observational studies employing longitudinal designs to examine either maintenance of CMD symptoms (Weich et al. 2005; Galea et al. 2007), social drift in individuals with CMD (Jokela 2014), or the association between the built environment and trajectories of depressive symptoms (Gariepy et al. 2015), as well a randomized controlled trial (Leventhal and Brooks-Gunn 2003), all neighborhood-level studies of CMD have employed cross-sectional designs. Place of residence has been exclusively measured at the time of data collection, although one longitudinal study was based on residence at multiple time-points (Jokela 2014).

Strengths and Limitations

Compared with studies of schizophrenia, sample sizes in CMD studies have generally been smaller, although these were generally adequately powered to detect any significant associations. Their reliance on cross-sectional survey-based designs is well suited to measuring prevalence in the general population, providing the sampling frame allows for adequate generalization. These study designs also mean a large number of confounding factors can be measured at the time of the survey, offering an advantage over register-based investigations which may be limited to routinely collected data. Such additional adjustment could explain the less pronounced association between urban environments and CMD than for schizophrenia, though further empirical work would be required in the schizophrenia literature to confirm this. Alternatively, studies of CMD typically use a broader variety of instruments to measure clinical disorder or general (often subclinical) psychopathology in their samples, which could lead to the observation of weaker associations overall, if exposure to social adversity was only causally important in the emergence of clinical disorders. Cross-sectional surveys have a number of inherent weaknesses; for example, people taking part in the surveys may be systematically different from those who do not, affecting (biasing) the results. In the studies we reviewed, participation rates varied from 54 (Ayuso-Mateos 2001) to 90.6% (Amoran et al. 2007) in urban-rural studies and between 38 (Guite et al. 2006) and 86% (Yen and Kaplan 1999) in neighborhood studies, suggesting that selective participation could be an issue in some studies. A further limitation of cross-sectional studies is that exposures (i.e., urban living) and confounders are measured concomitantly with the outcome (CMD), making it impossible to infer causality from any observed association; as the adage goes, correlation does not imply causation. While cross-sectional surveys can obtain measures which happened in the past, via interview or questionnaire, the answers elicited from this type of study may be prone to recall bias. If the probability of (mis)remembering an exposure or confounder in the past is related to the participant's mental health status, this type of bias can be particularly problematic, since any observed associations may be under- or overestimations of the true effect.

In summary, the evidence in favor of an association between contextual neighborhood factors and CMD was weaker than for schizophrenia and other non-affective psychotic disorders. The overall variance in the distribution of CMDs was generally less than 2%, with some indication that observed associations were due to residual confounding or household-level characteristics. These results suggest that compositional rather than contextual characteristics are driving any small area-level variation in CMD. Notwithstanding, most studies of CMD included here were cross-sectional, making it impossible to exclude selection biases as an explanation for the inconsistent or null findings observed. Further longitudinal research into the role of the neighborhood on CMD risk would therefore appear warranted.

Suicide

Each year more than 800,000 people die by suicide, making it the second leading cause of death among young people globally (World Health Organization 2014). Suicide therefore presents a major public health issue, particularly because a large amount of research suggests that rates are unequally distributed throughout the population. Thus, they vary by gender (Shiner et al. 2009; Langhinrichsen-Rohling et al. 2009; von Borczyskowski et al. 2010; Gibb et al. 2004; Hawton 2000), presence of an existing mental disorder (Beautrais 2000; Arsenault-Lapierre et al. 2004; Beskow 1979), chronic illness (Qin et al. 2014), unemployment (Page et al. 2013), low social support and social capital (Winfree and Jiang 2010; Kunst et al. 2013), and exposure to violence, abuse, or trauma (Fergusson et al. 2005; Fergusson and Lynskey 1995; Zapata et al. 2013; Rajalin et al. 2013; Pompili et al. 2009). While much of this research has focused on the individual level, the distribution of these risk factors, as well as suicide rates, varies according to characteristics of geographical areas, communities, and neighborhoods.

Investigation of the spatial distribution of suicide mortality dates back as far as the late nineteenth century, with Durkheim's seminal monograph (Durkheim 1897), in which he first noted that suicide rates clustered geographically. He proposed that individual acts of suicide were due to forces outside of the individual and that the geographical distribution of suicide rates could be explained by underlying patterns of social interaction and regulation. By contrast, Tarde (1903) argued that such patterns could be understood by imitation, where behaviors and beliefs are shared within a population. Numerous studies have subsequently investigated the spatial distribution of suicide, demonstrating that suicide mortality varies geographically. Such research allows appropriate investment in public mental health and prevention strategies and may give further insight into the aetiology of suicidal acts. In this section, we review the main studies on suicide risk in relation to both broad rural-urban variation and specific area-level or neighborhood differences in risk. We

Table 5 Summary of literature on rural-urban differences in suicide rates (From 2000 onward, chronologically ordered)

Study ID and setting	Diagnostic criteria and outcome	Case finding, age range, years	Sample size and no. of cases	Definition of spatial unit (rural-urban)	Timing of urban	Findings	Higher in	Confounders	Comments
(Wilkinson and Gunnell 2000) Australia	Suicide and undetermined deaths ICD-9 codes E950-9, E980-9 except E988.8	Australian Bureau of Statistics (ABS) Aged 15–34 years (1988–1997)		Nonmetropolitan classified as ≤20,000 people; metropolitan >20,000 people		Suicide rates in nonmetropolitan areas higher for men aged 15–24 years, but lower for women aged 25–34 years old compared to their counterparts in metropolitan areas. No other differences	Rural (men 15–24), equivalent (men, 25–34; women, 15–24), urban (women, 25–34)		
(Yip et al. 2005) China		China's Ministry of Health – death certificates issued by physicians (1991–2000)		Not specified	Not specified	Suicide rates consistently higher in rural cf. urban areas for men and women. Overall rates decreased over time, with an interplay of age, gender, and region driving change	Rural	Age, sex, region	Authors suggest that rapid modernization has corresponded to decreased suicide rates in China, due in part to increased standards of living, education, and medical care

(continued)

Table 5 (continued)

Study ID and setting	Diagnostic criteria and outcome	Case finding, age range, years	Sample size and no. of cases	Definition of spatial unit (rural-urban)	Timing of urban	Findings	Higher in	Confounders	Comments
(Singh and Siahpush 2002) USA		County mortality data – national mortality data files		Counties categorized using the US Department of Agriculture classification based on population size and proximity to metropolitan areas categorized into ten levels	At death (county-specific annual suicide deaths) $n = 3{,}101$ US counties	Male suicide rates were higher in rural areas. Gap increased over time and more pronounced in 15–24 men. For women, between 1970 and 1989, suicide rates were higher in urban areas, but this difference decreased over time, and rates were higher in rural areas by 1995–1997	Rural (men), urban (women, 1970–1989), rural (women, 1995–1997)	Age-, sex-, and county-specific deaths	Classification of rurality/ urbanicity was done in 1993 based on 1990 census. Criteria may not be appropriate across all time periods
		Aged 15+ years (1970–1997)						Adjusted for county-level variation in ethnic composition and divorce rate (as indicator of social disintegration)	
(Middleton et al. 2003) England and Wales	Suicide and undetermined death – ICD-9 codes E950-E959 and E980-E989, excluding E988.8	Office for National Statistics (ONS) (1981–1998)		9,264 electoral wards categorized by two indices: (Weich 2005) population density (quartiles); (Menezes 2014) population potential (quartiles) – a measure of geographic remoteness from large settlements	Postcode of usual residence at time of death, linked to electoral wards based on 1981 boundaries	Among 15–44-year-olds, suicides higher in remote areas. Rises in rates in rural areas were more related to population potential than population density	Rural	Townsend socioeconomic deprivation index and Congdon social fragmentation index	

(Caldwell et al. 2004) Australia	ICD-10 codes X60-X84	Australian Bureau of Statistics (ABS)	$N = 10{,}641$	Metropolitan centers (≥100,000), rural centers (10,000–99,999)		Rates higher for men in rural compared with metropolitan areas. For women, only those aged 30–44 had higher rates in rural areas	Rural (men, women aged 30–44), equivalent (women, other ages)		Young men with mental health disorders in rural areas less likely to seek professional help
		Aged 20+ years (1997–2000)		Population and rural areas (<10,000 people)					
(Levin and Leyland 2005) Scotland	Suicides and undetermined deaths	General Register Office for Scotland (GROS)		Four groups from the Scottish Household Survey rurality classification (Weich 2005); urban >10,000 people; (Menezes 2014); accessible rural = settlements <10,000 and within 30 min drive of settlement of ≥10,000; (Saha et al. 2005); remote towns = 3,000–10,000 and >30 min drive of a settlement of ≥10,000 (McGrath et al. 2004); remote rural = <3,000 population and >30 min drive from settlement of ≥10,000	Residence at death	Highest rates in remote rural areas. Age- and deprivation-adjusted models showed significantly greater risk of male suicide in remote and rural areas relative to urban areas, but lower risk of female suicide in accessible rural areas	Rural (men)	Carstairs deprivation indicator using car ownership, low social class, male unemployment and overcrowding, age, sex	High divorce rates and population loss between 1990 and 2000 predicted suicide rates
	ICD-9 codes E950-959.0 and E980-988.9	Ages 15+ years (1981–1999)							

(*continued*)

Table 5 (continued)

Study ID and setting	Diagnostic criteria and outcome	Case finding, age range, years	Sample size and no. of cases	Definition of spatial unit (rural-urban)	Timing of urban	Findings	Higher in	Confounders	Comments
(Yip et al. 2000) Australia and China (Beijing)	ICD-9 codes E950-E959	Australia – Australian Bureau of Statistics, Coroner's report. Beijing – death certificates and Public Security Office from the Ministry of Health (1991–1996)	Not specified	Not specified	Not specified	In Beijing suicide rates in rural areas were nearly three times higher than urban areas (~five times higher for women aged 25–34). In Australia, rural male suicide rate was higher than urban rate, but pattern reversed for women	Rural (Beijing: men and women, Australia: men) Urban (Australia: women)	Age, sex, region	Authors noted that the male/female suicide ratio in China was less than 1:1, but in Australia it was 4:1. In Beijing, elderly suicide rates were six times higher than the general population, but were not as increased in Australia
(Stark et al. 2007) Scotland	Suicide or undetermined cause ICD-9 codes E950-E959, ICD-10 codes X60-X84, E980-E989, Y10-Y34	General Register Office for Scotland (1981–1999)		Postcode sector used as proxy for rurality. Population density divided into quartiles	Death allocated to area of residence, rather than the area they died	Higher rate ratios in most and least densely populated quartiles. Association in women varied by age – no association under 25 years, similar pattern to men in middle age, lower rural rates in older women	Various (see left)		

(Page et al. 2007) Australia	Not specified	Australian Bureau of Statistics (ABU) Aged 15+ years (1979–2003)		Rural, remote, and metropolitan area classification system: (Weich 2005) metropolitan; (Menezes 2014) rural areas (large and small rural centers and other rural areas); (Saha et al. 2005) remote areas (remote centers and other remote areas)		Rural-urban differences reduced over the study period, but remained significant for men Female suicide rates were lower in rural and remote areas, particularly for those aged 25–34	Rural (men), urban (women)	Sex, age, country of birth, area SES	
(Pearce et al. 2007) New Zealand		Ministry of Health Aged 15+ years (1980–2001)	$N = 9{,}054$ suicide deaths	Census Area Unit (CAU) (~2,300 people) classified according to Statistics New Zealand's Urban-Rural Profile Classification to create a binary urban-rural variable	Domicile code (linked to CAU) at death	Higher rates in urban areas (1980–1982), but equivocal by the 1990s. Narrowing urban-rural differential may be due to raised suicide rates in more rural communities over time	Urban (1980–1982), equivalent (1990s)	Sex, residential domicile, age at death	

(*continued*)

Table 5 (continued)

Study ID and setting	Diagnostic criteria and outcome	Case finding, age range, years	Sample size and no. of cases	Definition of spatial unit (rural-urban)	Timing of urban	Findings	Higher in	Confounders	Comments
(Kapusta et al. 2008) Austria	ICD-8, ICD-9, ICD-10	Statistics Austria – death certificates (1970–2005)	99 districts	Five population density categories		Ratio of rural to urban suicide rates continuously increased over the last 30 years, indicating growing risk in rural areas	Rural	Age, sex, family status, and suicide method	
(Razvodovsky and Stickley 2009) Belarus		Belarus Ministry of Statistics Ages 15+ years (1990–2005)		Population density and structure of employment used. Classified as urban if >6,000 inhabitants or if >2/3 not employed in agriculture. All other settlements defined as "rural"		Higher in rural areas, although for those $\geq$75 years, urban rates were higher until 2000, but this reversed by 2005	Rural (men and women, except among oldest age groups by 2005)	Age, sex	Deteriorating social and economic situation may explain increasing suicide rates, with some rural-specific explanations, including increased social isolation and poor provision of medical services
(Kapusta et al. 2010) Austria	Not specified	Statistics Austria for mortality database	Not specified	Continuous measure of population density	Population density in 2001	Lower rates in urban areas	Rural	Population density, religion, unemployment rate, income	Access to mental health care was also related to rural-urban differences, with fewer facilities in rural areas

(Chang et al. 2011) Taiwan	Certified suicides, undetermined death, death by pesticide poisoning and suffocation: ICD-9 codes E950-E959, E980-E989, E863, and E913	1999–2007	N = 37,326 suicides	Deaths assigned to 358 districts according to registered address on death certificate	Residence at death	Rates highest in rural, East Taiwan. No evidence of above-average rates in large cities. In Taipei, rates increased toward suburbs. Income, population density, and lone-parent households associated with risk. Spatial patterning strongest in young people	Rural		
(Tak et al. 2012) Australia	ICD-10 codes X60-X84 and/or with the "intent" column registered as "2" (intentional self-harm)	National Coroners Information System (NCIS) (2004–2008)		Rural, Remote, and Metropolitan Classification (RRMA) and index of remoteness: (Weich 2005) metropolitan zone with population ≥100,000 (RRMA 1–2); (Menezes 2014) rural zones (RRMA 3–5); (Saha et al. 2005) remote zones (RRMA 6–7)	Post areas (PAs) ($n = 2{,}507$) assessed in 2006	Counties with low population density had higher suicide risk than heavily populated coastal cities. Rates higher for men in rural and remote areas. Spatial clusters identified in N. and W. areas. No rural-urban gradient for women	Rural (men), equivalent (women)	Age and sex standardized analyses	

(*continued*)

Table 5 (continued)

Study ID and setting	Diagnostic criteria and outcome	Case finding, age range, years	Sample size and no. of cases	Definition of spatial unit (rural-urban)	Timing of urban	Findings	Higher in	Confounders	Comments
(Park and Lester 2012) South Korea	Not specified	Korean Statistics Promotion Institute and census data Aged 15+ years (2005)	$N = 12{,}011$ suicides (pop. $n = 38.1$ m aged ≥ 15)	Korea's administrative districts. All townships without their own municipalities were classified as rural		Suicide rates higher in rural areas	Rural	Age, gender, marital status, month of suicide	Authors noted a high rate of pesticide and chemical use as method of suicide
(Patel et al. 2012) India	ICD-10 codes X60-X84	Registrar General of India Aged 15+ years (2001–2003)	$N = 2{,}684$ suicide deaths	6,671 small areas, randomly selected	Not specified	Age-standardized suicide rates were about two times higher in rural than urban areas. Rates varied between states, generally higher in south India	Rural	Age, sex, region, marital status, occupation, alcohol use, household fuel type, education	Higher rate in rural India may be linked to availability of pesticides
(Titelman et al. 2013) Nordic countries	Suicide and self-inflicted injury ICD-8 and ICD-9 codes E950-E959; ICD-10 codes X60-X84, Y87.0	Aged 15+ years (1980–2009)		No official definition. Rates in larger regions compared to each other		Rural areas in Norway, Finland, and Sweden had highest suicide rates, while capital cities had lower rates than national averages	Rural		
(Snider and Leo 2014) Australia	Cases of "possible suicide" and "undetermined causes" were not included	Queensland suicide register, aged 15+ years (2004–2008)	2,803 suicides	Dichotomous categorization of 38 rural-urban areas		Deprivation associated with age-standardized suicide mortality, particularly in men aged 15–59.	Deprivation increased suicide rates, particularly in urban areas	22 socioeconomic variables indexing material and social	Area-based indices of deprivation may produce inaccurate assessments of

						Deprivation factors had stronger association with suicide risk in urban areas, with nonsignificant or inverse effects in rural areas		conditions and relative socioeconomic disadvantage	rural socioeconomic status
(Ngamini Ngui et al. 2014b) Canada	ICD-10 codes X60-X84, Y87	Register Aged 15+ years (2004–2007)	$N = 2{,}951$	406 postcode areas known as "forward sortation areas"	Measured using postcode of place of death	Most likely clusters of suicide found in remote rural areas, lowest in metropolitan areas. Clusters likely driven by male rates; women more likely to die by suicide in urban areas	Rural (men), urban (women)	Age and sex	
(Qi et al. 2014) Australia	ICD-9, 950.0–959.9, and ICD-10, X60-X84	ABS Ages 15+ years (1986–2005)	$N = 45{,}293$ suicide deaths	Statistical local area	Place of suicide occurrence	Suicide rates higher in rural areas	Rural		
(Searles 2014) USA (16 states)		Deaths categorized as suicide in the National Violent Death Reporting System (2006–2008)		Based on population density, distance to metropolitan area, and employment to create *urban*, *rural adjacent*, and *rural nonadjacent* areas	County of residence	Demographic, socioeconomic, and mental health risk factors differed among rural and urban cases; in rural areas, cases were less likely to receive a psychiatric diagnosis and less likely to be in	Did not compare rural-urban differences directly	Age, sex, Hispanic ethnicity, marital status, homelessness, urban-rural residence, veteran status, mental health history	Risk factors for suicide, including substance abuse and availability of mental health services differ by urban-rural location

(*continued*)

Table 5 (continued)

Study ID and setting	Diagnostic criteria and outcome	Case finding, age range, years	Sample size and no. of cases	Definition of spatial unit (rural-urban)	Timing of urban	Findings	Higher in	Confounders	Comments
						treatment for mental health problems. History of substance abuse and job and financial problems were more prevalent among urban dwellers			
(Fontanella et al. 2015) USA	ICD-9 (clinical modification) codes E950-E959 and ICD-10 codes X60-X84, Y87.0, and U03	Aged 10–24 years (1996–2010)	$N = 66{,}595$ suicide deaths	2003 rural-urban continuum codes from the US Department of Agriculture, classifying 3,141 US counties into nine groups based on population size and adjacency to metropolitan areas		Rural suicide rates nearly double those in urban areas. Disparities persisted after controlling for confounders. Male rates declined in most urban areas over time, but remained stable in rural areas. Female rates increased in both rural and urban areas	Rural	Education, ethnicity, unemployment, poverty, female-headed households, unemployment, divorce, population density, health utilization, income	

briefly examine whether suicide methods differ between rural and urban settings and summarize the main hypotheses put forward to explain geographic differences in patterns of risk.

Urban Versus Rural Settings

To study how suicide rates vary between rural and urban areas, research studies have often used centralized registries, coroners' reports, and death certificates as a basis for investigation. Most studies have been conducted in high-income countries, including six from Australia (Caldwell et al. 2004; Tak et al. 2012; Snider and Leo 2014; Page et al. 2007; Qi et al. 2014; Wilkinson and Gunnell 2000), four from the USA (Fontanella et al. 2015; Hempstead 2006; Searles 2014; Singh and Siahpush 2002), two from Austria (Kapusta et al. 2008, 2010), three from the UK (Stark et al. 2007; Levin and Leyland 2005; Middleton et al. 2003), one from Canada (Ngamini Ngui et al. 2014a), and one from New Zealand (Pearce et al. 2007; Table 5). More recently, a handful of studies have been conducted in LAMIC, including Taiwan (Chang et al. 2011), India (Patel et al. 2012), Belarus (Razvodovsky and Stickley 2009), China (Yip et al. 2000), and South Korea (Park and Lester 2012).

Most studies have showed a consistently increased risk for suicide in rural compared with urban areas (Caldwell et al. 2004; Fontanella et al. 2015; Hempstead 2006; Kapusta et al. 2008; Chang et al. 2011; Razvodovsky and Stickley 2009). However, some studies have suggested that this effect may differ between men and women, as well as by age (Table 5), although such inconsistencies may be partly attributable to variation methodological approaches or differences over time. Variation in findings may also reflect the different characteristics of urban and rural places included in different studies and how the distribution of resources, including mental health care, varies by country.

Increased suicide risk was most consistently found for rural men (Caldwell et al. 2004; Tak et al. 2012; Taylor et al. 2005), with some evidence of a null effect (Tak et al. 2012) or reversed trends for women (Page et al. 2007; Ngamini Ngui et al. 2014b). It is currently unclear as to whether urban-rural differences in suicide rates are changing over time, with studies reporting that this mortality gap is both growing (Fontanella et al. 2015; Singh and Siahpush 2002; Kapusta et al. 2008) and narrowing (Page et al. 2007; Pearce et al. 2007; Razvodovsky and Stickley 2009); both are possible and may be context dependent and subject to a range of other social and economic determinants of health (Searles 2014) (see section "Neighbourhoods"). A Note on the Affective Psychoses (in some regions the rate is four times higher among men than women), the lower number of events for women may mean that some studies were underpowered to detect urban-rural variations in female suicide rates. Lower suicide rates in women, overall, may also be attributable to the fact that women are more likely to receive an open verdict than suicide on their death certificate (Linsley et al. 2001; Palmer et al. 2015), highlighting the need for careful definition of the outcome measure used in an epidemiological inquiry into variation in suicide rates.

Table 6 Summary of studies of suicide rates by area/neighborhood (From 1999 onward, chronologically ordered)

Study ID and setting	Diagnostic criteria and outcome	Case finding, age range, years	Definition of neighborhood	Timing of exposure	Area-level measures	Individual-level measures	Findings	Comments
(Neeleman and Wessely 1999) UK	Suicide verdict, suicide note found, method unambiguously indicated suicide, and/or suicidal intent recorded	Office for National Statistics and Coroner's reports $N = 329$ suicides in South London (population 902,008), 1991–1993	109 electoral wards in London boroughs Lewisham, Lambeth, Southwark, and Greenwich (mean pop. $N = 8{,}274$)	Postcodes linked to electoral wards	Small area ethnic densities and deprivation (Jarman index)	Age and sex	Minority suicide rates were higher in areas where minority groups were fewer. White suicide rates were higher in areas where more ethnic minorities live, independent of deprivation or age	Supports ethnic density hypothesis for completed suicide
(Whitley et al. 1999) UK	Deaths coded as suicide or open verdict: ICD-9 codes E950-E959, E980-E989	1981–1992		633 parliamentary constituencies of Great Britain	Townsend deprivation score, Congdon's anomie index, mean abstention rates from general elections in 1979, 1983, 1987, and 1992		Higher suicide rates associated with higher abstention, social fragmentation, and deprivation. Greatest for social fragmentation. Areas with greatest increases in social fragmentation over study period also had greatest increases in suicide	Ecological study unable to separate contextual from compositional effects

(Martikainen et al. 2004) Finland	ICD-9 codes E950-E959B, E959X and ICD-10 codes X60-X84, X870	Statistics Finland Ages 15–99 (1991–2001)	Functional regions ($n = 85$) were used, which are neighboring municipalities grouped according to travel-to-work areas and patterns of cooperation among municipalities	Area characteristics measured in 1990, 1992, 1993	Socioeconomic structure and deprivation, income inequality, social cohesion	Age, sex, SES, household income, housing tenure, economic activity, marital status, family type, household size, mother tongue	Suicide mortality varied between regions; larger for men than women. High mortality observed in deprived areas and areas with low family cohesion and voter turnout. Relationships were attenuated, but not fully accounted for after adjustment for individual characteristics. Area characteristics did not mediate/ modify effects of individual SES on suicide	Multilevel study
(Hempstead 2006) USA	ICD-9: E950-E959	Death certificates, medical examiner data 15 or older (1999–2001)	Municipality	Municipality of death (not residence) used	Population density, demographic structure, income, unemployment, and crime rates		Low population density and high % single-person households predicted suicide rates. More common in areas with declining population between 1990 and 2000 and where divorce rates were high	

(*continued*)

Table 6 (continued)

Study ID and setting	Diagnostic criteria and outcome	Case finding, age range, years	Definition of neighborhood	Timing of exposure	Area-level measures	Individual-level measures	Findings	Comments
(Agerbo et al. 2007) Denmark	ICD-8 codes E950-959 and ICD-10 codes X60-X84	Medical register on vital statistics Ages 25–60 (1982–1997)	Municipalities in Denmark ($N = 276$)	Year of suicide	% of people living alone, % employed individuals (based on male employment), median gross income (by male income)	Marital status, gross income, employment	When adjusted for individual measures, area-level associations were weak. Little evidence of cross-level interaction (that individual risk varied based on area characteristics)	Area-level associations explained by composition of high-risk individuals in some areas
(Stark et al. 2007) Scotland	ICD-9 codes E950-E959. ICD-10 codes X60-X84, E980-E989, and Y10-Y34	General Register Office for Scotland 1981–1999	Postcode sector (used as proxy for rurality)	Death allocated to area of residency, not area at death	Deprivation and population density		Deprivation was strongly associated with suicide rates in Scotland for all levels of population density for all ages	

(O'Reilly et al. 2008) Northern Ireland	ICD-10 codes X60-X84, Y10-Y34, Y87.0	Northern Ireland Statistics and Research Agency linked all those in 2001 census to deaths in the subsequent 5 years $n = 1,116,748$ ($n = 566$ suicides) aged 16–74		Census super-output area level (a standard government administrative area, with average population size 1,894)	Material deprivation (% receiving social security benefits), population density, social fragmentation (% in rented accommodation, unmarried, less than 65 living alone, population turnover in year preceding census)	Marital status, household size, SES, economic activity, self-reported health status	Higher suicide rates in more deprived and fragmented areas disappeared after adjustment for individual and household factors. No relationship between population density and suicide. Differences in rates of suicide between areas due to population composition than area-level factors	Area-level associations explained by composition of high-risk individuals in some areas
(Haws et al. 2009) USA	Not specified	Not specified (1990–1994)	States	Not specified	Elevation (i.e., altitude of state capital city)	Age, race/ ethnicity, sex	After adjustment, altitude of the state capital city was significantly correlated with suicide rates	

(*continued*)

Table 6 (continued)

Study ID and setting	Diagnostic criteria and outcome	Case finding, age range, years	Definition of neighborhood	Timing of exposure	Area-level measures	Individual-level measures	Findings	Comments
(Collings et al. 2009) New Zealand	ICD-9 codes E950-E959, E980-E989	2.8 million respondents in 1996 census, followed up for 3 years for mortality ($n = 1{,}101$ suicide deaths), aged 20–74 years at follow-up (1996–2000)	There were 1,775 area units (approximately 2000 people in each area) with borders based on locally recognizable communities		Index of neighborhood social fragmentation, Congdon index of fragmentation, neighborhood deprivation	Age, sex, ethnicity, educational qualification, marital status	No linear association between neighborhood fragmentation and suicide. However, a U-shaped association was suggested by Congdon index, indicating that living in the most or least fragmented decile of neighborhoods was associated with increased suicide rates. Neighborhood deprivation also predicted suicide rates. Authors concluded that neighborhood SES is a risk factor for suicide independent of individual-level socioeconomic position	

(Kelly et al. 2009) Europe		National suicide rates from the World Health Organization (2002–2004)	Comparing 11 European nations	2003–2004 variables measured in European Social Survey	Social trust, age, sex, marriage rates, income, and reported sadness	N/A	National suicide rates inversely related to social trust after controlling other factors	Ecological study
(Jagodic et al. 2013) Austria		Statistics Austria for mortality database (1991–2005)	Districts (90 in Austria)		Population density, religion (% Roman Catholic), mean income, unemployment rate, number of general practitioners, psychiatrists, and nonphysician psychotherapists		Neither density of general practitioners or psychiatrists associated with suicide rates, but weak association of association with psychotherapist density	Little variance in density of GPs in Austria may explain lack of association
(Tsai 2010) Taiwan		Population-based mortality database from the Department of Health in Taiwan (1998–2006)			Population characteristics (gender, age), socioeconomic status (marriage, unemployment, income), and climatic factors (temperature, atmospheric pressure, precipitation, duration of sunshine)	Not measured	Suicide was negatively correlated with temperature and positively correlated with sunshine. Socioeconomic and climatic factors contributed 52.7% and 6.8%, respectively, to variance in suicide mortality	

(*continued*)

Table 6 (continued)

Study ID and setting	Diagnostic criteria and outcome	Case finding, age range, years	Definition of neighborhood	Timing of exposure	Area-level measures	Individual-level measures	Findings	Comments
(Qi et al. 2010) Australia		Australian Bureau of Statistics (ABS) 1999–2003 $N = 2{,}445$ suicide deaths All ages		Statistical local area (SLA, $n = 452$) and local governmental area (LGA). Urban LGAs contained 2+ SLAs and rural/remote areas	Geographical variation	Age, sex, year and month of suicide, country of birth, statistical local area	Significant suicide clusters were discovered in NW and N Australia. These areas had very low population density	Higher rates may be due to social isolation and lack of mental health services. Compositional effects included high % minority groups and low SES
(Chang et al. 2011) Taiwan	Certified suicides, undetermined death, accidental pesticide poisoning, suffocation. ICD-9 codes E950-E959, E980-E989, E863, E913	(1999–2007)	358 districts – districts are the administrative unit below the county/city level	Residence assigned based on registered address on death certificate	Single-person households, people who moved to another district, marital status, lone-parent households, households not owner-occupied, nonemployed adults, population aged 15–17 not at school, adults with college or higher education, median household income, population with limiting long-term		After controlling for other area characteristics, increasing suicide rates were found in areas with high % lone-parent households, low household income, and low population density. Indicators of social fragmentation (i.e., population mobility and % living alone/ unmarried) were not strongly	

					illnesses, indigenous people, agricultural workers, population density		associated with suicide rates	
(Hooghe and Vanhoutte 2011) Belgium		Official death certificates (confirmed by a medical doctor) 1996–2005 $N = 12{,}000$ registered cases of suicide		308 communities in Flanders region	Social integration (religious participation, rate of single households, internal migration rate, external migration rate, presence of non-European inhabitants), socioeconomic deprivation (average income), rural-urban divide (population density – inhabitants/km^2), age, and sex		Community characteristics were significantly related to suicide rates. Measures of social integration were mixed: single-person households and deprivation were associated with higher suicide rates; religious participation was unrelated; population density, immigration, and presence of non-European inhabitants with lower suicide rates. Areas with older populations had higher suicide risks than expected	Ecological study Communities have an impact on suicidal behaviors, possibly via effects of social isolation

(*continued*)

Table 6 (continued)

Study ID and setting	Diagnostic criteria and outcome	Case finding, age range, years	Definition of neighborhood	Timing of exposure	Area-level measures	Individual-level measures	Findings	Comments
(Bando et al. 2012) Brazil	ICD-10	Brazilian and São Paulo State Death Registry Databases Ages 15 or older 1996–2008 $n = 98{,}904$ deaths (Brazil), 21,066 in São Paulo state, 5,589 in São Paulo city	Deaths assigned to states (27 Brazilian states and 558 micro-regions), 645 counties of SP state and 96 districts of SP city		Average income	Sex	Higher suicide rates found in wealthier areas of Brazil, São Paulo city, and poorer areas of São Paulo state. May reflect worldwide patterns, where direct associations with income are found in more equal areas, but inverse is found in areas with more inequality. Regional socioeconomic characteristics may moderate association between SES and suicide risk	Ecological study

(Congdon 2012) UK		Adult Psychiatric Morbidity Survey		Government office region (REG) and small area deprivation quintile (QIMD)	Age, ethnicity, social capital	Household income, education, marital status, employment, owner-occupied housing, receipt of income benefit, one-person household, migrant in previous year	Neighborhood perceptions associated with suicidal behaviors. However, area deprivation did not have a distinct significant influence, in contrast to a number of ecological studies. Deprivation may be partially mediated by social capital	
(Zhang and Wang 2012) China	Not specified	Part of a large psychological autopsy project Ages 15–34 $n = 392$ suicide, 416 controls			WHO community stress and problem scale – social and structural stresses, community behavioral problems	Age, sex, education, income, marital status, physical illness, mental disorder, and status in family	Neighborhood stresses and problems increase suicide risk in rural China. Health care, alcohol, job security, family disputes, and transportation found to have important roles in increasing suicide risks	

(*continued*)

Table 6 (continued)

Study ID and setting	Diagnostic criteria and outcome	Case finding, age range, years	Definition of neighborhood	Timing of exposure	Area-level measures	Individual-level measures	Findings	Comments
(Jagodic et al. 2013) Slovenia		Institute of Public Health of the Republic of Slovenia (mandatory registry of all deaths) (2000–2009)	Population divided into 12 statistical regions (administrative entities)	2000	GDP per capita, unemployment, marriage rates, divorce rates, mental health service availability (psychiatrists per 100,000, availability of psychological services and primary care doctors), prevalence of mental disorders, prescribed anxiolytics and antidepressants, age, and sex		Unemployment rate, marriage/ divorce ratio , psychiatrist availability, and quality of depressive disorder treatment predicted regional suicide rates	Ecological study

(Kunst et al. 2013) Netherlands	ICD-9 codes E950-E959 and ICD-10 codes X60-X84	Mortality record and population data from Statistics Netherlands (CBS) $n = 6{,}207$ suicide deaths $n = 117{,}569$ individuals 18+ years (1995–2000)	$N = 3{,}507$ neighborhoods. Place of residence defined by meaningful socioeconomic or geographical boundaries, corresponding to people's perceptions of their community	Place of residence by postcode data on January 1, 1995	Area income, population density, religious orientation, social capital (WBO Housing Demand Survey in 1998), and complimentary measure of social fragmentation based on Congdon index	Age sex, marital status, country of origin	After controlling for individual factors, suicide rates were 60% higher in areas with lowest income compared with highest income. Variations in suicide rates according to population density and cultural/religious variables were smaller and nonlinear. Suicide mortality rates were higher in areas with low social capital. Effect of social capital on suicide mortality more pronounced for some population subgroups (men, younger people, and unmarried individuals)	Cross-sectional registry-based study

(continued)

Table 6 (continued)

Study ID and setting	Diagnostic criteria and outcome	Case finding, age range, years	Definition of neighborhood	Timing of exposure	Area-level measures	Individual-level measures	Findings	Comments
(Okamoto et al. 2013) Japan		Data from previous survey, ages 20+ (2006–2007 for suicide and demographic variables, 2009 for social capital measures)	Administrative municipalities ($n = 20$)		Social capital (organizational membership, social trust, fairness, helpfulness, confidence in organizations), sex, age, population density, unemployment, primary industry workers, % elderly, psychiatrist availability		Suicide rates inversely related to social trust for men, but not other indicators of social capital	Cross-sectional ecological study

(Ngamini Ngui et al. 2014b) Canada	ICD-10 codes X60-X84 and Y87	Death registry 15+ years (2004–2007)	Forward sortation area (FSA) – first three letters of the postcode ($n = 406$)	Postcode of the place of death	% renters, % population without a diploma, unemployment rate, % agricultural workers	Age and sex	Areas with highest % single-person households more likely to contain a suicide cluster. Less likely in areas with high % 65+ years, single-parent families, and % without a diploma. For women, % single-parent families and agricultural workers increased likelihood of suicide cluster, whereas % with a university education and % unemployed decreased risk	Not able to test for individual characteristics Unknown how long people had lived in recorded location

(*continued*)

Table 6 (continued)

Study ID and setting	Diagnostic criteria and outcome	Case finding, age range, years	Definition of neighborhood	Timing of exposure	Area-level measures	Individual-level measures	Findings	Comments
(Zammit et al. 2014) Sweden	ICD-8 codes E950-E959, E980-E989; ICD-9, codes E950-E959, E980-E989; ICD-10 codes X60-X84, Y10-Y34	204,323 individuals born in Sweden in 1972–1977 up to age 26–31 ($N = 314$ died by suicide)			School level = average: foreign born, social fragmentation, low grade. Municipality = urbanicity, population density, index of social fragmentation (residential mobility, voting, % married/single households)	Sex, country of birth, history of mental illness, change of x between 8 and 16 years, parental SES, education, family income, marital status, school grade	Little evidence that municipality-level measures associated with suicide risk. Several school-level measures were associated with suicide risk. Suicide risk was higher for individuals who attended schools with higher % females. This applied equally to men and women. Finally, social/cultural distance may increase suicide risk	Municipalities are relatively large geographical areas which may not capture small area-level effects, which may be captured at school level

Neighborhoods

In an attempt to move beyond basic rural-urban gradients in suicide mortality, several studies have used small area investigations to examine neighborhood variation which may account for variation in suicide rates (Table 6), including social isolation and integration (Hooghe and Vanhoutte 2011). As before, this research has primarily been conducted in high-income countries, including Denmark (Agerbo et al. 2007), Australia (Qi et al. 2010), New Zealand (Collings et al. 2009), the UK (Stark et al. 2007; Congdon 2012; Neeleman and Wessely 1999; O'Reilly et al. 2008; Whitley et al. 1999), the USA (Hempstead 2006; Haws et al. 2009), Canada (Ngamini Ngui et al. 2014a), Belgium (Hooghe and Vanhoutte 2011), Austria (Kapusta et al. 2010), the Netherlands (Kunst et al. 2013), Finland (Martikainen et al. 2004), Taiwan (Chang et al. 2011; Tsai 2010), Brazil (Bando et al. 2012), Slovenia (Jagodic et al. 2013), and Japan (Okamoto et al. 2013). Both social fragmentation (Ngamini Ngui et al. 2014b; Hooghe and Vanhoutte 2011; O'Reilly et al. 2008; Whitley et al. 1999; Jagodic et al. 2013) and deprivation (Kunst et al. 2013; Snider and Leo 2014; Hooghe and Vanhoutte 2011; Collings et al. 2009; O'Reilly et al. 2008; Martikainen et al. 2004) have been associated with suicide mortality, as well as some evidence of a relationship with income inequality (Martikainen et al. 2004), low social capital (in particular social trust) (Kunst et al. 2013; Hempstead 2006; Kelly and Davoren 2009; Okamoto et al. 2013), and low population density (Hempstead 2006; Stark et al. 2007; Hooghe and Vanhoutte 2011; Qi et al. 2010). Nonetheless, these relationships have not been uniformly observed, with, for example, Chang et al. (2011) finding that social fragmentation was not strongly or consistently associated with suicide rates in Taiwan. A limited number of studies have also investigated climatic factors (Haws et al. 2009; Tsai 2010), including temperature, precipitation, sunshine, altitude, and atmospheric pressure, finding they only explained a small amount of variance in suicide mortality and were generally outweighed by socioeconomic factors.

Despite investigating similar area-level measures, a number of methodological approaches were employed in neighborhood-level inquiries of suicide risk (see Table 6), making direct comparisons challenging. For example, the spatial scale at which suicide rates were compared varied from locally recognizable communities to postcode areas, larger governmental administrative areas, or political constituencies. In a systematic review of this literature, Rehkopf and Buka (2005) found that studies conducted at smaller community levels were more likely to find lower suicide rates in higher socioeconomic areas than studies which used larger areas of aggregation (Rehkopf and Buka 2005). The authors recommend more attention to the size of region and measurement strategies in order to provide a clearer picture of how suicide rates vary by region.

Many studies were ecological – meaning that they only studied variance in rates between neighborhoods, regionals, and countries – and did not control for potentially relevant individual characteristics (Ngamini Ngui et al. 2014b; Whitley et al. 1999; Tsai 2010; Okamoto et al. 2013). This approach, while useful for hypothesis generation and further inquiry, makes it impossible to determine whether variation in suicide rates was due to compositional or contextual effects (O'Reilly et al. 2008). To partially mitigate the risk of ecological fallacy, several multilevel studies (combining both individual- and area-level risks) have been conducted. For example, following adjustment for a range of individual measures, Congdon et al. (Congdon 2012) found that area deprivation had no distinct influence on suicide rates, in contrast to other studies (Agerbo et al. 2001; O'Reilly et al. 2008; Martikainen et al. 2004). Nonetheless, simple control for individual factors also risks ignoring the complexity of how places make people and how certain high-risk individuals may be at more or less risk following exposure to further environmental factors (so-called cross-level interactions) (Macintyre and Ellaway 2003).

Several studies have investigated such interactions; Neeleman and Wessley (1999), for example, have shown that suicide rates in different ethnic groups, including the white majority, were higher in areas with a smaller population of one's own ethnicity (i.e., an ethnic density effect), as consistently shown for schizophrenia (section "Schizophrenia and Other Non-Affective Psychoses"). Similarly, Kunst et al. (2013) demonstrated that the effect of social capital on suicide mortality was more pronounced for certain population groups, including young people, unmarried individuals, and men. Finally, a systematic review conducted by Crawford et al. (2010), synthesizing data from 54 studies, found that established risk factors including unemployment, living alone, and ethnic minority status had a lower impact on suicide risk for individuals living in areas where these factors were more common (Crawford et al. 2010).

Variation in Risk by Suicide Methods

There is a strong evidence that the rural-urban gradient (higher rates in rural areas) for suicide mortality is strongly influenced by choice of method. Much of this literature points to the impact of accessibility when it comes to highly lethal means, including firearms, pesticides, and fertilizers, although this also appears to vary by setting. In countries including the USA, Canada, the UK, and Australia, there is consistent evidence showing that use of firearms for suicide is strongly spatially patterned, being more prevalent in rural and remote areas (Qi et al. 2014; Fontanella et al. 2015; Searles 2014; Kapusta et al. 2008; Ngamini Ngui et al. 2014b). Singh and Siahpush (2002) have noted that firearms accounted for 75% of all rural suicides, but only half of such incidents in urban areas, and accessibility to firearms has been cited as a major contributory factor for rural-urban differences in several studies (Fontanella et al. 2015; Singh and Siahpush 2002; Ngamini Ngui et al. 2014b; Judd et al. 2006; Hirsch and Cukrowicz 2014). Further, the changing availability of firearms may account for some time trends in suicide rates. Fontanella et al. (2015), for example, have noted that firearm ownership has decreased in urban areas, but remains stable in rural areas, potentially widening the disparity in both the accessibility of method and suicide rates between rural and urban communities. In urban areas, methods including jumping, poisoning, and hanging are more common

(Ngamini Ngui et al. 2014b). Pesticide poisoning is more common in some rural areas, including in Taiwan (Chang et al. 2011) and South Korea (Park and Lester 2012), with some evidence that hangings showed least geographic differences (Chang et al. 2011).

These distinct geographic patterns of suicide methods appear to be driven by the ready availability of method (Singh and Siahpush 2002), although may also be shaped by cultural norms, localized patterns (i.e., imitation; see Levin and Leyland 2005), and social expectations. Levin (Levin and Leyland 2005) also note that socioeconomic means may determine accessibility to certain more expensive methods, including the use of firearms or explosives. Changing attitudes toward firearms increased regulations on storage, and reduced access to firearms has been shown to reduce suicide rates (Wilkinson and Gunnell 2000; Rodríguez Andrés and Hempstead 2011). Large and Nielssen (2010) have used meta-analysis data to demonstrate that the decline in suicide in Australia coincided with a reduction in the availability of lethal means.

Main Hypotheses Explaining the Spatial Heterogeneity in Suicide Mortality

In addition to accessibility to methods and sociocultural influences, several other hypotheses have been proposed to explain the spatial patterning of suicide mortality, broadly organized into *compositional*, *contextual*, and *cross-level* explanations. We discuss each, in turn, below.

Compositional Hypotheses

Compositional explanations posit that the individual characteristics of people within certain locales, including the social drift of high-risk individuals into certain areas (Agerbo et al. 2007), account for the rural-urban and spatial differences in suicide (and other) rates. Low income and education, unemployment, and social isolation are all strongly related to suicide risk (Agerbo et al. 2001; O'Reilly et al. 2008) and may be more prevalent in some communities as a result of both structural and cultural influences maintained over successive generations. In many rural areas, these compositional effects may be compounded by the "out-migration" of young, educated individuals, which generates greater psychological distress or a sense of entrapment in those left behind (Judd et al. 2006; Alston 2012). It has been suggested that the subsequently increased levels of psychological distress may drive regional differences in suicide rates. However, there is no consistent evidence that differences in rates of psychiatric morbidity exist between areas with high and low suicide rates (Middleton et al. 2003; Judd et al. 2006). If compositional effects account for the majority of spatial variance in suicide risk, public mental health strategies to ameliorate this burden can focus on both indicated and selective prevention interventions which target high-risk individuals or subgroups of the population.

Contextual Hypotheses

Contextual effects refer to direct effects that characteristics of places have on suicide risk, which go beyond the totality of compositional effects. These contextual effects may operate either through increased likelihood of suicidal behavior or through low levels of social support (Agerbo et al. 2007). Contextual explanations also include how the organization, accessibility, and availability of mental health care vary between different areas. The evidence for contextual effects for suicide (Table 6) generally indicates that areas characterized by greater social fragmentation, isolation, and disintegration have higher mortality rates. This may reflect "differential changes over time in key social integration indicators," (Singh and Siahpush 2002) consistent with Durkheim's theory that *anomie* – low social integration – drives high suicide rates. Low levels of social integration may partly explain increased suicide rates in rural areas (Fontanella et al. 2015; Hirsch and Cukrowicz 2014), which may be confounded by changing demographic profiles in many rural and remote communities and deteriorating economic conditions (Middleton et al. 2003; Judd et al. 2006). It has been suggested, for example, that rurality per se may be less important than the differential effects of global economic trends on local regions, with certain rural communities particularly affected by downturns in economic activity. Further nuanced investigation of such intricate spatial variation of suicide is warranted (Judd et al. 2006).

Suicide rates may also be influenced by the availability and accessibility of mental health-care services and other public amenities in a given area (Fontanella et al. 2015; Middleton et al. 2003; Judd et al. 2006; Hirsch and Cukrowicz 2014). This may be compounded by issues of mental health literacy, less service utilization, concerns about stigma and discrimination, and lower willingness to seek help in some communities. Such barriers to care are often more common in rural communities, possibly due to cultural attitudes and shared norms (Singh and Siahpush 2002; Middleton et al. 2003; Judd et al. 2006; Hirsch 2006). Cultural factors and stigma can also present barriers to treatment, including the possible lack of anonymity in more rural, smaller communities (Fontanella et al. 2015; Hirsch and Cukrowicz 2014). In particular, community attitudes toward mental illness and help-seeking may contribute to elevated suicide rates, particularly among rural men (Judd et al. 2006). Other authors have noted that dominant masculine hegemony, which may be more common in some rural communities, tends to favor gendered expectations of stoicism and self-reliance, which may present a substantial barrier to help-seeking in the entire community (Judd et al. 2006; Alston 2012).

Cross-Level Hypotheses

Cross-level effects focus on how characteristics of geographical areas may have differential effects on individuals based on their particular characteristics. If prominent, these effects would make the debate about compositional versus contextual effects overly reductionist. Since few individual (i.e., compositional) risk factors are truly exogenous to the social environment, simple adjustment for these factors in the search for contextual mechanisms in suicide risk risks ignoring how the effects of place affect people's social and economic opportunities, including access to good education and high-income jobs. Much of the research reviewed above may thus have overlooked, or have been underpowered to detect, important cross-level interactions which may be vital for the development of putative prevention strategies in different contexts. Theoretical support for cross-level interactions can be traced back to Durkheim's proposal that the greater the dissonance between an individual and his/her social environment, the greater their risk of suicide (Agerbo et al. 2007). Likewise, Crawford and colleagues (2010) argued that the impact of unemployment on suicide risk may be exacerbated if unemployed individuals lived in area of high employment, although universal support for this has not been found (Martikainen et al. 2004). Combining the theoretical rational for cross-level interactions with empirical evidence is urgently warranted so we understand how "people make places and people make people" (Macintyre and Ellaway 2003) and permit appropriate public mental health responses in different populations, subgroups, and settings.

Conclusion

We have reviewed the literature on the association between urban living and three major sets of mental health outcomes, namely, psychotic disorders, common mental disorders, and suicide. Our comprehensive review of the current evidence, its strengths and limitations, and possible hypotheses to explain any variation should serve as a useful reference point for those working in epidemiology, public mental health, and mental health care and provision. But what do these findings say in regard to possible preventive strategies to reduce the global burden of mental disorders? What conclusions can we draw at this stage to promote the design and maintenance of healthy environments which promote positive mental health?

The reader will probably concur with our view that no universal answer exists to these questions. Common mental disorders show no consistent direction of association with urban or rural living, which in part might be due to heterogeneous study designs, difficulties with case detection, and other forms of biases inherent to studying prevalent conditions for which people in the community may rarely seek help. Psychotic disorders and suicide showed stronger, more consistent associations with the environment, but in opposite directions: psychotic disorders being more common in urban populations and suicide generally being elevated in more rural communities. The extent to which these patterns were causally determined, that is, due to the exposure rather than attributable to reverse causation, was considered, and for both outcomes, there is reasonable evidence to implicate social and environmental determinants in the aetiology of these mental health outcomes. These may occur at a variety of levels, including as a result of individual attributes, such as poverty, unemployment, and social class, as well as at household and neighborhood levels.

For both psychotic disorders and suicide, further research is required to investigate the extent to which social drift (of those with worse mental health into more deprived rural or urban communities) or upward social mobility (of those with better mental health away from such communities) may help to reveal whether patterns at the environmental level are purely compositional or at least partly contextual. Efforts to ameliorate exposure to adverse psychosocial adversities in the environment, including deprivation and social isolation, may help reduce the occurrence of both schizophrenia and suicide, albeit these efforts will need to be directed at different communities. For suicide the strong rural gradient associated with risk may be due in part to the availability of means, and more direct preventive strategies including tighter controls on firearms and pesticides may be worthwhile. For now, for all mental health outcomes, studies which can elucidate the multiple potential causal levels which drive increased risk of disorder are warranted, particularly those with a longitudinal element to their design which can tease out any sensitive windows to environmental factors over the life course.

References

Agerbo E, Torrey EF, Mortensen PB (2001) Household crowding in early adulthood and schizophrenia are unrelated in Denmark: a nested case-control study. Schizophr Res 47:243–246

Agerbo E, Sterne JAC, Gunnell DJ (2007) Combining individual and ecological data to determine compositional and contextual socio-economic risk factors for suicide. Soc Sci Med 64:451–461

Ahern J, Galea S (2011) Collective efficacy and major depression in urban neighborhoods. Am J Epidemiol 173:1453–1462

Akdeniz C, Tost H, Meyer-Lindenberg A (2014) The neurobiology of social environmental risk for schizophrenia: an evolving research field. Soc Psychiatry Psychiatr Epidemiol 49:507–517

Allardyce J (2001) Comparison of the incidence of schizophrenia in rural Dumfries and Galloway and urban Camberwell. Br J Psychiatry 179:335–339

Allardyce J et al (2005) Social fragmentation, deprivation and urbanicity: relation to first-admission rates for psychoses. Br J Psychiatry 187:401–406

Alston M (2012) Rural male suicide in Australia. Soc Sci Med 74:515–522

Amoran O, Lawoyin T, Lasebikan V (2007) Prevalence of depression among adults in Oyo State, Nigeria: a comparative study of rural and urban communities. Aust J Rural Health 15:211–215

Andrade L et al (2003) The epidemiology of major depressive episodes: results from the International Consortium of Psychiatric Epidemiology (ICPE) surveys. Int J Methods Psychiatr Res 12:3–21

Andrews G (2001) Prevalence, comorbidity, disability and service utilisation: overview of the Australian National Mental Health Survey. Br J Psychiatry 178:145–153

Arsenault-Lapierre G, Kim C, Turecki G (2004) Psychiatric diagnoses in 3275 suicides: a meta-analysis. BMC Psychiatry 4:37

Ayuso-Mateos J-L (2001) Depressive disorders in Europe: prevalence figures from the ODIN study. Br J Psychiatry 179:308–316

Bando DH, Brunoni AR, Benseñor IM, Lotufo PA (2012) Suicide rates and income in São Paulo and Brazil: a temporal and spatial epidemiologic analysis from 1996 to 2008. BMC Psychiatry 12:127

Beautrais A (2000) Risk factors for suicide and attempted suicide among young people. Aust N Z J Psychiatry 34:420–436

Beskow J (1979) Suicide in mental disorder in Swedish men. Acta Psychiatr Scand 277:1–138

Beyer KMM et al (2014) Exposure to neighborhood green space and mental health: evidence from the survey of the health of Wisconsin. Int J Environ Res Public Health 11:3453–3472

Bhavsar V, Boydell J, Murray R, Power P (2014) Identifying aspects of neighbourhood deprivation associated with increased incidence of schizophrenia. Schizophr Res 156:115–121

Blackwood DHR et al (2007) Are some genetic risk factors common to schizophrenia, bipolar disorder and depression? Evidence from DISC1, GRIK4 and NRG1. Neurotox Res 11:73–83

Blazer DG, Kessler RC, McGonagle KA, Swartz MS (1994) The prevalence and distribution of major depression in a national community sample: the National Comorbidity Survey. Am J Psychiatry 151:979–986

Bleuler E (1911) Dementia praecox or the group of schizophrenias. Translated by J. Zinken. New York: International Universities Press

Boydell J et al (2001) Incidence of schizophrenia in ethnic minorities in London: ecological study into interactions with environment. BMJ 323:1336–1336

Boydell J, van Os J, McKenzie K, Murray RM (2004) The association of inequality with the incidence of schizophrenia – an ecological study. Soc Psychiatry Psychiatr Epidemiol 39:597–599

Bresnahan M et al (2007) Race and risk of schizophrenia in a US birth cohort: another example of health disparity? Int J Epidemiol 36:751–758

Burns JK, Esterhuizen T (2008) Poverty, inequality and the treated incidence of first-episode psychosis: an ecological study from South Africa. Soc Psychiatry Psychiatr Epidemiol 43:331–335

Byrne M, Agerbo E, Eaton WW, Mortensen PB (2004) Parental socio-economic status and risk of first admission with schizophrenia- a Danish national register based study. Soc Psychiatry Psychiatr Epidemiol 39:87–96

Caldwell TM, Jorm AF, Dear KBG (2004) Suicide and mental health in rural, remote and metropolitan areas in Australia. Med J Aust 181:10–14

Chan KY et al (2015a) Prevalence of schizophrenia in China between 1990 and 2010. J Glob Health 5:010410

Chan KY et al (2015b) Urbanization and the prevalence of schizophrenia in China between 1990 and 2010. World Psychiatry 14:251–252

Chang SS et al (2011) Geography of suicide in Taiwan: spatial patterning and socioeconomic correlates. Health Place 17:641–650

Chiavegatto Filho ADP, Kawachi I, Wang YP, Viana MC, Andrade LHSG (2013) Does income inequality get under the skin? A multilevel analysis of depression, anxiety and mental disorders in Sao Paulo, Brazil. J Epidemiol Commun Health 67:966–972

Collings S, Ivory V, Blakely T, Atkinson J (2009) Are neighbourhood social fragmentation and suicide associated in New Zealand? A national multilevel cohort study. J Epidemiol Community Health 63:1035–1042

Congdon P (2012) Latent variable model for suicide risk in relation to social capital and socio-economic status. Soc Psychiatry Psychiatr Epidemiol 47:1205–1219

Craddock N, O'Donovan MC, Owen MJ (2005) The genetics of schizophrenia and bipolar disorder: dissecting psychosis. J Med Genet 42:193–204

Crawford MJ, Kuforiji B, Ghosh P (2010) The impact of social context on socio-demographic risk factors for suicide: a synthesis of data from case-control studies. J Epidemiol Community Health 64:530–534

Croudace TJ, Kayne R, Jones PB, Harrison GL (2000) Non-linear relationship between an index of social deprivation, psychiatric admission prevalence and the incidence of psychosis. Psychol Med 30:177–185

D'Souza DC, Sewell RA, Ranganathan M (2009) Cannabis and psychosis/schizophrenia: human studies. Eur Arch Psychiatry Clin Neurosci 259:413–431

Das-Munshi J, Becares L, Dewey ME, Stansfeld SA, Prince MJ (2010) Understanding the effect of ethnic density on mental health: multi-level investigation of survey data from England. BMJ 341:c5367

Dauncey K, Giggs J, Baker K, Harrison G (1993) Schizophrenia in Nottingham: lifelong residential mobility of a cohort. Br J Psychiatry 163:613–619

de Graaf R, Bijl RV, Ravelli A, Smit F, Vollebergh WAM (2002) Predictors of first incidence of DSM-III-R psychiatric disorders in the general population: findings from the Netherlands Mental Health Survey and Incidence Study. Acta Psychiatr Scand 106:303–313

de Graaf R, ten Have M, van Gool C, van Dorsselaer S (2012) Prevalence of mental disorders and trends from 1996 to 2009. Results from the Netherlands Mental Health Survey and Incidence Study-2. Soc Psychiatry Psychiatr Epidemiol 47:203–213

de Graaf R, ten Have M, Tuithof M, van Dorsselaer S (2013) First-incidence of DSM-IV mood, anxiety and substance use disorders and its determinants: results from the Netherlands Mental Health Survey and Incidence Study-2. J Affect Disord 149:100–107
De Silva MJ, McKenzie K, Harpham T, Huttly SRA (2005) Social capital and mental illness: a systematic review. J Epidemiol Community Health 59:619–627
Drukker M, Krabbendam L, Driessen G, van Os J (2006) Social disadvantage and schizophrenia. Soc Psychiatry Psychiatr Epidemiol 41:595–604
Dupéré V, Perkins DD (2007) Community types and mental health: a multilevel study of local environmental stress and coping. Am J Community Psychol 39:107–119
Durkheim E (1897) On suicide. London: Penguin Books
Eaton WW, Mortensen PB, Frydenberg M (2000) Obstetric factors, urbanization and psychosis. Schizophr Res 43:117–123
Erdem Ö, Prins RG, Voorham TAJJ, van Lenthe FJ, Burdorf A (2015) Structural neighbourhood conditions, social cohesion and psychological distress in the Netherlands. Eur J Public Health 25:995–1001
Faris REL, Dunham HW (1939) Mental disorders in urban areas. Chicago: Chicago University Press
Fergusson DM, Lynskey MT (1995) Suicide attempts and suicidal ideation in a birth cohort of 16-year-old New Zealanders. J Am Acad Child Adolesc Psychiatry 34:1308–1317
Fergusson DM, Horwood LJ, Ridder EM, Beautrais AL (2005) Suicidal behaviour in adolescence and subsequent mental health outcomes in young adulthood. Psychol Med 35:983–993
Fone DL, Dunstan F (2006) Mental health, places and people: a multilevel analysis of economic inactivity and social deprivation. Health Place 12:332–344
Fontanella CA et al (2015) Widening rural-urban disparities in youth suicides, United States, 1996–2010. JAMA Pediatr 169:466
Galea S, Ahern J, Rudenstine S, Wallace Z, Vlahov D (2005) Urban built environment and depression: a multilevel analysis. J Epidemiol Community Health 59:822–827
Galea S et al (2007) Urban neighborhood poverty and the incidence of depression in a population-based cohort study. Ann Epidemiol 17:171–179
Ganguli HC (2000) Epidemiological findings on prevalence of mental disorders in India. Indian J Psychiatry 42:14–20
Gariepy G et al (2015) The neighbourhood built environment and trajectories of depression symptom episodes in adults: a latent class growth analysis. PLoS ONE 10:e0133603
Gibb SJ, Beautrais AL, Fergusson DM (2004) Mortality and further suicidal behaviour after an index suicide attempt: a 10-year study. Aust N Z J Psychiatry 39:95–100
Giggs JA, Cooper JE (1987) Ecological structure and the distribution of schizophrenia and affective psychoses in Nottingham. Br J Psychiatry 151:627–633
Goldberg EM, Morrison SL (1963) Schizophrenia and social class. Br J Psychiatry 109:785–802
Grant BF et al (2009) Sociodemographic and psychopathologic predictors of first incidence of DSM-IV substance use, mood and anxiety disorders: results from the Wave 2 National Epidemiologic Survey on Alcohol and Related Conditions. Mol Psychiatry 14:1051–1066
Greiner KA, Li C, Kawachi I, Hunt DC, Ahluwalia JS (2004) The relationships of social participation and community ratings to health and health behaviors in areas with high and low population density. Soc Sci Med 59:2303–2312
Guite HF, Clark C, Ackrill G (2006) The impact of the physical and urban environment on mental well-being. Public Health 120:1117–1126
Halpern D, Nazroo J (2000) The ethnic density effect: results from a national community survey of England and Wales. Int J Soc Psychiatry 46:34–46
Hare E, Family H (1956) Setting and the urban distribution of Schizophrenia. Br J Psychiatry 102:753–760
Harrison G et al (2003) Association between psychotic disorder and urban place of birth is not mediated by obstetric complications or childhood socio-economic position: a cohort study. Psychol Med 33:723–731

Haukka J, Suvisaari J, Varilo T, Lönnqvist J (2001) Regional variation in the incidence of schizophrenia in Finland: a study of birth cohorts born from 1950 to 1969. Psychol Med 31:1045–1053
Haws CA et al (2009) The possible effect of altitude on regional variation in suicide rates. Med Hypotheses 73:587–590
Hawton K (2000) Sex and suicide. Gender differences in suicidal behaviour. Br J Psychiatry 177:484–485
Heinz A, Deserno L, Reininghaus U (2013) Urbanicity, social adversity and psychosis. World Psychiatry 12:187–197
Hempstead K (2006) The geography of self-injury: spatial patterns in attempted and completed suicide. Soc Sci Med 62:3186–3196
Hirsch JK (2006) A review of the literature on rural suicide: risk and protective factors, incidence, and prevention. Crisis 27:189–199
Hirsch JK, Cukrowicz KC (2014) Suicide in rural areas: an updated review of the literature. J Rural Ment Heal 38:65–78
Hooghe M, Vanhoutte B (2011) An ecological study of community-level correlates of suicide mortality rates in the Flemish region of Belgium, 1996–2005. Suicide Life Threat Behav 41:453–464
Jagodic HK, Rokavec T, Agius M, Pregelij P (2013) Availability of mental health service providers and suicide rates in Slovenia: a nationwide ecological study. Croat Med J 54:1198–1203
Jokela M (2014) Are neighborhood health associations causal? A 10-year prospective cohort study with repeated measurements. Am J Epidemiol 180:776–784
Jones-Rounds ML, Evans GW, Braubach M (2014) The interactive effects of housing and neighbourhood quality on psychological well-being. J Epidemiol Community Health 68:171–175
Judd F, Cooper AM, Fraser C, Davis J (2006) Rural suicide – people or place effects? Aust N Z J Psychiatry 40:208–216
Kapusta ND et al (2008) Rural-urban differences in Austrian suicides. Soc Psychiatry Psychiatr Epidemiol 43:311–318
Kapusta ND et al (2010) Availability of mental health service providers and suicide rates in Austria: a nationwide study. Psychiatr Serv 61:1198–1203
Kawachi I (2006) Commentary: social capital and health: making the connections one step at a time. Int J Epidemiol 35:989–993
Kelly BD, Davoren ÆM (2009) Social capital and suicide in 11 European countries: an ecological analysis. pp 971–977. doi:10.1007/s00127–009–0018-4
Kelly BD, Davoren M, Mhaoláin AN, Breen EG, Casey P (2009) Social capital and suicide in 11 European countries: an ecological analysis. Soc Psychiatry Psychiatr Epidemiol 44:971–977
Kelly BD et al (2010) Schizophrenia and the city: a review of literature and prospective study of psychosis and urbanicity in Ireland. Schizophr Res 116:75–89
Kessler RC et al (1994) Lifetime and 12-month prevalence of DSM-III-R psychiatric disorders in the United States. Results from the National Comorbidity Survey. Arch Gen Psychiatry 51:8–19
Kessler RC et al (2003) The epidemiology of major depressive disorder: results from the National Comorbidity Survey Replication (NCS-R). JAMA 289:3095–3105
Kirkbride JB, Jones PB (2011) In: Bhugra D, Gupta S (eds) Migration and mental health. Cambridge: Cambridge University Press, pp 15–31
Kirkbride JB et al (2006) Heterogeneity in incidence rates of schizophrenia and other psychotic syndromes: findings from the 3-center AeSOP study. Arch Gen Psychiatry 63:250–258
Kirkbride JB et al (2007a) Neighbourhood variation in the incidence of psychotic disorders in Southeast London. Soc Psychiatry Psychiatr Epidemiol 42:438–445
Kirkbride JB et al (2007b) Neighbourhood-level effects on psychoses: re-examining the role of context. Psychol Med 37:1413–1425
Kirkbride JB et al (2008) Testing the association between the incidence of schizophrenia and social capital in an urban area. Psychol Med 38:1083–1094

Kirkbride JB et al (2010) Examining evidence for neighbourhood variation in the duration of untreated psychosis. Health Place 16:219–225
Kirkbride JB et al (2011) Addendum. Psychol Med 41:2462
Kirkbride JB et al (2012) Systematic review of the incidence and prevalence of Schizophrenia and other psychoses in England. PLoS One. e31660
Kirkbride JB, Jones PB, Ullrich S, Coid JW (2014) Social deprivation, inequality, and the neighborhood-level incidence of psychotic syndromes in East London. Schizophr Bull 40:169–180
Kirkbride JB et al (2015) Social and spatial heterogeneity in psychosis proneness in a multilevel case–prodrome–control study. Acta Psychiatr Scand 132:283–292
Kovess-Masfety V, Lecoutour X, Delavelle S (2005) Mood disorders and urban/rural settings: comparisons between two French regions. Soc Psychiatry Psychiatr Epidemiol 40:613–618
Kunst AE, van Hooijdonk C, Droomers M, Mackenbach JP (2013) Community social capital and suicide mortality in the Netherlands: a cross-sectional registry-based study. BMC Public Health 13:969
Lane EA, Albee GW (1966) Comparative birth weights of Schizophrenics and their siblings. J Psychol 64:227–231
Langhinrichsen-Rohling J, Friend J, Powell A (2009) Adolescent suicide, gender, and culture: a rate and risk factor analysis. Aggress Violent Behav 14:402–414
Large MM, Nielssen OB (2010) Suicide in Australia: meta-analysis of rates and methods of suicide between 1988 and 2007. Med J Aust 192:432–437
Lasalvia A et al (2014) First-contact incidence of psychosis in north-eastern Italy: influence of age, gender, immigration and socioeconomic deprivation. Br J Psychiatry 205:127–134
Laursen TM, Munk-Olsen T, Nordentoft M, Bo Mortensen P (2007) A comparison of selected risk factors for unipolar depressive disorder, bipolar affective disorder, schizoaffective disorder, and schizophrenia from a danish population-based cohort. J Clin Psychiatry 68:1673–1681
Leventhal T, Brooks-Gunn J (2003) Moving to opportunity: an experimental study of neighborhood effects on mental health. Am J Public Health 93:1576–1582
Levin KA, Leyland AH (2005) Urban/rural inequalities in suicide in Scotland, 1981–1999. Soc Sci Med 60:2877–2890
Lewis G, Booth M (1994) Are cities bad for your mental health? Psychol Med 24:913–915
Lewis SW, Murray RM (1987) Obstetric complications, neurodevelopmental deviance, and risk of schizophrenia. J Psychiatr Res 21:413–421
Lewis G, David A, Andréasson S, Allebeck P (1992) Schizophrenia and city life. Lancet (London, England) 340:137–140
Linsley KR, Schapira K, Kelly TP (2001) Open verdict v. suicide – importance to research. Br J Psychiatry 178:465–468
Løberg E-M et al (2014) The cannabis pathway to non-affective psychosis may reflect less neurobiological vulnerability. Front Psychiatry 5:159
Löffler W, Häfner H (1999) Ecological pattern of first admitted schizophrenics in two German cities over 25 years. Soc Sci Med 49:93–108
Lofors J, Sundquist K (2007) Low-linking social capital as a predictor of mental disorders: a cohort study of 4.5 million Swedes. Soc Sci Med 64:21–34
Lofors J, Ramírez-León V, Sundquist K (2006) Neighbourhood income and anxiety: a study based on random samples of the Swedish population. Eur J Pub Health 16:633–639
Long J et al (2014) The prevalence of schizophrenia in mainland China: evidence from epidemiological surveys. Acta Psychiatr Scand 130:244–256
Lundberg P, Cantor-Graae E, Rukundo G, Ashaba S, Ostergren P-O (2009) Urbanicity of place of birth and symptoms of psychosis, depression and anxiety in Uganda. Br J Psychiatry 195:156–162
Macintyre S, Ellaway A 2003) In: Kawachi I, Berkman L (eds) Neighbourhoods and health. Oxford: Oxford University Press, pp 20–43
Marcelis M, Navarro-Mateu F, Murray R, Selten JP, Van Os J (1998) Urbanization and psychosis: a study of 1942-1978 birth cohorts in The Netherlands. Psychol Med 28:871–879

Marcelis M, Takei N, van Os J (1999) Urbanization and risk for schizophrenia: does the effect operate before or around the time of illness onset? Psychol Med 29:1197–1203
March D et al (2008) Psychosis and place. Epidemiol Rev 30:84–100
Martikainen P, Mäki N, Blomgren J (2004) The effects of area and individual social characteristics on suicide risk: a multilevel study of relative contribution and effect modification. Eur J Popul 20:323–350
Matheson FI et al (2006) Urban neighborhoods, chronic stress, gender and depression. Soc Sci Med 63:2604–2616
McCulloch A (2001) Ward-level deprivation and individual social and economic outcomes in the British Household Panel Study. Environ Plan A 33:667–684
McGrath J et al (2004) A systematic review of the incidence of schizophrenia: the distribution of rates and the influence of sex, urbanicity, migrant status and methodology. BMC Med 2:13
Mckenzie K (2008) Urbanization social capital and mental Health. Glob Soc Policy 8:359–377
Mednick SA (1970) Breakdown in individuals at high risk for schizophrenia: possible pre-dispositional perinatal factors. Ment Hyg 54:50–63
Menezes PR (2014) Commentary: epidemiological mental health research: contribution from low- and middle-income countries is essential. Int J Epidemiol 43:301–303
Middleton N, Gunnell D, Frankel S, Whitley E, Dorling D (2003) Urban–rural differences in suicide trends in young adults: England and Wales, 1981–1998. Soc Sci Med 57:1183–1194
Middleton N, Sterne JAC, Gunnell DJ (2008) An atlas of suicide mortality: England and Wales, 1988–1994. Heal Place 14:492–506
Mizrahi R (2016) Social stress and psychosis risk: common neurochemical substrates? Neuropsychopharmacology 41:666–674
Mohammadi M-R et al (2005) An epidemiological survey of psychiatric disorders in Iran. Clin Pract Epidemiol Ment Health 1:16
Moreno-Küstner B et al (2014) Prevalence of schizophrenia and related disorders in Malaga (Spain): results using multiple clinical databases. Epidemiol Psychiatr Sci: 1–11
Morgan C, Charalambides M, Hutchinson G, Murray RM (2010) Migration, ethnicity, and psychosis: toward a sociodevelopmental model. Schizophr Bull 36:655–664
Mortensen PB et al (1999) Effects of family history and place and season of birth on the risk of schizophrenia. N Engl J Med 340:603–608
Mortensen PB, Pedersen CB, Melbye M, Mors O, Ewald H (2003) Individual and familial risk factors for bipolar affective disorders in Denmark. Arch Gen Psychiatry 60:1209–1215
Myin-Germeys I, Krabbendam L, Delespaul PAEG, Van Os J (2003) Do life events have their effect on psychosis by influencing the emotional reactivity to daily life stress? Psychol Med 33:327–333
Neeleman J, Wessely S (1999) Ethnic minority suicide: a small area geographical study in south London. Psychol Med 29:429–436
Ngamini Ngui A et al (2013) Does elapsed time between first diagnosis of schizophrenia and migration between health territories vary by place of residence? A survival analysis approach. Health Place 20:66–74
Ngamini Ngui A et al (2014a) Disentangling the influence of neighborhood and individual characteristics on early residential mobility among newly diagnosed patients with schizophrenia: a multilevel analysis. Soc Psychiatry Psychiatr Epidemiol 49:1569–1578
Ngamini Ngui A, Apparicio P, Moltchanova E, Vasiliadis H-M (2014b) Spatial analysis of suicide mortality in Québec: spatial clustering and area factor correlates. Psychiatry Res 220:20–30
Norman RM, Malla AK (1993) Stressful life events and schizophrenia. I: a review of the research. Br J Psychiatry 162:161–166
O'Donoghue B et al (2016) Neighbourhood characteristics and the incidence of first-episode psychosis and duration of untreated psychosis. Psychol Med: 1–12. doi:10.1017/S003329171500286X
O'Reilly DO, Rosato M, Connolly S, Cardwell C (2008) Area factors and suicide: 5-year follow-up of the Northern Ireland population. Br J Psychiatry 192:106–111

Oher FJ et al (2014) The effect of the environment on symptom dimensions in the first episode of psychosis: a multilevel study. Psychol Med FirstView: 1–12
Okamoto M, Kawakami N, Kido Y, Sakurai K (2013) Social capital and suicide: an ecological study in Tokyo, Japan. Environ Health Prev Med 18:306–312
Omer S et al (2014) Neighbourhood-level socio-environmental factors and incidence of first episode psychosis by place at onset in rural Ireland: the Cavan-Monaghan First Episode Psychosis Study [CAMFEPS]. Schizophr Res 152:152–157
Page A, Morrell S, Taylor R, Dudley M, Carter G (2007) Further increases in rural suicide in young Australian adults: secular trends, 1979-2003. Soc Sci Med 65:442–453
Page A, Milner A, Morrell S, Taylor R (2013) The role of under-employment and unemployment in recent birth cohort effects in Australian suicide. Soc Sci Med 93:155–162
Palmer BS et al (2015) Factors influencing coroners' verdicts: an analysis of verdicts given in 12 coroners' districts to researcher-defined suicides in England in 2005. J Public Heal (UK) 37:157–165
Parikh SV, Wasylenki D, Goering P, Wong J (1996) Mood disorders: rural/urban differences in prevalence, health care utilization, and disability in Ontario. J Affect Disord 38:57–65
Park B, Lester D (2012) Rural and urban suicide in South Korea. Psychol Rep 111:495–497
Park S et al (2014) Increased use of lethal methods and annual increase of suicide rates in Korean adolescents: comparison with adolescents in the United States. J Child Psychol Psychiatry Allied Discip 55:258–263
Patel V et al (2012) Suicide mortality in India: a nationally representative survey. Lancet 379:2343–2351
Patten SB et al (2006) Descriptive epidemiology of major depression in Canada. Can J Psychiatr 51:84–90
Paykel ES, Abbott R, Jenkins R, Brugha TS, Meltzer H (2000) Urban-rural mental health differences in great Britain: findings from the national morbidity survey. Psychol Med 30:269–280
Pearce J, Barnett R, Jones I (2007) Have urban/rural inequalities in suicide in New Zealand grown during the period 1980–2001? Soc Sci Med 65:1807–1819
Pedersen CB (2006) No evidence of time trends in the urban-rural differences in schizophrenia risk among five million people born in Denmark from 1910 to 1986. Psychol Med 36:211–219
Pedersen CB, Mortensen PB (2001a) Family history, place and season of birth as risk factors for schizophrenia in Denmark: a replication and reanalysis. Br J Psychiatry 179:46–52
Pedersen CB, Mortensen PB (2001b) Evidence of a dose-response relationship between urbanicity during upbringing and schizophrenia risk. Arch Gen Psychiatry 58:1039–1046
Pedersen CB, Mortensen PB (2006a) Are the cause(s) responsible for urban-rural differences in schizophrenia risk rooted in families or in individuals? Am J Epidemiol 163:971–978
Pedersen CB, Mortensen PB (2006b) Urbanization and traffic related exposures as risk factors for schizophrenia. BMC Psychiatry 6:2
Pedersen CB, Mortensen PB (2006c) Urbanicity during upbringing and bipolar affective disorders in Denmark. Bipolar Disord 8:242–247
Phillips MR et al (2009) Prevalence, treatment, and associated disability of mental disorders in four provinces in China during 2001–05: an epidemiological survey. Lancet (London, England) 373:2041–2053
Pompili M et al (2009) Substance abuse, temperament and suicide risk: evidence from a case-control study. J Addict Dis 28:13–20
Probst JC et al (2006) Rural-urban differences in depression prevalence: implications for family medicine. Fam Med 38:653–660
Propper C et al (2005) Local neighbourhood and mental health: evidence from the UK. Soc Sci Med 61:2065–2083
Qi X, Tong S, Hu W (2010) Spatial distribution of suicide in Queensland, Australia. BMC Psychiatry 10:106
Qi X, Hu W, Page A, Tong S (2014) Dynamic pattern of suicide in Australia, 1986–2005: a descriptive-analytic study. BMJ Open 4:e005311

Qin P, Hawton K, Mortensen PB, Webb R (2014) Combined effects of physical illness and comorbid psychiatric disorder on risk of suicide in a national population study. Br J Psychiatry. doi:10.1192/bjp.bp.113.128785

Rajalin M, Hirvikoski T, Jokinen J (2013) Family history of suicide and exposure to interpersonal violence in childhood predict suicide in male suicide attempters. J Affect Disord 148:92–97

Razvodovsky Y, Stickley A (2009) Suicide in urban and rural regions of Belarus, 1990–2005. Public Health 123:27–31

Read J, van Os J, Morrison AP, Ross CA (2005) Childhood trauma, psychosis and schizophrenia: a literature review with theoretical and clinical implications. Acta Psychiatr Scand 112:330–350

Rehkopf DH, Buka SL (2005) The association between suicide and the socio-economic characteristics of geographical areas: a systematic review. Psychol Med 36:145

Reijneveld SA, Schene AH (1998) Higher prevalence of mental disorders in socioeconomically deprived urban areas in The Netherlands: community or personal disadvantage? J Epidemiol Community Health 52:2–7

Rodríguez Andrés A, Hempstead K (2011) Gun control and suicide: the impact of state firearm regulations in the United States, 1995–2004. Health Policy 101:95–103

Romans S, Cohen M, Forte T (2011) Rates of depression and anxiety in urban and rural Canada. Soc Psychiatry Psychiatr Epidemiol 46:567–575

Ross CE, Reynolds JR, Geis KJ (2000) The contingent meaning of neighborhood stability for residents' psychological well-being. Am Sociol Rev 65:581–597

Saha S, Chant D, Welham J, McGrath J (2005) A systematic review of the prevalence of schizophrenia. PLoS Med 2:e141

Saha S, Chant DC, Welham JL, McGrath JJ (2006) The incidence and prevalence of schizophrenia varies with latitude. Acta Psychiatr Scand 114:36–39

Sariaslan A et al (2015) Does population density and neighborhood deprivation predict schizophrenia? A nationwide Swedish family-based study of 2.4 million individuals. Schizophr Bull 41:494–502

Schelin EM, Munk-Jørgensen P, Olesen AV, Gerlach J (2000) Regional differences in schizophrenia incidence in Denmark. Acta Psychiatr Scand 101:293–299

Schofield P, Ashworth M, Jones R (2011) Ethnic isolation and psychosis: re-examining the ethnic density effect. Psychol Med 41:1263–1269

Searles VB, Valley MA, Hedegaard H, Betz ME (2014) Suicides in urban and rural counties in the United States, 2006–2008. Crisis 35:18–26

Selten J-P, Cantor-Graae E (2005) Social defeat: risk factor for schizophrenia? Br J Psychiatry 187:101–102

Selten J-P, Cantor-Graae E (2007) Hypothesis: social defeat is a risk factor for schizophrenia? Br J Psychiatry Suppl 51:s9–12

Selten J-P, van der Ven E, Rutten BPF, Cantor-Graae E (2013) The social defeat hypothesis of schizophrenia: an update. Schizophr Bull 39:1180–1186

Shaw RJ et al (2012) Impact of ethnic density on adult mental disorders: narrative review. Br J Psychiatry 201:11–19

Shiner M, Scourfield J, Fincham B, Langer S (2009) When things fall apart: gender and suicide across the life-course. Soc Sci Med 69:738–746

Silver E, Mulvey EP, Swanson JW (2002) Neighborhood structural characteristics and mental disorder: Faris and Dunham revisited. Soc Sci Med 55:1457–1470

Singh GK, Siahpush M (2002) Increasing Rural–Urban Gradients in US Suicide Mortality, 1970–1997. Am J Public Health 92:1161–1167

Skapinakis P, Lewis G, Araya R, Jones K, Williams G (2005) Mental health inequalities in Wales, UK: multi-level investigation of the effect of area deprivation. Br J Psychiatry 186:417–422

Snider CLA, De Leo D (2014) The influence of deprivation on suicide mortality in urban and rural Queensland: an ecological analysis. 1919–1928. doi:10.1007/s00127–014–0905-1

Song S, Wang W, Hu P (2009) Famine, death, and madness: schizophrenia in early adulthood after prenatal exposure to the Chinese Great Leap Forward Famine. Soc Sci Med 68:1315–1321

Sørensen HJ et al (2014) Population impact of familial and environmental risk factors for schizophrenia: a nationwide study. Schizophr Res 153:214–219
Stafford M, Marmot M (2003) Neighbourhood deprivation and health: does it affect us all equally? Int J Epidemiol 32:357–366
Stafford M, De Silva M, Stansfeld S, Marmot M (2008) Neighbourhood social capital and common mental disorder: testing the link in a general population sample. Health Place 14:394–405
Stark C, Hopkins P, Gibbs D, Belbin A, Hay A (2007) Population density and suicide in Scotland. Rural Remote Health 7
Steel Z et al (2014) The global prevalence of common mental disorders: a systematic review and meta-analysis 1980–2013. Int J Epidemiol 43:476–493
Stockdale SE et al (2007) The importance of social context: neighborhood stressors, stress-buffering mechanisms, and alcohol, drug, and mental health disorders. Soc Sci Med 65:1867–1881
Sundquist K, Frank G, Sundquist J (2004) Urbanisation and incidence of psychosis and depression: follow-up study of 4.4 million women and men in Sweden. Br J Psychiatry 184:293–298
Suvisaari JM, Haukka JK, Tanskanen AJ, Lönnqvist JK (2000) Decreasing seasonal variation of births in schizophrenia. Psychol Med 30:315–324
Szöke A et al (2014) Rural-urban variation in incidence of psychosis in France: a prospective epidemiologic study in two contrasted catchment areas. BMC Psychiatry 14:78
Tak Y et al (2012) Spatial analysis of suicide mortality in Australia: investigation of metropolitan-rural-remote differentials of suicide risk across states/territories. Soc Sci Med 75:1460–1468
Takei N, Sham PC, O'Callaghan E, Glover G, Murray RM (1995) Schizophrenia: increased risk associated with winter and city birth – a case-control study in 12 regions within England and Wales. J Epidemiol Community Health 49:106–107
Tarde G (1903) The laws of imitation. New York: Henry Holt
Taylor R, Page A, Morrell S, Harrison J, Carter G (2005) Social and psychiatric influences on urban-rural differentials in Australian suicide. Suicide Life-Threatening Behav 35:277–290
Thornicroft G, Bisoffi G, De Salvia D, Tansella M (1993) Urban-rural differences in the associations between social deprivation and psychiatric service utilization in schizophrenia and all diagnoses: a case-register study in Northern Italy. Psychol Med 23:487–496
Titelman D et al (2013) Suicide mortality trends in the Nordic countries 1980–2009. Nord J Psychiatry 67:414–423
Tomita A, Burns JK (2013) A multilevel analysis of association between neighborhood social capital and depression: evidence from the first South African National Income Dynamics Study. J Affect Disord 144:101–105
Tortelli A et al (2015) Schizophrenia and other psychotic disorders in Caribbean-born migrants and their descendants in England: systematic review and meta-analysis of incidence rates, 1950–2013. Soc Psychiatry Psychiatr Epidemiol 50:1039–1055
Tost H, Champagne FA, Meyer-Lindenberg A (2015) Environmental influence in the brain, human welfare and mental health. Nat Neurosci 18:1421–1431
Tsai JF (2010) Socioeconomic factors outweigh climate in the regional difference of suicide death rate in Taiwan. Psychiatry Res 179:212–216
Tweed DL et al (1990) Racial congruity as a contextual correlate of mental disorder. Am J Orthop 60:392–403
Vallée J, Cadot E, Roustit C, Parizot I, Chauvin P (2011) The role of daily mobility in mental health inequalities: the interactive influence of activity space and neighbourhood of residence on depression. Soc Sci Med 73:1133–1144
Van Os J (2000) Neighbourhood variation in incidence of schizophrenia: evidence for person-environment interaction. Br J Psychiatry 176:243–248
van Os J, McGuffin P (2003) Can the social environment cause schizophrenia? Br J Psychiatry 182:291–292
van Os J, Hanssen M, Bijl RV, Vollebergh W (2001) Prevalence of psychotic disorder and community level of psychotic symptoms. Arch Gen Psychiatry 58:663

van Os J, Hanssen M, Bak M, Bijl RV, Vollebergh W (2003) Do urbanicity and familial liability coparticipate in causing psychosis. Am J Psychiatry 160:477–482
van Os J, Krabbendam L, Myin-Germeys I, Delespaul P (2005) The schizophrenia envirome. Curr Opin Psychiatry 18:141–145
van Os J et al (2014) Identifying gene-environment interactions in schizophrenia: contemporary challenges for integrated, large-scale investigations. Schizophr Bull 40:729–736
Vassos E, Pedersen CB, Murray RM, Collier DA, Lewis CM (2012) Meta-analysis of the association of urbanicity with schizophrenia. Schizophr Bull 38:1118–1123
Vassos E, Agerbo E, Mors O, Pedersen CB (2015) Urban-rural differences in incidence rates of psychiatric disorders in Denmark. Br J Psychiatry. doi:10.1192/bjp.bp.114.161091 bjp.bp.114.161091
Veenstra G (2005) Location, location, location: contextual and compositional health effects of social capital in British Columbia, Canada. Soc Sci Med 60:2059–2071
Veling W et al (2006) Incidence of schizophrenia among ethnic minorities in the Netherlands: a four-year first-contact study. Schizophr Res 86:189–193
Veling W et al (2008) Ethnic density of neighborhoods and incidence of psychotic disorders among immigrants. Am J Psychiatry 165:66–73
Veling W, Hoek HW, Wiersma D, Mackenbach JP (2010) Ethnic identity and the risk of schizophrenia in ethnic minorities: a case-control study. Schizophr Bull 36:1149–1156
Veling W, Susser E, Selten J-P, Hoek HW (2015) Social disorganization of neighborhoods and incidence of psychotic disorders: a 7-year first-contact incidence study. Psychol Med 45:1789–1798
Veling W, Pot-Kolder R, Counotte J, van Os J, van der Gaag M (2016) Environmental social stress, paranoia and psychosis liability: a virtual reality study. Schizophr Bull. doi:10.1093/schbul/sbw031 sbw031
Vicente B et al (2006) Regional differences in psychiatric disorders in Chile. Soc Psychiatry Psychiatr Epidemiol 41:935–942
von Borczyskowski A, Lindblad F, Vinnerljung B, Hjern A (2010) Gender differences in risk factors for suicide: findings from a Swedish national cohort study. Can J Psychiatr 55:108–111
Wainwright NWJ, Surtees PG (2004) Area and individual circumstances and mood disorder prevalence. Br J Psychiatry 185:227–232
Wang JL (2004) Rural-urban differences in the prevalence of major depression and associated impairment. Soc Psychiatry Psychiatr Epidemiol 39:19–25
Wei G et al (2008) The first epidemiological study of mental disorders of Tibetans. World Cult Psychiatr Res Rev: 1–3
Weich S (2003a) Contextual risk factors for the common mental disorders in Britain: a multilevel investigation of the effects of place. J Epidemiol Community Health 57:616–621
Weich S (2003b) Geographic variation in the prevalence of common mental disorders in Britain: a multilevel investigation. Am J Epidemiol 157:730–737
Weich S (2005) Absence of spatial variation in rates of the common mental disorders. J Epidemiol Community Health 59:254–257
Weich S et al (2002) Mental health and the built environment: cross-sectional survey of individual and contextual risk factors for depression. Br J Psychiatry 180:428–433
Weich S, Twigg L, Lewis G, Jones K (2005) Geographical variation in rates of common mental disorders in Britain: prospective cohort study. Br J Psychiatry 187:29–34
Weich S, Twigg L, Lewis G (2006) Rural/non-rural differences in rates of common mental disorders in Britain: prospective multilevel cohort study. Br J Psychiatry 188:51–57
Weiser M et al (2007) Social and cognitive functioning, urbanicity and risk for schizophrenia. Br J Psychiatry 191:320–324
Werner S, Malaspina D, Rabinowitz J (2006) Socioeconomic status at birth is associated with risk of Schizophrenia: population-based multilevel study. Schizophr Bull 33:1373–1378
Whitley E, Gunnell D, Dorling D, Smith GD (1999) Ecological study of social fragmentation, poverty, and suicide. BMJ 319:1034–1037

Wilkinson D, Gunnell D (2000) Youth suicide trends in Australian metropolitan and non-metropolitan areas, 1988–1997. Aust N Z J Psychiatry 34:822–828

Winfree T Jr, Jiang S (2010) Youthful suicide and social support: exploring the social dynamics of suicide-related behavior and attitudes within a national sample of US adolescents. Youth Violence Juv Justice 8:19–37

World Health Organization (2014) Preventing suicide: a global imperative. Geneva: World Health Organisation

Xiang Y-T et al (2008) Prevalence and socio-demographic correlates of schizophrenia in Beijing, China. Schizophr Res 102:270–277

Yen I, Kaplan G (1999) Poverty area residence and changes in depression and perceived health status: evidence from the Alameda County Study. Int J Epidemiol 28:90–94

Yip PSF, Callanan C, Yuen HP (2000) Urban/rural and gender differentials in suicide rates: East and West. J Affect Disord 57:99–106

Yip PSF, Liu KY, Hu J, Song XM (2005) Suicide rates in China during a decade of rapid social changes. Soc Psychiatry Psychiatr Epidemiol 40:792–798

Zammit S et al (2010) Individuals, schools, and neighborhood: a multilevel longitudinal study of variation in incidence of psychotic disorders. Arch Gen Psychiatry 67:914–922

Zammit S et al (2014) Individual- and area-level influence on suicide risk: a multilevel longitudinal study of Swedish schoolchildren. Psychol Med 44:267–277

Zapata LB et al (2013) Orphaned and abused youth are vulnerable to pregnancy and suicide risk. Child Abuse Negl 37:310–319

Zhang J, Wang C (2012) Factors in the neighborhood as risks of suicide in rural China: a multilevel analysis. Community Ment Health J 48:627–633

Homicide and Suicide in Megacities

4

Carolina Ziebold Jorquera, Denisse Jaen-Varas, and
Jair de Jesus Mari

Abstract

There are 28 megacities in the globe and most are located in Asia. The rates of suicide and homicide of these cities are compared with national rates and country economic indicators such as gross domestic product (GDP), human development index (HDI), and inequity index (GINI). Concerning homicide, the highest rate (per 100,000 habitants) in megacities is in Rio de Janeiro, Brazil (29.9), and the lowest in Tokyo, Japan (0.2). Homicide rates above 10/100,000, regarded as of epidemic violence, are found in Rio de Janeiro, Brazil; São Paulo, Brazil (14.2); Karachi, Pakistan (12.3); and Lagos, Niger (12.2). Concerning suicide, the mean rate per 100,000 habitants in 22 of these megacities is 7.2, the highest in Osaka (20) and the lowest rate in Cairo (0.7). Epidemic levels of suicides are found only in megacities in Asia. There is a relationship between economic indicators with suicide (high GDP and HDI and low inequity) and homicide rates (low GDP and HDI and high inequity). The rates of suicide and homicide tend to show inverse correlation. Nine cities have at least one rate above 10/100,000 and nine cities have both rates below 5/100,000. The overall homicide rates found in megalopolises are lower than the global mean of rates of homicides found for the countries as a whole, except in Rio de Janeiro, Brazil, and Karachi, Pakistan. The megacities present lower suicide rates compared with the overall rates of the countries where they are placed. As a conclusion, there is no common pattern for homicide and suicide rates in megacities, but when homicide is high, suicide tends to be low. Higher country development and better social indicators are related to increased risk of suicide.

Keywords

Homicide • Suicide • Mental health • Megacities

C. Ziebold Jorquera (✉) • D. Jaen-Varas • J. de Jesus Mari
Department of Psychiatry, Universidade Federal de São Paulo, São Paulo, Brazil
e-mail: carolaziebold@gmail.com; deni_clau@hotmail.com; jamari17@gmail.com

N. Okkels et al. (eds.), *Mental Health and Illness in the City*, Mental Health and Illness Worldwide, https://doi.org/10.1007/978-981-10-2327-9_10

Contents

Introduction

The world is becoming increasingly urbanized. The United Nations (2015) estimates that 54% of the population now live in cities, nearly 3.9 billion inhabitants, and by the year 2050, 66% of the world's population is projected to be urban. Currently, close to half of the world's urban dwellers reside in relatively small settlements of less than 500,000 inhabitants, while only around one in eight live in one of the 28 megacities with more than 10 million inhabitants.

Globally, urban cities represent 2% of the earth's surface but use 75% of its resources. In consequence, megacities have the following three main dimensions as concerns: the social dimension (cultural diversity and variety, education, art, living conditions, transport, security, health care); the economic dimension (work, mass unemployment, improvement of infrastructure, new technologies, decentralization, repartition of wealth, capital equipment); and the ecological dimension (energy sources, sustainable development, air and water pollution, noise pollution, traffic jams, water supply, urban sprawl, urban environment protection, public transportation, and waste management) (European Association of National Metrology Institutes 2013; World Health Organization and United Nations Human Settlements Programme 2010).

A recent update by the World Health Organization and the United Nations Human Settlements Programme (2010) emphasized how cities may concentrate opportunities, jobs, and services for their citizens, but they also include risks and hazards for physical and mental health. In many cities around the world, health determinants are combined to create a triple threat of urban diseases and health conditions: infectious diseases, noncommunicable diseases, as mental health disorders and alcohol abuse, injuries, and violence.

As urban environments concentrate potentials risk factors for mental health and violence, it is relevant to explore which are the violence levels in megacities and to find out whether living in these human agglomerations would imply to have better or worse homicide and/or suicide indicators. Thus, the aim of this chapter is to examine the rates of suicide and homicide per 100,000 habitants in these cities, and to analyze whether they have a relation with national rates and country economic indicators as gross domestic product (GDP), human development index (HDI), and inequity index (GINI). Furthermore, possible risk factors associated with poor mental health in these megacities will be analyzed.

Violence

Violence can be understood as the intentional use of physical force or power, threatened or actual, against oneself, another person, or against a group or community who previously had lived "in peace" (Winton 2004), which either results in or has a high likelihood of resulting in physical, mental (psychological), or sexual injury (Winton 2004), death, maldevelopment, or deprivation (World Health Organization 2002, 2014). The World Health Organization (WHO) has defined epidemic violence as a homicide rate greater than 10 per 100,000 inhabitants in a city or a country (World Health Organization and United Nations Human Settlements Programme 2010).

The classification proposed by the *World Report on Violence and Health* in 2002 divides violence into three broad categories, arranged according to the features of the violent act: self-directed violence, interpersonal violence, and collective violence (World Health Organization 2002). Collective violence, in its multiple forms, receives a high degree of public attention. Violent conflicts between nations and groups, state and group terrorism, rape as a weapon of war, the movements of large numbers of people displaced from their homes, gang warfare, and mass hooliganism – all of these occur on a daily basis in many parts of the world. The health effects of these different types of events in terms of deaths, physical illnesses, disabilities, and mental anguish are massive (World Health Organization 2002).

Violence can have strong, negative impacts on economic development by drastically reducing growth and producing long-lasting detrimental social impacts (The World Bank 2011). During the past decade at least, urban violence is (or has become) the most serious form of lethal violence in the world. In 2000, an estimated 1.6 million people worldwide died as a result of self-inflicted (nearly half of these were suicides), interpersonal, or collective violence, for an overall age-adjusted rate of 28.8 per 100,000 population (World Health Organization 2002).

In 2012, the United Nations Office on Drugs and Crime (2013) estimated that deaths resulting from intentional homicide amounted to a total of 437,000 at the global level. The largest share of those deaths was registered in the Americas (36%) and large shares were also recorded in Africa and Asia (31% and 28%, respectively). Europe (5%) and Oceania (0.3%) accounted for the lowest shares of homicide by region.

Violence and Mental Health

The impact of violence on mental health is influenced by a range of factors. These include the psychological health of those affected prior to the event; the nature of the conflict; the form of trauma (whether it results from living through and witnessing acts of violence or whether it is directly inflicted, as with torture and other types of repressive violence); the response to the trauma by individuals and communities, and the cultural context in which the violence occurs (World Health Organization 2002).

The homicide rate is one of the ways to evaluate the level of violence in the cities. However, for every person who dies from violence, many more are injured and suffer a range of physical, mental, and other social and economic consequences. In addition, the fear of such violence further contributes to the social, economic, and political fragmentation of cities (World Health Organization and United Nations Human Settlements Programme 2010). Some of the psychological stresses related to conflicts are associated with or result from displacement, whether forced or voluntary, loss and grief, social isolation, loss of status, loss of community, and in some settings, acculturation to new environments (World Health Organization 2002). General manifestations of such stress, related to violence, can include depression and anxiety, psychosomatic ailments, suicidal behavior, intrafamilial conflict, alcohol abuse, antisocial behavior (World Health Organization 2002), insomnia, and substance abuse (World Health Organization and United Nations Human Settlements Programme 2010).

Although studies suggest a link between mental disorders and violence, the contribution of people with mental illness to overall rates of violence is small, and the magnitude of the relationship is exaggerated in the general population. The vast majority of people who are violent do not suffer from mental disorders (Stuart 2003).

The major determinants of violence are sociodemographic and socioeconomic factors such as being young, male, and of lower socioeconomic status, and lack of safe, stable, and nurturing relationships between children and their parents. Substance abuse appears to be a major determinant of violence, accounting for as much as a third of violent acts, and seven out of every 10 crimes of violence among mentally disordered offenders (Stuart 2003).

Homicide and Suicide Rates in Megacities

Most of the megacities are in Asia (16), six in America, three in Africa, and three in Europe. Tokyo is the city with the highest population, with 37 million inhabitants. Table 1 shows the 28 megacities, with homicides and suicide rates by cities wherever available and socioeconomic indicators by country: gross domestic product (GDP), human development index (HDI), and GINI index of economic inequality (zero expresses perfect equality, and 100 expresses maximal inequality).

Generally, cities with high homicides rates have low GDP and HDI and high GINI compared with cities with lowers levels of homicide. Conversely, for suicide rates, the GDP and HDI are higher in countries with high rates compared with low-rates countries. Inversely, the GINI is lower in the cities with epidemic level of suicide compared with those with low rate.

Nine of the cities show at least one of the rates of either suicide or homicide above 10 per 100,000 thousand, usually inversely correlated (when homicides are high, suicides tend to be low). Other nine megacities present lower rates than 5 per 100,000 thousand for both indicators: Jakarta, Manila, Cairo, Istanbul, Shanghai, Beijing, Guangzhou, Tianjin, and Shenzhen.

Table 1 List of megacities with sociodemographic indicators of their countries

City	Country	Continent	Population (United Nations 2015)	Homicide rate per 100,000 people 2012 (United Nations Office on Drugs and Crime 2013)		Suicide rate per 100,000 people		Gross domestic product (GDP) per capita current US$ 2014 (The World Bank 2016a)	Human development index (HDI) 2014 (United Nations Development Programme 2015)	Economic inequality, GINI index (The World Bank 2016b; Central Intelligency Agency 2016)
				City	Country	City	Country 2012 (World Health Organization 2012)			
Tokyo	Japan	Asia	37.833.000	0.2	0.3	18.7 (2014)	19.5 (2014)	36.194	0.89	37.9 (2011)
Delhi	India	Asia	24.953.000	3.2	3.5	10.3 (2014)	21.1	1.581	0.61	33.6 (2011)
Shanghai	China	Asia	22.991.000	0.9	1	4.4 (2013)	9.1 (2013)	7.590	0.73	46.9 (2014)
Mumbai (Bombay)	India	Asia	20.741.000	1.2	3.5	13.9	21.1	1.581	0.61	33.6 (2011)
Osaka	Japan	Asia	20.123.000	0.5	0.3	20 (2014)	19.5 (2014)	36.194	0.89	37.9 (2011)
Beijing	China	Asia	19.520.000	0.9	1	2.9 (2013)	9.1 (2013)	7.590	0.73	46.9 (2014)
Dhaka	Bangladesh	Asia	16.982.000	5.3 (2006)	2.7	–	10.6	1.087	0.57	32.0 (2010)
Karachi	Pakistan	Asia	16.126.000	12.3 (2013)	7.7	–	9.3	1.317	0.54	29.6 (2010)
Calcutta	India	Asia	14.766.000	1.3	3.5	15.5 (2014)	21.1	1.581	0.61	33.6 (2011)

(*continued*)

Table 1 (continued)

City	Country	Continent	Population (United Nations 2015)	Homicide rate per 100,000 people 2012 (United Nations Office on Drugs and Crime 2013)		Suicide rate per 100,000 people		Gross domestic product (GDP) per capita current US$ 2014 (The World Bank 2016a)	Human development index (HDI) 2014 (United Nations Development Programme 2015)	Economic inequality, GINI index (The World Bank 2016b; Central Intelligency Agency 2016)
				City	Country	City	Country 2012 (World Health Organization 2012)			
Istanbul	Turkey	Asia	13.954.000	4.7 (2008)	4	3.1 (2014)	7.9	10.515	0.76	40.2 (2012)
Chongqing	China	Asia	12.916.000	–	1	7.3 (2013)	9.1 (2013)	7.590	0.73	46.9 (2014)
Manila	Philippines	Asia	12.764.000	1.1	8.8	–	2.9	2.872	0.67	43.0 (2012)
Guangzhou	China	Asia	11.843.000	0.9	0.8	4.8 (2013)	9.1 (2013)	7.590	0.73	46.9 (2014)
Tianjin	China	Asia	10.860.000	–	1	2.5 (2013)	9.1 (2013)	7.590	0.73	46.9 (2014)
Shenzhen	China	Asia	10.680.000	0.9	1	4.8 (2013)	9.1 (2013)	7.590	0.73	46.9 (2014)
Jakarta	Indonesia	Asia	10.176.000	0.8 (2011)	0.6	1.2 (2011)	4.3	3.492	0.68	35.6 (2010)
Mexico City	Mexico	America	20.843.000	8.8	21.5	2.2 (2011)	4.2	10.325	0.76	48.1 (2012)
São Paulo	Brazil	America	20.831.000	14.2	25.2	4.7 (2012)	5.8	11.384	0.76	52.7 (2012)

New York-Newark	USA	America	18.591.000	5.1	4.7	6.0 (2009)	12.1	54.630	0.92	40.5 (2010)
Buenos Aires	Argentina	America	15.024.000	8.4	7.2	8.6 (2012)	8.3	12.509	0.84	42.5 (2012)
Rio de Janeiro	Brazil	America	12.825.000	29.9	25.2	2.9 (2012)	5.8	11.384	0.76	51.9 (2012)
Los Angeles	USA	America	12.308.000	6.5	4.7	7.1 (2009)	12.1	54.629	0.92	40.5 (2010)
Cairo	Egypt	Africa	18.419.000	2.4 (2011)	3.4 (2011)	0.7 (2009)	1.7	3.199	0.69	30.8 (2008)
Lagos	Nigeria	Africa	12.614.000	12.2	20	–	6.5	3.203	0.51	43.0 (2009)
Kinshasa	DR Congo	Africa	11.116.000	1.6	28.3	–	10.1	442	0.43	42.1 (2012)
Moscow	Russian Federation	Europe	12.063.000	3.8	9.2	–	19.5	12.735	0.80	41.6 (2012)
Paris	France	Europe	10.764.000	1.8	1	8.1 (2012)	12.3	42.732	0.89	33.1 (2012)
London	United Kingdom	Europe	10.189.000	1.3 (2011)	1 (2011)	7.9 (2013)	11.9 (2013)	46.332	0.91	32.6 (2012)

Suicide Rate by City sources: Japan (Portal Site List of Official Statistics of Japan 2016); India (National Crime Records Bureau of India 2014); China (Zhou et al. 2016); Istanbul (Turkish Statistical Institute 2014); Jakarta (Rozaki 2012); Mexico City (Instituto Nacional de Estadística y Geografía México 2013); Brazil (Ministerio da Saúde Brasil 2012); New York (New York City Department of Health and Mental Hygiene 2007); Buenos Aires (for homicide also) (Fleitas 2014); Los Angeles (Department of Public Health Los Angeles 2009); Cairo (Abdel Moneim et al. 2012); Paris (Observatoire National du Suicide 2016); and London (United Kingdom Government Office for National Statistics 2015). Suicide Rate by Country: Source WHO (World Health Organization 2012), except China (Zhou et al. 2016; National Bureau of Statistics of China 2014), Japan (Portal Site List of Official Statistics of Japan 2016), Argentina (for homicide also) (Fleitas 2014), and United Kingdom (United Kingdom Government Office for National Statistics 2015)

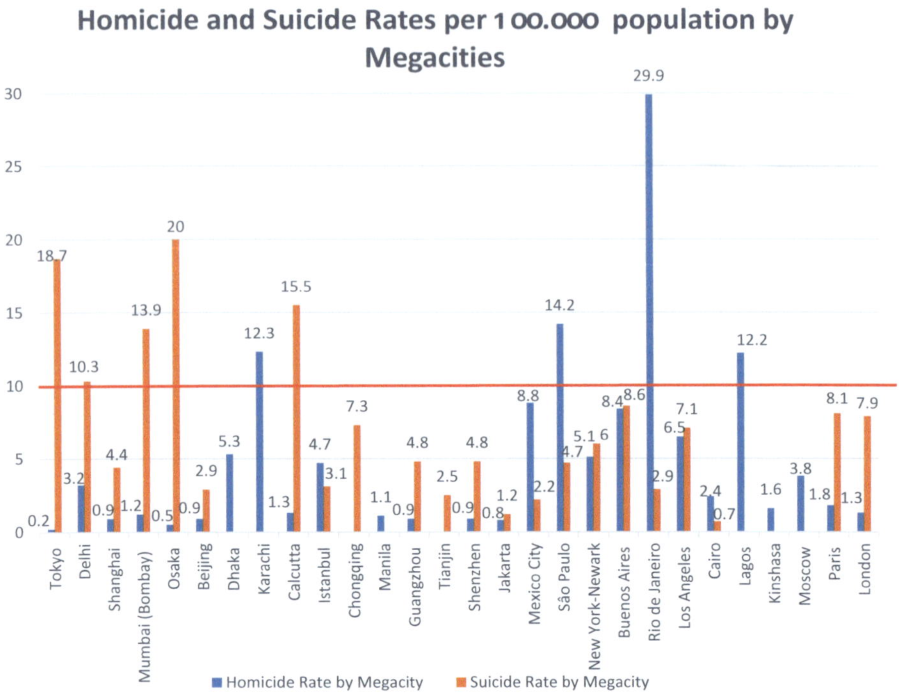

Fig. 1 Homicide and suicide rates per 100,000 populations by megacities (Using available data, sources in Table 1. The horizontal line represents the cutoff point for epidemic violence (rates above 10/100,000 population))

Figure 1 shows the homicide and suicide rates per 100,000 inhabitants by megacities and Fig. 2 shows the averages of homicide and suicide rates in these cities by geographical regions of the globe, compared with the overall data of the countries where the megacities are placed.

Homicide

While the global homicide rate is 6.9 per 100,000 population (United Nations Office on Drugs and Crime 2013), the homicide rate in these megacities is 5/100,000, fluctuating between 0.2 in Tokyo and 29.9 in Rio de Janeiro. The lowest homicide rate is found in Tokyo, and can be explained by economic stability, low social inequality, and high levels of development, together with low levels of weapons possession (Statistic Japan: Prefecture Comparisons 2016). Conversely, homicide rates considered as epidemic violence are found in Rio de Janeiro (29.9), São Paulo (14.2), Karachi (12.3), and Lagos (12.2). The overall homicide rates found in megalopolises are lower than the global mean of rates of homicides found for the countries as a whole, leading to the conclusion that homicides are not more common for those living in megalopolises except for places like Karachi and Rio de Janeiro.

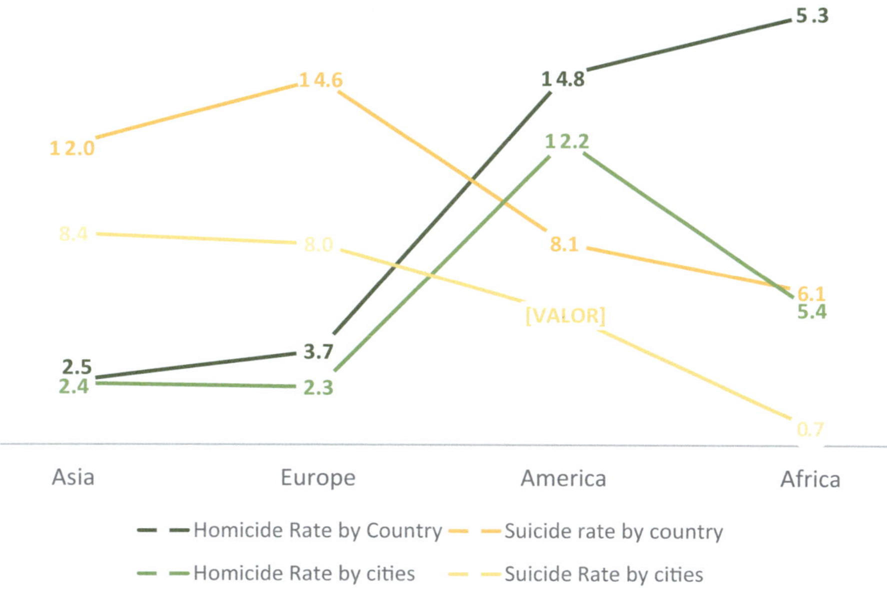

Fig. 2 Average of homicide and suicide rates per 100,000 habitants in megacities and their countries by geographical regions of the globe

The fact that homicide rates are overall lower in megacities than the whole country might be related to several factors such as the concentration of job opportunities, access to health, education, and safety (justice and police control). Nevertheless, these advantages can be extremely uneven within subpopulations of the megacity since homicide rates in deprived areas of the city can be worse than those rates found to the country as a whole. On the other hand, people living in rural areas are more isolated, with lower access to education and health opportunities, and more likely to be exposed to land disputes, violence against women, and fighting between groups belonging to different ethno-religious affiliations.

Understanding Epidemic Violence in Megacities

Near 90% of the populations of São Paulo and Rio de Janeiro present high levels of exposure to violence (Ribeiro et al. 2013). One-year prevalence estimates of any traumatic events are higher in Rio de Janeiro (35.1%) than São Paulo (21.7%). One-year prevalence estimates of exposure to assaultive violence are similar in the two cities, and exposure to other injury or shocking events is nearly twofold higher in Rio de Janeiro than in São Paulo (23.7% vs. 12.8%). In Rio de Janeiro accounts of witnessing atrocities, slaughters or massacres, and witnessing gang fights by criminal and drug trafficking organizations are very common. In addition to homicides by civilians, in 2010 the rate of persons killed by military or civil police in Brazil was 1.6/100,000 (Murray et al. 2013). Rio de Janeiro and São Paulo present high levels of socioeconomic inequality, as seen by the highest GINI found for the country. Social inequality, sexual abuse, domestic and community violence, early pregnancy, and low access to abortion lead to many neglected children who see in criminality and drug trading a good, and sometimes the sole, opportunity for social ascension. In addition to community violence, particularly in the outskirts and slums of these two cities, 20% of urban Brazilian children and adolescents are exposed to domestic violence in deprived areas, consisting of severe physical punishment by one of their parents (Bordin et al. 2009).

Homicide can be a common endpoint for many different behavioral pathways, including arguments between acquaintances, escalating domestic violence between spouses, robberies perpetrated by strangers, and, in some parts of the world, organized political violence. Organized violence is a violent behavior perpetrated by a social or political group that strives to achieve a specific political, economic, or social objective (Mian et al. 2002). This might explain the epidemic violence seen in Lagos and Karachi. In Lagos, Nigeria, factors related to high levels of homicide include illegal firearms possession, militancy activities to protest against high levels of poverty, political, and fighting between groups belonging to different ethno-religious affiliations (Pérouse 2011). There is more male participation in homicidal deaths than women, probably due to the fact that traditionally only males have been involved in militancy and communal clashes. In recent years, homicide rates in Nigeria have decreased, explained by cyclical political conflicts, and because the population does not trust security forces, reporting only a minority of crimes to the police. It is noteworthy that 70% percent of surveyed residents in Lagos report being fearful of becoming crime victims, with 90% fearing the prospects of being killed in a criminal attack, reflecting the repercussions of crime on citizens' quality of life, and their everyday concerns and mental health (World Health Organization and United Nations Human Settlements Programme 2010). A similar situation is seen in Karachi, where social upheaval, political instabilities, lawlessness, terrorism, economic disparity, and problems with security and safety are factors related to epidemic violence (Muhammad Gadit and Mugford 2007). In Karachi, victims of violence are mainly males under the age of 30, and almost half of the violent injuries are concentred in few neighborhoods, dominated by a single ethnic group, the Mohajirs, and their descendants, who had migrated to Pakistan at the time of partition from India in 1947 (Mian et al. 2002).

Suicide

The mean of suicide rate per 100,000 habitants in 22 of these megacities is 7.2, the highest shown to be in Osaka and the lowest rate in Cairo (0.7) (Table 1). Epidemic levels of suicides are found only in megacities in Asia: Osaka: 20; Tokyo: 18.7; Calcutta: 15.5; Mumbai: 13.9, and Delhi: 10.3. Thus, the highest rates of suicide are in Asia, and the lowest in Africa. As seen in Fig. 2, the megacities present lower suicide rates compared with the overall rates of the countries where they are placed.

Some factors related with committing suicide in Japan are (a) financial crisis (especially among male unemployed); (b) a social acceptance of suicide as a free act related to cultural themes such as honor, perfectionism, and guilt; (c) the rise of industrialization and isolation from family; (d) an extreme importance given to the workplace; and (e) presence of mental health disorder (mainly untreated depression) and/or a melancholic premorbid personality (Hayashi et al. 2010). According to the National Police Agency, the most common factors in triggering suicide in Japan are physical illness (47%), financial difficulties (23.8%), domestic problems (12.2%), and employment conflicts (7.2%) (Otsuka and Horita 2013).

Anomie is a term introduced by a French sociologist, Émilie Durkheim, to explain a condition of individuals or society characterized by a breakdown or absence of social norms and values, or governing rules that promote isolation, which could be related to one type of suicide (anomic) (Lenzi et al. 2012). Durkheim's suicide theory from 1897 establishes that social disintegration and social deregulation, and weakening of religious influence in modern societies, leading to individualism and anomie, increase risk of suicide. Based on Durkheim theory and modern sociological theories, a survey was conducted by collecting suicide rates from 87 nations, to study several economic and cultural factors related to suicide (Lenzi et al. 2012), and it has been found a positive relationship between increased suicide rates and levels of individualism, and secular-rational values. Also has been found correlation of suicide rates with years of schooling, divorce rates, and populations above 60 years of age. It is reasonable to suppose that high levels of development, individualism, and secular-rational values are related to an increased risk of suicide in more developed nations. Moreover, it has not been found highest suicide rates in urban population, giving rise to the hypothesis that living in a megalopolis does not make one more prone to suicide.

Different patterns related to suicide are presented in India and China. India presents a high suicide rate, with 21.1 per 100,000 inhabitants (World Health Organization 2012). Compared with other countries, India presents higher suicide rates among young women living in rural areas. Suicide in India is related to social and economic causes, such as alcohol consumption, unemployment, poverty, social inequities, and domestic violence. Alcohol consumption and agricultural work occupations are associated with risk of suicide in men, whereas being widowed, divorced, or separated is associated with a slightly decreased risk of suicide in women (Patel et al. 2012). Some predictors of attempted suicide in Indian women are common mental disorders, exposure to violence, physical illness, and experience of recent starvation (Maselko and Patel 2008). A similar pattern has been found in China, the only country where females have suicide rates similar to that of men (Zhang et al. 2011). Suicide is the leading cause of death among rural young women in China, who are frequently affected by violence. These factors could explain the lower rates found in the six megacities of China, when compared with the country's rural zones.

In another three megacities, there are a statistical combination of relatively low suicide and low homicides rates: Manila, Jakarta, and Cairo. One of the reasons might be under-reporting of statistics due to the nonacceptance of suicide and homicide by the Catholic Church (in the Philippines) or the Muslim religion (Indonesia and Egypt), associated with stigma for the family (Abdel Moneim et al. 2012; Radaniel et al. 2011; Suryani et al. 2009). Religion plays a major role at these locations, combined with strong family ties, and a belief that there is life after death, which averts feelings of hopelessness, a sentiment that has been described as an important predisposing factor for suicide (Abdel Moneim et al. 2012). In the Philippines, a predominantly Roman Catholic country, it is possible that reluctance to report deaths as suicide might contribute to biased low official rates. Nevertheless, the strong Roman Catholic culture could also contribute to preventing suicides, due to the beliefs and social norms associated with the religion. In the Philippines, another related factor was the social cohesion during the turbulent Martial Law era and its aftermath, because the reduction in suicide rates during periods of war and civil disturbance are well recognized (Radaniel et al. 2011). These low rates of suicides in Manila, Jakarta, and Cairo are in accordance with the idea that nations with very high traditional values have low rates of suicide.

Financial crises inevitably affect the mental health of the population. There is direct evidence regarding implications stemming from previous global financial crisis, such as the suicides seen in the Great Depression of 1929, and their implications for psychopathology, particularly depression, suicide, and alcohol and drug abuse. Direct socioeconomic consequences such as unemployment, material losses, debts, and social fragmentation can affect physical and mental health. A convincing association between unemployment and depression and suicide has been reported recently for Spain and Greece. Thus, one of the reasons implicated in suicide may be related to macroeconomic instability, particularly for developed countries, where material losses can lead to individual frustration, depression, and suicide.

As shown by these comparisons it was not found a common pattern between homicide and suicide rates in megacities. The other question is to find out whether local prevalence rates of mental disorders might be related to suicide and/or homicide rates in different social and cultural contexts.

Mental Health Risk in Megacities

In regards to severe mental illness, there is evidence of an increased risk of developing schizophrenia for individuals living in urban populations, compared with those born and/or raised in rural regions (McGrath et al. 2004). This effect on schizophrenia might be explained by environmental urban factors such as trauma experiences, and use of cannabis in interaction with personal proneness (van Os et al. 2010). Psychotic symptoms can be associated with growing up in an urbanized area and/or belonging to a minority group for both first- and second-generation immigrants. The effect of being a minority ethnic group on psychotic syndrome depends on the ethnic density of the area the person is living. The greater the proportion of the individual's own ethnic group in the area, the lower the risk for psychotic disorder. The differential sensitivity to the psychosis-inducing effects of environmental factors may be mediated by genetic factors (van Os et al. 2010).

Data on common mental health disorders prevalence in general population were found in five megacities of Asia: Beijing, Shanghai, Guangzhou, and Shenzhen, China (Kessler et al. 2015; Shen et al. 2006; Gu et al. 2013); Dhaka, Bangladesh (Islam et al. 2003); four mega cities of America: Mexico City, Mexico (Caraveo et al. 1999); Buenos Aires, Argentina (Kohn et al. 2005); São Paulo (Ribeiro et al. 2013; Andrade et al. 2012) and Rio de Janeiro (Ribeiro et al. 2013), Brazil; and one of Europe: Paris, France (Carta et al. 2002). Higher 1-year common mental health disorders prevalence is found in Brazilian cities compared with China: Anxiety disorders fluctuate between 18.8% in Rio de Janeiro and 20.8% in São Paulo and 2.7% in Beijing and Shanghai. Depression prevalence varies between 2% in Beijing and Shanghai to 9.4% in São Paulo; the levels of Posttraumatic Stress Disorder (PTSD) are much higher in São Paulo (5.0%) and Rio de Janeiro (3.3%), than those found in China for Beijing and Shanghai (0.2%). The prevalence of alcohol dependence is also higher in São Paulo and Mexico. One-year drug dependence in São Paulo (Andrade et al. 2012) is 0.5%, and 0% in Beijing and Shanghai (Shen et al. 2006). In Paris, 6-month prevalence rates are 5.9% of depression, 8.6% of anxiety, and 0.8% of substance-related disorder (Carta et al. 2002). Point-prevalence of depression has been examined in Guangzhou (Gu et al. 2013), with 0.8% and in Buenos Aires (Kohn et al. 2005) and Dhaka (Islam et al. 2003), both with 6%. In Dhaka, the point prevalence of anxiety disorder was 2.7% and substance-related disorder 2.4%. The Americas tend to present high levels of anxiety disorders, PTSD, and alcohol dependence.

Prevalence studies found in these megacities are insufficient to produce meaningful comparisons between mental health status and homicide and suicide rates due to the scarcity of epidemiological studies, where different sampling strategies and diagnostic criteria are applied. However, as it can be seen in Box 1 it is possible to describe risk factors related to mental health disorders, most related to social and demographic factors, violence, and trauma exposures (Ribeiro et al. 2013; Kessler et al. 2015; Andrade et al. 2012). The exposition to multiple traumatic events (especially traffic accidents, violence, and natural disasters), socioeconomic factors (neighborhood, low schooling and income), physical diseases, migration, and belonging to marginalized and discriminated groups increases the risk of presenting a mental disorder in megacities. Moreover, Box 1 summarizes risk factors for suicide and homicide (Ribeiro et al. 2013; Hayashi et al. 2010; Maselko and Patel 2008; Zhang et al. 2011).

Box 1 Factors related to mental health disorders, suicide and homicide

Mental health risk factors	Mental health protective factors
• Childhood sexual abuse • Childhood maltreatment, intimate partner violence and elder abuse • Psychopathology in parents • Alcohol and drug use	• Healthy family environment • Parents discipline language-based rather than physically-based • Cognitive and emotional regulation skills

(*continued*)

Mental health risk factors	Mental health protective factors
• Disability and Physical diseases • Exposure to chronic childhood adversity • Multiple traumatic events exposure • Loneliness and poor social relationships • Immigration with pre-existing vulnerability • Stigma and discrimination • Environmental stressors: overcrowding, deficient transport system, insecurity • Exposure to chronic community violence • Social and economic inequalities	• Good physical health • School involvement in childhood • Low traumatic events exposure • Healthy peer group • Healthy school/work environment • Pro-social activities participation • Healthy neighborhood environment • Social support • Social inclusion • Culture with gender equality • Spiritualism/religion • Job and educational opportunities

Suicide risk	Homicide risk
• Childhood sexual abuse • Cultural acceptance of suicide, social expectations • Alcohol and drug use • Gender: male • To be victim of partner violence • Previous intent and suicide ideation • Psychiatric disorder: current depression, anxiety disorder, personality disorder and psychosis • Stigma over mental health service use and mental health disorders • Unemployment, loss of social status • Lack of perceived emotional social support	• Alcohol use • Gender: male • Young aged 15 to 24 are the principal victims of homicide • Intimate partner violence • Community violence exposure • Culture of violence • Political/ethnic conflict • Association with drug-trafficking • Weapon access • Poverty • Poor housing conditions • Lack of social support • Unemployment, lack of job and educational opportunities • Social exclusion

As can been seen in Box 1, it is noteworthy that suicide, homicide, and mental health disorders share many common risk factors, mainly those influenced by exposure to violence and substance abuse. In this sense, it is a challenge for policy makers of megacities to implement preventive strategies to decrease violence and abuse of substances particularly directed for young adults. Although it can be concluded that homicide and suicide rates are usually inverse correlated, it is expected that factors which increase the rate of mental disorders in these megacities are similar to those related to violence (self or inter personal aggression). The direction of the violent behavior will be determined by social, cultural, and economic local factors.

Furthermore, urban populations tend to increase, there have been significant waves of war and economic refugees (within and inter-nation), and those are the people who will have to adapt to new cultural norms. Thus, large urban cities have to be prepared for diversity and integration, with a particular challenge in developing mental health services that are culturally sensitive for these vulnerable populations (Box 2).

Box 2 Highlight Findings

- There are 28 megacities in the world; most of them in Asia (16), 6 in America, 3 in Africa, and 3 in Europe.
- Highest homicides rates in megacities are related with lowest macroeconomic indicators: lows GDP and HDI, and higher levels of inequity (GINI).
- Highest suicides rates in megacities are positively related with GDP and HDI, and lowest levels of inequity (GINI).
- There are inverse correlation between homicide and suicide rates: when homicides are high, suicides tend to be low.
- Megacities with epidemic levels of violence –homicides rates above 10/10,000 population- are Rio de Janeiro, São Paulo, Karachi, and Lagos.
- In Brazil, homicides are related with community violence and high levels of socio-economic inequality. In Karachi and Lagos, homicides are related with ethnic fighting and organized political violence.
- The highest rates of suicide are in Asiatic megacities. Rates above 10/10,000 population of suicides are found in Osaka, Tokyo, Calcutta, Mumbai, and Delhi.
- Beyond of mental health problems, social risk factors for suicides in Japan are related with financial crisis and aspects related with Durkheim's suicide theory, which establishes that social disintegration in modern societies, leading to individualism and anomie, increasing risk of suicide. In other hand, suicide in India is related with alcohol consumption, unemployment, poverty, social inequities, and domestic violence.
- Homicides and Suicides rates in megacities are lower compared with national rates of the country where they are placed.
- Overall, living in megalopolis does not seem to make life either better or worse for mental health, but local risk factors in subpopulations of megalopolises can affect homicide, suicide, and mental health.

Conclusions

Overall, living in megalopolis does not seem to make life either better or worse for mental health. There is no evidence that homicides are more frequent in megalopolises; on the contrary, rates are lower than those expected for the countries where megalopolis are located, with exception of cities like Rio de Janeiro, where social inequalities play an important role in violence, and Karachi, where ethnic disputes are determinants of homicidal behaviors. Levels of development, individualism, secular-rational values, moral pressure, and intimate partner violence can be more important risk factors for suicide than simply being a dweller of a megalopolis.

Local risk factors in subpopulations within megalopolises define the adversities those individuals are exposed, determining their emotional development. Growing up in the wealthy areas of the city is completely different than being raised among violence and neglect in the slums and deprived violent areas in the outskirts of the city.

Cross-References

▶ Crime, Illegal Drugs, and Mental Health
▶ Discrimination in Urban Settings
▶ Institutions: Prisons and Asylums and the Deinstitutionalization
▶ Prospects and Challenges of Mental Health Services in a Major City in Latin America: Rio de Janeiro
▶ Urban-Rural Differences in Major Mental Health Conditions

References

Abdel Moneim W, Yassa H, George S (2012) Suicide trends in upper Egypt. J Forensic Sci 57(5). doi:10.1111/j.1556-4029.2012.02247

Andrade L, Wang Y, Andreoni S, Silveira C, Alexandrino-Silva C, Siu E, Nishimura R, Anthony J, Gattaz W, Kessler R, Viana M (2012) Mental disorders in megacities: findings from the Sao Paulo megacity mental health survey, Brazil. PLoS One 7(2):e31879. doi:10.1371/journal.pone.0031879

Bordin I, Duarte C, Peres C, Nascimento R, Curto B, Paula C (2009) Severe physical punishment: risk of mental health problems for poor urban children in Brazil. Bull World Health Organ 87(5):336–344

Caraveo A, Colmenares B, Saldívar H (1999) Morbilidad psiquiátrica en la Ciudad de México: prevalencia y comorbilidad en la vida. Salud Mental 22:62–67

Carta M, Kovess V, Hardoy C, Morosini P, Murgia S, Carpiniello B (2002) Psychiatric disorders in Sardinian immigrants to Paris: a comparison with Parisians and Sardinians resident in Sardinia. Soc Psychiatry Psychiatr Epidemiol 37:112–117

Central Intelligency Agency (2016) The World Factbook. Available via CIA. https://www.cia.gov/library/publications/the-world-factbook/rankorder/2172rank.html. Accessed 2 Feb 2016

Department of Public Health Los Angeles (2009) Los Angeles county suicide data. Available via LA county. http://publichealth.lacounty.gov/ivpp/injury_topics/Suicide/suicidedata_new.htm. Accessed 24 Mar 2016

European Association of National Metrology Institutes (2013) Mega cities. Available via EANMI. http://www.emrponline.eu/call2013/docs/MegaCities.pdf. Accessed 20 Oct 2015

Fleitas D (2014) Homicidios y suicidios en Argentina. Asociación de Politicas Publicas. Available via APP. http://www.app.org.ar/wp-content/uploads/2014/09/Homicidios-y-Suicidios-en-Argentina-2014.pdf. Accessed 24 Mar 2016

Gu L, Xie J, Long J, Chen Q, Chen Q, Pan R, Yan Y, Wu G, Liang B, Tan J, Xie X, Wei B, Su L (2013) Epidemiology of major depressive disorder in mainland China: a systematic review. PLoS One 8(6):e65356. doi:10.1371/journal.pone.0065356.g001

Hayashi N, Igarashi M, Imai A, Osawa Y, Utsumi K, Ishikawa Y, Tokunaga T, Ishimoto K, Harima H, Tatebayashi Y, Kumagai N, Nozu M, Ishii H, Okazaki Y (2010) Psychiatric disorders and clinical correlates of suicidal patients admitted to a psychiatric hospital in Tokyo. BMC Psychiatry 10:109. doi:10.1186/1471-244X-10-109

Instituto Nacional de Estadística y Geografía México (2013) Estadística de suicidios de los Estados Unidos Mexicanos 2011. Available via INEGI. http://internet.contenidos.inegi.org.mx/contenidos/productos//prod_serv/contenidos/espanol/bvinegi/productos/continuas/sociales/suicidio/2011/702825047436.pdf. Accessed 24 Mar 2016

Islam M, Ali M, Ferroni P, Underwood P, Alam F (2003) Prevalence of psychiatric disorders in an urban community in Bangladesh. Gen Hosp Psychiatry 25(5):353–357. doi:10.1016/s0163-8343(03)00067-7

Kessler R, Sampson N, Berglund P, Gruber M, Al-Hamzawi A, Andrade L, Bunting B, Demyttenaere K, Florescu S, de Girolamo G, Gureje O, He Y, Hu C, Huang Y, Karam E, Kovess-Masfety V, Lee S, Levinson D, Medina Mora M, Moskalewicz J, Nakamura Y, Navarro-Mateu F, Oakley Browne M, Piazza M, Posada-Villa J, Slade T, ten Have M, Torres Y, Vilagut G, Xavier M, Zarkov Z, Shahly V, Wilcox M (2015) Anxious and non-anxious major depressive disorder in the World Health Organization World Mental Health Surveys. Epidemiol Psychiatr Sci 24(3):210–226. doi:10.1017/S2045796015000189

Kohn R, Levav I, Caldas de Almeida J, Vicente B, Andrade L, Caraveo-Anduaga J, Saxena S, Saraceno B (2005) Los trastornos mentales en América Latina y el Caribe: asunto prioritario para la salud pública. Rev Panam Salud Publica 18(4/5):229–240

Lenzi M, Colucci E, Minas H (2012) Suicide, culture, and society from a cross-national perspective. Cross-Cult Res 46(1):50–71. doi:10.1177/1069397111424036

Maselko J, Patel V (2008) Why women attempt suicide: the role of mental illness and social disadvantage in a community cohort study in India. J Epidemiol Community Health 62(9):817–822

McGrath J, Saha S, Welham J, El Saadi O, MacCauley C, Chant D (2004) A systematic review of the incidence of schizophrenia: the distribution of rates and the influence of sex, urbanicity, migrant status and methodology. BMC Med 28:2–13

Mian A, Mahmood S, Chotani H, Luby S (2002) Vulnerability to homicide in Karachi: political activity as a risk factor. Int J Epidemiol 31(3):581–585

Ministerio da Saúde Brasil (2012) Taxa de mortalidade específica por causas externas. Available via DATASUS. http://tabnet.datasus.gov.br/cgi/deftohtm.exe?idb2012/c09.def. Accessed 23 Mar 2016

Muhammad Gadit A, Mugford G (2007) Prevalence of depression among households in three capital cities of Pakistan: need to revise the mental health policy. PLoS One 2(2):e209. doi:10.1371/journal.pone.0000209

Murray J, Cerqueira D, Kahn T (2013) Crime and violence in Brazil: systematic review of time trends, prevalence rates and risk factors. Aggress Violent Behav 18:471–483. doi:10.1016/j.avb.2013.07.003

National Bureau of Statistics of China (2014) China Statistical Yearbook. Available via NBS. http://www.stats.gov.cn/tjsj/ndsj/2014/indexeh.htm. Accessed 26 Mar 2016

National Crime Records Bureau of India (2014) Accidental deaths & suicides in India –2014-maps. Available via NCRB. http://www.ncrb.gov.in/StatPublications/MAPS-2014/adsi-2014%20maps/all%20map%20adsi-2014.pdf. Accessed 24 Mar 2016

New York City Department of Health and Mental Hygiene (2010) Injury Statistics Brief. Suicides in New York City, 2007–2009. Available via NYChealth. https://www1.nyc.gov/html/doh/downloads/pdf/ip/ip-suicides-in-new-york-city.pdf. Accessed 24 Mar 2016

Observatoire National du Suicide (2016) Suicide. Connaître pour prévenir: dimensions nationales, locales et associatives. Available via Ministère des Affaires sociales et de la Santé, France. http://drees.social-sante.gouv.fr/IMG/pdf/ons2016_mel_010216.pdf. Accessed 26 Mar 2016

van Os J, Kenis G, Rutten B (2010) The environment and schizophrenia. Nature 468. doi:10.1038/nature09563

Otsuka Y, Horita Y (2013) Statistics on suicides of Japanese workers. Japan Labor Rev 10(4):44–54

Patel V, Ramasundarahettige C, Vijayakumar L, Thakur J, Gajalakshmi V, Gururaj G, Suraweera W, Jha P (2012) Suicide mortality in India: a nationally representative survey. Lancet 379:2343–2351

Pérouse M (2011) Nigeria Watch: third report on violence (2006–2011). Available via Nigeria Watch. http://www.nigeriawatch.org/media/html/NGA-Watch-Report11.pdf. Accessed 24 Mar 2016

Portal Site List of Official Statistics of Japan (2016) Social indicators by prefecture 2016. Health and medical care. Available via e-Stat. http://www.e-stat.go.jp/SG1/estat/ListE.do?bid=000001068038&cycode=0. Accessed 27 Mar 2016

Radaniel M, Lebanan-Dalida M, Gunnell D (2011) Suicide in Philippines; time trend analysis (1974–2005) and literature review. BMC Public Health 11:536. doi:10.1186/1471-2458-11-536

Ribeiro W, Mari J, Quintana M, Dewey M, Evans-Lacko S, Vilete L, Figueira I, Bressan R, de Mello M, Prince M, Ferri C, Coutinho E, Andreoli S (2013) The impact of epidemic violence on the prevalence of psychiatric disorders in Sao Paulo and Rio de Janeiro, Brazil. PLoS One 8(5): e63545. doi:10.1371/journal.pone.0063545

Rozaki A (2012) Suicide among Indonesian children/teenagers. Available via Kyoto Review. http://kyotoreview.org/issue-12/suicide-among-indonesian-childrenteenagers/. Accessed 26 Mar 2016

Shen Y, Zhang M, Huang Y, He Y, Liu Z, Cheng H, Tsang A, Lee S, Kessler R (2006) Twelve-month prevalence, severity, and unmet need for treatment of mental disorders in metropolitan China. Psychol Med 36(2):257–267. doi:10.1017/S0033291705006367

Statistic Japan: Prefecture Comparisons (2016) Murder Victims. Available via Statistic Japan. http://stats-japan.com/t/kiji/10567. Accessed 2 Mar 2016

Stuart H (2003) Violence and mental illness: an overview. World Psychiatry 2(2):121–124

Suryani L, Page A, Lesmana C, Jennaway M, Basudewa I, Taylor R (2009) Suicide in paradise: aftermath of the Bali bombings. Psychol Med 39(8):1317–1323. doi:10.1017/S0033291708004893

The World Bank (2011) Violence in the city. Understanding and supporting community responses to urban violence. Available via UNICEF. http://www.unicef.org/protection/Violence_in_the_City.pdf. Accessed 20 Oct 2015

The World Bank (2016a) GDP at market prices (current US$). Available via WB. http://data.worldbank.org/indicator/NY.GDP.MKTP.CD. Accessed 29 Mar 2016

The World Bank (2016b) GINI index (World Bank estimate). Available via WB. http://data.worldbank.org/indicator/SI.POV.GINI. Accessed 29 Mar 2016

Turkish Statistical Institute (2014) Turkey in statistics 2014. Available via TURKSTAT. https://biruni.tuik.gov.tr/bolgeselistatistik/sorguSayfa.do?target=degisken. Accessed 27 Mar 2016

United Kingdom Government Office for National Statistics (2015) Suicides in the United Kingdom, 2013. Registrations. Available via Office for National Statistics. http://www.ons.gov.uk/ons/dcp171778_395145.pdf. Accessed 29 Mar 2016

United Nations (2015) World urbanization prospects: the 2014 revision. Available via UN. http://esa.un.org/unpd/wup/publications/Files/wup2014-Report.pdf. Accessed 29 Mar 2016

United Nations Development Programme (2015) Human development report 2015: work for human development. Available via UNDP. http://hdr.undp.org/sites/default/files/hdr_2015_statistical_annex.pdf. Accessed 29 Mar 2016

United Nations Office on Drugs and Crime (2013) Global study on homicide: trends, contexts, data 2013. Available via UNODC. https://www.unodc.org/documents/gsh/pdfs/2014_GLOBAL_HOMICIDE_BOOK_web.pdf. Accessed 29 Mar 2016

Winton A (2004) Urban violence: a guide to the literature. Environ Urban 16(2):165–184. doi:10.1177/095624780401600208

World Health Organization (2002) World report on violence and health. Available via WHO. http://apps.who.int/iris/bitstream/10665/42495/1/9241545615_eng.pdf. Accessed 24 Oct 2015

World Health Organization (2012) Global Health Observatory data repository. Suicide rates. Data by country. Available via WHO. http://apps.who.int/gho/data/node.main.MHSUICIDE?lang=en. Accessed 29 Mar 2016

World Health Organization (2014) Global status report on violence prevention. Available via WHO. http://www.who.int/violence_injury_prevention/violence/status_report/2014/en/. Accessed 29 Mar 2016

World Health Organization & United Nations Human Settlements Programme (2010) Hidden cities: unmasking and overcoming health inequities in urban settings. World Health Organization, Switzerland

Zhang L, Li Z, Li X, Zhang J, Zheng L, Jiang C, Li J (2011) Study on the trend and disease burden of injury deaths in Chinese population, 2004–2010. PLoS One 9(1):e85319. doi:10.1371/journal.pone.0085319

Zhou M, Wang H, Zhu J, Chen W, Wang L, Liu L, Yu S, Tan F, Barber R, Coates M, Dicker D, Fraser M, González-Medina D, Hamavid H, Hao Y, Hu G, Jiang G, Kan H, Lopez A, Phillips M, She J, Vos T, Wan X, Xu G, Yan L, Yu C, Zhao Y, Zheng Y, Zou X, Naghavi M, Wang Y, Murray C, Yang G, Liang X (2016) Cause-specific mortality for 240 causes in China during 1990–2013: a systematic subnational analysis for the Global Burden of Disease Study 2013. Lancet 387 (10015):251–272. doi:10.1016/s0140-6736(15)00551-6

Part III

Urban Mental Health: Challenges

Institutions: Prisons and Asylums and the Deinstitutionalization

5

Sean Z. Kaliski

Abstract

As noted by Penrose in the mid-twentieth century the numbers of psychiatric hospital beds and prisoners in most countries seem to be almost consistently inversely related. Superficially this implies that provision of institutional psychiatric care somehow ameliorates the crime rate. Conversely deinstitutionalization, which primarily emptied and reduced beds in psychiatric hospitals, should have led to increased rates of crime and incarceration of mentally ill individuals. The reality is somewhat complicated; many discharged patients are leading more satisfying lives, but significant numbers of people with serious mental illness are homeless, receive inadequate care, misuse substances, and have other risk factors associated with criminal behavior. More than expected numbers of these people are in prisons or are being admitted into ever-expanding forensic mental health facilities, which can be characterized as a process of reinstitutionalization by stealth. A future program should perhaps recognize that a certain number of mentally ill people do require long-term institutional care within the current context of creating adequate community care for most people with psychiatric disorders.

Keywords

Deinstitutionalization • Transintitutionalization • Homelessness • Serious mental illness • Community care • Rehabilitation • Prisons

Contents

S.Z. Kaliski (✉)
Department of Psychiatry and Mental Health, University of Cape Town, Cape Town, South Africa
e-mail: sean.kaliski@uct.ac.za

N. Okkels et al. (eds.), *Mental Health and Illness in the City*, Mental Health and Illness Worldwide, https://doi.org/10.1007/978-981-10-2327-9_17

Introduction

Almost from their inception in the eighteenth century psychiatric hospitals were built in lush countryside settings outside cities. Their original purpose was to provide a refuge, an asylum, from the hurly burly of city life that was presumed to cause or aggravate mental illness (Shorter 1997). There were no effective treatments then and rural tranquility was believed to offer the best hope for recovery. Most became self-contained communities that cultivated their own food and manufactured small goods. Some asylums were located within cities, such as Bethlem in London, that were regularly visited by the local population in order to ogle the inmates' bizarre and uncontrollable behaviors. It was not surprising that madness and crime were easily conflated.

Almost 80 years ago Penrose (1939) documented that in European countries there were inverse correlations between the numbers of inmates in psychiatric (then called "mental") hospitals and the numbers of prisoners and deaths due to murder. Almost as an aside he also noted that there seemed to be a positive correlation between the birth rate and number of inmates in psychiatric hospitals. He concluded that if the mental health needs in a country were addressed the prevalence of serious crimes would decrease. His observations, which became known as "Penrose's Law or Hypothesis," were subsequently taken to mean that this was an invariant inverse relationship, which implied that reducing psychiatric beds anywhere must always lead to an increase in the numbers of prisoners (Gunn 2000). Consequently, deinstitutionalization, the process that discharged large numbers of chronically institutionalized psychiatric patients mostly into facilities within cities and accordingly reduced the number of beds available for long-term care should be blamed for increases in prisoners and crime. The USA, one of the prime instigators of deinstitutionalization, is now renowned for having the highest incarceration rate in the world and superficially appears to confirm Penrose's hypothesis.

Another view, also long regarded as a truism, was that there has always been a constant group of individuals who repeatedly move between psychiatric and penal institutions, as Gunn (1977) almost 40 years ago pointed out:

> "....there is an important group of patients, mainly men, whose basic problem is that their disorder, whether it is alcoholism, personality disorder or schizophrenia, leaves them vulnerable and in need of asylum in the broadest sense. These men constitute the 'social litter' of our modern cities, and they move from hospital to prison, to doss house

and back again, like a stage army tramping round and round, making a much greater impression than their numbers warrant simply because we have no facilities for them" (p.321)

The possible relationships between psychiatric and penal institutions are complicated by the obvious observation that most individuals who have psychiatric disorders do not commit crimes and many prisoners do not require psychiatric treatment. In fact, there are no authorities who contend that serious mental illness is an important cause of crime, other than in a very small proportion of offenders (especially if substance misuse and personality disorders are excluded). Undoubtedly there are offenders who belong in both psychiatric and penal settings, and often are ultimately admitted to long-term forensic mental health facilities, but the dispute really centers on how many of these individuals actually fall into this overlap and whether there is any compelling evidence that altering provisions in either penal or psychiatric services cause changes in the other.

Deinstitutionalization

People with chronic serious mental illness used to be long-term residents in publicly funded psychiatric hospitals. Originally these hospitals were intended to be asylums where psychotic and demented people could be humanely removed from society. Chaimowitz (2012) has aptly noted "as time went on, those asylums became overcrowded and institutional. Rather than places of sanctuary and treatment, some became places to warehouse people with mental illness" (p. 2).

In 1960 there were about 563,000 psychiatric beds (i.e., 314 beds per 100,000) in the USA. By 1990 this had dwindled to 98,000 (i.e., 40 beds per 100,000) and to just over 57,000 beds by 1998 (21 beds per 100,000) (Lamb and Bachrach 2001; Markowitz 2006). Likewise the average length of stay also decreased from 6 months in the 1960s to the current average of 15 days. Similar drastic reductions in beds have been recorded in Canada, Denmark, and the United Kingdom (Kramp and Gabrielsen 2009; Munk-Jorgensen 1999; Sealy and Whitehead 2004). In 1978, Law 180 was passed in Italy that effectively closed all psychiatric hospitals in that country at a stroke. Bed ratios in the Western Cape, South Africa, a developing country, declined from 61 per 100,000 in 1995 to 49 per 100,00 in 2000 and 39 per 100,000 by 2005, a reduction of 36% over 10 years (Lund et al. 2010a, b). Valkenberg Mental Hospital in Cape Town had more than 2,000 beds in 1980. Currently the hospital has 300 beds (of which 150 belong to the forensic mental health service), with an average length of stay of 40 days compared to many years previously.

A combination of factors conspired to produce this effect: the introduction of new psychotropic drugs, especially the neuroleptics, that seemed to obviate the need for hospitalization; a growing dismay at the abysmal conditions in psychiatric hospitals and apparent maltreatment that was conducted within their wards, as exemplified in films such as *One Flew Over the Cuckoo's Nest*; and the influence of the anti-psychiatry movement that gained momentum during the 1960s (Chaimowitz 2012; Szasz 2011; Turner 2004).

Deinstitutionalization was primarily a politically motivated process that assumed that management and quality of life for those with chronic serious mental illness would be superior if they lived in the community. There was also a supposition that outpatient care with drug treatment was less expensive than long-term inpatient hospitalization (Sisti et al. 2015). Virtually no pilot studies were conducted beforehand to investigate whether this was feasible or actually affordable (Fakhoury and Priebe 2002).

Three parallel processes defined deinstitutionalization. Firstly, the discharge of long stay patients into the community (dehospitalization); secondly, the provision of residential care in the community, which has often been called trans-institutionalization because many of these community facilities resembled long stay hospitals, and now also included newly diagnosed cases that previously would have been admitted to chronic wards; and thirdly the creation of specialized noninstitutionalized services for those with chronic psychiatric disorders (Carlier et al. 1996; Sealy and Whitehead 2004). If any of these people required admission to stabilize their mental state, this was supposed to be to an acute facility in a general hospital. Only intractable or complicated cases were to be referred to psychiatric hospitals, and not for long-term care. All commentators agreed that emptying psychiatric hospitals had to be matched with the provision of adequate community care and resources (Fakhoury and Priebe 2002). An unnoticed irony was that deinstitutionalized patients were mostly discharged into facilities within urban areas, such as halfway houses and boarding rooms, in apparent contradiction with the original purpose of providing them with refuge from supposed toxic city environments.

Not everyone was entranced with deinstitutionalization (Chaimowitz 2012). A study conducted in two provinces in South Africa on 425 institutionalized psychiatric patients concluded that 61% had high support needs (mostly as they were poor functioning) and belonged in communities that had scanty resources (Dartnall et al. 1999). They therefore had a low discharge potential, and there was a low likelihood that the state would ever provide adequate resources in the community to enable their successful discharge. In fact, they concluded that there should not be any decrease in psychiatric beds, but that existing facilities were needed to deal with acute admissions. It was arguably not too much of a stretch to extrapolate this reservation to the developing world.

The Consequences of Deinstitutionalization

Many former chronic inpatients settled well into their communities and were much better off than when previously institutionalized, even though they seldom achieved full integration in their communities (Fakhoury and Priebe 2002; Lamb and Bachrach 2001; Leff 2001).

Progress in implementing the ideals of deinstitutionalization has been remarkably uneven in developed countries, let alone in lower and middle income countries. In the Netherlands, community-based care has expanded far more than hospital care was reduced, probably because more citizens with relatively minor disorders started using the service. Hospital care in Germany has been replaced with institutionalized homes that are staffed and almost resemble the old chronic hospitals that still provide residents with limited access to the wider community, and in Sweden an insufficient reallocation of resources to community services has resulted in high rates of hospitalization (Fakhoury and Priebe 2002). Similar scenarios have occurred in Asia and South America. In the Western Cape, South Africa, which is the best resourced province in the country, discharged psychiatric patients were not provided with residential care and mostly only had access to primary health care clinics that were already overwhelmed with having to deal with the myriad of medical illnesses, especially HIV/AIDS that afflicted the impoverished population. Consequently, it has been difficult to assess, treat, and monitor those with serious mental illnesses, especially high frequency users who tend to be more behaviorally disturbed and to misuse substances (Lund et al. 2010b).

For a distressing large number of people, deinstitutionalization has been followed by inadequate provision of community facilities that has had many dire consequences:

- *Homelessness*. Inadequate, and sometimes inappropriate, residential facilities in the community have led to distressingly high numbers of impoverished people with psychiatric disorders living on the streets (Turner 2004). Markowitz's (2006) survey of 81 US cities with populations of more than 50,000 demonstrated a sturdy inverse correlation between the number of public psychiatric hospital beds and rates of homelessness. High numbers of private psychiatric beds did not affect the relationship, presumably because people with serious mental illness do not have access to these facilities. Although homelessness was associated with higher rates of arrests, he was not able to ascertain whether homeless mentally ill people were more likely to be arrested (see "Homelessness and challenges of community care").
- *High arrest rates*. In many cities police officers arrest mentally ill vagrants not only because they have committed crimes but in order to provide care, sometimes called "mercy bookings" (Markowitz 2006), or because their behavior disturbs others. A recent meta-analysis of longitudinal studies that followed up deinstitutionalized patients concluded that very few members of these cohorts were subsequently imprisoned (Winkler et al. 2016). They did concede that there are a rising number of people with mental health problems who are homeless or in prison but attributed this to other criminogenic causes. Their conclusions were somewhat undermined by the high numbers of unreported or untraced subjects in many of the studies because people are often lost to follow up because they are in prison or have moved elsewhere.
- *Substance and alcohol abuse*. It is more difficult to get alcohol and substances in psychiatric institutions and living in the community offers almost limitless

opportunities to obtain them. It is well known that psychotic disorders are highly comorbid with substance abuse, and that substance abuse constitutes a high risk of committing violent crimes (Grann and Fazel 2004; Van Dorn et al. 2012).

- *Shortage of psychiatric hospital beds*. Even where community resources are good there are not enough beds for acute admissions (Chaimowitz 2012). Emergency rooms, local clinics, and shelters, which are now overwhelmed with psychiatric patients, can only offer temporary respite, while clinicians often frantically seek beds in nearby psychiatric hospitals.
- *Transintitutionalization*. Instead of being treated and housed within the walls of a psychiatric hospital, mentally ill individuals are instead being admitted to general hospitals, living in structured residences, being certified to forensic psychiatric hospitals or incarcerated (Fakhoury and Priebe 2007; Priebe et al. 2005; Primeau et al. 2013; Sealy and Whitehead 2004). In the USA a survey of former patients from 162 VA hospitals found that following discharge 86% were mostly in other inpatient units, nursing homes, or organized residences in the community (Irmiter et al. 2007). The term transinstitutionalization implies that almost by sleight of hand psychiatric patients have been moved sideways from a psychiatric institution into another institutional setting in the community that almost seems to be pretence of reintegrating them into the community.
- *Inadequate care for those newly diagnosed with a serious mental illness*. There is some speculation, but unfortunately not with enough empirical support, that the prevalence of serious mental illness together with substance abuse is expanding in all countries. Many of these new cases have poor social supports and struggle to find adequate long-term care (Priebe et al. 2005; Sisti et al. 2015).
- *Urbanization of mental health care*. Asylums were originally placed in rural areas because the stress of living in cities, where overcrowding, crime, substance abuse, and poverty were rampant, was assumed to aggravate or even cause mental illness, and conversely mentally ill people were feared and believed to be dangerous (Cohen and Freeman 1945; Shorter 1997). It seemed logical to remove them to the tranquility of the countryside both for their rehabilitation and to remove them from their fellow citizens. As psychiatry developed into a recognized medical speciality that began to use current medical technologies and biological treatments, asylums were transformed into hospitals that accordingly needed facilities, such as radiological and laboratory services, access to drugs, and a variety of highly trained personnel who were reluctant to live far away from cities. Psychiatric hospitals also became integral teaching platforms for universities. Furthermore, as deinstitutionalization gained traction it became obvious that the urban areas best provided for the range of necessary services that discharged patients required. These included well-resourced clinics, social services, adequate housing, and employment opportunities (Fakhoury and Priebe 2007; Lamb and Bachrach 2001; Munk-Jorgensen 1999). Ironically two centuries after asylums transported psychiatric patient to rural refuges they are now believed to be better off living in cities. History will have to judge this assumption.

Criminalization of the Mentally Ill – Is This a Consequence of Deinstitutionalization?

Crime rates vary hugely between countries. Comparisons are difficult because definitions of offences, arrest, and conviction rates depend on the different laws and resources in each country. Even a seemingly straightforward offence such as murder surprisingly does not have a universal definition (Taylor et al. 2009). Nevertheless, criminal activity generally has a relatively high base rate in most countries and accordingly is caused by a myriad of motivations, some of which may flow directly from psychiatric symptoms. Developing countries, such as South Africa, may have even higher rates, especially of violent crime (Stein et al. 2008).

Similarly epidemiologic surveys have confirmed that psychiatric disorders, especially if substance misuse and mood disorders are included, also have high base rates in most countries with lifetime prevalences that range from 13% to 30% (Swanson et al. 1990). As both mental illness and crime have relatively high base rates, they can be expected to co-occur commonly without being causally related. Also the relative number of offences, particularly violent crimes, committed by mentally ill people is always a small fraction of total offences that are mostly committed by nonpsychotic individuals. It may therefore be very difficult to detect whether there has been an increase in offending behavior and determining its cause in this group. There do not seem to be any large-scale epidemiologic studies that have determined how many mentally ill people actually commit crimes, especially violent crimes, and what their outcomes were.

An important requirement for the involuntary admission of psychotic individuals in South Africa under the Mental Health Care Act of 2002 requires that the user constitutes a danger to others. Therefore, most are admitted after committing violent acts for which they are not charged and therefore do not contribute to the overall crime data. Some are indicted and then referred to a forensic mental health observation unit for determination of triability and criminal responsibility. Those who are found not guilty by reason of insanity (or not triable) are admitted to forensic hospitals and, again, their crimes are not recorded in official statistics. In other words, the true prevalence of those with serious mental illness who are diverted to mental health facilities after committing offences without being charged is not really known. There are also few data on why some are indicted instead of being diverted.

Nevertheless, it is now accepted that people with schizophrenia are significantly more likely to be convicted for crimes, especially for crimes of violence. If these individuals have other risk factors associated with criminal behavior, this likelihood increases markedly. Some insist that this occurs independently of deinstitutionalization and have demonstrated that the rate of criminality in people with serious mental illness has always been roughly consistent over time; when the periods of pre- and post-deinstitutionalization were compared there was not a spike in convictions in the post-deinstitutionalized-group, despite the paucity of data from the long era of chronic hospitalization(Wallace et al. 2004). These authors do acknowledge that many persons with serious mental illness are in prisons but inexplicably blame policies in the criminal justice system rather than deinstitutionalization. An undeniable fact is that our cities now have large numbers of people with serious mental illness, who are often homeless, that would not commit crimes if they were institutionalized.

Are Prisons Repositories of Mentally Ill People Who Should Be in Psychiatric Hospitals?

Rising numbers of mentally ill individuals are being imprisoned. It has been noted that probably the largest mental institution in the USA is the Los Angeles County jail (Chaimowitz 2012). A now incontrovertible finding is that there is a high prevalence of psychiatric disorders in both remand and convicted prisoners. The rates of psychotic illnesses range from 3.5% to 5.5%, compared to 1% to 2% in the general population, with lifetime prevalences of up to 8% (Birmingham et al. 1996; Fazel and Seewald 2012). When substance abuse is added to the list of diagnoses, the rates rise to between 60% and 70%. In Fazel and Seewald's meta-analysis (Fazel and Seewald 2012), the rates of psychosis among prisoners in low- and middle-income countries were considerably higher than those in high-income countries, and that the rates had not differed significantly from those reported 10 years earlier. Their conclusion was that the higher rates in low- and middle-income countries was a consequence of their poor mental health resources, implying that prisons were being used to house behaviorally disturbed psychotic individuals, and that the rates in high-income countries were stable because of improved psychiatric care in their prisons, increased diversion from prison to hospital, and generally better living conditions.

In Denmark, where mentally ill offenders are not diverted to psychiatric hospitals but are always indicted, there seems to be a strong trend that initially closure of general psychiatric beds had no noticeable effect on the numbers of forensic admissions, but over time the decreased inpatient psychiatric care was associated with a significant growth in the numbers of mentally ill offenders, especially for violent offences (Kramp and Gabrielsen 2009). Surprisingly intensive community interventions (including psychiatric emergency services) could not compensate for the loss of psychiatric beds. A recent 13 year survey in Sweden revealed that whereas in the general population there were 45 violent crimes per 1,000 people, there were 215 violent crimes per 1,000 patients. According to their data about 5% of all violent crime were committed by people with severe mental illness, which seemed to imply that if these individuals had been residing in psychiatric institutions the rate of violent crime would have diminished by 5% (Fazel and Grann 2006).

These studies do not differentiate between those who have become mentally ill because of the stress of imprisonment, those for whom their mental illness was incidental to their incarceration, and those who committed their offences because of their symptoms.

Nevertheless, studies continue to demonstrate that the inverse relationship between psychiatric beds and prison populations remains sturdy. A recent longitudinal study in six South American countries over 20 years confirmed that the rates of increase in prisoners tracked the decrease in psychiatric beds; it seemed that there was on average an increase of five prisoners for every one psychiatric bed lost. This relationship persisted even when economic factors were controlled (in order to exclude impoverishment as an interfering variable) (Mundt et al. 2015). But they did concede that this does not establish a definite causal link between the two phenomena and that both may be influenced by unidentified factors. No one can determine how many of the increase five prisoners (for each psychiatric bed lost) actually have a mental illness. Nevertheless, there does seem to be some connection between the numbers of psychiatric patients in the community and their rate of offending.

Is There a Relationship Between Psychiatric Symptoms and Criminal Behavior?

A meta-analysis of 20 studies concluded that individuals diagnosed with psychotic disorders, especially with schizophrenia, were at high risk of committing violence (Fazel et al. 2009). Similarly the rate of convictions for violence crime among those with bipolar disorder was 8.4% compared to 3.5% in the general population (Fazel et al. 2010). Comorbid substance abuse heightened the risks considerably in both disorders, although those with schizophrenia continued to be at risk when substance abuse was excluded from the analysis. Offenders with bipolar disorder tend to derive from criminogenic families and abuse substances. Van Dorn et al. (2012), in their reanalysis of the National Epidemiologic Survey on Alcohol and Related Conditions findings, contended that all studies use lifetime diagnoses to account for specific events, such as violent crime, which would suggest the possibility that at the time of the offence the perpetrator may not have been actively disordered. Nevertheless, they did find that a diagnosis of a serious mental illness over the previous year was strongly associated with violence, which they labeled as "spatio-temporal contiguity." But this still fails to reassure that active symptoms, which surely would be a core reason for admitting such individuals to a psychiatric institution, actually accounts for their criminal behavior.

A longitudinal study that compared 221 mentally ill and control parolees over 1 year did confirm that those with mental illness were more likely to violate parole (and to be rearrested) but for reasons similar to reasons that applied to the control group too, and not for any factors directly related to their psychiatric disorder (Skeem et al. 2013). They concluded that offenders with psychiatric disorders were more likely to have longstanding antisocial personality and that only a small minority had actually committed crimes because of their symptoms. Others have concluded that perhaps one in ten mentally ill offenders commit an offence because of psychotic symptoms (Peterson et al. 2010).

A multisite comparison of patients that had been discharged from general and forensic psychiatric hospitals demonstrated that the general psychiatric patients were more likely to have psychotic symptoms, lower functioning and over a 2 year period a higher likelihood of violent behavior (Hodgins et al. 2006). This seemed to confirm that partially treated psychotic individuals were more likely to be violent possibly because of ongoing symptoms, and also provided some evidence that long-term care that stabilized symptoms may be beneficial.

Reinstitutionalization by Stealth: The Growth of Forensic Mental Health Services

The numbers of beds in forensic psychiatry hospitals have risen dramatically since 1990 (Priebe et al. 2005). Increasing numbers of people with serious mental illness, who otherwise may have been in long stay psychiatric wards, are being charged with crimes and subsequently (after a forensic assessment) hospitalized indefinitely under court order in an ever expanding forensic mental health service (Chaimowitz 2012). Between 1990 and 2003 forensic psychiatry beds throughout Europe increased by up to 143%, and these high admission rates correlate with the reduced number of general psychiatric beds available in those regions as well as increases in community facilities and prisoners (see Fig. 1; Coid et al. 2001; Hodgins et al. 2006; Priebe et al. 2005). The number of forensic patients (who have been found not guilty by reason of insanity) in the Western Cape has tripled over the last 20 years, 80% of whom have had previous admissions to the general psychiatric system. The bed occupancy in this forensic mental health service is consistently above 110% (Kaliski 2013).

This is very likely another manifestation of transinstitutionalization, in that chronically ill but behaviorally disturbed individuals are now being diverted and institutionalized in forensic psychiatry hospitals. Unfortunately, given the current fearfulness for mentally ill offenders that has been bolstered by a reliance on risk assessment tools that are mostly not valid, relatively very few are ever discharged fully back into the community. This is occurring ironically while many are rejoicing at the apparent successes of community care. Forensic hospitals almost by stealth have become the new chronic institutional care system.

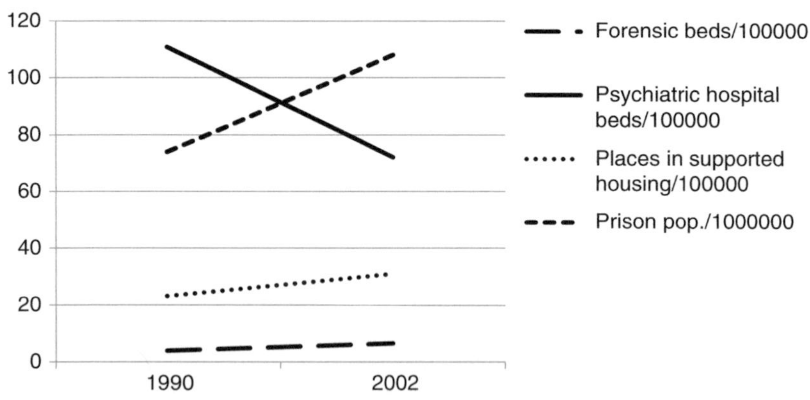

Fig. 1 Composite of trends in six European countries between 1990 and 2002. Derived from data provided by Priebe et al. (2005)

This is not necessarily a bad development. Hodgins et al. (2006) in their multisite comparison between general and forensic psychiatric patients found that the former had a significantly higher level of psychotic symptoms and lower levels of psychosocial functioning than the latter on discharge, even though the forensic patients were much more likely to have been admitted following violent behavior. The general psychiatric patients were also more likely to assault others in the 2 years following discharge, which ultimately led to admission to a forensic psychiatry hospital. This is surely an indication that the lack of long-term treatment programs for people with serious mental illness, a deficiency that has followed deinstitutionalization, does result in violent behavior.

Should the Asylums Be Reopened?

Community programs have mostly resulted in inadequate care for people with serious mental illness even where there are well-resourced community facilities. Most of these people are chronically ill, low functioning, often poorly compliant with their medication, susceptible to substance abuse and victimization. As Sisti et al. (2015) opine

> "For persons with severe and treatment-resistant psychotic disorders, who are too unstable or unsafe for community-based treatment, the choice is between the prison-homelessness-acute hospitalization-prison cycle, or long term psychiatric institutionalization. The financially sensible and morally appropriate way forward includes a return to psychiatric asylums that are safe, modern and humane" (p.244)

A new model of care is obviously needed. It would surely comprise of comprehensive community care and support for the majority of those with psychiatric disorders, but also provide long-term institutional care for those who cannot care for themselves, are unable to use available community resources properly and need long-term care. This scenario will work best in developed countries. Low- and middle-income countries face four major barriers to upscaling their mental health services (which would be necessary to provide adequate mental health care in the community), namely, (1) the absence of financial resources and government commitment; (2) overcentralization of all services; (3) ongoing problems with integration of mental health care into community and primary health care settings; (4) scarcity of trained personnel and general shortage of mental health expertise (Eaton et al. 2011) Therefore, in the developing world there are few expectations that community resources will in the foreseeable future significantly improve to make wholesale deinstitutionalization in their cities feasible. But no one is advocating reopening asylums in the countryside. Even in developed countries rural areas are often inaccessible, poorly funded and offer fewer mental health resources, especially qualified personnel and support services (Moore and Walton 2013). For the foreseeable future most mental health care will therefore be better delivered in urban settings.

Conclusion

Penrose's hypothesis continues to have continues to have validity as there still does seem to be an inverse relationship between the numbers of state funded psychiatric beds and numbers of prisoners. The many causal links that connect psychiatric beds to prison capacities remain elusive. Many people with serious mental illness who previously would have been languishing in soulless psychiatric institutions are now leading more satisfying lives somewhere in mostly urban communities, even if in similarly structured facilities. But there remain an unknown number of people, either newly diagnosed or discharged, for whom community supports are either absent or inadequate for their special needs who would benefit from long-term care. Too many of them are suffering in prisons.

Cross-References

▶ Crime, Illegal Drugs, and Mental Health
▶ History of Urban Mental Illness
▶ Homelessness and Challenges of Community Care
▶ Homicide and Suicide in Megacities

References

Birmingham L, Mason D, Grubin DH (1996) Prevalence of mental disorder in remand prisoners: consecutive case study. Br Med J 313:1521–1524

Carlier VE, Lamberts RD, Fouwels AJ et al (1996) PTSD in relation to dissociation in traumatized police officers. Am J Psychiatr 153:1325–1328

Chaimowitz G (2012) The criminalization of people with mental illness. Can J Psychiatr 57:1–7

Cohen L, Freeman H (1945) How dangerous to the community are State hospital patients? Conn State Med J 9:265–276

Coid J, Kahtan N, Cook A et al (2001) Predicting admission rates to secure forensic psychiatry services. Psychol Med 31:531–539

Dartnall E, Modiba P, Porteus K et al (1999) Is deinstitutionalization appropriate? Discharge potential and service needs of psychiatric inpatients in KwaZulu-Natal and the Eastern Cape. Centre for Health Policy, University of Witwatersrand, South Africa

Eaton J, Mccay L, Semrau M et al (2011) Scale up of services for emtnal health in low-income and middle-income countries. Lancet 378:1592–1603

Fakhoury W, Priebe S (2002) The process of deinstitutionalization: an international overview. Curr Opin Psychiatry 15:187–192

Fakhoury W, Priebe S (2007) Deinstitutionalization and reinstitutionalization: major changes in the provision of mental healthcare. Psychiatry 6:313–316

Fazel S, Grann M (2006) The population impact of menal illness on violent crime. Am J Psychiatry 163:1397–1403

Fazel S, Seewald K (2012) Severe mental illness in 33 588 prisoners worlwide: systematic review and meta-regression analysis. Br J Psychiatry 200:364–373

Fazel S, Gulati G, Linsell L et al (2009) Schizophrenia and violence: systematic review and meta-analysis. PLoS Med 6:e1000120

Fazel S, Lichtenstein P, Grann M et al (2010) Bipolar disorder and violent crime. Arch Gen Psychiatry 67:931–938
Grann M, Fazel S (2004) Substance misuse and violent crime: Swedish population study. BMJ 328:1233–1234
Gunn J (1977) Criminal behaviour and mental disorder. Br J Psychiatry 130:317–329
Gunn J (2000) Future directions for treatment in forensic psychiatry. Br J Psychiatry 174:332–338
Hodgins S, Muller-Isberner R, Allaire J (2006) Attempting to understand the increase in the numbers of forensic beds in Europe: a multi-site study of patients in forensic and general psychiatric services. Int J Forensic Men Health 5:173–184
Irmiter C, Mccarthy JF, Barry KL et al (2007) Reinstitutionalization following psychiatric discharge among VA patients with serious mental illness: a national longitudinal study. Psychiatry Q 78:279–286
Kaliski S (2013) Reinstitutionalization by stealth: the forensic mental health service is the new chronic system. Afr J Psychiatry 16:13–17
Kramp P, Gabrielsen G (2009) The organization of the psychiatric srvice and criminality committed by the mentally ill. Eur Psychiatry 24:401–411
Lamb HR, Bachrach LL (2001) Some perspectives on deinstitutionalization. Psychiatr Serv 52:1039–1046
Leff J (2001) Why is care in the community perceived as a failure? Br J Psychiatry 179:381–383
Lund C, Kleintjes S, Kakuma R et al (2010a) Public sector mental health systems in South Africa: inter-provincial comparisons and policy implications. Soc Psychiatry Psychiatr Epidemiol 45:393–404
Lund C, Oosthuizen P, Flisher AJ et al (2010b) Pathways to inpatient mental health care among people with schizophrenia spectrum disorders in South Africa. Psychiatr Serv 61:235–240
Markowitz FE (2006) Psychiatric hospital capacity, homelessness, adn crime and arrest rates. Criminology 44:45–72
Moore MA, Walton BA (2013) Improving the mental health functioning of youth in rural communities. Contem Rural Soc Work 5:85–103
Mundt AP, Chow WS, Arduino M et al (2015) Psychiatric hospital beds and prison populations in South America since 1990. Does the Penrose hypothesis apply? JAMA Psychiatry 72:112–118
Munk-Jorgensen P (1999) Has deinstitutionalization gone too far? Eur Arch Psychiatry Clin Neurosci 249:136–143
Penrose L (1939) Mental illness and crime:outline of comparative study of European statistics. Br J Med Psychol 18:1–15
Peterson AL, Skeem J, Hart EL et al (2010) Analyzing offence patterns as a function of mental illness to test the criminalization hypothesis. Psychiatr Serv 61:1217–1222
Priebe S, Badesconyi A, Fioritti A et al (2005) Reinstitutionalization in mental health care: comparison of data on service provision from six European countries. BMJ 330:123–126
Primeau A, Bowers TG, Harrison MA et al (2013) Deinstiutionalization of the mentally ill: evidence for transinstitutionalization from psychiatric hospitals to penal institutions. Compr Psychol 2:1–10
Sealy P, Whitehead PC (2004) Forty years of deinstitutionalization of psychiatric services in Canada: an empirical assessment. Can J Psychiatr 49:249–257
Shorter E (1997) A history of psychiatry: from the era of the asylum to the age of Prozac. Wiley, New York
Sisti DA, Segal AG, Emanuel IZJ (2015) Improving long-term psychiatric care. Bring back the asylum. JAMA 313:243–244
Skeem J, Winter E, Kennealy P et al (2013) Offenders with mental illness have criminogenic needs, too: toward recidivism reduction. Law Hum Behav 38:212–224
Stein DJ, Seedat S, Herman A et al (2008) Lifetime prevalence of psychiatric disorders in South Africa. Br J Psychiatry 192:112–117
Swanson JW, Holzer CE, Ganju VK et al (1990) Violence and psychiatric disorder in the community: evidence from the Epidemiologic Catchment Area surveys. Hosp Community Psychiatry 41:761–770

Szasz T (2011) The myth of mental illness: 50 years later. Psychiatrist 35:179–182
Taylor PJ, Dunn E, Ogloff JRP et al (2009) Offenders with mental disorder on five Continents: a comparison of approaches to treatment and demographic factors relevant to measurement of outcome. Int J Forensic Ment Health 8:81–96
Turner T (2004) The history of deinstitutionalization and reinstitutionalization. Psychiatry 3:1–4
Van Dorn R, Volavka J, Johnson N (2012) Mental disorder and violence: is there a relationship beyond substance abuse? Soc Psychiatry Psychiatr Epidemiol 47:487–503
Wallace C, Mullen PE, Burgess P (2004) Criminal offending in schizophrenia over a 25-year period marked by deinstitutionalization and increasing prevalence of comorbid substance use disorders. Am J Psychiatr 161:716–727
Winkler P, Barrett B, Mccrone P et al (2016) Deinstitutionalised patients and homelessness and imprisonment: systematic review. Br J Psychiatry 208:421–428

Crime, Illegal Drugs, and Mental Health

6

Joana Corrêa de Magalhães Narvaez, Caroline Galli Moreira, and Fernando Rosa da Rosa

Abstract

This chapter is about the complex connections between drugs, violence, criminality, and mental illness. To approach this theme, we recall the history of drug from the perspective of the amplified social context involving illness and public policies that guide the discussion about violence and drugs insertion in urban setting. Drug use, manifestations of crime, and the development of mental illness are socially produced symptoms and the way they are expressed in our current society indicate vulnerabilities and dysfunctionalities of the urban organization. The etiology of the problem is complex and based on a society with marked differences. Drugs bring to light a circuit of illegality and criminality, forming a parallel system of laws and punishment. The mental illness can add damages to inhibitory control, increase impulsivity and emotional instability, leaving the users more exposed to violence outcomes. The criminality can designate a stigma without a clear distinction between the evolutionary phases of substance dependence, which inhibits the development of therapies directed to different profiles.

J.C. de Magalhães Narvaez (✉)
Laboratório de Psiquiatria Molecular, National Institute for Translational Medicine, Hospital de Clínicas de Porto Alegre, Federal University of Rio Grande do Sul, Porto Alegre, RS, Brazil

Center for Drug and Alcohol Research, Hospital de Clínicas de Porto Alegre, Federal University of Rio Grande do Sul, Porto Alegre, RS, Brazil
e-mail: jonarvaez@gmail.com

C. Galli Moreira
Center for Drug and Alcohol Research, Hospital de Clínicas de Porto Alegre, Federal University of Rio Grande do Sul, Porto Alegre, RS, Brazil

Brasilian Psychiatric Association, Florianópolis, SC, Brazil
e-mail: carolinegmoreira@hotmail.com

F. Rosa da Rosa
Social Service for the Industry (SESI-RS), Porto Alegre, RS, Brazil
e-mail: nandorosa@gmail.com

N. Okkels et al. (eds.), *Mental Health and Illness in the City*, Mental Health and Illness Worldwide, https://doi.org/10.1007/978-981-10-2327-9_3

Prevention and treatment of these conditions are as broad and multicausal as the problem itself. So, we proposed a theoretical model of functional staging of drug users to identify a clinical expression associated with vulnerabilities and violence expression. Anyway, the treatment to substance abuse should not configure an exclusion of these aspects that can make up the user profile and its relationship with society. The analysis of violent contexts should include socioenvironmental factors that should not be reduced to the social group where it is manifested. The association of drugs with violence may be an aspect that highlights the emphasis it has on the current social context. It also indicates the need to develop public health policies that address drugs, violence, and mental illness in a more holistic way.

Keywords
Drugs • Crime • Violence • Public policies • Treatment

Contents

Introduction

Drugs and violence have always been parts of the human experience and their impacts are expressed in various forms, all over the world. According to the latest World Report of the United Nations on Drugs and Crime published in 2015, one in every 20 people aged between 15 and 64 years, has used illicit drugs (UNODC – United Nations Office on Drugs and Crime 2015). Drugs represent a cost to society which is estimated to more than $200 billion dollars a year (Sadock et al. 2007). In addition to the impact on health, drugs are also associated with significant levels of social violence. In fact, here is another problem of global dimensions: the high rates of violence prompted the World Health Organization (WHO) to consider it as a public health problem. According to the WHO report on violence and health (Krug et al. 2002), every year more than one million people lose their lives, and many more suffer nonfatal injuries as a result of interpersonal, collective, or self-inflicted violence. Overall, violence is among the leading worldwide causes of death for people aged 15–44 years. Although, precise estimates are difficult to obtain, the cost of violence translates into billions of dollars invested every year in health care around the world, and billions more for national economies in terms of lost working days, law enforcement, and wasted investment.

It is in the modern urban environment of a large number of cities that one can find an inseparable relation between the use of illicit drugs and violence, verified in the dimensions of the vast and complex ecosystem generated in the interrelation of drug trafficking, users' behaviors, and police action.

Drugs and Violence in the City

We can't talk about violence and drugs without understanding the cultural role of drugs in society. Marlene Xiberras (2007), a French sociologist, has researched the historical association between different cultures and certain drugs, tracing what might be called a cultural epidemiology. She summed up the history of civilization of psychoactive products and the significance that those drugs have for each society. In its early days, the Western history shows that some individuals began to adopt alcohol as a way to access the divine intoxication. In spite of the consumption of alcohol and other drugs has lost much of its religious content, profane and recreational practices never ceased.

Drug experimentation is associated, since the midtwentieth century, to the idea of engaging in illegal activities. The Brazilian context is particularly related to violence not only because of its increasing prevalence shown by national surveys (Carlini et al. 2011), but also because of entrenched social and environmental vulnerabilities, such as drug use. At the end of the century, Brazil's "drug issue" is based on urban violence. From an anthropological point of view, we emphasize that while the use of illicit drugs among the middle class of the 1960s and 1970s was related to pleasure, self-discovery, and accomplishment, the association of drugs and weapons and the increasing control of illicit substances trade by drug trafficking, drug use began to be seen as urban violence (Velho 1998). Indeed, there is the establishment of trafficking and consumption groups, as occurs in "crackolândia" in São Paulo, Brazil, where high rates of substance dependence are found. The violence between drug traffickers and the repression by police added to the social prejudice make up a framework where rights violations are frequent. Such drug-related lifestyles exist throughout the world and may be seen as microcontexts where subcultures related to drug dependence are created. At the same time, the urban space where this subcultures interaction occurs is also relevant, and is an important focus point intertwined with psychological, social, and biological dimensions.

Drug issues are permeated by social, cultural, and geographical aspects with the varying prevalence of specific drugs in different countries being a factor that also adds different features to the policies. Population studies (CEBRID 1997; SAMHSA: Substance Abuse and Mental Health Services Administration 2001) show that lifetime use of alcohol is discrepant between countries such as Brazil (74.6%), Chile (86.5%), and the USA (82.4%). Similarly, lifetime use of marijuana has closer result in various countries such as Greece (8.9%), Brazil (8.8%), Poland (7.7%), although, in disagreement with data coming from countries such as the USA (40.2%) and the UK (30.8%). The prevalence of lifetime cocaine use is similar in Germany (3.2%) and Brazil (2.9%), well below the USA (14.2%) and Chile (5.3%). Lifetime crack cocaine use is 0.7% in Brazil, which was about half of that value shown in an American study (SAMHSA: Substance Abuse and Mental Health Services Administration 2001), has been growing exponentially and was found to be 1.4% in the last national survey (Carlini et al. 2011). Therefore, there is a significant variance in prevalence and drug type in different countries.

The issue of drugs and violence reveals a complex and circular interaction that can only be attained with a contextualized overview and with a deep understanding of meanings that occurs in what Fernandes (1998) calls psychotropic territories. With the addition of drug effects in terms of increased paranoia and deregulation of inhibitory control (as it dilutes and reduces components of self-control), the user may be more likely to express an aggressive pattern that replicates what he was frequently exposed to in its original environment. Crack cocaine users manifest self-destructive behaviors related to risk exposure and environment-directed aggression. This validates the evidence that internal and external environments are seen as threatening. Thus, there is a subjective search for security through arming in a culture where guns are valued as an instrument of power and affirmation (Horta et al. 2010). This is a constitutional issue; the more vulnerable the subject feels in relation to its self-protection and social integration abilities, the more probable is its appealing to a poor defensive, projective, and externalizing system.

Social Vulnerabilities Associated to Drug Use in the City: A Culture of Violence

Nowadays, especially regard to low- and middle-income countries, drugs such as cocaine show the dead serious impact of violence surrounding the drug issue. However, drugs should not only be referred as a cause of violence and crime outcomes once it may be entangled in a complex sociocultural plot. Contexts of violence preceding substance use may easily allude to drug use. A sociodemographic analysis shows that assault is more frequent in males, in black, brown, yellow, and indigenous adolescents, children of mothers with low education, kids studying in public school, or working. Among individual risk behaviors, the use of tobacco, alcohol, and illicit drugs, as well as variables related to the mental health domain such as impaired socialization, were associated to a positive history of aggression (Malta et al. 2014).

Socialization and family structure seem to impact on outcomes such as drug use. Studies with homeless children and adolescents in several countries have been conducted to investigate the factors associated with substance use. These factors include parental drug use, poor integration in school activities and learning difficulties, changes in family structure, domestic violence, and pressure from peers (Olley 2006). On the other hand, the maintenance of family bonds, the involvement in school, negative history of domestic violence, spending less time (daily and total years) in the streets, and social links seem to be protective factors among adolescents living on the streets who do not use drugs. The misfortunes to which homeless teenagers are exposed such as hunger, domestic violence, or cold, combined with the lack of support and supervision of a parent or guardian and the easy access to various substances, predisposes them to a frequent and heavy drug use (Rachlis et al. 2009).

Public policies and the way the government reaches this population should be considered and weighted. Some studies (Werb et al. 2011) have established that policing practices may contribute to adverse outcomes related to health and social issues among these structurally vulnerable populations, including drug users, who are marginalized as a result of the intersection of sociostructural inequalities, such as the criminalization of drugs, poverty, and homelessness. Epidemiological studies have linked repressive strategies to increased drug-related risks and harms, including the risk of HIV exposure, overdoses and decreased search for and entry in health services (Friedman et al. 2006; Beletsky et al. 2014).

Drug users have illustrated how racism intersects with poverty and the stigma associated with drugs, and also how it can make them disproportionately vulnerable to surveillance in public spaces that sounds like discriminatory. In a Canadian study (Markwick et al. 2015), almost all users that were followed reported having experienced or witnessed violence by public or private security which was often based on suspicion of theft (whether real or unfounded).

The daily violence linked to a structure that should provide security instead of brutality strengthens social barriers to access public or private health care areas and thus reinforces the structural vulnerability of drug users. These experiences of exclusion, violence, and aggression increment social marginalization (Markwick et al. 2015).

Concerning ethnicity, Latin American adolescents have a higher annual prevalence of drug use compared to non-Hispanic whites and African Americans (Johnston et al. 2015). Cultural stressors such as those related to acculturation, discrimination, immigration, poverty, and violence in the community, have shown to be implicated in these results. Stress is commonly related to a wide range of risk behaviors predisposing to substance use, and it also applies to social stressors. About 30% of young Latin Americans showed association with risk factors for the use of multiple drugs and alcohol such as family stress, financial difficulty, "acculturative gap", and contact with gangs and drugs (Cardoso et al. 2016). Additionally, it is known that adolescents exposed to household use of drugs such as marijuana, tobacco, and alcohol may have reckless attitudes regarding substance use in addition to an easier access to drugs and the development of drug consumption behaviors.

The Impact of Early Drug Use

The incorporation of drugs in a given culture has an impact that transposes subjective and social barriers. It brings physiological manifestation with behavioral impact which, in turn, brings new vulnerabilities to social life. An important aspect in the context of violence insertion is the earliness of substance experimentation. In Brazil, for example, drugs considered legal – alcohol and tobacco – are those with the lowest age of first use (average around 12 years) (CEBRID 1997), while the experimentation of marijuana and cocaine occurs around 13 and 14 years old, respectively. The drug experimentation in U.S. corroborates these data, once it is equally premature (SAMHSA: Substance Abuse and Mental Health Services Administration 2001). The precocity of drug experimentation, in addition to affirming a possible context of low social restraint, also covers impairments from the point of view of brain development with a potential impact in increasing impulsiveness and reducing the inhibitory control in decision-making. For example, studies with crack cocaine users show that, most of the times, drug experimentation takes place quite early in this population, usually in the range from 12 to 24 years of age (SAMHSA: Substance Abuse and Mental Health Services Administration 2001). Initiating drug use during adolescence may influence the development of the prefrontal cortex, the brain area responsible for neuropsychological functions and impulse control, and whose maturation extends to 20 years of age approximately (Autry and Monteggia 2012).

Therefore, the sooner drug use initiates, the earlier it acts on an immature brain impacting on inhibitory control and rising the risk of launching a regular, problematic, and addicted use (Boys et al. 2001). On the other hand, the defective social structure does not provide enough family support, making young people seek in dealers the boundaries they lack in family environment. A population study (Narvaez et al. 2015) that deepened the evaluation of these aspects, revealed that subjects with lifetime crack cocaine use report some specificities regarding parental structure that includes greater maternal presence and paternal absence at home, even when controlled for the use of cocaine and social group. That same population also shows a higher prevalence of aggressive episodes and firearm possession, as well as risky sexual behavior (Narvaez et al. 2014a).

Gender and Violence in the City

Gender violence is one of the most expressive violence manifestations related to drug use in the urban setting. Although, data linking high rates of violence to drug users, little is known about gender differences associated with violence among this population. A prospective cohort study in the population of injected drug users in Canada noted that, while binge drug use and drug trafficking were associated with violence in women, early experimentation and incarceration were significant for males. Women were more likely to be attacked by acquaintances, partners, and commercial sex customers, while men were more likely to suffer violence from strangers and police (Marshall et al. 2008). These results indicate significant gender differences with respect to the predictors and characteristics of violence between genres.

Many explanations have been proposed to the fact that women who inject drugs may be at greater risk of becoming infected with HIV than their male partners. Women are more susceptible to stigmatization and marginalization by society and thus are more likely to hide their behavior of injecting drugs. Unsafe injecting practices may be more common among women given the difficult access to treatment and needle/syringe programs, as well as the lack of services targeting women (UNODC/UN-Women/WHO/International Network of People who Use Drugs 2014). High rates of prostitution – whether to maintain their own consumption or to support drug use from their sexual partner – and infrequent condom use have been reported among female users of injected drugs. At the same time, those who are not involved in prostitution show higher rates of needle and syringes sharing (Croxford et al. 2015). The combined risks of unsafe injecting practices and unprotected prostitution substantially increase the risk of acquiring HIV for women. Women notably seem to face more barriers in accessing treatment opportunities and while one in three drug users all over the world are female, only one in five drug users under treatment are women (UNODC – United Nations Office on Drugs and Crime 2015).

An important aspect regarding violence is the self-inflicted aggression. In this sense, an indicator of social impact among drug users is the exposure to risks through sexual behavior, which is often analyzed in its gender-dependent specificities. However, this kind of behavior shows a trend towards a uniform distribution among users of hard drugs with sex work being the most evenly distributed behavior between the genres, as well as risk exposure. In this sense, studies with crack cocaine users have discussed the issue of sexual violence and sexual risk behaviors, such as prostitution and irregular use of condoms (Santos Cruz et al. 2013; Inciardi and Surratt 2001). These actions obviously have consequences such as the enhanced prevalence of HIV and hepatitis in subjects using crack cocaine or other routes of administration of cocaine when compared to the general population (FIOCRUZ 2013). It has also been demonstrated that even when they are informed about risk behaviors for HIV, those users still maintain actions that can expose them to HIV infection and spread the virus (de Azevedo et al. 2007). Another study focusing on female drug users described high rates of infrequent condom use (49.3%), exchanging sex for drugs (17.8%), and the prevalence of HIV infection (37%) (von Diemen et al. 2010). These results suggest that drug use is associated with risky sexual behavior in users of both genres.

Domestic violence is another gender issue which is growing in the urban context between drug users. Crack users with episodes of domestic violence experienced more aggressions and beatings as victim outside their home relationship, and also showed greater difficulty in controlling impulsivity in regards to assaulting or hurting someone, more involvement in episodes of physical aggression and they are more charged with violent crimes than crack users without domestic violence. Domestic violence seems to be a possible predictor for the expression of social violence in general.

Drugs and Crime

Even though there was a decline in crime rates in European countries in the postwar period, on the other hand, USA and Brazil showed an increase in crime from the 1980s on which was seen by the number of crimes against property and violent crimes such as robbery, kidnapping, and murder which rose. In 1985, for example, trafficking and drug use caused three times more condemnations than 20 years before, which may have been exacerbated by the impact of repressive and prohibitive policies.

In the mid-1980s, studies on incarcerated populations found a 64% rate for the abuse of alcohol and illegal drugs prior to committing an offense (Jeffrey et al. 1988). Newer studies about the prevalence of mental illnesses in different nationalities corroborated that the abuse of alcohol or other substances was the most frequent diagnosis in samples of defendants and arrested individuals, corresponding to 60% (Telles 2004). In addition, a survey with subjects convicted of murder found that 35.9% of perpetrators were under the influence of alcohol at the moment of crime and 24% of victims had consumed a psychoactive substance (73% alcohol and 27% cocaine) (Chabur et al. 2002).

In 2010, a study conducted in ten major cities in the USA revealed that, on average, about 70% of arrested men had used an illicit drug (Office of National Drug Control Policy 2011), while drug use in general population of males accounted for 11.2% (SAMHSA: Substance Abuse and Mental Health Services Administration 2001). Similar results were shown in Australia in 2008, where a study based on data from ten locations around the country found that 65% of all prisoners, including drug law offenders, tested positive for illicit drug use (Gaffney et al. 2010). In the UK, similar results were found (Gordon et al. 2006).

These data are supported by a meta-analysis with 30 studies showing that the chances of committing injuries and illegal acts in illicit drug users were between 2.8 and 3.8 times greater than for nonusers, thereby strengthening the evidence of an association between drug use and crime. These results indicate that the correlation of drug and crime varies according to the type of substance (Bennet et al. 2008). Table 1 summarizes the risk of committing illegal acts in users of different substances based on the aforementioned study.

Regardless of cultural diversity, drug and crime indicators point to a growing emphasis, which stresses the need for considering the impact of drug policy. If on one hand, drugs are expressed as a factor associated with increased crime, on the other hand, they may act as mediators with an indirect association to factors linked to expression of violence.

Drugs and Homicide

The number of murders is considered as the best international indicator to scale violence anywhere in the world. Naturally, mortality data are shown to be associated to areas with some of the highest level of consumption and commerce of drugs.

Table 1 Relation between substance use and illegal acts

Substance of use	Risk of committing illegal acts
Crack cocaine	Six times higher
Heroin	Three times higher
Cocaine	2.5 times higher
Marijuana	1.5 times higher
Amphetamines	1.9 times higher

However, drug-related deaths may be due to secondary variables such as homicides and crime, and not due to physiological effect. Again crack cocaine, a drug that scores high in violence ranks and that is highly addictive due to its plasma peak, lends itself as a good example. An evidence of this is the way in which drug-related deaths are established: they are more commonly associated with trafficking elements or clashes with police than with the damage caused directly by the drug (Kessler and Pechansky 2008). Mortality among crack cocaine users is estimated to be 20%, and the main cause is homicides (Ribeiro et al. 2004). Obviously, this is a complex discussion when taken in more detail, once it is possible that subjects become additionally exposed to situations of risk of death in terms of impulsivity and impaired decision-making caused by drug effects.

Crime and Crack-Cocaine

A Brazilian cohort who followed crack cocaine users for 5 years showed that 17.6% died during the study and murder was the most prevalent cause (Ribeiro et al. 2006). Therefore, the mortality among crack cocaine users is predominantly due to violent causes (56.5%) and complications related to HIV or cytomegalovirus infection (30.4%) (Ribeiro et al. 2004). Even those users seeking health services to treat their addiction show high rates of mortality (10%) and other negative outcomes like imprisonment (7%) (Laranjeira et al. 1998). In this sense, it is estimated that crack cocaine users are 57.4% more likely to be arrested (Ferreira Filho et al. 2003). What is true is that literature converges in affirming that crack cocaine users have problems regarding crimes.

In relation to violent behavior, it is known that crimes are common in this population: studies have reported that 23–80% of crack cocaine users engage in illegal activities to obtain drugs such as theft and drug trafficking (Kessler et al. 2012), and that 40% of them report the use of weapons for self-protection which, in turn, generates high rates of detention in this population (Santos Cruz et al. 2013). According to some studies, the increased prevalence of violence observed among crack cocaine users when compared to users of inhaled cocaine could be a result of secondary variables such as psychiatric comorbidities, multiple drug use, and socio-demographic features (Vaughn et al. 2010). Nevertheless, a study by Kessler et al. (2012), shows higher rates of legal problems – such as theft and robbery (23%) and assault (32%) – in crack cocaine users, even after controlling for potential factors of confusion, including psychiatric comorbidities. A population study (Narvaez et al. 2014a) demonstrated that individuals with history of crack cocaine use are 30 times more likely to possess blade weapons and firearms when compared with general population, even in a conservative statistical model controlling for sociodemographic variables such as social class and sex. Concerning firearm possession, the indicators showed up to be significant among crack cocaine users, even after adjustment for use of inhaled cocaine, indicating that stronger drugs may trigger some behaviors that are potentially more aggressive.

Substance Abuse and Mental Illness

Epidemiological studies show that individuals who meet criteria for abuse and dependence of alcohol and other drugs (excluding tobacco dependence) are much more likely to meet the criteria for other psychiatric disorders. Depressive symptoms are common among people diagnosed with abuse and substance dependence. Between a third and half of all those with abuse or addiction to opiates and around 40% of those with alcohol abuse or dependence meet the criteria for major depressive disorder at some point in their lives. Substance use is also an important precipitating factor in suicide, which is an important indicator of mental illness and self-inflicted violence. People who consume substances in abusive levels are 20 times more likely to die from suicide than the general population. Although, comorbidity between the disorders related to alcohol and mood disorders is widely recognized, between 20% and 50% of all individuals with disorders related to alcohol also meet the diagnostic criteria for anxiety disorder, with phobias and panic disorder particularly frequent comorbid diagnoses (Sadock et al. 2007).

Crack-Cocaine and Mental Illness

Evidences show that the relevant comorbidity in the context of crack cocaine use is the presence of antisocial personality disorder (Falck et al. 2002). A multicenter study conducted in Brazil has shown an important prevalence of antisocial personality disorder among crack cocaine users, which accounted to 25% in comparison to a 9% rate among nonusers of cocaine. However, that work included only drug users and broader data obtained from population studies have shown a lighter perspective, although, still alarming. Population data from Brazil show the prevalence of antisocial personality disorder in crack cocaine users accounting to 15.4%, while it is found to be 1.3% in nonusers (Narvaez et al. 2014b). From this perspective, illegal acts committed by crack cocaine users may not always be driven directly by the drug (Guimarães et al. 2008). In this sense, some subjects may have a previous comorbidity that, when added to the vulnerabilities of their social scenario, pushes them to the drug context searching for references of social transgression such as the drug dealer which, in the slum, has an authority role.

Crime and Mental Illness in the City

Relating also to the use of drugs and other psychiatric comorbidities, studies have shown a significant association between antisocial personality disorder and violence in patients with schizophrenia. The Community study on the association between mental disorders and violence of the Epidemiologic Catchment Area (Swanson et al. 1990), 8–10% of individuals with schizophrenia reported that they had been violent in the past 12 months, compared to 2% of the general population. The individuals that report drug abuse – presenting or not mental disorders – had a higher prevalence of violent behavior. Another study (Eronen 1995) found that homicidal women had a 70 times increased risk of antisocial personality disorder, alcohol dependence/drugs with personality disorder and schizophrenia with addiction to alcohol/drugs, compared to women in the general population.

Studies comparing criminal activity of people discharged from psychiatric hospitals with people without mental disorders have found higher rates of crime among individuals with severe mental disorders (such as schizophrenia, bipolar disorder, major depression, and delusional disorder). In a study (Tengstrom and Hodgins 2002) of male patients released from forensic hospitals in Canada, Finland, Germany, and Sweden diagnosed with schizophrenia, schizoaffective disorder, or schizophreniform disorder, 22.7% of forensic patients had been convicted of murder. And the evaluation of the group of patients from general psychiatric hospitals found that those with previous conviction had a significantly higher prevalence of diagnosis of substance abuse and antisocial personality disorder. People with substance abuse have a risk 12–16 times more likely to engage in violent behavior than others who do not use substances (Nestor 2002).

For Hodgins (2001), violent behavior is associated with contextual factors – including police repression and availability and adequacy of psychiatric treatment – and individual factors – presence of comorbid personality disorders and disorders related to alcohol and/or drugs, and lack of adherence to treatment and lack of insight regarding the disorder. In this sense, it is that individuals with mental disorders in combination with antisocial personality disorder and substance abuse/chemical dependency are often rejected by clinicians and more oriented towards the criminal justice system than to the mental health system (Taylor and Gunn 1984). However, the impact of antisocial personality disorder in discriminating aggressive behaviors is still controversial (Guimarães et al. 2008).

As we seen above, substance abuse is both a cause and symptom of mental illness. The relationships between geography and substance abuse are complex, including cultural traditions and ease of obtaining drugs. The concentration of substance abuse can create a self-reinforcing cycle as that area becomes more tolerant, and attracts more users. In most of the urban situations, this causes a constant increase in volume and frequency of crimes, and the presence of mental disorders among those users only bring more fuel to an already highly explosive environment. Drug and mental illness may add less inhibitory control in the decision making process, increase impulsivity and emotional instability, leaving the subjects more exposed to violence outcomes.

Drugs and Crime: A Social Hypothesis

Functional impairments and difficulties in social integration are two features which increase the vulnerability of drug users, bringing them closer to violent contexts. There are high levels of unemployment and persistent challenges in achieving and maintaining financial stability among this population since they have limited opportunities and significant barriers to achieve a stable job. This way, drug users are more prone to engage in prohibited activities such as drug dealing, prostitution, and theft, all of this in the search for money to support the addiction. So, these subjects end up associated with negative outcomes to health and welfare such as exposure to infectious diseases and incarceration (Richardson et al. 2015). Crack cocaine, as a hard drug with social consequences, is once again interesting to this discussion. Despite the population of crack cocaine users is composed of young adults in working age, these subjects are disconnected from the formal labor market or unemployed in most cases. Besides, 80% of crack cocaine users reported early substance use – between 16 and 26 years old – suggesting the need to approach this problem from adolescence (Guimarães et al. 2008), a time when they have not entered the labor market yet but had probably interrupted their education process. Studies show that this population is exposed to early traumas without a protective context, which therefore induce a neuropsychological impact that hampers school entrance and will keep hampering afterwards (Narvaez et al. 2012). Therefore, it is observed that these subjects are exposed to various risk situations and social vulnerabilities, which reveals a serious public health problem and contributes to increase violence (Guimarães et al. 2008) and dysfunctional outcomes.

About one-third of drug users report finding regular barriers to employment such as discrimination by the employer, drug testing requirement in the workplace, employers' unwillingness to take on those subjects with criminal records, limited access to education skills or formal jobs, without mentioning homelessness and the lack of previous employment opportunities in the formal labor market. These data come from a study conducted with drug users living in Vancouver – Canada that showed the relationship between income generation and high rates of violence exposure: the chances of being exposed to violence were significantly higher in all prohibited and illegal income generating activities including theft, drug trafficking, activities in the street (car wash, informal recycling and begging), and sex work. The most frequently described kinds of violence were beatings (which account for over two-thirds of the reported episodes), followed by gun attack and sexual violence (Richardson et al. 2015).

In this context of informal income generation, including drugs and prostitution, physical violence sometimes acts as an instrument of regulation and punishment and may even become amplified by police enforcement. In addition, income-generating activities such as begging, informal recycling (*binning*) and cleaning car windows (*squeegeeing)*, are often targets of policing. The need for money is then worsened by the high cost of drugs and by policies hindering the access to opportunities and traditional institutions to drug users, increasing their dependence on alternative economies (Richardson et al. 2015). Being on the street means experiencing

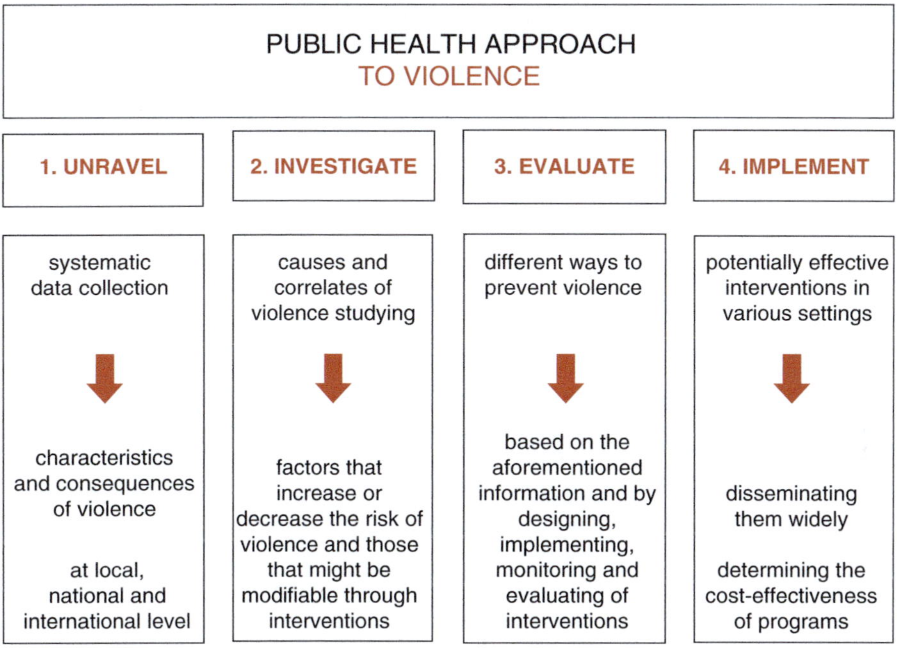

Fig. 1 Public Health Approach to Violence

capitalist society in another way, since the recognition of the subjects is not necessarily related to their productive capacity but to the development of survival strategies appropriate to each urban context. This does not mean not to need money, but to target it as a way of acquiring the essentials for each day.

Public Policies, Violence, Crime, and Drugs

Criminality is a complex phenomenon with multiple biopsychosocial determinants and is part of the human condition from the earliest times, though, is currently considered as a public health problem. The public health approach to violence is based on scientific requirements, which involves four steps (BJS – Bureau of Justice Statistics 2001) (Fig. 1).

The war on drugs: the case of USA

Speaking of measures to fight violence and crime necessarily include a debate on drug-related policies, which is a globally controversial issue covering different approaches. In this sense, the USA is very illustrative in the way it has dealt with and rethought this issue.

The USA spends about 26 billion US dollars a year in its war on drugs whose purpose is, in part, capturing and restraining drug dealers and users. This, therefore, leads to the following scenario: drug users account for 25% of the US prison population; 10% of all arrests are due to drug-related nonviolent offenses; 40% of drug arrests are due to marijuana possession; 20% of prisoners are underage teenagers, and the actual number of young prisoners has risen more than 80% between 1993 and 1997. Although, ethnical patterns in drug use do not differ markedly, nonwhites represent almost 75% of drug users inside prisons. The number of prisoners per capita has more than doubled between 1985 and 1997 largely because of the drug war. The US rate of incarceration for drug offenses (149 per 100,000 inhabitants in 1995) exceeds the rates of most Western European nations for all crimes (for e.g., 95 per 100,000 for France in 1995) which in most part are due to nonaggressive causes (BJS – Bureau of Justice Statistics 2001; MacCoun and Reuter 2001).

In the USA, prohibitionist ideas and repressive approaches prevail in criminal control over drugs, and about 80% of that country's drug policy budget is used in repression (Caballero and Bisiou 2000). The severity of the US legislation mainly concerns the trafficking of imported substances such as cocaine and crack cocaine. The structure of criminal control over drugs in USA was foreseen by the Controlled Substances Act (CSA) in 1970, which in accordance with international conventions, established repression based on criminalizing both possession for personal use and trafficking. Expanding the CSA, other laws were enacted subsequently such as the Controlled Substances Penalties Amendment Act 1984 and the Anti-Drug Abuse Act of 1986, which increased penalties, created new charges, and even predicted the intervention of the army in the fight against trafficking. Despite being inserted in a democratic country, the American criminal drug system is considered very authoritarian when compared to the civil law system, which is adopted by other countries (Rodrigues 2006). The 1990s were then marked by an unprecedented expansion of criminal control over drugs in the USA. More repressive laws were created, which increased the number of controlled substances without, however, succeeding in achieving the goals of reducing demands. Targeting the ideal of abstinence – tough assuming the risk of reducing individual freedom – the Violent Crime Control and Law Enforcement Act of 1994 authorized the establishment of Drug Courts, specialized courts for drug use offenses, whose aim is to submit the indictee for drug use or possession to involuntary treatment. The goal pursued by the American legislator to deter criminals through harsh penalties provided by law, however, has not been achieved, given the figures released by the last world drug report released by the UN, which shows that it has not reduced the offer and demand for drugs, despite the huge number of prisoners.

There is a constant debate about the most effective way to deal with drug issues all over the world and there are significant variations considering a combative approach, a preventive one and decriminalization. In the USA, a still incipient group – although, a growing one – composed by government officials, commentators, and academics have argued that the current policy to offensively prosecute drug dealers and users should be reconsidered. In this sense, the current drug policy is compared with the prohibition of alcohol from 1920 to 1934, and it is argued that abolishing drug laws would eliminate the profit, gangs, and traffickers. Although, without approving such a radical reversal of the national policy on the subject, there are already studies on the possibility of legalizing drugs of abuse in order to reduce the incidence of violent crimes (Sadock et al. 2007). The fact is that there is a growing tendency to view drug abuse as a disease to be treated and prevented rather than being controlled and managed by criminal justice. US initiatives to decriminalize the possession of small amounts of marijuana in the 1970s were followed by actions to legalize the use of marijuana for medical purposes in the 1990s (Grossman et al. 2002).

The legalization process of marijuana that has been taking place in recent years in some countries – especially in some states of the USA – has generated controversy, since with legalization, more people have access to the drug, which can generate an increase in the prevalence of consumption and a more pathological development, such as we observed in relation to alcohol and tobacco. On the other hand, legalization allows reducing the circulation of the drug by the traffic, generating a legal control of the production and sale – and, consequently, a reduction of the social violence that the drug traffic generates.

Treatment and Prevention

The preventive approach of treatment brings out harm reduction strategies, which aim to prevent and treat drug addiction without requiring users to stop their use. This approach recognizes abstinence as a goal and not a condition to receive treatment (Caballero and Bisiou 2000), since the sobriety can be reached in the medium and long term. In this sense, harm reduction constitutes a secondary step in preventing drug abuse. In opposition to a policy with a prohibitionist nature, preventive strategy defends the control on drugs through a preventive and public health approach. It consists in a nonrepressive alternative that may coexist with the prohibition, since it admits a greater medical and social interference when dealing with drugs issues. For example, heroin treatment programs with methadone aimed at increasing the self-esteem of addicted individuals through medical care, avoiding contact with the illicit market, and preventing crimes against property to support heroin addiction. In the mid-1990s, the preventive model in the form of replacement therapy, was already implemented in all European countries, with the Netherlands and Switzerland considered the most advanced countries in this policy. Germany, Spain, Italy, Austria, and Luxembourg are already applying this model for some time. Even the criminal law of most repressive countries in Europe, such as Greece, Finland, and Norway, already have some programs like these, though with a limited coverage, which can be explained by the strong influence of the prohibitionist model in these countries (Rodrigues 2006).

A Theoretical Staging Model to Violence Expression and Risk Exposure Among Drug Users

The dimensional and dynamic diagnosis of substance use is a current trend in addition to specifying therapeutic approaches to these patients. The functional staging of drug users can help to identify a clinical expression associated with the social context in which they live in, besides identifying vulnerabilities that may influence the expression of their disease (Cosci and Fava 2013). We proposed a theoretical staging model (Narvaez 2014) conceptually organized in the criterion of clinical plausibility. The staging model in its entirety systematically identifies several functional patterns among drug users (more specifically crack cocaine), progressively demonstrated in terms of severity, as they impact on admission, progression, prognosis, and treatment. The model is composed of impairments and functional features, associated to clinical prognostic variables, behavioral, neuropsychological, and response to treatment aspects. These elements make up the model, aiming to contemplate the diversity of aspects involved in the use of substances, without incurring a reductionist classification of users, solely focused on the pattern of consumption. Here, we restrict ourselves to exposing the variable linked to violence and exposure to risks.

This model implies the presence of "latent" factors that may exist previously to drug experimentation, such as trauma and mental illness, which can either increase the risk of trying a drug (self-medication), stimulating the progression to substance abuse and also worsens prognosis with violence expression. Profile features such as socioeconomic status and early age experimentation also comprise some of those latent factors and seem to be associated with a profile more prone of changing from drug use initiation to drug abuse.

The mental illness while can drive to the use of drugs, insofar as they are more vulnerable to drug exposure, can also be aggravated or triggered in the course of chronic drug use. If, on one hand, comorbidity with mental illness can indicate a poor prognosis and a greater propensity for exposure to violence, on the other hand, treating it can leverage results, protecting the subject from the impulse to self-medicate through the drug.

The staging phases can denote a progressive character of self-directed aggression, whether a passive aggression – by risk exposure – or a more active, outward, and hetero-aggressive outburst, culminating in a behavior with social impact. We then propose this progression according to the different phases, as follows (Fig. 2):

The clinical expression of poor functionality in drug users may occur through social outwardly expressed markers such as violence, impulsive, and destructive behaviors. These markers, whether or not prior to drug use, are all indicators of a poor prognosistic (Narvaez 2014). Crack cocaine users, which are widely referred to as representative of exposure and promotion of violence, show distinct phases and expressions that run adjacent to violent behavior. Scientific literature demonstrates that crack cocaine is associated with greater exposure to sexual risk behavior, violence, and misdeeds (Inciardi and Surratt 2001; Guimarães et al. 2008; Nappo et al. 1994). In addition to the indexes of impact on society, crack cocaine users may

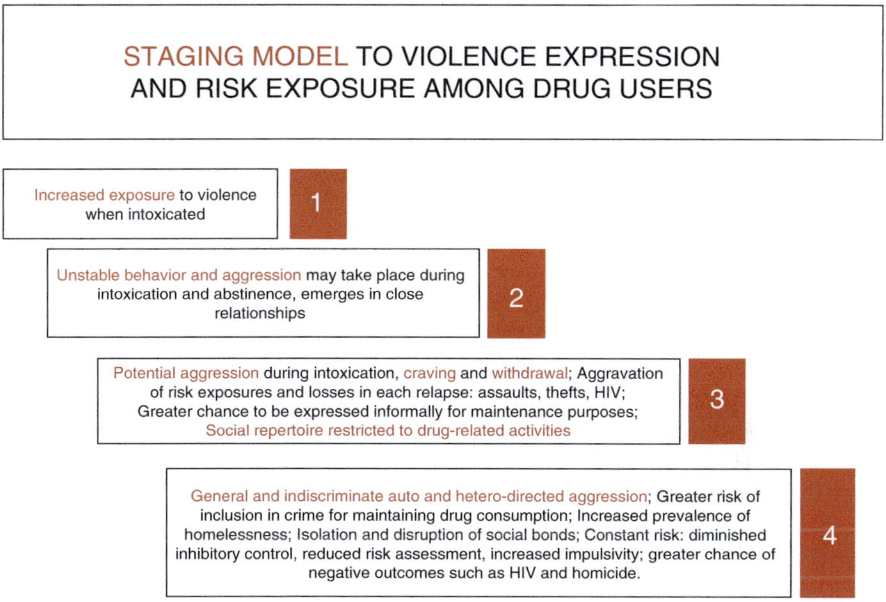

Fig. 2 Data based on: Narvaez J. O amor nos tempos do crack. Porto Alegre: Universidade Federal do Rio Grande do Sul, 2014.

be impaired in multiple critical areas such as: quality of life, ability to work/study, and satisfactory maintenance of social bonds. Besides, those subjects are usually polydrug users and show important rates of psychiatric comorbidities, and all of these aspects that will directly influence the treatment to be offered.

Another aspect we consider important when proposing the staging of substance use is to take into account social functionality, which usually includes history of dysfunctional families with relational and legal problems. It presents a context that is vulnerable to the development of aggressive behavior and also results from explicit social violence and traumas in the primary family core. The latter, in turn, is marked by domestic violence and traumas that drive the search for anesthetic substances and self-medication (Narvaez 2014), a scenario that favors new traumatization and the maintenance of violent behavior, through acting out.

Both social capacity and social losses should be considered when composing the treatment. Social reintegration should be a priority if it is intended to counterbalance the drug use with other life course opportunities. Moreover, the criminality can designate a stigma without a clear distinction between the evolutionary phases of substance dependence, which inhibits the development of therapies directed to the different phases and profiles (Narvaez 2014). It must be remembered that problems associated with violence add more difficulties in terms of resolution of conflicts and strains. Strengthening and rescuing links with the local culture and territory may promote self-esteem and psychosocial possibilities, in addition to impact in treatment. Authors suggest that the state and its health mechanisms must be present in the community preventing illegal access to guns, encompassing the reduction of repressive, or punitive measures in familiar environments (Horta et al. 2010). Anyway, the treatment should not configure an exclusion of these aspects that can make up the user profile and its relationship with society.

Conclusion

Drug use, manifestations of crime and the development of mental illness are socially produced symptoms that indicate vulnerabilities and dysfunctionalities of our current social organization. Considering the history of mankind, none of these are new. However, the way drug use, crime, and mental illness are expressed in today's society indicates the weaknesses of our urban organization. Analyzing them separately is reductionist, since these phenomena are only superficial manifestations of a deeper layer. How do we educate people? Why do drug anesthesia, break with reality or disruption through violence become attractive? What kind of subjectivation is expressed through these manifestations?

Questioning is the answer, since people who grow up in scarce socio-affective contexts tend to charge society's bill, overflowing their suffering through disruptive manifestations. Urban organizations demarcate these distinctions in a more compressed space, where comparison with the status of others is inevitable, even if unconscious. As long as we look only at the symptom and use repressive means, we will still have to contain the social manifestation that does not stop. The etiology of the problem is complex and based on a society with a lot of social inequalities and marked differences. The drug culture demarcates a system of laws and codes, outside normal social laws. More inclusive public policies and investment in education and health are necessary actions to deal with a part of the population that, when feeling neglected, defends itself by bouncing back.

Drug issue shift from self-inflicted violence to violence inflicted to others, possibly replicating beyond the individual realm what they experienced in early contexts of life. In this sense, drug use is strongly associated with manifestations of social violence, however, it is important to warn that the causal chain through which these variables are interconnected is complex and may be multidimensional. Violence can be attributed to secondary variables such as sociodemographic, psychiatric disorders, personality, and family structure. It is not known whether these aspects behave as moderators, mediators, or confounders. Violence may not only be a consequence of substance use, but also when in a broader context, it may consist as a cause to early onset experimentation of drugs. Subsequently, as long as drug actions reduce self-control, this aggressive pattern can be more easily replicated in the society. In this sense, drug and mental illness may add less inhibitory control in the decision making process, increase impulsivity and emotional instability, leaving the subjects more exposed to violence outcomes. Also, despite the obvious impact of drugs in relation to risk behaviors and self-directed violence, it may act as markers for these behaviors and not necessarily as a causal factor. This reinforces the need for longitudinal studies that actively investigate the chains that lead to such associations.

The analysis of violent contexts should include socio-environmental factors that should not be conditioned to the social group where it is manifested. The association of drugs with violence may be an aspect that highlights the current social context. From the therapeutic point of view, discriminating drugs, mental illness, and social suffering is a purely didactic task. Where does one start and where does the other end? Who produces what? Comorbidity with mental illness may boost the pathological progression of drug use but may also be aggravated or triggered in the course of chronic drug use. For treatment purposes, the subject must be integrated. Prevention and treatment of those conditions are broad as multicausal is the problem. A broader contextualization of the subjective and social environments can set the basis for the establishment of policies that address drugs, violence, and mental illness from a public health point of view, thus directing this social demand for treatment areas in the urban context.

Cross-References

- ▶ Development of Mental Health Services in a Major City in Latin America: Santiago
- ▶ Homicide and Suicide in Megacities
- ▶ Mental Health and Urban Life: Experiences from Colombia
- ▶ Prospects and Challenges of Mental Health Services in a Major City in Latin America: Rio de Janeiro

References

Autry AE, Monteggia LM (2012) Brain-derived neurotrophic factor and neuropsychiatric disorders. Pharmacol Rev 64(2):238–258

Beletsky L, Heller D, Jenness SM, Neaigus A, Gelpi-Acosta C, Hagan H (2014) Syringe access, syringe sharing and police encounters among people who inject drugs in New York City: a community-level perspective. Int J Drug Policy 25(1):105–111

Bennet T et al (2008) The statistical association between drug misuse and crime: a meta-analysis. Aggress Violent Behav 13:107–118

BJS – Bureau of Justice Statistics (2001) U.S. Department of Justice, Sourcebook of Criminal Justice Statistics 2000. U.S. Government Printing Office, Washington, DC

Boys A, Marsden J, Strang J (2001) Understanding reasons for drug use amongst young people: a functional perspective. Health Educ Res 16(4):457–469

Caballero F, Bisiou Y (2000) Droit de la drogue. Dalloz, Paris, p 744

Cardoso JB, Goldbach JT, Cervantes RC, Swank P (2016) Stress and multiple substance use behaviors among Hispanic adolescents. Prev Sci 17(2):208–217

Carlini EA et al (2011) III Levantamento Domiciliar Sobre o Uso de Drogas Psicotrópicas no Brasil – 2010. CEBRID – Centro Brasileiro de Informações sobre Drogas Psicotrópicas – Universidade Federal de São Paulo, São Paulo

CEBRID (1997) IV Levantamento Sobre o Uso de Drogas entre Estudantes de 1° e 2° graus em 10 Capitais Brasileiras 1997. http://www.cebrid.epm.br/estudantes/intro.htm; http://www.cebrid.epm.br/estudantes/intro.htm; http://www.cebrid.epm.br/estudantes/intro.htm. Accessed 25 Feb 2017

Chabur JE, Escobar-Cordoba F, Martinez JW, Garzon A, Gomez D (2002) Trastorno de personalidad antisocial en condenados por homicídio en Pereira, Colombia. Inv Salud 4 (2):89–94

Cosci F, Fava GA (2013) Staging of mental disorders: systematic review. Psychother Psychosom 82 (1):20–34

Croxford S et al (2015) Sex work amongst people who inject drugs in England, Wales and Northern Ireland: findings from a National Survey of Health Harms and Behaviours. Int J Drug Policy 26 (4):429–433

de Azevedo RC, Botega NJ, Guimarães LA (2007) Crack users, sexual behavior and risk of HIV infection. Rev Bras Psiquiatr 29(1):26–30

Eronen M (1995) Mental disorders and homicidal behavior in female subjects. Am J Psychiatry 152 (8):1216–1218

Falck RS, Wang J, Carlson RG, Eddy M, Siegal HA (2002) The prevalence and correlates of depressive symptomatology among a community sample of crack-cocaine smokers. J Psychoactive Drugs 34(3):281–288

Fernandes L (1998) O sítio das drogas. Aveiro, Editorial Notícias

Ferreira Filho OF, Turchi MD, Laranjeira R, Castelo A (2003) Perfil sociodemográfico e de padrões de uso entre dependentes de cocaína hospitalizados. Rev Saude Publica 37(6):751–759

FIOCRUZ (2013) Estimativa do número de usuários de crack e/ou similares nas Capitais do País. http://www.icict.fiocruz.br/sites/www.icict.fiocruz.br/files/livreto_epidemiologico_17set.pdf. Accessed 25 Feb 2017

Friedman SR, Cooper HLF, Tempalski B, Keem M, Friedman R, Flom PL, Des Jarlais DC (2006) Relationships of deterrence and law enforcement to drug-related harms among drug injectors in US metropolitan areas. AIDS 20:93–99

Gaffney A et al. (2010) Drug use monitoring in Australia: 2008 annual report on drug use among police detainees, monitoring report no. 9. Institute for Criminology, Canberra

Gordon L et al (2006) The economic and social costs of Class A drug use in England and Wales, 2003/04. In: Singleton N, Murray R, Tinsley L (eds) Measuring different aspects of problem drug use: methodological developments. Home Office Online Report, vol 16, 6th edn. Home Office, London

Grossman M, Chalouopka FJ, Shim K (2002) Illegal drug use and public policy. Health Aff 21 (2):134–145

Guimarães CF, Santos DVV, Freitas RC, Araújo RB (2008) Profile of crack users and factors related to criminality at the detoxication ward at Hospital Psiquiátrico São Pedro. Rev Psiquiatr Rio Gd Sul 30(2):101–108

Hodgins S (2001) The major mental disorders and crime: stop debating and start treating and preventing. Int J Law Psychiatry 24(4-5):427–446

Horta RL, Horta BL, Pinheiro RT, Krindges M (2010) Violent behavior in adolescents and parent–child cohabitation. Rev Saude Publica 44(6):979–985

Inciardi JA, Surratt HL (2001) Drug use, street crime and sex-trading among cocaine-dependent women: implications for public health and criminal justice policy. J Psychoactive Drugs 33 (4):379–389

Jeffrey RW, Pasewark RA, Bieber S (1988) Insanity plea: predicting no guilty by reason of insanity adjudication. Bull Am Acad Psychiatry Law 16(1):35–39

Johnston LD, O'Malley PM, Miech RA, Bachman JG, Schulenberg JE (2015) 2013 Overview: key findings on adolescent drug use. Institute for Social Research, University of Michigan, Ann Arbor

Kessler F, Pechansky F (2008) A psychiatric view on the crack phenomenon nowadays. Rev Psiquiatr Rio Gd Sul 30(2):96–98

Kessler FH, Barbosa Terra M, Faller S, Ravy Stolf A, Carolina Peuker A, Benzano D, Brazilian ASI Group, Pechansky F (2012) Crack users show high rates of antisocial personality disorder, engagement in illegal activities and other psychosocial problems. Am J Addict 21:370–380

Krug EG et al (2002) World report on violence and health. World Health Organization, Geneva

Laranjeira R, Dunn J, Rassi R, Mitsushiro SS, Fernandes MS (1998) Seguimento de usuários de crack após dois anos. Rev Bras Psiquiatr 47(5):233–236
MacCoun R, Reuter P (2001) Drug war heresies: learning from other vices, times, and places. Cambridge University Press, Cambridge
Malta DC, Prado RR, Caribe SSA, Silva MMA, Andreazzi MAR et al (2014) Factors associated with injuries in adolescents, from the National Adolescent School-based Health Survey (PeNSE 2012). Rev Bras Epidemiol 17(Suppl 1):183–202
Markwick N, McNeil R, Small W, Kerr T (2015) Exploring the public health impacts of private security guards on people who use drugs: a qualitative study. J Urban Health 92 (6):1117–1130
Marshall BDL, Fairbairn N, Li K, Wood E, Kerr T (2008) Physical violence among a prospective cohort of injection drug users: a gender-focused approach. Drug Alcohol Depend 97:237–246
Nappo SA, Galduróz JC, Noto AR (1994) Uso do "crack" em São Paulo: fenômeno emergente? Rev Bras Psiquiatr 16:75–83
Narvaez J (2014) O amor nos tempos do crack. Universidade Federal do Rio Grande do Sul, Porto Alegre
Narvaez JC, Magalhães PV, Trindade EK, Vieira DC, Kauer-Sant'anna M, Gama CS, von Diemen L, Kapczinski NS, Kapczinski F (2012) Childhood trauma, impulsivity, and executive functioning in crack cocaine users. Compr Psychiatry 53(3):238–244
Narvaez JC, Jansen K, Pinheiro RT, Kapczinski F, Silva RA, Pechansky F, Magalhães PV (2014a) Violent and sexual behaviors and lifetime use of crack cocaine: a population-based study in Brazil. Soc Psychiatry Psychiatr Epidemiol 49(8):1249–1255
Narvaez JC, Jansen K, Pinheiro RT, Kapczinski F, Silva RA, Pechansky F, Magalhães PV (2014b) Psychiatric and substance-use comorbidities associated with lifetime crack cocaine use in young adults in the general population. Compr Psychiatry 55(6):1369–1376
Narvaez JCM, Pechansky F, Jansen K, Pinheiro RT, Silva RA, Kapczinski F, Magalhães PV (2015) Quality of life, social functioning, family structure, and treatment history associated with crack cocaine use in youth from the general population. Rev Bras Psiquiatr 37(3):211–218
Nestor PG (2002) Mental disorder and violence: personality dimensions and clinical features. Am J Psychiatry 159(12):1973–1978
Office of National Drug Control Policy (2011) ADAM II: 2010 annual report – Arrestee Drug Abuse Monitoring program II. Office of National Drug Control Policy, United States, Executive Office of the President, Washington, DC
Olley BO (2006) Social and health behaviors in youth of the streets of Ibadan, Nigéria. Child Abuse Negl 30:271–282
Rachlis BS, Wood E, Zjanga R, Montaner JSG, Kerr T (2009) High rates of homelessness among a cohort of street-involved youth. Health Place 15:10–17
Ribeiro M, Dunn J, Laranjeira R, Sesso R (2004) High mortality among young crack cocaine users in Brazil: a 5-year follow-up study. Addiction 99(9):1133–1135
Ribeiro M, Dunn J, Sesso R et al (2006) Causes of death among crack cocaine users. Rev Bras Psiquiatr 28:196–202
Richardson LA et al (2015) Socioeconomic marginalization in the structural production of vulnerability to violence among people who use illicit drugs. J Epidemiol Community Health 69:686–692
Rodrigues LBF (2006) Controle penal sobre as drogas ilícitas: o impacto do proibicionismo no sistema penal e na sociedade. Faculdade de Direito da Universidade de São Paulo, São Paulo
Sadock BJ, Sadock VA, Ruiz P (eds) (2007) Kaplan & Sadocks comprehensive textbook of psychiatry, 9th edn. Artmed, Porto Alegre
SAMHSA: Substance Abuse and Mental Health Services Administration (2001) U.S. Department of Health and Human Services. Summary. http://www.samhsa.gov. Accessed 25 Feb 2017
Santos Cruz M, Andrade T, Bastos FI, Leal E, Bertoni N, Villar LM, Tiesmaski M, Fischer B (2013) Key drug use, health and sócio-economic characteristics of young crack users in two Brazilian cities. Int J Drug Policy 24(5):432–438

Swanson JW, Holzer CE, Ganju V, Jono RT (1990) Violence and psychiatric disorder in the community: evidence from the Epidemiologic Catchment Area Surveys. Hosp Community Psychiatry 41(7):761–770

Taylor PJ, Gunn JC (1984) Violence and psychosis. I. Risk of violence among psychotic men. Br Med J 288(6435):1945–1949

Telles LE (2004) Características sociodemográficas, patológicas y delictivas de reos sometidos a examen de Responsabilidad Penal, em el Instituto Psiquiátrico Forense, en Porto Alegre, Brasil, durante el año 2000. Universidad Nacional de La Plata – Argentina, La Plata

Tengstrom A, Hodgins S (2002) Criminal behavior of forensic and general psychiatric patients with schizophrenia: are they different? Acta Psychiatr Scand Suppl 412:62–66

UNODC – United Nations Office on Drugs and Crime, World Drug Report (2015) Sales No. E.15. XI.6. United Nations Publication, New York

UNODC/UN-Women/WHO/International Network of People who Use Drugs (2014) Policy brief on "Women who inject drugs and HIV: addressing specific needs". Vienna International Centre, Vienna

Vaughn MG, Fu Q, Perron BE, Bohnert AS, Howard MO (2010) Is crack cocaine use associated with greater violence than powdered cocaine use? Results from a national sample. Am J Drug Alcohol Abuse 36(4):181–186

Velho G (1998) Nobres e Anjos. Um Estudo de Tóxicos e Hierarquia. Fundação Getulio Vargas Editora, Rio de Janeiro

von Diemen L, De Boni R, Kessler F, Benzano D, Pechansky F (2010) Risk behaviors for HCV- and HIV-seroprevalence among female crack users in Porto Alegre. Arch Womens Ment Health 13 (3):185–191

Werb D, Rowell G, Guyatt G, Kerr T, Montaner J, Wood E (2011) Effect of drug law enforcement on drug market violence: a systematic review. Int J Drug Policy 22(2):87–94

Xiberras M (2007) A sociedade intoxicada. Instituto Piaget, Lisboa

Discrimination in Urban Settings

7

Soumitra Pathare, Antonio Ventriglio, Rubens Dantas, and Dinesh Bhugra

Abstract

Stigma against people with mental illness, mental health professionals, and mental health institutions is common around the globe and has been described for a long time. There is no doubt that stigma takes different forms from active to passive discrimination. Causes of stigma are many and stigma is related to what people see as cause of mental illness. It also depends upon how individuals and the society at large see specific mental illnesses, their causes, interventions, and outcomes. Inevitably, levels of stigma will vary across cultures and how cultures influence presentation of mental illness and how and where help is sought from. Stigma is also related to perceived and real identity of the individuals who experience mental illness. There is often wide-spread stigma against mental illness, those with mental illness, and also those who work in the field. Stigma needs to be differentiated from prejudice and from discrimination. Prejudice is also about cognitive and motivational processes which allow an individual to hold negative views about others or situations. Stigma is about creating the "other" which confirms the identity of individuals who are different from us. Discrimination against people with mental illness on the other hand works within the framework which may well be legally sanctioned. There are a number of international and regional human rights enshrined in law and also agreed

S. Pathare
Centre for Mental Health Law and Policy, Indian Law Society, Pune, India

A. Ventriglio
Department of Mental Health, Department of Clinical and Experimental Medicine, University of Foggia, Foggia, Italy

R. Dantas
Fundação do ABC, Santo André, SP, Brazil

D. Bhugra (✉)
Institute of Psychiatry, Psychology and Neuroscience, King's College London, London, UK
e-mail: dinesh.bhugra@kcl.ac.uk; andrea.livingstone@kcl.ac.uk

N. Okkels et al. (eds.), *Mental Health and Illness in the City*, Mental Health and Illness Worldwide, https://doi.org/10.1007/978-981-10-2327-9_8

between nation states. These incorporate rights such as inability to marry, inability to vote, and not being allowed to inherit property or deal with contractual matters. In this chapter, we will illustrate examples from four areas of discrimination – right to property, right to employment, right to vote, and right to marry – for which considerable evidence is available.

Keywords
Stigma • Discrimination • Human rights • Right to vote • Right to marry • Right to employment

Contents

Introduction

Stigma against people with mental illness, mental health professionals, and mental health institutions reflects negative attitudes and behaviors. Different cultures see mental illness and those with mental illness in different ways. Stigma is related to personal identity as well. Stigma needs to be differentiated from prejudice and from discrimination. Prejudice against an individual or situation is also about cognitive and motivational processes which allow an individual to hold negative views about others or situations. Stigma is about against the "other." This creation of "the other" confirms the identity of individuals who are different from us, and thus stigma validates our identities. How and when people with mental illness and their families and carers experience prejudice and how they express this exposure and how they deal with it are important for a number of reasons. This will influence engagement with services and therapeutic alliances but will also affect service planning and delivery. Discrimination against people with mental illness on the other hand works within the framework which seems to be legally sanctioned. There are a number of international and regional human rights agreements on various individual rights, and these include a number of basic human rights. These incorporate rights such as inability to marry, inability to vote, and not being allowed to inherit property or deal with contractual matters.

Cities demonstrate different rates of psychiatric disorders in comparison with rural areas. However, stigma and discrimination have not been studied separately in cities. Our perception is that stigma may be different as rural families may be more sociocentric, leading to perhaps lower expectations from those with mental illness. However, legal system does not make any distinction between rural and urban areas.

In this chapter, we will illustrate examples from four areas of discrimination – right to property, right to employment, right to vote, and right to marry – for which considerable evidence is available.

Attributes of Right to Property

The ownership of property is important to individuals for its economic benefits, a sense of ownership and security, and its contribution to the social status of the owner. Historically, in some countries only landowners enjoyed the right to vote thus the right to property and vote were strongly interlinked. In many countries, women never had full ownership rights in respect of property. In patriarchal societies, male members of family are placed as preferred heirs in law of succession. In Kerala, for example, till the middle of the nineteenth century property was inherited through the matrilineal line and polyandry was the norm in some groups and consequently women had the power in the family. The right to property is therefore an important right to enable individuals to participate fully in social and community life. In rural areas, land has a particular value especially as it is handed down from generation to generation.

The concept of right to property is not limited to holding a property in one's name. Right to property in its effective sense means the right to acquire, hold, enjoy, and dispose of the property based on one's own wishes. Thus, right to consent and contract becomes a significant part of this right. Right to property also means the right to receive economic benefits out of it. A person can acquire/ own property by making a contract to purchase the property or by succession. Similarly, the freedom to dispose of the property can be exercised by entering into a contract to sell, by gifting it, or by making a will. There are additional problems in our understanding of differences based on legal reasoning. This can be attributed to variations in terms used in law which vary from *insanity, madman, mental derangement or imbecility, mental disorder, mental disturbance, lunatic, usual idiocy (*sic*), madness, mentally incompetent, and idiots* to name a few. In most cases, these terms are not defined in the law and are left open to interpretation by both the lawyers and clinicians which in itself further contributes to discrimination. Some of the terms are a result of colonial rule, and laws have not been changed for centuries. The use of these pejorative terms in legislation also play an important part in increasing stigma against persons with mental health problems and, conversely, stigma associated with mental health problems is likely to mean that these terms are often interpreted in a manner which is detrimental to the interests of persons with mental health problems, thereby setting up a vicious cycle.

Definitions: it may be helpful to clarify some of the definitions in the context of various conventions and treaties before we discuss the areas of discrimination.

A "Treaty" is an international written agreement between the sovereign states. It is governed by international law. By "Ratification," a country accepts to be legally bound by the principles of treaty. By "Accession," a country accepts an offer to become a party to a treaty already negotiated and signed by other states. It has the same legal effect as ratification. By "Succession" of treaty, a successor country consents to be legally bound by the obligations of predecessor country to the international treaty and has same effect as ratification.

Social Model of Disability Versus a Medical Model of Disability

The medical model of disability essentially focuses on individual bodily impairments, and disability is seen as a deficit of functioning due to this physical or mental impairment. In stark contrast, the social model of disability outright rejects this conceptualization of disability. Social model theorists conceptualize "impairment" as a loss of functioning of some physical or mental organ or the loss of a body part. However, they conceptualize "disability" as a society that discriminates, disadvantages, and excludes people with impairments, as it does not make appropriate accommodations and gives preference to those without impairments (Anastasiou and Kauffman 2013). Anastasiou and Kauffman (2013) define it: "*Disability* is the loss or limitation of opportunities to take part in the normal life of the community on an equal level with others due to physical and social barriers" (p. 442). The Convention on Rights of Persons with Disabilities (CRPD) (UN General Assembly 2007) also uses a social model to understand disability. It too conceptualizes disability as a social construct and says "disability results from the interaction between persons with impairments and attitudinal and environmental barriers that hinders their full and effective participation in society on an equal basis with others" (preamble *e*, p. 1).

International and Regional Human Rights Conventions and the Right to Property

The Universal Declaration of Human Rights (UDHR) (UN General Assembly 1948). Article 17 of the UDHR recognizes the right of the individual to own property when it states that "everyone has the right to own property alone as well as in association with others" and "no one shall be arbitrarily deprived of their property." The right to property is not specifically protected in either the International Covenant on Economic, Social and Cultural Rights (ICESCR) (UN Committee on Economic, Social and Cultural Rights (CESCR) 2006) or the International Covenant on Civil and Political Rights (UN General Assembly, International Covenant on Civil and Political Rights 1966) (ICCPR). However, other rights such as legal capacity which affect the right to ownership of property are protected by these conventions. Later international human rights instruments protect the right to property more specifically. The right to own property as well as to inherit property is protected under Article 5 of the International Convention on Elimination of All Forms of Racial Discrimination (UN General Assembly 1965). Article 15 of the Convention on Elimination of All

Forms of Discrimination against Women (UN General Assembly 1979) also recognizes women's right to enter into contracts. Article 16 of the same Convention recognizes women's right to ownership, enjoyment, and disposal of property on an equal basis with others. The Convention on Rights of Persons with Disabilities (CRPD) (2014) specifically calls upon countries to "take all appropriate and effective measures to ensure the equal right of persons with disabilities to own or inherit property, to control their own financial affairs and to have equal access to bank loans, mortgages and other forms of financial credit, and shall ensure that persons with disabilities are not arbitrarily deprived of their property" [Article 12 (5) of the CRPD] (Convention on the Rights of Persons with Disabilities 2014).

Regional human rights conventions do recognize the right to property. For example, Article 1 of the European Convention on Human Rights (Council of Europe 1950) recognizes a person's "right to peaceful enjoyment of possessions." Similarly, Article 23 of the American Declaration on the Rights and Duties of Man (American Declaration 1948) states that "Every Person has the right to own such private property as meets the essential needs of decent living and helps to maintain the dignity of the individual and of the home." Article 14 of the African Charter on Human and Peoples' Rights (Organization of African Unity (OAU) 1981) explicitly states "The right to property shall be guaranteed."

Although in theory, most member states recognize the right of persons with mental health problems to succession (inheritance); however, it has been shown that over one third do not recognize the right of persons with mental health problems to enter into contracts or write their own will and testament (Bhugra et al. 2016a). The apparent nondiscrimination in the right to succeed or inherit property is of little benefit to persons with mental health problems as they are unable to enter into contracts (to sell/dispose of the property) or give it away in a will or testament. Hence, effectively they are unable to enjoy the inheritance in many member states.

Laws dealing with contracts in many countries use the term "unsound mind" or "sound mind" to debar or allow, respectively, persons from entering into valid contracts. A plain reading of the text does not necessarily equate the term "unsound mind" with mental illness; however, courts in countries have equated these two terms and treated them as interchangeable (Goldman 1990) and thus led to denial of legal capacity for persons with mental health problems. In its General Comment 1 on Article 12 of the CRPD, the Committee on Rights of Persons with Disabilities has categorically stated that "The CRPD (Article 12) now makes it clear that 'unsoundedness of mind' and other discriminatory labels are not legitimate reasons for the denial of legal capacity (legal standing and legal agency)," and it further goes on to reiterate that "Under article 12 of the Convention, perceived or actual deficits in mental capacity must not be used as justification for denying legal capacity" (Convention on the Rights of Persons with Disabilities 2014).

There is little doubt that although some laws may result from stigma, many result from lack of knowledge, whereas others are meant to protect the most vulnerable individuals in the society thereby going to an extreme assuming that all individuals with mental illness have similar problems in severity as well as type. Policy makers need to take into account that nondiscrimination is guaranteed under international law (e.g., ICESCR) and their countries have an obligation to ensure that laws do not discriminate against individuals with mental health problems. Thus, it is critical to understand how persons with mental health problems may be enabled and indeed empowered to enter into valid contracts and enabled so that they can exercise their testamentary capacity. The CRPD recommends that a supported decision-making paradigm be developed and provided rather than current prevalent substituted decision-making paradigm in country legislation (Convention on the Rights of Persons with Disabilities 2014).

Presently, in most member states, a person with mental disability is routinely judged as not being competent in making decisions and a guardian is appointed to make decisions on their behalf. The major problem with this substituted decision making paradigm is that the guardian is assumed to have best interests at heart of the individual with mental illness and also in the process excluding individual from the decision-making process.

Right to Marry

Marriage is one of the oldest social institutions which is culturally approved and supported although types and contracts within marriage may well change across cultures. In many cultures, there is an almost universal expectation that individuals will get married so that they can procreate and bring family up together as a unit. Depending upon whether the individuals come from egocentric societies or sociocentric societies or from sex-positive or sex-negative societies their own expectations from marriage as an institution may well differ from those of the kinship or society they belong to. In sociocentric societies, often the marriage is seen as occurring between families and not individuals and these are most likely in rural areas rather than urban areas in many countries. Marriage of individuals who have mental illness or those who go on to develop mental illness once married may face legal obstacles in specific ways. "Married" status is associated with greater longevity while those unmarried, divorced, or widowed people particularly especially men show higher than expected mortality rates (UN General Assembly, International Covenant on Economic, Social and Cultural Rights 1966). There is no doubt that realization of right to marry by a person is not an important part of one's life choice but also an exercise in personal liberty. It is clearly evident that every individual's right to marry or remain married is regulated by his or her country's family codes, civil codes, marriage laws, or divorce laws. However, social expectations also play a role. For example, in some countries parents want to get their children with mental illness married off to ensure there is a care-giver after their death. Often this is seen in cases of intellectual disability.

Internationally, like the right to property, right to marry, right to remain married, and dissolution of marriage have been enshrined in several human rights instruments. These include Article 16 of Universal Declaration of Human Rights (1948) and Article 10 of International Covenant on Economic, Social and Cultural Rights (1966) (ICESCR) (Bhugra et al. 2016b). For persons with mental disabilities, provisions of Article 23.1(a) of United Nations Convention on the Rights of Persons with Disabilities (2007) (UN General Assembly 2007) (CRPD) explicitly state that the state parties (countries) shall ensure "The right of all persons with disabilities who are of marriageable age to marry and to found a family on the basis of free and full consent of the intending spouses is recognized." State parties (countries) are also required to take "effective and appropriate measures to eliminate discrimination against persons with disabilities" in relation to protection of right to marry and found a family. In the context of Article 4 (b) of CRPD, these effective and appropriate measures include legislative measures "including legislation, to modify or abolish existing laws, regulations, customs and practices that constitute discrimination against persons with disabilities." Taken together, these obligations require member states to bring their marriage regulating laws in line with the principles of CRPD, so that persons with mental health problems or disabilities can marry on the basis of free and full consent of the intending spouses.

In a study of domestic laws in 193 countries to explore whether laws affected the rights of people with mental illness with respect to marriage, it was noted that 37% of countries explicitly prohibit marriage by persons with mental health problems. In 11% (21 countries) the presence of mental health problems can render a marriage void or can be considered grounds for nullity of marriage. In 6% (11 countries) persons with mental health problems are allowed to marry subject to approval from others. Of these 11 countries, 4 countries allow marriage with permission from appropriate court or family council; in 4 countries persons with mental health conditions may marry subject to consent from a parent, guardian, or custodian; and in 3 countries marriage is only permitted subject to approval from an expert or an institution. Thus, in many countries basic human rights related to marriage are being flouted (Lauber and Rössler 2007). Thus, discrimination against people with mental illness is rife in many countries. The increased likelihood of getting divorce on grounds of mental illness is another major problem. This may be used in domestic disputes to find an easy solution thus isolating and discriminating further against individuals with mental illness. The implications of such discrimination are as follows:

Society holds "disapproving and devaluing attitudes" towards those who are divorced because of mental illness (Thara et al. 2003a). This is more glaring in the case of women who are thus more likely to be ridiculed by the society if nuptial ties are broken due to the woman's mental illness. In India, Thara et al. (2003b) found that women and family members experience extreme stigma and distress due to divorce because of mental illness than due to mental illness alone.

Mental health care in the community with traditional gender roles and gender role expectations may well contribute to social exclusion as much as legal provisions of divorce on the basis of mental illness. Marriage laws however institutionalize the stigma of mental health problems and lead to downstream violation of multiple human rights particularly for women with mental health problems.

Although provisions in the law regulating annulment or divorce on the basis of mental health problems are theoretically gender-neutral in most countries, in practice these appear to have more antifemale effects in some countries. As noted above, use of nonstandardized terms to describe mental health conditions can contribute to individual interpretation as well as add to further stigma as this indicates stereotypes. While decision-making capacity is task-specific, ill-defined terms such as "unable to reason"and "unsound mind" often get equated with mental health condition. Committee on the Rights of Persons with Disabilities has expressly stated in General Comment No. 1 (Convention on the Rights of Persons with Disabilities 2014) that "perceived or actual deficits in mental capacity must not be used as justification for denying legal capacity." Moreover, use of terms like "madman," "demented," or "deranged" perpetuate stigma of mental illness. Laws in some countries require an expert such as a trained psychiatrist to decide whether a person with mental health problems can perform and fulfill duties in a marriage.

Right to Vote

In any democratic country, the right to vote is an important right. This indicates and signifies a right to choose and select and participate in the civil society as a full citizen. This right is also enshrined in international human rights treaties. The right of "everyone" to take part in the political process and elections is based on universal and equal suffrage. Although these International Conventions have been ratified by large majority of United Nations member states, their application across the globe is by no means universal and remains incredibly patchy.

In any civil society, democratic principles are based on individual's right to participate in the decision-making processes which may affect all citizens. Thus, each vote has the same value irrespective of gender, religion, or sexual orientation. This principle of voting rights dates back to ancient times. Although there have been changes to the actual voting systems such as proportional representation, the basic principle of active democratic participation through casting the vote remains the same. The right to vote is an important right signifying freedom and full citizenship and is protected by international human rights treaties. Article 21 of the Universal Declaration of Human Rights (1948) (UDHR) recognizes the right of "everyone" to take part in the political process and elections should be based on universal and equal suffrage. This right to vote and participate in the political process is also protected under Article 25 of the International Covenant on Civil and Political Rights (1966) (ICCPR). More recently, the Convention on Rights of Persons with Disabilities (2007) (CRPD) in Article 29 reinforces the right of disabled persons (including those with mental health problems and disabilities) to participate in political and public life on an "equal basis with others." Article 29 of the convention also specifically talks about providing persons with disabilities the opportunity and the right to vote and be elected. These International Conventions have been ratified by large majority of United Nations member states. The ICCPR adopted by the United Nations General Assembly in 1966 has been ratified by 167 countries, and the CRPD adopted by the United Nations General Assembly in 2006 has been ratified by 157 countries to date. There is a responsibility on UN member states ratifying these conventions to ensure that citizens in their countries are able to enjoy these treaty based rights by taking "all appropriate legislative, administrative and other measures for the implementation of the rights recognized in the present Convention" [Article 4 (1) (a) of the CRPD]. This recognizes that rights protected in international human rights documents will only be realized if these are translated into domestic legislation by the member states ratifying these conventions and treaties.

Bhugra et al. (2016c) found that after studying the laws of 193 member states only 11% ($n = 21$) placed no restrictions on the right to vote by persons with mental health problems. Thirty-six percent member states ($n = 69$) deny all persons with any mental health problems a right to vote without any qualifier. Again varying and stigmatizing terminology is used in legislation to describe persons with mental health problems, e.g., insanity, weakness of mind, unsound mind, lunatic, mental disability, mentally ill, psychiatric patient, "certified insane," "medically proven total mental incapacity," "mental derangement," "notoriously insane," mental incapacity, mentally incompetent, "lost his mind," "demented," seriously weakened mental state, mentally deficient, "insane or imbecile," certified to be insane, and "mental ineptitude." In all the above cases, these terms are not defined in the law and are left open to interpretation. Another 11% member states ($n = 21$) denied the right to vote to detained persons; of these nine member states specifically denied the right to vote to persons who were detained under the mental health law, while the remainder denied the right to vote to all those who were interdicted or judicially interdicted. These authors reported that in 29% member states ($n = 56$) a restriction on right to vote for persons with mental health problems required an order (declaration) from a court or magistrate. Thirty-four percent of UN member states ($n = 57$) who have ratified the ICCPR and 36% of member states ($n = 57$) who have ratified the CRPD deny persons with mental health problems the right to vote and are failing their obligation under these international conventions.

Sometimes even within the same country different states disallow voting rights. Hessler and Buchanan 2012 looked at the laws of 50 US states and found that 19 states (40%) denied all persons with mental illness the right to vote on the basis of mental illness and 12 states did so on the grounds of incompetence or lack of capacity.

The chief argument often made for disallowing persons with mental ill-health from voting is the need to maintain integrity of the electoral process, which requires a certain level of informed choice by individuals who have the capacity to make such choices. Often there is a general stereotypical assumption that all individuals with mental illness lack capacity to make an informed choice and thus should not be allowed to participate in the electoral process. However, it is important to remind ourselves that capacity is task-specific (i.e., while a person may not necessarily have the capacity to make decisions on their finances or treatment, but may retain the capacity to make other decisions, including voting choices) and varies over time.

Right to Employment

Work is an important aspect of human functioning. In addition to providing financial security work and employment also lead to social connections, self-esteem, and improved self-image. Dunn et al. (2008) in a qualitative study found that there were two streams for understanding the positive role of work in an individual's life – first of these was personal meaning and the second was the understanding that work promotes recovery. Developing a sense of pride as well as financial reward was seen as important results of work. Furthermore, right to work and employment is indispensable for social integration of persons with mental health problems.

Mental health shares a two-way relationship with work and social integration. Self-esteem and social identity are closely tied with the ability to work and earn livelihood (Tiggemann and Winefield 1984; Winefield and Tiggemann 1990). Work also provides an opportunity to integrate meaningfully with the society. Yet, persons with mental disabilities face direct discrimination by employers and colleagues, and indirect discrimination through structural disadvantages such as lack of incentives to foster employment of persons with mental disabilities coupled with a generalized neglect by policy makers (Stuart 2006). Universally, participation of persons with mental disabilities in the labor market has been strikingly low.

Stigma of mental illness, whether internalized or not, has been known to pose the most prominent individual-level as well as structural barrier to the entry of persons with mental disabilities in competitive labor market and earn wages on an equal basis with others. It is entirely feasible that in joint families in rural areas where individuals with mental illness are not the sole earners, even menial tasks keep them occupied and grant small income.

International law has adopted a human-rights-based perspective for promoting participation of persons with mental disabilities in labor market. In particular, Article 23 of the Universal Declaration of Human Rights (1948); Article 6 and Article 7 of the International Covenant on Economic, Social and Cultural Rights (1976) (ICESCR), and Article 27 of United Nations Convention on the Rights of Persons with Disabilities (2006) (CRPD) protect the rights of persons with mental disabilities to work and employment. CRPD holds historical significance for bringing paradigm shift in the perception of persons with mental disabilities from objects of charity to people with equal human rights. Article 27 of the CRPD expressly proclaims that states parties recognize that persons with mental disabilities have equal rights to work in open, inclusive, and accessible environment and earn livelihood by the work of their free choice. It requires states parties to ensure that persons with mental disabilities are protected from harassment and slavery in the labor market and promote realization of right to work through "legislation and other appropriate measures" (paragraph 1 of Article 27). Moreover, equal access to employment opportunities for persons with mental disabilities also reinforces their equal recognition before the law and enjoyment of legal capacity as per Article 12 of CRPD.

Although international conventions and instruments protect these rights for persons with mental health problems, the realization of these rights remains a prerogative of domestic laws. Member states which ratify international conventions are legally bound to amend and/or abolish existing noncompliant legislation (see Article 4 (1) of CRPD). Unless domestic laws are made compliant to the CRPD and other international conventions, the expressive power of these human rights treaties is undervalued.

These steps include being aware of workplace environment, learning from motivations of organizational leaders, learn from others, understand and explore opportunities, and look at sources and resources which are available and make a start. Nardodkar et al. (Bhugra et al. 2016d) reported that discrimination against people with mental illness is at many different levels from being excluded at the time of short-listing to being excluded at place of work where they may not receive any legal support. Fourteen percent (28 member states) of countries do not have laws which explicitly prohibit discrimination on the grounds of mental disability at the time of recruitment, and another 13% (26 member states) of countries have laws with conflicting provisions. In 30% (58 member states) of the countries, there are no explicit or implied provisions in the law for providing reasonable accommodation to persons with disabilities, including mental disabilities.

Negative discrimination in the laws leads to denial of equal opportunities to access economic resources. For instance, mention of "mental impairment" in the interpretation of the term "disability" in laws assures that persons with mental health problems have equal access to disability associated rights protected by law in a particular state. Laws of nearly 56 countries which do not explicitly define disability or mention "mental impairment/disability/illness" in the definition of disability leave scope for uncertain interpretation, thus leading to likelihood of unequal access to rights and resources for persons with mental disabilities and mental illness. One third of countries deny equality to persons afflicted with mental illness by not prohibiting discrimination on the ground of mental disability during recruitment. Laws of another 61% countries which do not explicitly prohibit discontinuation of employment on the basis of health grounds including mental disability pose barriers to retaining employment. Work reduces incapacity and fosters recovery from mental illness. Therefore, provisions of laws to give access to reasonable accommodations at workplace also play an important role in rehabilitation process. Services required to deliver support for management of illness at workplace, networking of mental health care professionals with employers, can be adopted as part of reasonable accommodations at workplace. It has been argued that affirmative action strategies such as a quota system, giving incentives to employers to recruit persons with mental health problems, should be used to eliminate stigma of mental illness and can end discrimination towards persons afflicted with mental illness (Hessler and Buchanan 2012).

It is important to emphasize that laws in low and lower middle income countries are less likely to protect against inequality, discrimination, and human rights violations. In many countries, these laws date back to colonial times and have not changed at all. Often, protection of human rights through legislative and policy measures requires government spending on mental health systems and parallel community and human resources.

Conclusions

Just looking at four areas of discrimination, it is clear that only about 10% countries have laws which are nondiscriminatory against people with mental illness. Although we did not look at specific discrimination issues between rural and urban areas, it is imperative that further qualitative work is undertaken to explore the impact of these laws in various settings.

The dialogue with policymakers and stakeholders are critical in this case. Policymakers often argue that ICESCR rights such as right to work are subject to progressive realization (see General Comment No. 18, paragraph 19). However, 50 years after the ICESCR was adopted, countries have failed to amend or replace legislation which fail to halt discriminatory practices. Policymakers must be held responsible and regularly reminded that discrimination against people with mental illness is unlawful and legal system needs to change to ensure that those with mental illness can lead a full and fulfilling life. A majority of countries are failing in their obligations under international human rights law when they fail to provide legal protection against discrimination in employment, inheriting property, for right to vote, and right to marry. The jumble of definitions must be sorted out as a matter of urgency.

Cross-References

▶ Mental Health Care in 11 Major Cities Over Two Decades

References

American Declaration of the rights and duties of Man. Adopted by the ninth international conference of American States, Bogotá, 1948. Available at: https://www.cidh.oas.org/Basicos/English/Basic2.American%20Declaration.htm. Accessed 11 Sept 2016.

Anastasiou D, Kauffman JM. The social model of disability: dichotomy between impairment and disability. J Med Philos; 2013;38(4):441–459. doi: 10.1093/jmp/jht026.

Bhugra D, Pathare S, Joshi R, Nardodkar R, Torales J, Tolentino EJ, Dantas R, Ventriglio A (2016a) Property inheritance and contract and persons with mental illness. Int Rev Psychiatry 28:402–408

Bhugra D, Pathare S, Nardodkar R, Gosavi C, Ng R, Torales J, Venriglio A (2016b) Legislative provisions related to marriage and divorce of persons with mental health problems: a global review. Int Rev Psychiatry 28:386–392

Bhugra D, Pathare S, Gosavi C, Ventriglio A, Torales J, Castaldelli-Maia J, Toelntino EJ, Ng R (2016c) Mental illness and the right to vote: a review of legislation across the world. Int Rev Psychiatry 28:395–399

Bhugra D, Ventriglio A, Pathare S (2016d) Freedom and equality in dignity and rights for persons with mental illness. Lancet Psychiatry 3(3):196–197

Convention on the Rights of Persons with Disabilities (CRPD/C/GC/1), 11 Apr 2014

Council of Europe, European Convention on Human Rights, as amended by Protocol 14, 4 Nov 1950, ETS 5. Available at: http://www.refworld.org/docid/3ae6b3b04.html. Accessed 27 June 2016

Dunn E, Wewiorski N, Rogers E (2008) The meaning and importance of employment to people in recovery from serious mental illness: results of a qualitative study. Psychiatr Rehabil J 32 (1):59–62

Goldman N (1990) Mortality differentials by marital status: an international comparison. Demography 27(2):233–250. doi:10.2307/2061451

Hessler K, Buchanan A (2012) Equality, democracy, and the human right to health care. In: Medicine and social justice: essays on the distribution of health care. Oxford University Press, Oxford. p 97
Lauber C, Rössler W (2007) Stigma towards people with mental illness in developing countries in Asia. Int Rev Psychiatry 19(2):157–178
Organization of African Unity (OAU), African Charter on Human and Peoples' Rights ("Banjul Charter"), 27 June 1981, CAB/LEG/67/3 rev. 5, 21 I.L.M. 58 (1982). Available at: http://www.refworld.org/docid/3ae6b3630.html. Accessed 27 June 2016
Stuart H (2006) Mental illness and employment discrimination. Curr Opin Psychiatry 19 (5):522–526
Thara R, Kamath S, Kumar S (2003a) Women with schizophrenia and broken marriages – doubly disadvantaged? Part II: family perspective. Int J Soc Psychiatry 49(3):233–240
Thara R, Kamath S, Kumar S (2003b) Women with schizophrenia and broken marriages – doubly disadvantaged? Part I: patient perspective. Int J Soc Psychiatry 49(3):225–232
Tiggemann M, Winefield AH (1984) The effects of unemployment on the mood, self-esteem, locus of control, and depressive affect of school-leavers. J Occup Psychol 57(1):33–42
UN Committee on Economic, Social and Cultural Rights (CESCR), General comment no. 18: the right to work (Art. 6 of the Covenant), 6 Feb 2006, E/C.12/GC/18. Available at: http://www.refworld.org/docid/4415453b4.html. Accessed 1 July 2016
UN General Assembly, Convention on the Elimination of All Forms of Discrimination Against Women, 18 Dec 1979, United Nations, Treaty Series, vol 1249, p 13. Available at: http://www.refworld.org/docid/3ae6b3970.html. Accessed 27 June 2016
UN General Assembly, Convention on the Rights of Persons with Disabilities: resolution/adopted by the General Assembly, 24 Jan 2007, A/RES/61/106. Available at: http://www.refworld.org/docid/45f973632.html. Accessed 27 June 2016
UN General Assembly, International Convention on the Elimination of All Forms of Racial Discrimination, 21 Dec 1965, United Nations, Treaty Series, vol 660, p 195. Available at: http://www.refworld.org/docid/3ae6b3940.html. Accessed 27 June 2016
UN General Assembly, International Covenant on Civil and Political Rights, 16 Dec 1966, United Nations, Treaty Series, vol 999, p 171. Available at: http://www.refworld.org/docid/3ae6b3aa0.html. Accessed 27 June 2016
UN General Assembly, International Covenant on Economic, Social and Cultural Rights, 16 Dec 1966, United Nations, Treaty Series, vol 993, p 3. Available at: http://www.refworld.org/docid/3ae6b36c0.html. Accessed 27 June 2016
UN General Assembly, Universal Declaration of Human Rights, 10 Dec 1948, 217 A (III). Available at: http://www.refworld.org/docid/3ae6b3712c.html. Accessed 27 June 2016
Winefield AH, Tiggemann M (1990) Employment status and psychological well-being: a longitudinal study. J Appl Psychol 75(4):455–459. doi:10.1037/0021-9010.75.4.455

Loneliness and Single-Person Households: Issues of *Kodokushi* and *Hikikomori* in Japan

8

Takahiro A. Kato, Naotaka Shinfuku, Norman Sartorius, and Shigenobu Kanba

Abstract

Traditionally, Japanese life used to be based on village communities and was a society that highly valued neighborhood relations, but in modern urban life, it is becoming increasingly difficult to establish and maintain close human connections, especially in the city. This creates a situation in which people are likely to become lonely. In other words, people who had hitherto lived in traditional family groupings and communities have increasingly come to spend more time as lone individuals with many actually living alone in the city. Some people living in single-person households in the city tend to feel loneliness, and sometimes this develops into a variety of mental health and psychiatric illnesses based on loneliness. In this chapter, we introduce issues pertaining to single-person households in Japan, especially focusing on kodoku-shi (lonely death) and hikikomori (social withdrawal). Regarding hikikomori, we discuss its psychopathology and propose a stage- and/or condition- oriented therapeutic approach. These issues related to urban single-person households are not merely Japanese or one nation issues but are in fact increasingly global phenomena and as such require breakthrough measures based on worldwide research.

Keywords

Loneliness • Isolation • Kodoku-shi (lonely death) • Hikikomori • Amae • Shame

T.A. Kato (✉) • S. Kanba
Department of Neuropsychiatry, Graduate School of Medical Sciences, Kyushu University, Fukuoka, Japan
e-mail: takahiro@npsych.med.kyushu-u.ac.jp; t.a.kato@live.jp

N. Shinfuku
Department of Social Welfare, School of Human Sciences, Seinan Gakuin University, Fukuoka, Japan

N. Sartorius
Association for the Improvement of Mental Health Programmes (AMH), Geneva, Switzerland

N. Okkels et al. (eds.), *Mental Health and Illness in the City*, Mental Health and Illness Worldwide, https://doi.org/10.1007/978-981-10-2327-9_9

Contents

Introduction

As modernization and urbanization have progressed, people who had hitherto lived in traditional family groupings and communities have increasingly come to spend more time as lone individuals with many actually living alone. This situation of single-person households tends to induce isolation and/or loneliness and sometimes leads to a variety of mental health and psychiatric issues. Here, we introduce issues pertaining to single-person households in Japan, especially focusing on kodoku-shi (lonely death) and hikikomori (social withdrawal).

The Elderly and Kodoku-Shi (Lonely Death)

In Japan's past, there was a strong custom of adult children living with their elderly parents in extended families, but due to urbanization and the nuclear family, many elderly couples and individuals have come to live alone. Often there is no problem while both partners are healthy, but after the loss of a partner, elderly people tend to fall into a solitary state. In most cases, all is good while activities of daily life (ADL) are intact, but as elderly people living alone develop cognitive decline, dementia and/or physical disability, such single elderly are forced into loneliness with many having their lives fall apart.

The problem of solitary elderly is a serious social issue. Often with almost no social contact with anyone, weeks or sometimes months pass and even in the case of death, nobody notices for a few weeks or even a few months. In Japan, this outcome of a sad lonely death is called "kodoku-shi" (Brasor 2014; Nobel 2010). According to data released by the Tokyo Metropolitan Medical Examiner's Office, the number of deaths at home of people over 65 living alone in the Tokyo metropolitan area (the 23 wards of Tokyo) is increasing every year from 1451 in 2003 to 2869 in 2013 (Japan Cabinet Office 2015). This constitutes a doubling over the last decade.

Furthermore, in an opinion poll conducted by the Japan's Cabinet Office in 2012, the proportion of elderly aged 60 years or over who considered dying alone to be an issue of concern (very concerning + somewhat concerning) was less than 20% overall, while the proportion was over 40% for elderly single-person households (Japan Cabinet Office 2015). Such data strongly suggests that loneliness is becoming a close concern among elderly people especially for single-person households in urban area of Japan.

In addition, kodoku-shi occurs among not only the elderly but also middle-aged people. Japan's Urban Renaissance Agency (URA) reported that the total number of kodoku-shi cases of URA apartments occupants who were discovered more than a week after death was 194 in 2013, and among these cases, 65 people were less than 65 years old (Japan Cabinet Office 2015). Thus connected with such assisted living, it is important to note the social issue of middle-aged and elderly people who had been receiving financial social welfare protection subsequently succumbing to kodoku-shi once such protection has been terminated. Recently, the issue of kodoku-shi in areas afflicted by major disasters such a large earthquakes has also been highlighted (Kako and Ikeda 2009).

In order to avoid social isolation of the elderly and kodoku-shi, further improvement of senior citizens' financial and other welfare services is indispensable (▶ Chap. 13, "Implementing Community Care in Large Cities and Informal Settlements: An African Perspective" by Robertson, L., Paul Szabo, C., in this book). In recent years, welfare services have expanded from their core role of providing financial/human support to offering regular home visits for people living alone and increased community spaces allowing for increased social interactions. Such services remain inadequate and need to be further expanded. This issue now common in Japan, especially in urban areas, is likely to occur in other countries that will face an aging society in the future and as such constitutes an emerging global issue of importance (▶ Chap. 3, "Urban-Rural Differences in Major Mental Health Conditions" by Solmi, F., Dykxhoorn, J., Kirkbride, J., in this book). Evidence concerning kodoku-shi is extremely limited, and therefore new and in-depth epidemiological and intervention research is call for moving forward.

Hikikomori

In Japan, one of the most highlighted issues of loneliness faced by many younger people is the problem of hikikomori (social withdrawal), which was originally reported in Japan. The word "hikikomori" has been listed in the Oxford Dictionary since 2010 as "(*in Japan*) *the abnormal avoidance of social contact, typically by adolescent males/a person who avoids social contact [Origins] Japanese, literally 'staying indoors, (social) withdrawal*'." In the guideline of hikikomori (Japanese Ministry of Health, Labour and Welfare), hikikomori has been defined as a social withdrawal condition (or, persons) staying almost all time and every day in their own home and preventing school/job participation for 6 months and more, and schizophrenia is not included in this condition (Saito 2010). Hikikomori negatively impacts not only the affected individual's mental health but also population-level education and workforce stability and as such is an urgent issue in the administration of Health, Welfare and Labor (Kato et al. 2011, 2016). We have recently developed a new hikikomori diagnosis based on the above diagnosis standards in Japan (Saito 2010) as well as suggestions from Teo and Gaw (2010). We have proposed that the following four criteria should be included and observed for 6 months or more: (1) the person stays at home almost all day, almost every day; (2) the person avoids nearly all social situations such as school, work, etc.; (3) the person avoids direct social interaction with family or acquaintances; and (4) the above hinders the individual's social life. Individuals who fulfil all four of these criteria will be defined as "hikikomori." People with severe hikikomori cannot leave their homes at all, while majority of people with hikikomori can occasionally go out for shopping and the like.

Hikikomori Persons Who Live Alone

In Japan, many persons of hikikomori live with their families, but a portion of such individuals live alone. Our recent survey has shown that the majority of persons with hikikomori live with their families; however, a small number live alone (11%) (Teo et al. 2015a). The following case vignette is a typical case of a hikikomori sufferer who lives alone:

> Mr H is a single man in his early thirties living alone in an apartment in a big city. Raised in a regional city by working parents and with a sister 5 years his senior. As his father lived in other cities for his work, he saw him only about once every 2–3 months and as such did not have much of a relationship with his father. During early childhood and schooling, H experienced no major difficulties and despite being quiet he had a few good friends. H had always for some reason dreamed of living in a large cosmopolitan city. An average student academically, he was accepted to the economics faculty of a metropolitan university, and started to live alone. Unfortunately, he could not make many friends, and his university life was rather lonely. H had hoped to become a civil servant but failed the examination and instead found employment at a middle-size trading company. After a year of training, H was assigned to the sales division but found the work difficult and came to be constantly reprimanded by his supervisors and by his 4th year was often absent, felt depressed and finally took leave from his position at his 5th year.
>
> While on leave, he did not return to the family home but spent his days alone in his apartment on his computer. Finally coming to the conclusion by himself that 'I'm not made out for sales' and 'This company is not for me', he resigned his position. Afterwards, H lived on his savings but running out of money started to work short-term manual labor jobs thinking 'this is not the type of work I'm meant to be doing'. This also did not last. Once again feeling fatigued and depressed, he visited a nearby psychiatric clinic where social welfare protection was recommended for him. He applied and began to receive payments. After this, H does not look for reemployment but spends most of the day on the Internet and playing online games. For food, everyday he buys convenience store 'lunch boxes' or pre-prepared meals from supermarkets. H has no friends nearby, has lost contact with his old friends and basically has no friendships.
>
> With this situation having continued for 5 years, his parents urge him to return home, which he does not do, but rather continues to live in this solitary manner. At a complete loss for what to do, his parents seek advice at the Hikikomori Support Center…

As shown in this case, some hikikomori persons living without families are supported by the welfare system in addition to parental support and dependence. In Japan, even when unemployed, it is possible to obtain full economical support to live alone, if certain conditions are satisfied (such as poverty and mental/physical illnesses) by a governmental social welfare system called "seikatsu-hogo" and "shogai-nenkin." Against this backdrop, unemployed single-person households are increasing not only amongst the elderly but also among younger and middle-aged people. In many cases, young people start living alone when entering university or finding employment, but even if they drop out of school or quit their jobs and become unemployed, many continue to receive financial assistance from their parents and/or the social welfare system enabling them to remain shut in alone.

In such situations, human relationships are diluted, as is contact with family members, friends, and colleagues resulting in a solitary condition. When this situation lasts more than 6 months, the individual may be included within the definition of hikikomori. Such individuals sometimes go out to shop at convenience stores and supermarkets and occasionally go out for fun, but they usually stay at home and spend most of their time watching TV and surfing the Internet. The existence of such lone hikikomori has been facilitated by the development of information technology (IT). With the development of Internet, most shopping can be done via the net (and "net shopping" coupled with advanced delivery networks is especially advanced in Japan), making it possible to live without going out. Furthermore, even when withdrawn from society and shut in at home, it has become possible to enjoy various types of entertainment such as online games in complete isolation. Indeed, a universal world can be achieved by living alone.

Psychopathological Understandings of Hikikomori Living with and Without Families

While hikikomori persons who live alone share some commonality with hikikomori persons living with family members, there are also important differences. Both types of hikikomori sufferers often have a present/past history of psychiatric comorbidity such as avoidant personality disorder, major depressive disorder, paranoid personality disorder, post-traumatic stress disorder, social anxiety disorder, dysthymic disorder, obsessive–compulsive personality disorder, and specific phobia (Teo et al. 2015b). Although there is yet to a comprehensive epidemiological survey, from the viewpoint of the authors' clinical experience, narcissistic personality disorder seems to be especially prevalent among hikikomori persons who live alone. Many of those who are suffering in families often have problems at an earlier stage of separation and independence from their parents. On the other hand, hikikomori persons who live alone tend to have achieved separation to some extent, but there is a tendency toward insufficient independence. In addition, hikikomori people who live alone often experience strong discord with their families, refuse to live with their families, and many have the desire to run away from their families. Interestingly, such hikikomori youth are often found to live in relative close proximity to their parents, suggesting the existence of internal conflict between independence and dependence. Such behaviors may be considered the result of Japanese "*amae*" culture. Japanese "*amae*"-related dependent behaviors were originally described by the late Takeo Doi, a psychiatrist and psychoanalyst (Doi 1973). To a great extent, dependent behaviors related to "amae" are conducted with the belief that the parent will forgive all. Doi believed that Western societies tend to consider such dependence in children to be something that should be overcome or corrected, while in Japan "*amae*" remains an acceptable mode of behavior even in adult life (Doi 1973). Doi discussed "sullenness" or a sullen withdrawal as one transformation of "*amae*"; thus the behaviors of hikikomori may be seen to be close relation to the classic behavior of "*amae*." Hikikomori persons living with families are suggested to be indirectly induced by "*amae*" to the extent that parents accept their child staying at home for prolonged periods of time (Kato et al. 2012), while hikikomori persons living alone also have some similar tendencies because majority of such people do not live a great distance from their families.

Internationalization of Hikikomori Phenomenon

Recent studies have strongly suggested that cases of hikikomori exist not only in Japan but also many other countries. Our previous case vignette survey among psychiatrists in Australia, Bangladesh, India, Iran, Japan, Korea, Taiwan, Thailand, and the United States has suggested that hikikomori cases are observed in all countries especially in urban areas (Kato et al. 2012). In addition, hikikomori-like cases have been reported in France, Hong Kong, Oman, South Korea, and Spain (Furuhashi et al. 2012; Garcia-Campayo et al. 2007; Lee et al. 2001, 2013; Malagon-Amor et al. 2015; Sakamoto et al. 2005; Teo and Kato 2015; Wong et al. 2015). However, with the exception of Japan, this phenomenon has long been ignored. To our knowledge, alternative terms for hikikomori do not exist in Western countries, while hikikomori-like phenomena have recently been termed "운둔형외토리; 隠遁型" in Korea and "宅男/宅女" in China. We have conducted the first international clinical survey focusing on hikikomori using the above evaluation tool for hikikomori and revealed that persons with hikikomori exist in the United States, South Korea, and India (Teo et al. 2015a).

Hikikomori was first identified in Japan, and it was previously regarded as a culture-bound syndrome unique to Japan (Kato et al. 2012). The Japanese tend to socially interact in a manner that prioritizes indirect communication, and this is because of a unique sense of values involving concepts such as the above-introduced "*amae*" and "shame." Our series of international surveys have shown that hikikomori exist in countries with different cultural and social backgrounds (Kato et al. 2016, 2012; Teo et al. 2015a, b). Why did hikikomori manifest as an international mental illness? Even now, compared to Western societies, young people in Asian societies, including South Korea, Japan, and Taiwan, tend to be more economically dependent on their parents, and this phenomenon seems to be one of the expressed forms of "*amae*" (Kato et al. 2012). Even though the concept of "*amae*" was originally considered to be uniquely Japanese, contemporary views suggest that "*amae*" is actually more universal in nature (Niiya et al. 2006). Thus, there is an interesting parallel to the hikikomori phenomenon that has been thought of as unique to Japan but, as our preliminary results show, is perceived by psychiatrists as occurring in a variety of other countries. "Shame" is another factor to consider in our understanding of the psychopathological mechanisms of hikikomori (Kitayama 1998). There has been a general tendency within Japanese society to have a strong sense of shame in front of others, which might compound attitudes that allow for hikikomori in Japanese society. However, this shame consciousness might not be something uniquely limited to Japanese society. In the famed American musical "*Beauty and the Beast*," the selfish prince is magically turned into a beast by a witch. Believing that he cannot venture outside in his ugliness, and with a deep shame consciousness, he withdraws in his castle. This exactly equates to some hikikomori youth in Japan today. Many hikikomori sufferers harbor an ideal (narcissistic) self-image – "I should be an executive," or "I should become a great person," and when they are faced with the fact that this ideal self-image has not been actualized in the community, it is the Japanese "shame society" of "I'm not worth to go out (in public), I'm ashamed," and the belief that going out in public will lead to a miserable and shameful experience that has led many young hikikomori sufferers to cut ties with society at large. What they fear is that in society, they will be treated as "ugly people"; in other words, what they fear most is a denial of their very own narcissistic self-image.

Stage- and Condition-Oriented Therapeutic Approaches for Hikikomori

Therapeutic interventions for hikikomori are difficult to achieve, and this is particularly so for hikikomori persons living alone when compared to those living with their families. Hikikomori persons living with family members are occasionally introduced to some kinds of support services when their family seeks help even without their own help-seeking behaviors. However, in the case of hikikomori persons living alone, there is a tendency not to seek these support services which would potentially free the individual from their socially withdrawn situation. In such situations, it is often only when for economic reasons, solitary life becomes difficult leading to a breakdown in life that help is sought. This is also the case in the situation of a mental crisis.

As shown in Fig. 1, we herein propose that individuals with hikikomori can be classified into the following five stages based on living with families [A category] and living alone [B category]: (*1A*) *rarely leaves own room and rejects cohabitating family*, (*2A*) *rarely leaves own home but with some communication with cohabitating family*, (*3A*) *sometimes leaves own home and/or some interaction with others*, (*2B*) *lives alone and mostly does not interact with others, and* (*3B*) *lives alone but with some interaction with others.*

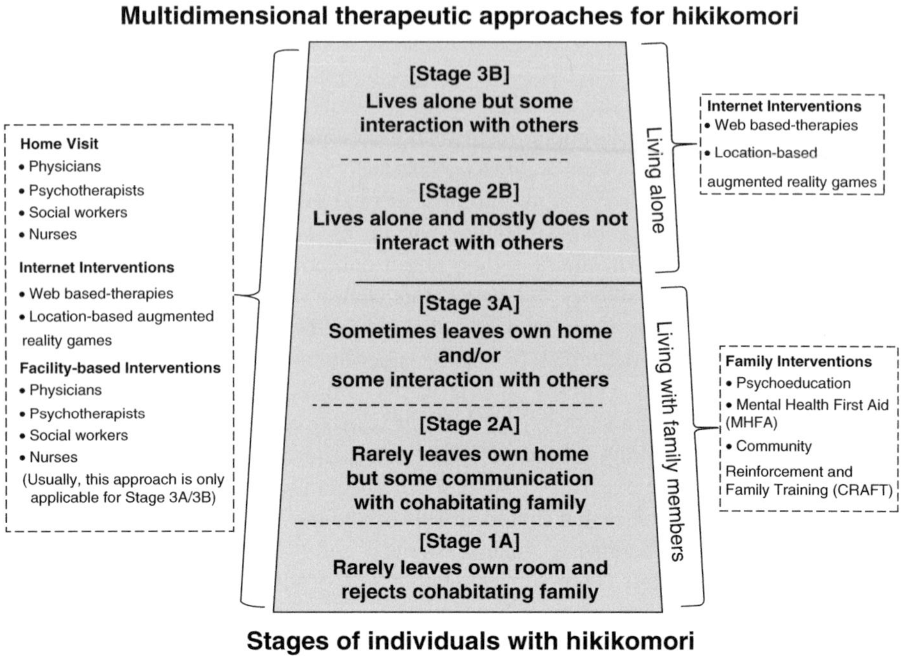

Fig. 1 Multidimensional therapeutic approaches for hikikomori
Individuals with hikikomori can be classified into the following five stages based on living with families (A category) and living alone (B category): (*1A*) *rarely leaves own room and rejects cohabitating family*; (*2A*) *rarely leaves own home but some communication with cohabitating family*; (*3A*) *sometimes leaves own home and/or some interaction with others*; (*2B*) *lives alone and mostly does not interact with others*; *and* (*3B*) *lives alone but some interaction with others*
Family interventions and direct interventions with the person themselves are both important. Initially, it is unlikely that the persons themselves would seek treatment, and therefore family interventions are crucial in case of living with family members. It is important for families to acquire the appropriate knowledge and techniques for dealing with hikikomori individuals. The Australian-developed Mental Health First Aid (MHFA) may be useful for educating the actual skills to contact the individuals appropriately. Furthermore, home visits by physicians, nurses, and psychologists also play an important role.
For individuals who live alone, a different intervention method may be necessary than for those living with their families. Early intervention methods utilizing technologies such the Internet and *Pokémon Go* may be particularly important in guiding such individuals to outside support organizations. Once in contact with support organizations, such as medical facilities (psychiatric hospitals/clinics), psychotherapy rooms, mental health support centers, and hikikomori support centers, various types of psycho-bio-social interventions can be applicable in the same way as other psychiatric disorders. In any case, a combination of various approaches is called for.

Interventions with the hikikomori persons themselves and family interventions are both important. Initially, it is unlikely that the persons themselves would seek treatment, and therefore family interventions are crucial in case of living with family members. In clinical practice dealing with hikikomori, the first consultation is often made by family members (especially parents). However due to a lack of knowledge (about mental illness in general and hikikomori in particular) and prejudices against such mental conditions, in many cases, family members cannot respond directly to individuals with these ailments, are unable to intervene at all, and waste many years without seeking help. As a result, we are now facing the new issue of longer-term hikikomori of ever increasing age. Thus, it is important for families to acquire the appropriate knowledge and techniques for dealing with hikikomori individuals. Mental Health First Aid (MHFA) may be useful for educating actual skills for family members in how they may interact with hikikomori sufferers appropriately. MHFA was originally developed as a 12-hour educational course that teaches participants (mainly laypeople) how to identify, understand, and respond to signs of mental illnesses, which gives participants the skills needed to reach out and provide initial help and support to someone who may be developing a mental health or experiencing a crisis (Kitchener and Jorm 2002, 2006). The five steps of the MHFA (third version) are as follows: step (1), approach the person, assess, and assist with any crisis; step (2), listen nonjudgmentally; step (3), give support and information; step (4), encourage the person to get appropriate professional help; and step (5), encourage other support (Kitchener et al. 2013). We have been promoting usage of the MHFA in Japan (Hashimoto et al. 2016; Kato et al. 2010; Suzuki et al. 2014) and are now in the process of developing an educational support model that enables families (especially parents) of persons with hikikomori to obtain specific skills and knowledge in dealing with hikikomori based on MHFA. Community Reinforcement and Family Training (CRAFT), a disorder-specific and partner-assisted intervention, that was originally developed for family members of individuals with substance use disorders (Meyers et al. 1998, 2002) has also been highlighted as another powerful interventional candidate for family members dealing with hikikomori individuals (Sakai and Nonaka 2013). Furthermore, home visits by physicians, nurses, psychologists, and social workers also play an important role.

For individuals who live alone, a different intervention method may be necessary than for those living with their families. Early intervention methods utilizing technologies of the Internet and mobile tools may be particularly important in guiding such individuals to outside support organizations. Very recently, the appearance of "*Pokémon Go*" – the location-based augmented reality game, which connects the Internet society with real road maps, has brought us to a new phase. Some people with hikikomori who have not gone out for years are leaving their homes in search of *Pokémons* (Kato et al. 2017). To our amazement, a single male patient who previously was barely able to go out who live alone has begun to venture out daily with the emergence of *Pokémon Go*. His main destination is public parks filled with other people in search of *Pokémons*, though they do not converse with each other. No matter what the reason, for a hikikomori suffer who rarely goes out, any "adventure" outside is important as the first step in the treatment. Notwithstanding obvious limitations, we can perhaps expect much from this as an evolutionary therapeutic tool enhancing motivation toward the outside. Such new location-based augmented reality games that can connect the Internet society with real roadmaps, offer a tempting new approach to treating hikikomori. It is important to utilize such technology to successfully attract people with hikikomori from their isolated room to places where they may be treated. Looking to the future with the aim of a fundamental breakthrough combating the hikikomori condition, we need to develop an online intervention system using e-learning materials that allows easy usage for persons with hikikomori themselves and also their family members. Once in contact with support organizations, such as medical facilities (psychiatric hospitals/clinics), psychotherapy rooms, and/or mental health support centers, and hikikomori support centers, various types of psycho-bio-social interventions can be applicable in the same way as other psychiatric disorders. In any case, a combination of various approaches is called for.

Psychotherapeutic Approaches Against Hikikomori

In an international hikikomori survey, we previously asked persons with hikikomori in Japan, India, South Korea, and the United States whether they themselves want to have treatment such as psychotherapy and/or pharmacotherapy with direct methods or indirect online methods (Teo et al. 2015a). This international survey revealed for the first time that most hikikomori persons would like to undergo some form of treatment to resolve their hikikomori status (Teo et al. 2015a). Most desired psychotherapy (counseling) rather than pharmacological therapy. Significantly more persons desired direct psychotherapy and pharmacological therapy than online treatment using webcams and the like. The most popular form of treatment was direct (face-to-face) personal psychotherapy by a mental health professional. With regard to comparisons between different countries, American hikikomori persons had a more pronounced tendency than Japanese counterparts to not desire treatment from their primary care physician.

Our international research has supported the finding that the majority of hikikomori sufferers feel a strong sense of loneliness and have functional disabilities. The fact that most people with hikikomori desire therapeutic interventions also deserves attention. At the time the survey was initiated, we expected that because most people with hikikomori do not go outside, they would be likely to actively search out nondirect forms of treatment over the Internet; however, in contrast to this hypothesis, most respondents desired direct treatment. We believe that the results of this survey will be beneficial in considering treatment strategies for people with hikikomori. Based on our clinical observations (no survey exists to our knowledge), we suppose that most persons are extremely isolated from venues of social interaction in an attempt to be freed from annoying human relationships; however, at the same time, they desire direct interpersonal interaction at a preconscious and/or unconscious level especially as their shut-in durations are prolonged. Psychotherapy is useful in handling these ambivalent sentiments, and the reason why most persons with hikikomori desire direct psychotherapy may lay within this internal motivation.

Psychotherapy has long been one of the main approaches for the treatment of hikikomori in Japan (Kano and Kondo 2000; Kitayama et al. 2001; Nakamura and Shioji 1997), and various psychotherapeutic methods, including the psychoanalytical approach, the group psychotherapy approach, and the house call approach, have been practiced (Kano and Kondo 2000). In terms of psychoanalysis, it has been pointed out that psychopathologies such as schizoid and narcissistic personalities underlie the hikikomori phenomenon (Kano and Kondo 2000). Among various types of psychotherapy, psychoanalytic psychotherapy conducted over a long duration plays a significant role in treating these personality pathologies. Ambivalent sentiments such as "I want to engage with this person" and "I want to let him/her in peace" emerge among analysts engaged in psychoanalytic treatment, particularly non-face-to-face psychoanalysis employing couches. Winnicott D.W. proposed such a treatment approach involving an ambivalent paradoxical attitude and venue, where a patient can be "alone but not alone," that can cultivate a patient's ability "to be alone" (Winnicott 1958). Another important point that psychoanalysis alerts us to is the fact that even individuals living out their lives in perfectly healthy ways have some degree of hikikomori-like qualities (such feelings as "want to be alone" and/or "let me alone"). This understanding helps us to develop a sense of empathy with hikikomori and to build a unified treatment approach. Fairbairn W.D. proposes the "exciting object" and the "rejecting object" as the two main elements of personality (Fairbairn 1952). Hikikomori may be an extreme manifestation of the rejecting object (Kato et al. 2015). Dynamic group psychotherapy based on psychoanalytic theories is effective in allowing people with hikikomori to observe each other and achieve mutual understanding by sharing experiences with the above pathologies in groups (Kubo et al. 2017). Additionally, in many cases, people with hikikomori have not been examined at medical institutions, and in such cases, in-home counseling is effective; recent research has suggested the usefulness of house calls among hikikomori in South Korea (Lee et al. 2013).

Such clinical experience resulting from the practice of Japanese psychotherapists should be widely introduced overseas, and at the same time, we should attempt to build evidence of the effectiveness of psychotherapy.

Future Perspectives

The main causes of the internationalization of hikikomori may be urbanization, globalization, and in particular the international spread of nondirect communication due to the popularization of the Internet (Kato et al. 2011, 2015). We have recently been proposing a concept of "boundless syndromes" in modern society (Kato and Kanba 2016). A more interconnected world due to globalization and the growing importance of "(Inter)net society" may be major contributing factors behind the hikikomori phenomenon. An interconnected world is resulting in a boundless society and building a boundless psychological world in modern persons, which may form a variety of novel psychopathological conditions including hikikomori. Behind the increase of people with hikikomori, there may be an interconnection with societies that are increasingly boundless. Ironically, this very interconnectedness may spur an increase in the number of people who need to drastically "disconnect" (withdraw) from society. As information technology continues to evolve the social issues brought about regarding single-person households will be of increasing importance. We should consider an interconnected world producing novel psychopathology in the twenty-first century.

Hikikomori, Kodoku-Shi, and Suicide as Urban Loneliness-Related Phenomena

As modernization and urbanization have progressed, people who had hitherto lived in traditional family groupings and communities have increasingly come to spend more time as lone individuals with many actually living alone in the city. Traditionally, Japanese life used to be based on village communities and was a society that highly valued neighborhood relations, but in the modern urban life, it is becoming increasingly difficult to establish and maintain close human connections. This creates a situation in which people are likely to become lonely. As shown in this chapter, this situation of single-person households in the city tends to induce isolation and/or loneliness and sometimes leads to a variety of mental health and psychiatric issues based on loneliness. In other words, we should not overlook the fact that as individuals in single-person households in the city experiencing loneliness, they also develop associated mental health issues, including hikikomori, kodoku-shi, and also suicide. One of the major background factors of suicide is loneliness (▶ Chap. 4, "Homicide and Suicide in Megacities" by Ziebold, C., Jaen-Varas, D., Mari, J., in this book), and, interestingly, an American expatriate living in Japan remarked that hikikomori seems to be an alternative suicidal behavior. Even though no epidemiological data exists, there have been many case reports of hikikomori persons who committed suicide. Further studies between hikikomori and suicide should be conducted. Hikikomori was first reported in Japan, and recently Japan's Cabinet Office reported that the number of sufferers presently is in excess of 500,000 (age 15–39). This number does not take into account a much larger population when including middle- and senior-aged hikikomori people (Jozuka 2016). Although there has yet to be any epidemiological research, we have seen an increase in visits of hikikomori sufferers aged over 40 years old and their parents at our hikikomori advice facilities; thus it has been predicted that the aging of such individuals will continue to develop into a serious social issue in the not-too-distant future. Especially in urban areas where we see little to no community bonds, it is more and more likely that lone hikikomori suffers will one day also become victims of kodoku-shi.

Conclusion

As shown above, the hikikomori-like phenomenon is not limited to Japan but also exists in many other countries; thus we have warned that hikikomori and kodoku-shi may not simply be Japanese cultural/social phenomena but actually indicators of a pandemic of psychological problems that a continually globalizing society will have to face in the near future (Kato et al. 2011). The presented topics related to single-person households in the city are not merely Japanese or one nation issues but are in fact increasingly global phenomena and as such require breakthrough measures based on worldwide research. Precisely as we are living in such a lonely era, especially for people who live alone in cities, many of us are beginning to seek human connectivity in new ways as the emergence and popularity of Facebook and Instagram exemplify. Meanwhile for those who are not satisfied by the emotional connection brought about by such social network services, a new kind of deep loneliness is being experienced. While a fundamental breakthrough to these loneliness-related situations is yet to be thought of, perhaps it is by utilizing such global networks that we may achieve one.

Conflict of Interests

In this review paper, all the authors report no financial relationships with commercial interests.

Cross-References

▶ Homicide and Suicide in Megacities
▶ Implementing Community Care in Large Cities and Informal Settlements: An African Perspective
▶ Urban-Rural Differences in Major Mental Health Conditions

References

Brasor, P (2014) Japan's lonely people: where do they all belong? The Japan Times. http://www.japantimes.co.jp/news/2012/03/04/national/media-national/japans-lonely-people-where-do-they-all-belong/#.WLJr4RjCP_C. 22 June 2014

Cabinet Office, Government of Japan (2015) Heisei 26 Nendo Koureika-no-Zyoukyou oyobi Kourei-syakai-taisaku no Jisshi-Zyoukyou (2014 Annual report about the situation of aging society and strategies against the aging society) from Cabinet Office, Government of Japan. http://www8.cao.go.jp/kourei/whitepaper/w-2015/html/zenbun/s1_2_6.html

Doi T (1973) The anatomy of dependence. Kodansha International, Tokyo

Fairbairn WD (1952) Psychoanalytic studies of the personality. Tavistock Publications Limited, London

Furuhashi T, Figueiredo C, Pionnie-Dax N, Fansten M, Vellut N, Castel PH (2012) Pathology seen in French "Hikikomori". Seishin Shinkeigaku Zasshi 114(10):1173–1179

Garcia-Campayo J, Alda M, Sobradiel N, Sanz Abos B (2007) A case report of hikikomori in Spain. Med Clin (Barc) 129(8):318–319

Hashimoto N, Suzuki Y, Kato TA, Fujisawa D, Sato R, Aoyama-Uehara K, Fukasawa M, Asakura S, Kusumi I, Otsuka K (2016) Effectiveness of suicide prevention gatekeeper-training for university administrative staff in Japan. Psychiatry Clin Neurosci 70(1):62–70

Jozuka E (2016) Why won't 541,000 young Japanese leave the house? CNN. http://edition.cnn.com/2016/2009/2011/asia/japanese-millennials-hikikomori-social-recluse/

Kako M, Ikeda S (2009) Volunteer experiences in community housing during the great Hanshin-Awaji earthquake, Japan. Nurs Health Sci 11(4):357–359

Kano R, Kondo N (2000) Seinen no Hikikomori (adolescent social withdrawal) (in Japanese). Iwasaki gakuzyutsu shuppansha, Tokyo

Kato TA, Kanba S (2016) Boundless syndromes in modern society: an interconnected world producing novel psychopathology in the 21st century. Psychiatry Clin Neurosci 70(1):1–2

Kato TA, Suzuki Y, Sato R, Fujisawa D, Uehara K, Hashimoto N, Sawayama Y, Hayashi J, Kanba S, Otsuka K (2010) Development of 2-hour suicide intervention program among medical residents: first pilot trial. Psychiatry Clin Neurosci 64(5):531–540

Kato TA, Shinfuku N, Sartorius N, Kanba S (2011) Are Japan's hikikomori and depression in young people spreading abroad? Lancet 378(9796):1070

Kato TA, Tateno M, Shinfuku N, Fujisawa D, Teo AR, Sartorius N, Akiyama T, Ishida T, Choi TY, Balhara YP, Matsumoto R, Umene-Nakano W, Fujimura Y, Wand A, Chang JP, Chang RY, Shadloo B, Ahmed HU, Lerthattasilp T, Kanba S (2012) Does the 'hikikomori' syndrome of social withdrawal exist outside Japan? A preliminary international investigation. Soc Psychiatry Psychiatr Epidemiol 47(7):1061–1075

Kato TA, Teo Alan R, Tateno M, Choi Tae Y, Balhara Yatan Pal S, Kanba S (2015) Introducing the international research of hikikomori in Japan, India, Korea and the United States. Jpn J Clin Psychiatry (Jpn) 44(12):1625–1635

Kato TA, Kanba S, Teo AR (2016) A 39-year-old "Adultolescent": understanding social withdrawal in Japan. Am J Psychiatry 173(2):112–114

Kato TA, Teo AR, Tateno M, Watabe M, Kubo H, Kanba S (2017) Can Pokemon GO rescue shut-ins (hikikomori) from their isolated world? Psychiatry Clin Neurosci 71(1):75–76

Kitayama O (1998) Transience: its beauty and danger. Int J Psychoanal 79(Pt 5):937–953

Kitayama O, Saito T, Watanabe T, Muto S (2001) Zadankan Hikikomori Ni Tsuite (Forum Hikikomori) (in Japanese). In: Watanabe T, Muto S (eds) Gendai No Esprit 403-Hikikomori. Shibundo, Tokyo, pp 5–34

Kitchener BA, Jorm AF (2002) Mental health first aid training for the public: evaluation of effects on knowledge, attitudes and helping behavior. BMC Psychiatry 2(1):1

Kitchener BA, Jorm AF (2006) Mental health first aid training: review of evaluation studies. Aust N Z J Psychiatry 40(1):6–8

Kitchener BA, Jorm AF, Kelly CM (2013) Mental health first aid manual, 3rd edn. Mental Health First Aid Australia, Melbourne

Kubo H, Kato TA, Kanba S (2017) Group psychotherapy for Hikikomori: consideration of "Difficulty in communicating with others" in the introduction period of group. Jpn J Adolesc Psychiatry (Jpn) 26(2):197–198

Lee KM, Koo JG, Kim EJ, Lee SH (2001) The psychosocial characteristics of Oiettolie adolescents (in Korean). Korean J Counsel Psychother 13(1):147–162

Lee YS, Lee JY, Choi TY, Choi JT (2013) Home visitation program for detecting, evaluating and treating socially withdrawn youth in Korea. Psychiatry Clin Neurosci 67(4):193–202

Malagon-Amor A, Corcoles-Martinez D, Martin-Lopez LM, Perez-Sola V (2015) Hikikomori in Spain: a descriptive study. Int J Soc Psychiatry 61(5):475–483

Meyers RJ, Miller WR, Hill DE, Tonigan JS (1998) Community reinforcement and family training (CRAFT): engaging unmotivated drug users in treatment. J Subst Abuse 10(3):291–308. PMID: 10689661

Meyers RJ, Miller WR, Smith JE, Tonigan JS (2002) A randomized trial of two methods for engaging treatment-refusing drug users through concerned significant others. J Consult Clin Psychol 70(5):1182–5. PMID: 12362968

Nakamura K, Shioji R (1997) Taijin kyoufushou to hikikomori (in Japanese). Rinshou Seishin Igaku (Clin Psychiatry) 26:1169–1176

Niiya Y, Ellsworth PC, Yamaguchi S (2006) Amae in Japan and the United States: an exploration of a "culturally unique" emotion. Emotion 6(2):279–295

Nobel J (2010) Japan's 'Lonely Deaths': a business opportunity. TIME. http://content.time.com/time/world/article/0,8599,1976952,00.html?xid=rss-topstories. 06 Apr 2010
Saito, K (2010) Hikikomori No Hyouka-Shien Ni Kansuru Gaido-Rain [Guideline of Hikikomori for their evaluations and supports]. Ministry of Health, Labour and Welfare, Tokyo. http://www.ncgmkohnodai.go.jp/pdf/jidouseishin/22ncgm_hikikomori.pdf
Sakai M, Nonaka S (2013) CRAFT Hikikomori no Kazoku-Shien workbook (CRAFT workbook for supporting family members of hikikomori). Kongoshuppan, Tokyo
Sakamoto N, Martin RG, Kumano H, Kuboki T, Al-Adawi S (2005) Hikikomori, is it a culture-reactive or culture-bound syndrome? Nidotherapy and a clinical vignette from Oman. Int J Psychiatry Med 35(2):191–198
Suzuki Y, Kato TA, Sato R, Fujisawa D, Aoyama-Uehara K, Hashimoto N, Yonemoto N, Fukasawa M, Otsuka K (2014) Effectiveness of brief suicide management training programme for medical residents in Japan: a cluster randomized controlled trial. Epidemiol Psychiatr Sci 23 (2):167–176
Teo AR, Gaw AC (2010) Hikikomori, a Japanese culture-bound syndrome of social withdrawal?: a proposal for DSM-5. J Nerv Ment Dis 198(6):444–449
Teo AR, Kato TA (2015) The prevalence and correlates of severe social withdrawal in Hong Kong. Int J Soc Psychiatry 61(1):102
Teo AR, Fetters MD, Stufflebam K, Tateno M, Balhara Y, Choi TY, Kanba S, Mathews CA, Kato TA (2015a) Identification of the hikikomori syndrome of social withdrawal: psychosocial features and treatment preferences in four countries. Int J Soc Psychiatry 61(1):64–72
Teo AR, Stufflebam K, Saha S, Fetters MD, Tateno M, Kanba S, Kato TA (2015b) Psychopathology associated with social withdrawal: idiopathic and comorbid presentations. Psychiatry Res 228 (1):182–183
Winnicott DW (1958) The capacity to be alone. Int J Psychoanal 39(5):416–420
Wong PW, Li TM, Chan M, Law YW, Chau M, Cheng C, Fu KW, Bacon-Shone J, Yip PS (2015) The prevalence and correlates of severe social withdrawal (hikikomori) in Hong Kong: a cross-sectional telephone-based survey study. Int J Soc Psychiatry 61(4):330–342

Part IV

Urban Mental Health Around the World

Mental Health Care in 11 Major Cities Over Two Decades

9

David Goldberg, Graham Thornicroft, Maryam Abbasinejad, Francesco Amaddeo, Gavin Andrews, Dinarte Ballester, Stephen T. Baron, Stuart Bell, Ellis Busnello, Marcel Buster, Jack J. M. Dekker, Klement Dymi, Ahmad Hajebi, Mathew James, Marianne Kastrup, David LeCount, Loiva dos Santos Leite, Ahmad Mohit, R. Srinivasa Murthy, Shinfuku Naotaka, Sara Jane Escouto dos Santos, Norman Sartorius, Crista Taylor, Jan Theunissen, Ana Cristina Tietzmann, Wilco Tuinbreijer, Kathleen Westcoat, and Matty A. S. de Witt

Abstract

This chapter investigates how mental health services have changed over the past two decades in a selection of 11 major cities from four different continents. Overall, deinstitutionalization has moved patient care away from hospitals and into community settings where mental health centers are now developing outpatient services. The cities presented are Amsterdam (the Netherlands), Baltimore (the United States), Bangalore (India), Copenhagen (Denmark), Kobe (Japan), London (United Kingdom), Porto Alegre (Brazil), Sydney (Australia), Madison (the United States), Tehran (Iran), and Verona (Italy). These cities have been selected to represent a global perspective including different continents, cultures,

D. Goldberg (✉)
Institute of Psychiatry, King's College London, London, UK

G. Thornicroft
Institute of Psychiatry Psychology and Neuroscience of King's College, London, UK

M. Abbasinejad
Department for Mental Health and Substance Abuse, Roozbeh Hospital, Tehran, University of Medical Sciences, Tehran, Iran

F. Amaddeo
Department of Neurosciences, Biomedicine and Movement Sciences, University of Verona, Verona, Italy
e-mail: francesco.amaddeo@univr.it

G. Andrews
CRUfAD, UNSW at St Vincent's Hospital, Darlinghurst, NSW, Australia
e-mail: gavina@unsw.edu.au

N. Okkels et al. (eds.), *Mental Health and Illness in the City*, Mental Health and Illness Worldwide, https://doi.org/10.1007/978-981-10-2327-9_19

and religions. This chapter reports of several promising community care initiatives and evidence that the era of institutionalization may be ending.

Keywords
Mental health services • Community health services • Institutionalization • Mental Health in Primary Care

D. Ballester
Psychiatry Service, Hospital Mãe de Deus, Porto Alegre, Brazil
e-mail: ballester.dinarte@gmail.com

S.T. Baron • C. Taylor • K. Westcoat
Behavioral System Baltimore, Inc. (BHSB), Baltimore, MD, USA

S. Bell
Oxford Health NHS Foundation Trust, Northampton, UK

E. Busnello
Universidade Federal do Rio Grande do Sul, Porto Alegre, Brazil
e-mail: ebusnello32@terra.com.br

M. Buster
Department of Epidemiology and Health Promotion, Public Health Service, Amsterdam, The Netherlands

J.J.M. Dekker
Arkin Mental Health Care, Research Department, Amsterdam, The Netherlands

K. Dymi • M. Kastrup
Psychiatric Centre Ballerup, Copenhagen University Hospital, Copenhagen, Denmark
e-mail: klementdymi@yahoo.com; d121304@dadlnet.dk; marianne.kastrup@dadlnet.dk

A. Hajebi
Department of Psychiatry, Tehran Psychiatry Institute, University of Medical Sciences, Tehran, Iran

M. James
Mental Health Service, St Vincent's Hospital, Sydney, Australia

D. LeCount
Field Faculty Associate School of Social Work, University of Wisconsin, Madison, USA
e-mail: lecount@wisc.edu

L. dos Santos Leite
Municipal Health Authority, Porto Alegre, Brazil
e-mail: loiva@sms.prefpoa.com.br

A. Mohit
Tehran University of Medical Sciences and Health Services, Tehran, Iran
e-mail: mohidahmad@gmail.com

R.S. Murthy
National Institute of Mental Health and Neurosciences (NIMHANS), Bangalore, India
e-mail: smurthy030@gmail.com

S. Naotaka
International Center for Medical Research and Treatment, Kobe University, Kobe, Japan
e-mail: shinfukunaotaka@gmail.com

Contents

S.J.E. dos Santos
Municipal Health Authority, Porto Alegre, Brazil
e-mail: saraes@sms.prefpoa.com.br

N. Sartorius
Association for the Improvement of Mental Health Programmes (AMH), Geneva, Switzerland
e-mail: sartorius@normansartorius.com

J. Theunissen
GGZ in Geest Research Department, Amsterdam, The Netherlands

A.C. Tietzmann
Medical Residency Program in Psychiatry – Universidade Federal de Ciências da Saùde de Porto Alegre, Porto Alegre, Brazil
e-mail: actietzmann@gmail.com

W. Tuinbreijer
Department of Public Mental Health Care, Municipal Health Service, Amsterdam, The Netherlands

M.A.S. de Witt
Department of Epidemiology and Health Promotion, Public Health Service, Amsterdam, The Netherlands

Introduction (David Goldberg and Graham Thornicroft)

In 1997, the Bethlem Royal Hospital in London celebrated its 750th birthday, and as part of these we invited 11 cities to describe what they were doing faced with the challenge of past or recent deinstitutionalization and how they planned to improve care in the future. They were chosen because they had either published interesting work on the path toward community care or had been involved in improving the training offered to primary care physicians in mental health skills. The proceedings were published as a Maudsley Monograph book titled *Mental Health in Our Future Cities*. The purpose of this review is to answer the question: what has happened, in terms of mental health care, to these 11 cities in the last 20 years? In each chapter about one of the eleven cities, the opening paragraph called "The Situation in 1996" has been added by David Goldberg and Graham Thornicroft to give readers the context of the original chapter.

Mental Health and Urbanization (Norman Sartorius)

The rate of urbanization over the past century has remained steady, regardless of wars, famines, natural or human-made catastrophes, forced migration, or speed of industrial development. Nearly two thirds of the population of China lives in urban areas and at least half of the population of India is now urbanized. Other low and middle income countries follow the same trend. The proportion of the population living in urban areas in the highly industrialized countries is now approximately 80% and growing (Figs. 1 and 2).

Urbanization has a variety of consequences for mental health and for mental health care. The most important of these is undoubtedly the change of the nature of communities in which people live. Villages were geographical areas housing groups of people who were related by family or knew each other for many years. Certain rules of behavior were accepted by all villagers. Modern villages are different: in many a number of houses are deserted and others belong to people living in towns and spending (often only a few) weekends in their weekend residence. The population of villages in developing countries has also changed: in many the young and able have left the village to go and work in towns in the same or in another country leaving behind wives with children, the elderly, and the disabled.

Growing at a fast rate towns in many of their parts do not have any "village" communities (in the sense described above). Recent migrants, the poor and the unemployed inhabit the same high rise dwellings not knowing each other and often not wanting to know anyone living in the same compound. In addition, rapid urbanization has not only changed the structure of towns but also led to a rapid increase of their size reaching up to 15 or 20 millions inhabitants in which the rules of managing smaller towns are no longer valid.

The patterns of urban morbidity have also changed. In addition to a high prevalence of substance abuse and interpersonal violence which have characterized urban slums in the past century, towns of the twenty-first century have populations with a significant proportion of people of higher age which contributes to increasing prevalence of various forms of comorbidity of mental and physical disorders.

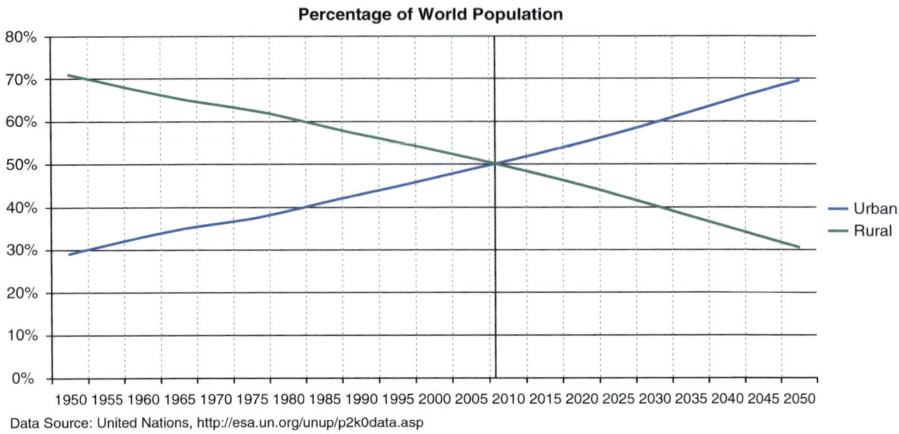

Fig. 1 Proportions of the world's population living in urban and rural areas 1950–2050

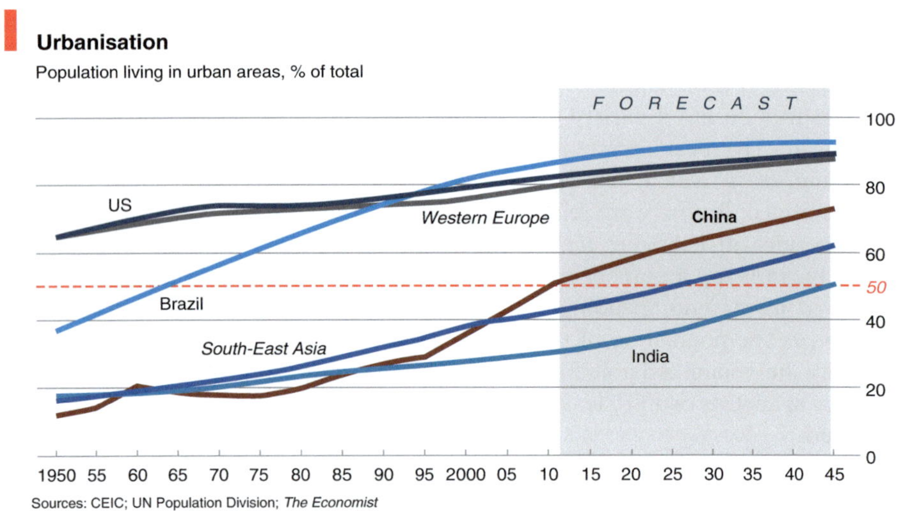

Fig. 2 Proportions of population living in cities in different parts of the world, 1950–2050

It is amazing that until now there are no universally agreed guidelines nor policies concerning the organization of mental health care (or indeed health care in general) in cities. The Alma Ata declaration, which was based on an international consensus about the organization of health care, was to a large extent applicable to the development of health care systems in rural areas with fairly well-defined communities. There is no international declaration nor agreement about urban health services. Saying that this is not possible or not necessary because urban health strategies are very complex is nonsense: it is precisely because of the complexity of health care in urban settings that a consensus on principles of providing care should have been sought by the World Health Organization or some other of the H8 organizations (H8 refers to the eight major international organizations dealing with health. The World Health Organization is one of them, supposedly the "primus inter pares".). The recently adopted Sustainable Development Goals have a goal that deals with sustainable and healthy cities: but there is no cross reference between that goal and the goal that deals with health care and wellbeing.

If I were invited to propose planks for a platform on which to base the urban mental health care strategy, I would suggest the following:

1. Stigma of mental illness – particularly when it is combined with poverty and other problems that mentally ill people often have – is often worse than the illness itself because it leads to an array of disadvantages ranging from difficulties of finding or maintaining housing and employment to experiencing violence. Mental health services should therefore include in their programs and budgets continuing and vigorous action against the stigmatization and discrimination of people with mental illness and their families.
2. The general health care personnel should be educated and encouraged to deal with common mental disorders with guidance and advice of specialists if these are available. The care for the severely mentally ill should be in the hands of specialized personnel working closely with carers – family members or peers ("specialized personnel" are not necessarily psychiatrists and could be nurses, social workers or psychologists with training in the management of severe mental disorders.).
3. Staff working in the mental health service should develop criteria which will allow them to (a) assess whether the family or other carers are fit to provide care for a person who has a severe mental illness and (b) assess whether a person who has mental illness is fit to be placed in the family or other care unit.
4. Education of family and other carers as well as the provision of moral and material support to them while caring for persons with mental illness should be budgeted and given high priority. Programs such as the FIRST AID and others helping in the management of illness and its symptoms can be useful in making care more efficient and humane (Kitchener et al. 2015). The education of families should use family members with experience in caring as well as people with the experience of mental illness as faculty in teaching other families.
5. As resources become available, it is important to develop different types of service which will be suited to the needs of patients and their carers. This should include in-patient and out-patient facilities as well as various forms of transition services, nursing homes, peer apartments, and other services developed in harmony with the sociocultural setting in which the services are to function.
6. General health care services should be urged to use methods of prevention of mental disorders ranging from appropriate perinatal care and correction of iodine

deficiency in women in fertile age to the correction of sensory deficits in children and measures reducing substance abuse problems. The recent trend of leaving the responsibility for primary prevention of mental disorders with parents led to the reduction of their use particularly in families of low education and stricken by poverty.

7. The problems related to the comorbidity of mental and physical disorders should be given special attention and lead to a coordination of primary care agents and the specialized services concerned. Unless people with mental illness are given comprehensive care, they will continue to have significantly higher mortality and poorer quality of life than the rest of the population. Suggestions about the organization of services for people with mental illness will have to be reviewed and revised as time goes on. The speed of urbanization and the changes it introduces in the structure and functioning of large cities in the undoubtedly hides surprises for the organization of health care and the main principle of services organization will have to be flexibility and readiness to change the organization and structure of services so as to make them appropriate for settings in which they will function.

Overview of the Changes in the 11 Cities Over the Past 20 Years (David Goldberg and Graham Thornicroft)

There have been widespread increases in the populations of all the cities, accompanied by clear movements towards care in community settings in those cities where there was still reliance on large mental hospitals, such as Amsterdam (Holland), Copenhagen (Denmark), and Porte Alegre (Brazil). In these cities, there has been widespread creation of mental health centers and a corresponding shift of resources to them away from the mental hospital. In Copenhagen, there are now more people in touch with the service despite fewer beds. In Porto Alegre these centers have become the main focus of care, and where specialized centers for child psychiatry have been created, they have provided such care for the first time, and the transformation has been facilitated by social cash transference programs aimed at the very poor, such as "Bolsa Família," with the resources managed mainly by women, which has had a huge impact, moving millions of people to a higher social class.

Those cities whose services in the community were most advanced in 1996, like Baltimore (USA), Madison (USA), and Verona (Italy), have all developed their services still further, either by refining the specialist community teams in the US cities or by continuing to produce research into the basis of a community service in Verona.

Effects of Funding Changes on Mental Health Systems

Some cities, for example, Copenhagen and Baltimore, report increases of funding devoted to mental health, Copenhagen by transforming their mental health system, and Baltimore by greatly improving the administrative machinery for managing the system.

But the reverse has occurred in London, where the government instituted severe austerity measures which have meant that mental health services have been cut back to a far greater extent than general medical services, and practices like assertive community treatment abandoned despite evidence of its efficacy. In Copenhagen, by the way of contrast, there has been an expansion of mental illness services with more people receiving care, which has been largely shared between mental health centers and small units in local areas. The mental health services in Madison claim that only 9% of their total budget is spent on hospital services, but admit that their in-patient service is very far from satisfactory.

The Location of Mental Health Services and Stigmatization

Having to go to a Mental Health Center (where only mentally disordered people have treatment) may be seen as stigmatizing, while strengthening the ability of primary care physicians to recognize and treat common mental disorders, and for those with severe disorders to receive care alongside everyone else in general hospitals, may well be the least stigmatizing place of treatment of all.

If mental illness services are administered by administrations operating their own separate services, this can reinforce the view that psychological problems are more easily seen as something apart from physical diseases. In England and Wales, sometimes separate provider organizations (called mental health trusts) operate alongside those for general health. Sometimes this makes some sense, where social care is an important component of desirable services, for example, in services for dementia, for learning disability, and for those with long-term severe mental disorders: in these cases social services need to work closely with medical services to achieve the best solution. Another potential advantage of separate trusts is that they may minimize asset stripping from mental to physical health care, although any such protection has been hard to detect in recent years. In Copenhagen, mental health has been separated from general health to produce a completely separate service, which is likely to be seen as something quite different from other disorders, a perception which might be encouraged by the decision to locate the administrative headquarters in the building of the old mental hospital.

New Services for People with Psychological Disorders

New services for people with the mental illness have been reported in many cities during the past 20 years. There has been a profusion of new programs in Sydney (Australia), ranging from the Program for Early Intervention & Prevention of Disability to services for distinct patient groups, such as the services for people living with HIV/AIDS or hepatitis or those needing Dialectical Behaviour Therapy. A further example is a combination of assertive outreach to local NGOs and street corners, along with assertive case management, which has been successful in linking clients with stable accommodation and mainstream services.

The Impact of the Digital Revolution

Help Initiated by Experts

In Sydney, participation in the eHealth initiatives has produced Internet-delivered CBT with a cost comparable to 2 months of medications and shown equal efficacy, wide availability, and acceptability to consumers, but most importantly that quality can be assured. There are now a dozen different online CBT programs catering for a raft of anxiety and affective disorders. It also now focuses on the top causes of disability, for example, chronic pain, arthritis, chronic obstructive pulmonary disease/asthma, diabetes, and cancer, in which anxiety and depression play such a significant part. In Bangalore (India), voluntary organizations working in different parts of the city to provide basic mental health care to the population. At YourDOST, people seeking support are connected to the right experts – psychologists, psychotherapists, counselors, life coaches, and career guides – who help them through free and anonymous online chats. New services using the Internet include several in Bangalore focusing on self-evaluation, self-care, and graded support with emotional counseling.

Services Initiated Within the Community

In Bangalore, a group of parents (AMEND) have not only provided support to each other, but taken up advocacy activities like the organization of emergency ambulance support and welfare benefits to persons/families with chronic mental illness. Other specific initiatives address the needs of students and cancer survivors (CIPHER, CARER program) aimed at providing online support to specific health groups. It is to be hoped that this exciting work will be developed in other places as well.

A Downside to the Internet

In both Kobe (Japan) and Bangalore, there have been increasing numbers of young people who avoid social contact altogether and stay at home. They often communicate only through the Internet and play Internet games.

The Role of Primary Care in Services for Psychological Disorders

In London (United Kingdom), the National Service Framework for Mental Health laid down two standards for care for mental disorders in primary care dealing with both treatments which should be available without referral and to access to specialist mental illness services. More recently, the government's Improving Access to Psychological Therapies (IAPT) has trained psychology graduates to work with primary care physicians across the country in a range of evidence-based treatments.

They have substantially improved the capacity of the National Health Service to provide treatment to people with depression and anxiety-related disorders. In Tehran, in order to respond to the enormous increase in the urban population, additional modifications have been made at the community level to ensure that everyone has access to mental health as well as care for their physical health. A new group of worker, with M.Sc. degree in psychology, has joined primary care physicians at the second level of service, in order to provide psychological treatments in primary care.

In Porto Alegre and Bangalore, where important work was done in the past between mental health services and primary care, better integration between the two is introduced as "work in progress" in the former and is omitted in the latter. The enormous burden of both common mental disorders and unrecognized severe mental disorders is often not seen by psychiatrists as part of their task.

Research Findings About Community Care from Some of the Cities

In Australia, a "National Survey of Mental Health and Wellbeing" was undertaken in 1997 and 2007, which demonstrated the high prevalence of states of anxiety and depression in the Australian population. This has been followed by the Internet-based treatment packages already described.

In Verona, diagnosis alone accounts for only 6% of variance of length of stay in hospital, but including socioeconomic status accounts for 43% of the variance. Other projects have shown that patients are at almost twofold higher risk for death than the general population and this becomes extremely higher for those diagnosed with drug addiction or personality disorders. Finally, when service satisfaction was measured with the Verona Service Satisfaction Scale, weaknesses were identified in the physical layout of facilities, in the lack of involvement of relatives and in the information provided.

In London it was shown that community treatment orders (CTOs) did not reduce the rate of readmission of people with psychosis, and the trial found no reduction in overall hospital admission support that might justify the significant curtailment of patients' personal liberty. Despite evidence that assertive outreach for severe mental illness is effective, it was discontinued in London in the interests of austerity, and the patients re-assigned to services called "recovery and support teams." In Sydney the liaison service for primary care physicians and the 24 h Acute Care Service were both abandoned. In Japan the fact that costs of care in private hospitals were covered by public funds created perverse incentives to allow longer in-patient stays.

Increasing Urbanization Across the World

It is much easier to provide good community services in middle sized, relatively well-integrated cities like Madison and Verona than in large overpopulated cities like London and Baltimore – not to mention cities the size of Cairo or Tokyo. Sartorius argues that when cities reach the size of 15 or 20 million "the rules of managing smaller towns are no longer valid."

We would argue that special problems arise at much smaller populations – London, with a population of 8.7 million, being an example. Other large cities in the UK also have problems of high rates of admissions of those with severe mental illness, high rates of homelessness and crime, and people a greater degree of complexity in the combination of mental illness and substance use disorders – but London always tops the list. Stuart Bell, a Chief Executive of the South London and Maudsley Trust during these years, describes the possible advantages of having an overall Mayor of large cities – and this indeed can be very effective in achieving at least some degree of autonomy, provided that the Mayor, or the equivalent top level for the city government, has a clear interest in services for people with mentally illness, is guided by the evidence of effective practice, and invests in proportion to the level of need in the population. If such an overarching authority is either not available or uninterested, the only advice we can offer is to regard the big city as a set of smaller units of between a quarter to one-third of a million, depending on the administrative size of provider authorities. Within each of these needs should be assessed and national guidelines should be followed.

Conclusion

There are promising signs that the era of institutionalization has either ended or is soon coming to an end, in all of the 11 cities – even in Japan, the government has ordered all regions to move away from hospital-based services to develop more community-based mental health services. It is clearly much easier to move to satisfactory community services in smaller cities with a tradition of relatively stable family life – but the influential work at Madison shows that much can be done elsewhere. In low- and medium-income countries, the work at Bangalore and Porte Alegre shows that much can be achieved by cash payments to those in greatest need – namely people with severe mental illness and their family members.

The City of Amsterdam (Jan Theunissen, Matty A. S. de Witt, Wilco Tuinbreijer, Marcel Buster, and Jack J.M. Dekker)

Situation in 1996 (This paragraph is written by D. Goldberg and G. Thornicroft)

The "Amsterdam model" emerged in 1987, as an attempt to modernize and co-ordinate mental health care in Amsterdam, from the previous model where in-patient care was from a huge distant mental hospital and care in the city mainly from "RIAGGS," which provided out-patient and community care from a wide range of mental health and other professionals. They were preparing to divide the city into three sectors and to gradually move in-patients into their appropriate sector, in to various forms of residential accommodation, and with each center connected to an existing hospital in the city. At the time of our conference, they planned to close the big hospital finally in 2001. Meanwhile, the RIAGGS for Amsterdam were five in number, and co-ordination of services between them and the other forms of service for mentally ill people was incomplete.

Acute Psychiatry (Short and Medium Care) in Amsterdam

In the period 1975–1985, most psychiatric patients from Amsterdam were not admitted in Amsterdam, but in the Psychiatric Hospital Santpoort (approximately 1100 beds), 30 km outside of Amsterdam. This hospital had four short stay wards (approx. 150 beds; duration of treatment 3–6 months) and three long stay wards (approx. 300 beds; duration of treatment 6 months to 3–4 years). Apart from these, there were wards for youth and adult patients with personality disorders (approx. 65 and 90 beds; treatment duration 3 months to 2 years) and wards for adult and elderly chronic patients (approx. 500 beds; treatment duration 3–25 years). In addition to Santpoort, there was the Valerius Clinic in the center of Amsterdam with approx. 120 beds (treatment duration 2 weeks to 6 months) and two long stay units with both 50 beds.

In the period 1980–1995, the psychiatric hospital Amsterdam closed all its beds in Santpoort and created new institutions within the city of Amsterdam. The underlying thought in 1990 was to reshape the wards for short and long stay in Santpoort and the Valerius Clinic into five so-called Social Psychiatric Service Centers (SPDCs). Day-treatment should substitute psychiatric admissions and was to be central element of treatment. That should make a swift reintegration to the patient's own house possible to keep the connection of patients to their home environment intact with as little seclusion and stigma as possible.

In the period 1995–2005, five SPDCs existed with teams including psychiatrists, psychologists, and social psychiatric nurses (SPVs) and inpatient facilities. They provided the regular social psychiatric care and were aimed at preventing hospitalization and providing home treatment. But still vast numbers of patients received their care at the office (Dekker 1996). Anno 2015, three SPDC's still exist; however, they are now called clinics like in the old days. The capacity of short and long stay treatment for psychiatric patients in 1990 of 640 beds is reduced to 300 beds in 2015. So the targets in 1990 for treating the patients in their home environment with as little seclusion and as much autonomy as possible is more than achieved. A negative side effect of this development is the increase in the percentage involuntary admissions from 30% to 70% in the period 1990–2015 (van der Post 2012).

Long Term Care for Persons with Severe Mental Illness

In 1995 the removal operation from long-term psychiatric patients that hitherto had lived for years in Santpoort in the dunes, back to the city of Amsterdam, was completed.

Most patients had moved to the locations in Amsterdam. Duurkoop (1995) investigated the clinical and social functioning and the quality of life of these patients before and after the removal. He concluded that all long-stay chronic patients could be placed from the hospital to alternative housing facilities in Amsterdam. However, their psychiatric symptoms did not change, nor did their social functioning. There was no increase in the use of neuroleptics, a decrease in seclusions and time of seclusion and in the most severe group of patients, daily living skills improved.

In 1997 it was estimated that the city of Amsterdam counted about 4700 people who suffered from severe and long-term mental health problems and were in care in one of the specialized mental health services in the city. It was estimated that another 2400 patients with severe mental illness did not receive help for their problems or received help from a general health practitioner (Kroon et al. 1998).

In those years, local spokesmen and media proclaimed that experiences of increased inconveniences and hassle in the city could be linked to people with severe mental problems that before were long-term custodised in psychiatric hospitals far from the city. In a survey study in 2005, Theunissen et al. (2013) concluded that about 4700 patients that met the criteria for severe long-term mental illness were in care in one of the psychiatric facilities in Amsterdam. Sixty percent of them suffered from schizophrenia, their societal participation was low, and they had many unmet needs, especially in the social domain. But in spite of that they rated their quality of life as rather positive. With regard to the perceived annoyance in the city, it was concluded that the inconvenience could not be attributed to this group of patients in care; in most of the psychiatric crises this group experienced, there was an almost direct response from the mental health system.

It became also clear that most of the patients, who in the psychiatric hospital, as a complete care institution, received a daily activity program, now lacked meaningful day-to-day activities as they were living more or less independently in the city. Loneliness and lack of social contacts were commonly mentioned by patients as the main problems in daily living. Between 1990 and 2000, nine day-care centers were established, some in the facilities of the psychiatric hospital and some as independent units that after some time were run by clients themselves. Theunissen (2006) investigated the outcome of four day-care and rehabilitation-centers in terms of clinical and social functioning, needs for care and quality with a controlled design. He found no difference in outcome in terms of psychiatric symptoms and social functioning between patients that received these services, compared with patients who received the regular mental health care.

One of the main changes in care for the severe mental illness-group in the last decade is the acknowledgment of the importance of the user-perspective in treatment and care. User-knowledge, based on former experiences with disease and treatment, is seen as a necessary addition in treatment.

Public Mental Health Care

Treatment of severe mental illness clients living in society proved to be difficult, as a result of a lack of illness-insight and of motivation for and compliance with treatment. This group increased in the period of de-institutionalization in Amsterdam described above and needs to be approached and engaged actively. The reason to offer them care was sometimes a public one and not an individual one, and public mental health care therefore plays an important role.

Patients often received care from several institutions, resulting in overlapping care or patients falling out of the system, because of a lack of coordination. In response, the public health service in Amsterdam started a person-centered approach: for every patient one care-coordinator was appointed and an integrated care plan with common goals on the domains of care, income, daytime activities, and housing was formulated. From 2007 to 2014, in Amsterdam, over 5000 homeless persons were included in this project and had an integrated care plan. About 73% reported to have achieved their goals at least partly after 2.5 years and 72% achieved the outcome measure of a stable mix, which was first defined as a stable situation on care, income, and housing.

The public health service also provided harm reduction and care for chronic heroin addicts. This developed from a low threshold methadone-bus in 1979 into three integrated care facilities in 2013, where the public health service and the social welfare agency work closely together. Addiction care, mental health care, somatic care, social work, welfare, and reintegration were offered in an integrated way. To those for whom methadone treatment provided insufficient results, additional heroin-supplementation is offered to achieve a better quality of life.

The "Safetynet department" was introduced as a response to an increase of worries about nuisance and concerns from police, housing corporations, other professionals, and citizens. It consisted of social psychiatric nurses, psychologists, and psychiatrists who make house calls in which they make an inventory of the problems and the support and care systems and direct patients (back) to the appropriate care and support. Safetynet also developed interventions for upcoming problematic populations. An example is the Top600, aimed at youth that commit violent high-impact crimes. Based on the assumption that many psychosocial problems underlie the criminal activities, Safetynet has implemented a social psychiatric screening, resulting in referral to appropriate care into the city's approach that before mainly focused in punishment and repression. This intervention is now also implemented in other target population such as citizens suspected of radicalization.

Discussion: Complexities and Solutions

The main change in the funding of care for patients with severe mental illness has taken place in 2015. The funding of all nonmedical treatment and support, also the sheltered living facilities, till then funded by the general health insurance companies according the AWBZ, was then transferred to the municipal government. The municipalities have, next to the responsibility for the welfare and health of their citizens, other tasks like reducing annoyance in the neighborhoods and reducing the dependence of citizens on welfare. Sometimes these tasks do not concordate. To avoid collisions of interests, with the patient as the victim, complex networks around the patient have been implemented with representatives of benefit agencies, the employment office, the housing association, social welfare, sometimes the police and the psychiatric nurse around the table.

The other main change is the transference of care for patients with severe mental illness to basic care. The part of the population with stable functioning and mild psychiatric symptoms is transferred from the FACT teams to the general practitioner who is supported by a nurse trained in psychiatry.

The future will tell how these complexities will influence the life and recovery of the citizens with severe mental disorders.

The City of Baltimore: A Strong Commitment to Community Based Services (Stephen T. Baron, Kathleen Westcoat, and Crista Taylor)

Situation in 1996 (This paragraph is written by D. Goldberg and G. Thornicroft)

Baltimore was included because it offers a radically different model from Madison – it operates through a Board consisting of government officials, community leaders, and users of mental health treatments and is responsible for regulating the system in all of the three sectors of the city, but not for providing services. Its goals are to ensure a high quality of care, of continuity of care, and the efficient use of scarce resources. Care is provided by seven Lead Agencies, of a sector of Baltimore serving between 200 and 250 K residents. They previously described the expansion of the range of services available to the Case Management Units in each sector, the provision of crisis care, and improvements in the continuity of care.

National and Federal Developments

The first major change was initiated on July 1, 1997, when the state of Maryland received federal approval to have all Medicaid recipients placed in Managed Care Organizations (MCOs) for their primary health care. The federal approval allowed Maryland to carve out specialty mental health services from the MCOs and have the mental health system managed by the State's Mental Hygiene Administration. Specialty mental health care was defined as state certified mental health agencies (primarily private nonprofits) or a licensed mental health professional (psychiatrist, psychologist, nurse practitioner, professional counselor, or social worker). At the time of the change, about 50% of the revenue to support the Public Mental Health System's community-based services was awarded through contracts to the state certified mental health agencies with the other 50% coming from reimbursement fees primarily from the Federal Medicaid program. Medicaid funds are a joint federal/state match and in Maryland the match is 50% federal and 50% state dollars.

This change moved the system into predominately a fee for service structure with less reliance on grant funding and much more emphasis on reimbursement for services, primarily from Medicaid. To implement this change, the State of Maryland contracted with a private sector Behavioral Health Organization to serve as an Administrative Service Organization (ASO) to manage the system by paying claims, credentialing individual practioners, performing utilization management, as well as issuing regular data reports. The State retained authority for determining eligibility for services, identifying the services to be reimbursed and the reimbursement levels, and setting the parameters for utilization management. The early years of transitioning the system were difficult on a number of levels. Many providers lacked internal billing and claims systems, the State had limited experience in developing the new policies and procedures needed to manage the fee for service structure, and the ASO had difficulty managing the claims payment structure. It took a good 5 years for the system to mature where the new structure was well understood and was performing at an acceptable level.

Maryland eagerly participated in President's Obama's Affordable Care Act (ACA) which expanded Medicaid eligibility. The expansion included single adult males many of whom are dealing with substance use issues. With the increase in the number of individuals eligible for Medicaid, the state expanded the role of the ASO to also manage reimbursement for substance use disorder services. Similar to the financing structure of the public mental health system, there was a reduction in grants to providers and an increase reliance on reimbursing services through a fee for service structure.

The City of Baltimore

Baltimore City is home to about 10% of the state's population and, due to the disproportionate number of individuals living below poverty, represents over 30% of the users of the state's public mental health system. Over the past 20 years, there have been a number of changes at both the City and State levels. These changes have included the following.

- The organization of the funding structure for the Public Mental Health System
- The increase in availability of data
- The incorporation of Evidence Based Practices into the public system
- The merger of the management of the mental health and substance use disorder systems into a public behavioral health authority
- An increase in collaborations with the criminal justice system
- Recognition of the value of peer support and the creation of peer driven services.
- The development of a full continuum of integrated, crisis response services

The system has stabilized and supported an increase of utilization over the years. In Baltimore City from 2011 to 2015 the number of individuals seen in a year grew from 41,492 individuals to 53,370, a 29% increase over that 5 year period. In addition, the average cost of care in the City went down over that 5-year period from an average cost of $5438 in FY' 2011 to $4.939, a 9% decrease for that time period (BHSB website, annual reports).

The change has enabled the state to generate and use much more reliable data on utilization and cost information since much of these data are obtained from claims paid. Data on cost and utilization did not exist before July 1997 change. Another example of using data to better manage the public behavioral health system has been the state's initiation of its Outpatient Management System (OMS). OMS requires providers to report on a number of indicators every 180 days. The OMS has the capacity to report on areas such as:

- Reduction in thoughts of self-harm
- Reduction in depressive symptoms
- Reduction in consumers smoking
- Reduction in consumers' incarceration
- Reduction in homelessness

The next significant change occurred in 2013 for Baltimore City, when the City merged the two entities that managed the City's mental health system (Baltimore Mental Health Systems (BMHS)) and it substance use disorder system, Baltimore Substance Abuse Systems into a new entity Behavioral Health System Baltimore (BHSB) to form a single integrated behavioral health authority. The motivation for this change was the recognition of the large number of individuals with co-occurring disorders.

In July 2014, the State of Maryland merged its substance use and mental health treatment systems to create an integrated behavioral health authority at the state level. At both the city and state levels, much time has been spent on organizing the administrative functions of the system. The next step of making changes to assessment and treatment protocols to ensure that the integration of the administrative structure benefits individuals with co-occurring disorders is still in process.

Changes to the Mental Health Service

Another change over the past 20 years has been the incorporation of Evidence-Based Practices (EBPs) in the State and City's public system. Beginning in the late 1990s in partnership with the University of Maryland and Dartmouth College, BMHS began looking at methods to incorporate EBPs in its public system. The State's Mental Hygiene Administration shared in this commitment and worked to bring EBPs state wide. Since the early 2000s, the state has contracted with the University of Maryland to expand evidence based practices and to assist with training and monitoring of the fidelity of the practice. Over the past decade, the State has incorporated EBPs into the public behavioral health system including the following:

- Assertive Community Treatment (ACT)
- Supported Employment
- Trauma-Focus Cognitive Behavioral Therapy (TF-CBT)
- Functional Family Therapy (FFT)

- Integrated Dual Diagnosis Treatment (IDDT)

To support the expansion of EBPs, the State has employed strategies that include the following:

- Measuring adherence to the EBPs fidelity scale
- Reimbursing at a higher rate for EBPs
- Providing ongoing technical assistance

These strategies have led to an increase in the use of EBPs in the City and State.

Since the publication of the "Mental Health in Our Future Cities," there have been a number of challenges for public behavioral health systems at the City, State, and Federal levels. There is a growing recognition that a large number of individuals with mental illness are in contact with the criminal justice system. In response to this, BHSB has implemented a number of interventions which have been guided by the sequential intercept model. The sequential intercept model provides a conceptual framework for communities to organize targeted strategies for justice-involved individuals with behavioral health disorders. The sequential intercept model targets the areas of law enforcement, initial detention/initial court hearing, jails/courts, re-entry, and community corrections (https://www.prainc.com/wpcontent/uploads/2015/10/SIMBrochure.pdf).

Initially begun by both BMHS and BSAS, BHSB has maintained a focus and commitment to these areas. BHSB has developed a close relationship with the Baltimore City Police Department that provides active training to new police officers, it has a close relationship with the judges and other court personnel and actively supports specialized drug and mental health courts, assists court-ordered individuals to assess community services, and works with its provider network to support individuals returning from incarceration as well as assisting with transition planning from state hospitals for individuals found incompetent to stand trial or not criminally responsible and being conditionally released.

Baltimore has over the past 20 years has made a strong commitment to peer activities. BHSB and its predecessor agency established three Wellness and Recovery Centers. These centers are operated by primary consumers and offers peer-led support services. Each of the centers has a unique focus as one was established to support homeless individuals with a mental illness and another has a primary focus on supporting individuals with a mental illness who are lesbian, gay, bi-sexual, transgender, and questioning (LGBTQ) and another supports the Transitional Age Youth (TAY) population.

A comprehensive, integrated (substance use and mental health) crisis response system is the backbone of any successful behavioral health system. A good crisis system reduces harm including death and reduces overall costs for the behavioral health system. The Crisis system in Baltimore City consists of a 24/7 crisis line for both mental health and substance use disorder calls and has an entity that provides mobile crisis services, a community-based residential crisis facility and a residential detoxification program. A separate crisis response system for children and youth also exists, which provides mobile crisis services and community-based urgent care for youth and their families.

The Baltimore crisis system has become a valuable component of the City's community based system as demonstrated by its overall growth in the following areas:

- Over the past 10 years, the hotline has seen a 73% increase in calls. In FY'06 there were 26,833 calls and by the end of FY'16 there were 46,435 annual calls.
- The number of crisis bed admissions many of which are divergence from an inpatient admission has grown from 593 admissions in FY'08 to 669 admissions in FY'15, a 13% increase.
- During the past 7 years, mobile crisis calls have shown an 18% increase from 2285 annual deployments in FY'09 to 2793 in FY'15. (BHSB website, annual reports)

The Future

Many challenges remain in ensuring that the Baltimore City residents have access to a rich range of community services. The experiences of the past 20 years have taught us that we must continue to keep our focus on a number of areas including the following:

- Maintaining a strong local behavioral health authority with the authority and responsibility to manage the public system
- Ensuring services are easily accessible, individualized, and flexible
- Ensuring that data are collected and used
- Developing strong collaborative relationship with other agencies including but not limited to: Housing, Education, Police, Corrections, Courts, and State and Federal governments
- Ensuring there is accountability present in the system
- Creating a financing structure that includes incentives and risks and supports strong clinical care
- Ensuring that the system has well-defined outcomes and knows if individuals are benefiting from receiving services

Bangalore City, India (1996–2016) (R. Srinivasa Murthy)

Situation in 1996 (This paragraph is written by D. Goldberg and G. Thornicroft)

Bangalore was included because of its participation in research into the role of primary care in providing mental health services. It is the home of the National Institute of Mental Health and Neurosciences (NIMHANS), but services are also provided by the major general hospitals in the city, all of which also have both in- and out-patient services. The role of NIMHANS has been to provide training for primary care staff, who are likely to produce the least stigmatizing form of service for those with psychological illnesses.

National Changes

During the two decades, the National Mental Health Programme, formulated in 1982, has seen increased support. From only one district mental health program, there are more than 200 districts covered by the program of integration of mental health care with general health care. The rapid growth of the private sector mental health care with little regulation is a matter of importance. The fact that the health is a state subject makes the growth and development of mental health care across states an uneven one. This is clearly seen in the recent two reports. The National Mental Health Survey (NIMHANS 2016) and review of the mental health care by the National Human Rights Commission present a mixed picture of progress and problems. The finding of 13% of the population with a mental disorder and over 80% not under care presents a continuing challenge. The other observation of the unequal spread of services across the more than 30 states of India and the lack of mechanisms to monitor the mental health program are areas of concern. The suggestion to form a National Mental Health commission is timely and can address the barriers to care identified during the last two decades. The other initiatives to support mental health needs of the population of significance are (i) in October 2014, a new National Mental Health Policy of India was accepted; (ii) the Mental Health act, 1987, is under revision with a more progressive Mental Health Care Bill, 2016; and (iii) there have been other legislative changes relating to persons with disabilities, women, sexual violence, and juvenile justice.

Bangalore City

The progress of the last two decades in Bangalore city (with about 1% of the population of the country) illustrates both the mental health challenges of a rapidly growing population and the innovative approaches that can address the emerging mental health needs of the population. India entered an important period of economic and trade liberalization in 1991, and after 25 years, there is full evidence of the benefits of "opening up" of the economy as well as the growing "inequity" in the society. During the period of 1996–2016, in Bangalore city, there were dramatic changes in the physical growth of the city, population characteristics, along with rapid industrialization and growing mental health needs of the population.

The last review of mental health city services was around the time of the millennial years. Of the many changes around 2000, the most important was the growth of the information technology (IT) industry in the city. Bangalore has come to seen as the "silicon valley" of India. There are estimated over 1000 IT companies in the city. This has meant the internal migration of professionals from all over the country and from abroad. The population itself has grown from 4 million in 1991 to 11.5 million in 2016 (Bangalore district is growing at twice the rate of Delhi, 7 times the rate of Chennai, and 10 times the rate of Mumbai and Hyderabad cities). There is commensurate growth of high rise apartments and increase in the number of vehicles (5.5 million, with an annual 10% increase). The living conditions have both improved with the addition of a large number of fly-overs, metro services for mass mobility of the population, a new airport, along with an increase in populations living in slums (about 10%). A visitor coming to the city, after 20 years, would feel lost with the changes associated with the growth and development of the city.

The above changes have resulted in newer mental health needs of the population and novel approaches to deliver mental health care to the population.

Of the many changes at the level of individuals, families, and communities, three changes are significant. The youth of today (over 50% of population is less than 35 years of age) is an aspirational generation. They are more inward looking, success oriented, and are willing to challenge the traditional values and relationships. This change leads to greater potential for achievement but also exposes them to risks of not being able to depend on traditional supports in times of crises. A reflection of this change is seen in the increasing use of intoxicants/prescription drugs in the city, suicides, and Internet addition (Shepherd-Smith 2012). At the family level, the move from joint/extended families to nuclear families is the most striking change. The positive aspect is the opportunities for growing beyond the "controls" of the joint families. However, in times of crisis, there is less of supports either for the self or to growing children, elderly persons or in situations of marital problems and crisis situations (Srinivasa Murthy 2003). The large number of alternative community care facilities, from day care facilities, short and long stay facilities in the private sector is a reflection of the lack of families to care for patients with chronic illnesses. The internal migration of young people and lack of assimilation of the migrants to the city are also credited with the growing violence, substance abuse, and homelessness.

Of the many special population groups, the wide range of new groups of workers has attracted the attention of mental health professionals. The artificial work of "call centers" and the competitive IT industry are associated with higher rates of mental disorders in this population (Kalyanasundaram et al. 2007). The growth of slum population having limited accesses to services is an another mental health challenge.

New Mental Health Initiatives

There have been a number of mental health initiatives to address the emerging needs. Further efforts have been made to integrate mental health with general health, both in the public and voluntary sectors (Pradeep et al. 2014). Towards promotion of mental health and prevention mental disorders, life skills education in schools has been implemented and demonstrated to be effective (Bharath and Kumar 2010). Wellness initiatives of National Mental Health and Neurosciences (NIMHANS), initiative with an IT company, titled "White Swan" is an attempt to both make mental health the focus rather than only mental illness. This has resulted in a number of activities like a newsletter, public lectures, empowering youth, and caregiver trainings. There is also an attempt to share mental health skills with the general population aimed at prevention of mental disorders and promotion of mental health using mass media. Another important development is the large number of voluntary organizations working in different parts of the city to provide basic mental health care to the population. These agencies are also training general population as "counselors" to provide mental health first aid. One other initiatives is the use of the information technology and social media to reach the population with wide range of mental health needs. There are a number agencies ("YourDOST"; nSmiles, etc.) working in the city using this approach. Driven by the personal loss of a friend by suicide, the founder of YourDOST initiative provides free and anonymous online counseling. At YourDOST, people seeking support are connected to the right experts – psychologists, psychotherapists, counselors, life coaches, and career guides – who help them through free and anonymous online chats. The website operates round the clock. It has also motivational literature and provides counseling in several Indian languages. "nSmiles" focuses on self-evaluation, self-care, and graded support with emotional counseling. There are also specific initiatives address the needs of students and cancer survivors (CIPHER, CARER program) aimed at providing online support to specific health groups. These are significant developments and needs to be understood to see what section of the population it will cater to and what mental health issues are feasible to be addressed by these approaches.

Another initiative of the last two decades is the family support group, AMEND. This group of parents have not only provided support to each other, but taken up advocacy activities like the organization of emergency ambulance support and welfare benefits to persons/families with chronic mental illness (Srinivasa Murthy 2003).

Stigma is an important barrier to care (Loganathan and Srinivasa Murthy 2011). In the city for over 10 years, there are regular public mental health education programs over the television, radio, and other mass media. Specific studies to evaluate some of the interventions have been undertaken. Fighting stigma will have to be a long-term goal and should use a wide range of initiatives.

The other major development is the wider use yoga and meditation as part of psychiatric services. To date, there is evidence of use in specific mental disorders either as a standalone treatment or as an adjunct to psychiatric treatment (Verambally and Gangadhar 2016). This development has important implications:

First, these measures are accepted by the community.

Second, they are easy to deliver.

Third, these measures can be practiced by the individuals and thus can lead to empowerment of the population. Fourthly, the growing scientific evidence of the value of yoga and meditation gives credibility to their value in mental health care.

Recognition of the importance of mental health needs of the persons diagnosed with diabetes and cancer, efforts have been initiated towards self-care and emotional health of caregiver. Common to all the new initiatives, is the need for quality control and monitoring of the work of large number of counselors and care providers with varying degrees of skills.

Conclusion

During the 20 years of review, there have been dramatic changes in the mental health in the country. There is greater recognition of the importance of the mental health of the population. There is now a mix of both public sector and private sector initiatives. There is growth of the Bangalore city along with recognition of emergence of new mental health needs of specific population groups. The growing need has been addressed both by local small scale initiatives and by use of traditional knowledge of the society, along with policy changes at the national level. Recent reviews also highlight the urgent need for greater leadership of academic institutions and greater use of indicators of care, especially the outcome indicators to monitor the changes. The new openings utilizing the information technology hold promise for the future to reach care to all those who need mental health care.

Copenhagen Mental Health Services (Marianne Kastrup and Klement Dymi)

Situation in 1996 (This paragraph is written by D. Goldberg and G. Thornicroft)

The mental health service in Copenhagen was influenced by the Psychiatric Plan of 1987: this aimed to promote collaboration with primary care physicians and hospital services, to proceed with de-institutionization, and in-patient beds were to be reduced. Services at a given general hospital were to have the same catchment area as other departments, and these were to be the same as that of a community mental health center. These were to be established over the whole city, dealing with populations of between 25,000 and 55,000 people. All departments should have a comprehensive responsibility and be able to deal with forensic problems, as well as drug and alcohol dependency.

In 1997 there was a general principle of Catchment areas, meaning that each psychiatric department had in general total responsibility for its catchment area. There were exceptions that Copenhagen still had a psychiatric hospital (Sankt Hans) for forensic patients, long-term care, and special long-term treatment of abuse, drawn from more than one catchment area.

National Developments

National Board of Health Report on Psychiatry

The National Board of Health established a committee to produce recommendations for the field of psychiatry (Kastrup 1998). In their report, they list a number of challenges: the significant excess mortality among psychiatric patients, the uneven quality reported in the assessment and treatment of psychiatric patients, the extensive use of coercion, and the lack of continuity of care between regional and local municipal care.

The aim of the plan is to ensure and prioritize the field of psychiatry – both the promotion of mental health and the treatment of psychiatric illness – so there will be the same standards and rights for psychiatric and somatic patients. As of 2015, all the different levels of specializations from basic to highly specialized functions, with one exception (dangerous forensic patients), are represented in the Capital Region.

National Strategy for Psychiatric Research

In 2015, three ministries: for health and prevention, for children, and for education and research published a joint plan for psychiatric research (Ministeriet for sundhed og forebyggelse 2015). The aim is to promote research on early intervention and the causes and mechanisms leading to mental illness that increases our understanding of the mental disorders and to promote interventions that may prevent mental illness, reduce stigmatization and the excess mortality due to physical illness, and support that mentally ill may obtain a meaningful and satisfying life.

The City of Copenhagen Today

The city has undergone significant demographic changes during the last decades as it has become increasingly fashionable among young, well-to-do, to live in the city center. We see a continuous influx of inhabitants and an increasing number of students to the many educational institutions. Further, the migrant population is increasing, and the city is becoming more ethnically diverse, with the proportion of ethnic minority children in the public schools increasing.

Mental Health Services in Copenhagen

Administrative Developments

Since 1997, the mental health services in Denmark have been completely reorganized (Kastrup 1998). In Denmark, psychiatric treatment is primarily a public task, and health services are still publicly funded via the taxation system. With the structural reform in 2007, the country became divided into five regions, and as health authorities, they have the overall responsibility for all health services.

The structural reform meant that the Copenhagen Hospital Corporation that comprised both somatic and mental health services was dissolved. Instead, a new organization, the Mental Health Services of the Capital Region of Denmark, was created, comprising the entire Greater Copenhagen with approximately 1.6 million inhabitants. Today the mental health services of the Capital Region are administratively and economically completely separated from somatic health services. Mental health is separated from general health: nonmedical staff members have been much in favor of these changes, while many psychiatrists have felt it is a loss that the health services have been divided. The new organization that is one administrative authority constitutes the largest psychiatric hospital in Denmark (see Table 1).

Table 1 Copenhagen Mental Health Services data comparing the situation in 1997 and 2015–2016

Description of the situation in the 1997 book	The current situation
Cph population 471,300 total foreign population in 1995: 44,661	Cph population 594.245 (2016) total foreign population, non-Danish citizens (2016) 93.320; non-western migrants 58.261 (2015) second generation, non-western migrants 28.034
In 1997 the psychiatric services of the Copenhagen Hospital Corporation had a total of 993 beds At the Copenhagen Hospital Corporation, the number of psychiatric beds was approximately 1.7 per 1000. In 1995 all Danish psychiatric departments had 34,248 admissions and 33,912 discharges. Of these, 4895 admitted and 5936 discharged from the municipality of Copenhagen	Mental health services in the capital region provide treatment to 32% of all mental patients in Denmark (2015) and consists of 10 Mental health centers (each serves a catchment area) providing: • Inpatient units • Outpatient clinics • Community mental health services • Outreach services • Specialised treatment 46,568 patients in treatment 544,939 outpatient appointments (2015): 1176 inpatient beds 13 community mental health teams 23 assertive outreach teams 8 mental health care of older people teams 12 early intervention in psychosis teams 37 outpatient clinics (2015): Approx. 4400 staff (full-time equivalent) https://www.psykiatri-regionh.dk/english/Sider/default.aspx

There has been a substantial increase in the absolute amount of money allocated to mental health. A press release from Danish Psychiatric Association stated “that there has been a reduction of beds by 22% during the last 8 years despite an increase by 48% of patients in contact with psychiatry” (January 2015). The professional background of the staff has also changed, with fewer psychiatrists, and more non-psychiatrists.

Therapeutic Developments

In parallel with the rest of Denmark, the Capital Region has during the last decades focused on community treatment and outpatient care. During the same period, the number of inpatient beds has decreased.

Community mental health centers were already established in 1997, but since then assertive community teams (ACT teams) have been established at all psychiatric centers, aiming to provide outreach care to the severest, primarily psychotic patients. Further, there has been established a number of early intervention of psychosis teams that focus on young persons with newly diagnosed psychosis.

Emergency care is part of the psychiatric centers, but rationalization of services has resulted in the number of emergency clinics being reduced, and today there are only a few left with a 24-h service.

The mental health centers have to a large extent, each been concentrating on a particular catchment area, providing comprehensive services to the population from that area, and citizens were referred to the local service of their catchment area and only exceptionally were they to receive care from other centers. These changes have been accompanied by the introduction of assertive teams and teams for early intervention at each psychiatric center.

This policy has over the recent years been questioned, in particular by academic psychiatry, as there have been strong voices advocating for more specialized treatment options across the catchment areas, instead of all centers providing all kinds of care. As a result we today see a number of "competence centers" that receive patients from the entire region. Thus, there is presently a competence center for affective disorders, for transcultural psychiatry, for forensic psychiatry, for psychotherapy, for recovery and rehabilitation, for dual diagnoses, and for suicide prevention. These competence centers are located at the different psychiatric centers and are part of them.

The Hospital Plan for the Capital Region (2015)

The vision for psychiatry is to build psychiatric services that are person-centered, based on recovery and rehabilitation, and provide a variety of treatment options. To provide psychiatric care that is easily accessible, with a particular focus on outpatient care, with focus on prevention and continuity of care, and with minimal use of coercion. This is a continuation of the "Vision for future psychiatry" from 2011, emphasizing the need for precise diagnoses, new types of medication, new treatment options, but also a wish to treat patients referred, as fast as possible with the assumption that the earlier treatment starts, the more likely is the positive outcome. Long-term admissions have negative consequences for the patients' self-perception (Region Hovedstaden 2011).

Digital Communication

The potential for better coordination and collaboration among the various stakeholders in Copenhagen is enhanced further in the last decade, especially because of the new digitalization of health information at national level, the electronic patient journals (EPJs) that started in 2007, and other programs/platforms related with health/mental health.

The aim is to merge all health information in one file system that is accessible to health professionals as well as patients, in order to reduce mistakes and increase security.

Assessment of Strategies for the Future

Increasing attention is being paid to quality assessment, and the Mental Health Services of the Capital Region has been accredited by the International Joint Commission. Lean initiatives are introduced at all levels in order to increase productivity and minimize unnecessary activities. A 3-year plan "Focus, simplification, improvement" has as one of its objectives to improve the measurement of quality and outcome of treatment (Lau 2016)

Coercive Measures

Denmark has been criticized internationally for its extensive use of physical restraints. Several initiatives have been taken to reduce coercion. Increased resources and projects directed at coercian have resulted in a drastic reduction in the use of coercion in the Capital Region. This focus has had a clear impact on the perception of coercive measures and interest in finding alternatives (Region Hovedstadens Psykiatri 2016).

Mental Health Promotion

The municipality of Copenhagen has produced for 2015–2018 a paper describing factors promoting and inhibiting mental health outlining a series of possible interventions. The action plan for mental health describes prioritized areas and invites to a dialogue with other authorities in order for Copenhagen to achieve the goals for the city's health policy. The health policy of Copenhagen (2015–2025) has as a primary goal to ensure all its citizens a longer life and with a better quality of life. The current social inequity in health should be abolished and all citizens given possibilities to enjoy a healthy life (Københavns Kommune 2015a, b).

Challenges

The Mental Health Services of the Capital Region are facing a number of challenges including:

- There is among many psychiatrists a concern that the number of beds has decreased too drastically with serious consequences for the provision of satisfactory treatment.
- A large number of forensic patients occupy the limited number of available psychiatric beds, which may change the atmosphere on the psychiatric wards.
- Health services have been increasingly criticized to focus too much on control, checklists, bureaucratic guidelines, and so on, at the expense of focusing on the contact with patients as the core performance for health professionals.
- Many experience that the health system is governed by an economic agenda and a demand for increased productivity at the expense of the academic interest and scientific development of the discipline.
- Psychiatric patients have still an excess mortality due to, for example, insufficient treatment of their physical disorders, unhealthy lifestyle, and side-effects of medication.

- There is a need to reduce any overuse of psychiatric drugs. There is a concern among lay and professional media on the increased use of psychotropic drugs

Conclusion

The Mental Health Services are undergoing constant changes and the goals are ambitious including a wish to implement a culture of constant improvement for the benefit of the mentally ill, to work towards more equality with other health services, to reduce stigmatization and work for better quality of care and patient satisfaction, and to let patients and their relatives increasingly have influence on the care provided.

Mental Health and Illness in Kobe, Japan (Naotaka Shinfuku)

Situation in 1996 (This paragraph is written by D. Goldberg and G. Thornicroft)

Kobe had partially modernized their mental illness services by lowering the number of psychiatric beds relative to the national average. Day care and some consultation services are provided by the Prefectural Mental Health Center, which also plays an administrative role in promoting prevention and rehabilitation. Local health centers had recently begun to concentrate on geriatric and mental health services in 1996. Kobe has encouraged family training programs in most of its 41 Health Centers and responded to the challenge of the Hanshin Awaji earthquake by setting up a Psychological Care Center to deal with the many people with posttraumatic stress disorder. Full modernization was hampered by a government policy of stimulating the building of psychiatric hospitals by the private sector, resulted in an enormous increase in bed numbers over the whole country. Fees for treatment in private hospitals are covered by national insurance schemes, so length of stay is very much longer in Japan than in other high-income countries. Ambulant treatment is largely by private psychiatric clinics. The need to increase the number and range of residential care facilities if a large number of long-stay psychiatric patients are to be discharged from hospital is acknowledged.

National Developments in Mental Health Policies

Several changes took place in developing psychiatric services in the past 22 years from 1995 to 2016. The Ministry of Health, Labor and Welfare (MHLW) has ordered all the municipalities in Japan to move away from hospital-based services to community-based mental health services. In 1995, new "Mental health and welfare law" was enacted to merge health and welfare services for the persons suffering from physical, developmental, and mental disabilities.

Mental health and welfare law promoted the construction of a variety of community mental health facilities such as day care, group home, sheltered workshop, and so on. At the same time, psychiatric social workers have been recognized as important disciplines for community psychiatry and they have been licensed for legalizing their contribution.

Changes of Financing System in Psychiatry

The MHLW introduced the financing system to pay less for long-stay patients. This has decreased the population of schizophrenic inpatients in Japan. In 1995, schizophrenic patients occupied more than 80% of the inpatient population in psychiatric hospitals. In 2015, the proposition is less than 60%.

Instead, the number of persons with dementia has increased as psychiatric inpatients. The MHLW planned to decrease 70,000 psychiatric beds in the 5 year plan. But the decrease was minimal as private psychiatric hospitals were reluctant to reduce the beds, contributing their major source for revenue. The MHLW introduced the financing incentive to cover psychiatric consultation fee and encouraged the setup of private psychiatric clinics. This policy has succeeded in opening up of private mental clinics which has mushroomed in cities. Within 15 years, the number of private psychiatric clinics increased to more than 6000. They are treating persons suffering from depression, neurosis, and stress-related problems.

Depression and Suicide Prevention

From 1998 to 1999, Japan experienced the sharp increase of suicide from 24,000 to 31,000. This was due to the change of traditional employment system.

Globalization affected the middle aged workers who lost their long-term jobs. The high rate of suicide continued more than 10 years and the government promulgated "Suicide prevention law." The law highlighted the importance of the prevention and early treatment of depression.

Social Withdrawal and Hikikomori

There have been increasing number of young people who avoid social contact and stay home. They often communicate only over internet and spend times in playing Internet games. According to the survey of the national broadcast in 2005, so-called hikikomori was reported more than 1.6 million. Their pathologies are wide ranged. Our survey revealed Hikikomori is not limited to Japan. Support or help by ex-hikikomori persons is the most effective to save them from the state of hikikomori (Kato et al. 2012).

Senile Dementia

Japan is experiencing the rapid aging of the population. In 2015, the elderly population over 65 years was more than 26%, while the young population under 15 years was less than 16% and the number is shrinking. Age is the important risk factor for cognitive impairment. The biggest health and social concern at present is how to cope with the increases of persons with dementia.

Nursing Care Insurance

In 2000, MHLW started the nursing care insurance system to provide care at home and in the communities. This was in addition to the national health insurance system covering all medical services at hospital and clinics. The nursing care insurance has enabled to provide comprehensive medical and care service at home and in the communities for patients suffering from all kind of disabilities from physical frailties, mental disabilities to cognitive impairment. Persons with disabilities can receive necessary services at home and nearby communities. These services include the home visits by nurses, rehabilitation specialists, and care providers.

Forensic Psychiatry

In 2003, the MHLW passed the law to set up special wards for psychiatric patients who committed serious crimes. The enactment of this law was facilitated by a social event of mass killing of primary school students by a person with a history of psychiatric hospitalization. The law ended the custodial role of psychiatric hospitals. A number of private psychiatric hospitals had functioned as private prison for mental patients who committed crimes and were hospitalized as involuntary admission.

Changing Attitude of Psychiatrists

The Japan society of Psychiatry and Neurology (JSPN) was occupied by internal fighting since 1970s. The annual congress of the JSPN had no scientific program for more than 20 years. The official journal was also stopped publishing. The organization of the 12th Congress of the 2002 World Psychiatric Association (WPA) in Yokohama ended the internal fighting. The event opened up the eyes of Japanese psychiatrists to the world (Shinfuku 2012). One of the big changes of Japanese psychiatry from 1995 to 2016 was the change of attitude of Japanese psychiatrists. They are now trained to pay more attention to the human right of inpatients. They became aware of the trends in psychiatric services given from foreign countries. They consider community psychiatry as the norm for the future psychiatric care in Japan.

The City of Kobe

Kobe is the capital of Hyogo Prefecture and one of Japan's ten largest cities. Kobe is a city on Osaka bay in central Japan close to Osaka and Kyoto located between the sea and the Rokko mountain range. Kobe is well known for its signature marbled beef and scenic setting of mountains framing the harbor. Kobe is considered one of Japan's most attractive cities. Kobe has been an important port city for many centuries. Kobe port was among the first to be opened to foreign trade in the nineteenth century along with other ports including Yokohama, Nagasaki, Hakodate, and Niigata.

The Great Hanshin Awaji Earthquake

The Great Hanshin Awaji Earthquake, with a magnitude of 7.2 on the Richter scale, hit the southern part of Hyogo Prefecture and neighboring areas at 5.46 a.m. on January 17, 1995. Kobe was in most damaged area where, houses, buildings, and railway were collapsed and where "life line" services such as water, gas, electricity, and telephone were interrupted for a long period. The great quake killed more than 5500 people on the spot and left about 310,000 people homeless.

I personally experienced the Great Hanshin Awaji Earthquake on spot at that time. I have also been witnessing the recovery process of victims in the Kobe area over 20 years. As psychiatrist, I have also been observing the changes of psychiatric problems of the Earthquake victims over the time (Shinfuku 1997, 1999).

Mental Health and Illness in the City of Kobe

Mental health policy of the Kobe city has been adapted in the planning in national mental health policy. From 1995 to 2016, major changes took places in mental health of the population of Kobe city. Also new developments were observed in psychiatric services reflecting the changes in national mental health policies.

Mental Health Problems Resulting from the Hanshin Awaji Earthquake

Hanshin-Awaji earthquake made a big damage to the city and the population in Kobe in 1995. But Kobe city recovered from the damage within a few years. The achievement were the results of hard works and concerted efforts of Kobe citizens, volunteers, central and local governments officers as well as many supporting agencies. More than 100,000 habitants lost their housings and had to live in shelters and then temporary housings. Many victims have suffered from long term physical and psychological consequences due to long term stressful life situation. The wide variety of physical and psychological problems appeared soon after the disaster. The first symptoms were emotional agitation, anxiety, panic attack, and sleep problems. Those symptoms and signs were gradually replaced by depressive symptoms. After 1 year, alcohol-related problems among victims became prevalent, whereas post-traumatic stress disorder (PTSD) was rarely observed among those victims (Shinfuku 2002). Lessons learned from the Hanshin Awaji earthquake were used in the planning of disaster prevention and care for the disaster victims in Japan and in other countries. The experience was transferred to support the victims of the Great Tohoku Tsunami disaster in 2011.

Conclusion

Kobe city experienced the Great Hanshin Awaji Earthquake in 1995 which devastated the city and its population. The city has recovered from the damage in the past

20 years. Now it is hard to recognize its damage for outsiders. But the memory of the tragedy still remains in the mind of the affected population. The increase of the elderly population and needs for services to them including for the persons with cognitive impairment are now the big challenges to Kobe.

Services in Major Cities: Focus on London, UK (Stuart Bell)

Situation in 1996 (This paragraph is written by D. Goldberg and G. Thornicroft)

In 1995, mental health services in greater London were provided by 27 health care provider organizations. *Hospital services*: London shares problems with other big UK cities, but to an even greater extent: bed occupancy rates as high as 125%, delays and long journeys for admission, assaults and sexual harassment onwards, high thresholds for admission, use of Accident and Emergency departments at night, and staff shortages of all kinds. Between 1969 and 1991, in-patient resident populations steadily fell by 66%, but numbers of admissions were rose by 8%. *Community services*: limited to home treatment/crisis resolution services in many parts of London, long delays for many services, services orientated to the needs of ethnic minority scanty, and London's primary care services - these services are significantly worse than those in other UK cities. In the original book, the Chief Executive of the Maudsley also had a chapter, and we have invited Stuart Bell to give an updated account of how things seem now from his viewpoint.

A Chief Executive's View of Services in Major Cities

Since the publication in 1998 of "Mental Health in Our Future Cities" cities has if anything assumed even greater importance in political and economic discourse across the world. In the United Kingdom, London has become so dominant that deliberate efforts have been made to bolster the power of other major regional centers to counterbalance its influence. In particular Manchester has emerged as an exemplar of major regional devolution. Several local authorities have pooled their sovereignty, accepted a single elected mayor, and in return have received unprecedented degrees of devolved authority from national government over significant components of the public realm, including healthcare.

In some respects, this represents a major challenge to the concept of a "national" health service, but for mental health services can be seen as an acknowledgment of the principle that good care is dependent as much upon the effective coordination of housing, employment, social care, education, and the criminal justice system as upon the delivery of health services by themselves. Interestingly much of the case for "DevoManc" (Developments in Manchester) is couched in terms which focus upon the importance of improving the life chances and educational attainment of young people for the long-term prosperity of the local economy. One could not look for a better opportunity to make the case for investment in good mental health care to offer a valuable return to the wider community, the local economy, and the next generation.

Over the course of the last 18 years, mental health services have also made significant progress in improving and demonstrating their ability to bring better "value" to healthcare, in the meaning of the term developed by Michael Porter. Cognitive behavioral therapy for anxiety and depression, early intervention in psychosis, treatments for drug and alcohol addiction, a range of interventions for the more effective treatment of mental disorders in young people, and the importance of liaison psychiatry in improving the outcomes of treatment for physical ill-health and maternity care now all have a sound and compelling economic case behind them. The National Health Service "Five-Year Forward View for Mental Health" (2016) drew together and set out many of these arguments.

The combination of these two developments – the significance of cities as major economic powerhouses (dependent in so many respects upon the flourishing of industries of the mind – in education, in academia, and in the financial, service, and creative industries) and the growing evidence base for investment in good mental health care – ought to have led to a renaissance in mental health services. In reality however the pressures upon them feel as acute as they were back in 1998. The very same National Health Service document, which draws together the compendium of evidence for the value of investments in mental health care in 2016, acknowledges at the same time that appropriate investment has failed to materialize over the course of the last two decades. Mental health remains a poor relation, and while the arguments for its importance are acknowledged, it still remains very difficult to shift the overall balance of investment in its favor across health care systems worldwide.

Traditionally so much of health care is devoted to the diseases of old-age, and as life expectancy increases, health care systems, particularly those which incentivize the consumption of interventions, will inevitably find it more and more difficult to give priority to resources for illnesses which affect predominantly young people and adults of working age. But the age profile in cities is often very much younger than that of the hinterland which surrounds them, and for them the case for protecting the mental health and resilience of their young people and of their workforce will be an important and competing priority. That is at the heart of the case which Manchester has made. It is possible therefore that our future cities may become the places which see the most developed arguments for better mental health care not just as manifestations of traditional health planning, but as economic necessities rooted within a wider social milieu. It will be interesting to see whether their ability to rise to that challenge depends on the extent to which they are able to exercise greater autonomy in relation to their national governments. Perhaps the success of cities in improving the mental health of their populations will become one of the key tests of their ability truly to take charge of their own destinies over the course of the next couple of decades.

Mental Health Services in London 1997–2016 (Graham Thornicroft and David Goldberg)

Key Policy Changes Affecting Mental Health Services in London

The main policy mental health issues which have affected London are summarized in Table 2.

National Service Framework for Mental Health, 1999–2008

One of the most important policies that changed the provision of care in London was the National Service Framework for Mental Health (NSFMH), which drew upon a review of the vast array of relevant evidence, including related information from other countries (Thornicroft 2000). The framework set seven standards for local services to achieve:

- Mental health promotion (standard 1)
- Primary care and access to services (standards 2 and 3)
- Effective services for people with severe mental illness (standards 4 and 5)
- Caring about carers (standard 6)
- Preventing suicide (standard 7).

London's services also needed to comply with the 1999 National Health Service Plan which added more specific detail in describing what services should be provided in each local area. Specifically it required that three types of community mental health teams should be provided in each borough (local government area)

Table 2 Key mental health policy changes in England 1997–2016

1999	National Service Framework for Mental Health and NHS Plan: national policy
2000–3	Policy Implementation Guidelines on: early intervention; home treatment; crisis intervention; community mental health teams; suicide prevention; women's mental health; black and ethnic minority mental health
2005	Mental Capacity Act
2007	Mental Health Act 2007 – amendment of the Mental Health Act 1983 with the introduction of supervised community treatment
2008	Refocusing the Care Programme Approach
2009	New Horizons – a Shared Vision For Mental Health: national policy
2011	No Health Without Mental Health – A Cross-Government Mental Health Outcomes Strategy For People Of All Ages
2013	Whole-Person Care: From Rhetoric To Reality. Achieving Parity Between Physical and Mental Health
2014	Closing The Gap: Priorities For Essential Change In Mental Health

in London, as well as across the whole of England for: (i) early intervention, (ii) assertive outreach, and (iii) crisis resolution/home treatment. Interestingly, although no independent evaluation of the NSF or the National Health Service Plan was conducted, some data are available of the impact of these policies, showing that admission rates appear to have decreased to inpatient wards after the introduction of Home Treatment Teams and that suicide rates fell when many of the provisions of the suicide reduction policy were implemented (While et al. 2012).

The impact of the NSFMH and the National Health Service Plan, over this first decade of the review period, was substantial extra investment in mental health. During the main period of implementation (1999/2000–2005/06), expenditure on adult mental health services increased in real terms by 47% – an annual rise of 6.7%. Across England as a whole, most funding for the NSF-MH/National Health Service Plan was from central government. The funds available were earmarked as usable only for mental health. They included £700 million over 3 years to implement the NSFMH and £120 million distributed via a Mental Health Modernisation Fund. These funds and a strong performance management program to implement the new teams meant that mental health services across London improved substantially during this period.

Introduction in London of Community Treatment Orders

The Mental Health Amendment Act in 2007 enabled the introduction of supervised community treatment, a type of community treatment order (CTOs). Subsequently a large scale evaluation of the impact of CTOs, called the OCTET trial, showed that the imposition of compulsory supervision did not reduce the rate of readmission of psychotic patients, and the trial found no reduction in overall hospital admission support that might justify the significant curtailment of patients' personal liberty (Burns et al. 2013). London then needed to respond to a series of further policies, including New Horizons in 2009, and No Health Without Mental Health in 2009, Closing the Gap in 2014, and the Five Year Forward View for Mental Health in 2016.

Mental Health Changes Specific to London

When thinking of such a populous and complex city as London, a number of contextual factors need to be borne in mind: the astonishing ethnic diversity across the city, with over 100 languages spoken for example; the wide disparities between the most affluent and the most economically and socially deprived, often in adjacent neighborhoods or streets; the challenges of maintaining an effective health care workforce given the rising price of accommodation and cost of living, and the large numbers of homeless people and of young and often socially isolated people, both students and internal migrants, all at risk for substance use disorders in particular.

The second decade of this review period was one of the disinvestments rather than investment. For example, one large London mental health service provider reported that over the period from 2009/10 to 2013/14 it had seen a net reduction in funding of £12 m, "with the pace of net disinvestment being accelerated and set to do so further." They report that "If the funding provided by the Department of Health and which passes through Clinical Commissioning Groups (CCGs) then the net gain to local CCGs from disinvestment and efficiency in local mental health services has been approximately £50 m. In each of our local CCGs we have seen net savings from mental health services of at least 32% over the last 7 years." (Docherty and Thornicroft 2015). These reductions in the National Health Service budget for mental health care in London primed a series of rapid reconfigurations of, especially, the community mental health teams, with the aim of reducing costs (Docherty and Thornicroft 2015; Centre for Economic Performance's Mental Health Policy Group 2012; McManus et al. 2009; Green et al. 2005).

The first casualty in London of cost-cutting was the Assertive Outreach teams, which were decommissioned in most parts of London in the decade following 2000. While the patients served by these teams continued to need care, they were for the large part re-distributed to generic community mental health teams treating people with severe mental illness, often termed "Recovery and Support Teams."

A series of other key features have been the hallmarks of mental health needs and services in the capital city in the last decade. There is continuing evidence of higher rates of compulsory treatment for nonwhite patients, whose service satisfaction is understandably less than for others (Kmietowicz 2003). Indeed overall admission rates in the capital continue to be higher than for other parts of Britain. There has been an exception to service reductions, namely the establishment across London, and indeed across all of England, a new set of psychological treatment teams, called Improving Access to Psychological Therapies, which have substantially improved the capacity of the National Health Service to provide treatment to people with depression and anxiety-related disorders.

Almost £7.5 billion is spent each year to address mental ill health in London, including spending on health and social care, welfare benefits and costs to education services, and the criminal justice system. However, these costs are only a fraction of the total £26 billion lost to London each year through lost productivity and reduced quality of life (Mayor of London 2014).

The service changes described here have taken place at the same time that the consequences of mental ill health upon London and Londoners are increasingly well understood. Indeed the wider impacts of mental ill health result in about £26 billion each year in total economic and social costs to London (Mayor of London 2014). Young people are also affected with at least 1 in 10 children having a clinically significant mental health problem, meaning 111,000 young people in London. The impacts of childhood psychiatric disorders cost London's education system approximately £200 million per year (Mayor of London 2014).

There are also substantial costs to social care with 45% of looked after children aged 5–17 experiencing mental disorder, and 65,000 older Londoners experience dementia, a figure that is expected to almost double over the next 30 years. In social care costs, London boroughs spend about £550 million a year treating mental disorder, and another £960 million is spent annually on welfare benefits to support people with mental ill health. These costs do not even tackle the problem. Informal and unpaid carers are left to contribute £1.2 billion worth of support each year to people with mental disorders.

How can one summarize all these changes over two decades in a city of 8.7 million inhabitants? It is fair to say that for London therefore, a city of great football teams, the last 20 years has been a "game of two halves," with investment, provision, and outcomes improving during the first decade, and many of these gains lost during the subsequent period of austerity and disinvestment. This perhaps demonstrates that when "push comes to shove," and when difficult healthcare funding decisions need to be made, that cutting mental health care continues to be a prime manifestation of structural/systemic stigma and discrimination against people with mental illness.

Madison Model: Highlights (1996–2016) (David LeCount)

National Changes

The Patient Protection and Affordable Care Act (Obamacare), enacted in 2010, increased those with health insurance in the state of Wisconsin by 200,000 (U.S. Department of Health & Human Services 2016) (by 211,000 and nationally by an estimated 20 million. Most with private insurance receive traditional mental health services through the private sector, such as health maintenance organizations (HMOs) or private practice agencies. There still are many people without health insurance in the United States. As a result of the Mental Health Parity Act (1996), private insurance pays for mental health services more equal to physical care, covering longer-term psychiatric inpatient treatment episodes, psychotherapy, and psychotropic medications.

The Madison Model

After 20 years (1996–2016) the Madison Model comprehensive system of care remains intact, with primarily positive service embellishments. The culture of caring also continues, with services more integrated, consumer centered, strength based, and recovery oriented. During this time, the population of Dane County increased from 390,000 to 523,600 (7/1/2015 data), with approximately half living primarily in the City of Madison. The number of people with serious and persistent mental illness receiving services has increased from approximately 1500, to an estimated 3000 people. Services to people with less serious mental illnesses, many without the ability to pay, increased from an estimated 3000, to 6000.

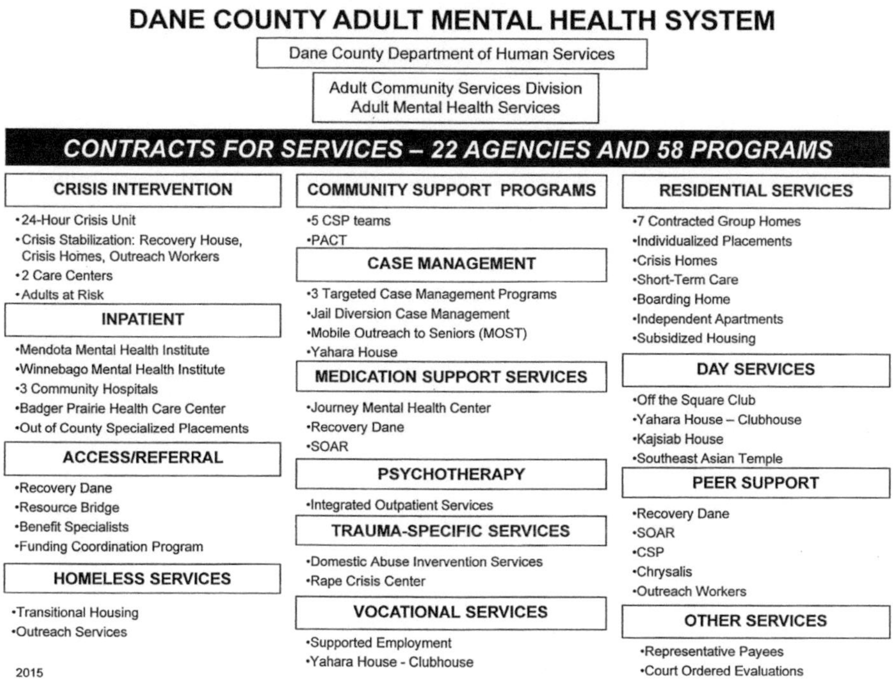

Fig. 3 Dane county adult mental health system overview chart. Date: 2015

For over 40 years, the County has been the state mandated lead agency for service oversight and coordination, performing planning, development, delivery, budgeting, and evaluation of services. In Dane County (Madison, Wisconsin), this is accomplished by public sector contracting for services with NPOs in the community, with the number of agencies increasing from 18 to 22 and programs increasing from 40 to 58 (Fig. 3).

The array of services addressing varying levels of need include:

- Comprehensive Community Support Programs – ACT Model addressing highest needs, patterned after Program of Assertive Community Treatment (PACT) demonstrated efficacy in the 1970s
- Varying levels of Supervised Living Arrangements
- Inpatient – Voluntary and Involuntary, for highest levels of acuity
- Case Management & Service Facilitation
- Emergency Services, including crisis stabilization
- Day services, patterned after Clubhouse International
- Psychotherapy & Psychotropic Medications
- Work Related Services – primarily supported employment
- Outreach and other Outpatient Services
- Centralized Intake, including Resource Center for service related questions

Funding Changes

The doubling of services coincides with funding increasing from $13.2 M to $25.3 M. The 2016 county budget for adult mental health services, including inpatient, comes to $27.9 M.

- Approximately 85% of funding continues to be for services to people with the most severe psychiatric disorders.
- Funding distribution for community services is 91%, vs. hospital at 9%, remains basically the same.

While State and County direct funding paid for the majority of services in the past, approximately 68% (County 42%, State 26%), government now pays around 43%, with the remaining being primarily Medicaid fee for services (FFS) revenues. Consumers continue to receive Supplemental Security Income (SSI), a monthly stipend to pay for basic needs while living in the community.

Programmatic Changes and Expansions

A comprehensive continuum of services has been operative since 1986, when the Madison Model was rated number one in the country by Public Citizen Health Research Group (Washington, DC). While service expansions and innovations have continued, this section will address initiatives occurring the past 20 years. Consumers as providers, staff intensive crisis stabilization services, comprehensive community support, earlier levels of intervention, and a consumer centered (integrated, holistic, trauma informed) recovery emphasis have been some of the major themes during this time. Most of these outcome-based program innovations have been fostered through state/federal Medicaid funding initiatives, under the umbrella of psychosocial rehabilitation.

SOAR Case Management Services: Starting in 1997, as a consumer service provider program, SOAR now provides a variety of services to consumers across the needs spectrum. While the primary emphasis is providing case management services, SOAR has a weekly psychotropic medication clinic and provides a central entry point into the system through Recovery Dane, a phone-based referral service for people requesting mental health services. A parallel service at the County level, called Aging and Disabilities Resource Center, was more recently created to provide answers to service needs and inquiries. More recently, SOAR developed Solstice House, a peer respite center. There, consumers can stay for brief periods of time in a home like setting, taking a time out as needed within this self-directed care facility operated by Certified Peer Specialists. SOAR provides training for consumers with the "lived experience" to be certified as Peer Specialists. To date, 80 consumers have been certified in Dane County, and many are providing support and skill training within existing programs.

Crisis stabilization started in 2000, to provide alternatives to hospitalization.

People at risk for hospitalization are afforded surround staff support services in the community for brief periods of time. Clients receiving this Medicaid billable service are generally referred through the Emergency Service Unit, the inpatient authorization agent. More recently, two residential centers have been added (20 slots), providing staff intensive services in lieu of hospitalization.

Community Recovery Services (CRS): Medicaid pays for peer supports, provided by Certified Peer Specialist, supported employment, based on Individual Placement and Services (IPS) evidence-based practice (EBP), and some Community Living Supportive services, such as adult family homes (AFH) and group homes (CBRFs).

Comprehensive Community Service (CCS) is a State Medicaid initiative Dane County has recently implemented. Eligible participants must meet a functional screen, and a level of need less than assertive community treatment (ACT model), but more than traditional outpatient services (psychotherapy and psychotropic medications). Payment is significantly higher, as it pays for real cost of services. CCS is consumer directed and includes core mental health services and holistic services, such as yoga and equestrian therapies. In addition to CCS incorporating into existing programs for consumers who meet functional criteria, several other programs have been developed through this funding to address more immediate needs. As more people are found eligible, this program will continue to expand.

Assertive Community Treatment: More than 500 people with the highest mental health needs receive services through PACT and three other programs operated by Journey Mental Health Center in locations throughout Madison. One program, Community Treatment Alternatives (CTA), works with people with criminal backgrounds, coming out of jail or prison. They provide similar ACT services to people in lieu of jail and coordinate the conditional release (CR) program for forensic clients discharged from Mendota Mental Health Institute. The most recent program implemented through Corrections funding is Opening Avenues to Reentry Success (OARS), which enrolls people with severe psychiatric disorders coming out of the state prison system. Hopefully, CTA will model for corrections with what has been accomplished in the psychiatric inpatient deinstitutionalization movement, i.e., development of responsible comprehensive community-based mental health services for people in lieu of jail and prison.

Promoting Recovery from Onset of Psychosis (PROPs): This grant funded program, operated by Journey Mental Health Center, (formerly Mental Health Center of Dane County) targets people age 15–25, who are experiencing early symptoms of psychosis. The goal is to provide mental health intervention close to the illness onset, to maximize treatment effectiveness. Certified Peer Specialists are a part of this program. Additionally, *Journey Mental Health Center* has procured funding for primary care and holistic services to provide better-integrated services to people with severe psychiatric disorders in an attempt to reduce much earlier death rates and increase quality of life.

Three Negative Developments

- Involuntary psychiatric inpatient is no longer provided at the local state hospital, Mendota Mental Health Institute, which now primarily provides forensic inpatient services. People going through civil commitments must be transported to the other state hospital, Winnebago Mental Health Institute, which is 80 miles northeast of Madison.
- Jail mental health services are now provided by a national corporation (vs. the local mental health center), resulting in less efficiency in the implementation of aftercare services.
- Journey Mental Health Center is providing much less community-oriented psychiatric residency training, a program they excelled in, and that resulted in hiring many residents in the public sector, thereby continuing the Madison Model tradition of excellence.

What We Have Learned and Performance Outcomes

Resulting from the early evidence based practice demonstrated through PACT and other innovative programs, the Madison Model provided national and international training for mental health professionals to learn evidenced based and best practices.

Some lessons learned are:

- A centralized agency (County) overseeing and coordinating services avoids duplication and provides cost effective and clinically sound treatment resulting in a well-balanced system of care that benefits the consumer and community.
- How the money is spent, approximately 90% for community-based services and 10% for psychiatric inpatient treatment, represents a well-balanced system of care, with services provided in the least restrictive environment.
- Program and system outcomes can be enhanced through best practices, such as fixed points of responsibility within programs and systemically, along with EBP such as Assertive Community Treatment, Individual Placement and Services work model, Clubhouse International, Case Management, and Consumers as providers.
- Minimizing psychiatric hospitalizations:
  - Psychiatric inpatient care is used primarily for short-term acute treatment when the symptoms are most pronounced
  - Psychiatric inpatient care can be minimized when responsible community-based alternatives have been developed, such as staff intensive crisis stabilization facilities that are more community centered and consumer friendly
  - Decision making from a community vs. hospital perspective minimizes psychiatric inpatient episodes and keeps the emphasis on community, the most facilitating environment.
    - Crisis oversight has provided community based stabilizing alternatives for approximately 50% of people presenting for hospitalization
    - Crisis unit initial authorization, ongoing monitoring of inpatient services, and community crisis stabilization alternatives, has kept average lengths of hospital episodes at approximately 15 days for involuntary, and 3–5 days for voluntary.

- Excellent functional community-oriented psychopharmacological psychiatry allowing for teaming and interdisciplinary decision making input with mutual respect at all levelsmaximizes cost-effective and clinically sound utilization of resources in keeping with community based practices.
- Recovery and psychosocial rehabilitation can be supported through concerted funding initiatives described above, further enhancing consumer well-being.
- Laws and legal processes are important in protecting human rights and minimizing more restrictive approaches and provide due process for involuntary hospitalizations.
- Supervised living arrangements should be transitional, teaching self care skills for consumers to live in more independent living settings.
- Well-incorporated working relationships along with fixed points of responsibility throughout the system of care, including related networks such as law enforcement and basic needs (food, clothing, and shelter), are essential to consumer growth, development, and quality of life.
- Integrated services to address the multiple needs (mental health, substance misuse, and physical) in a clinical, holistic, trauma-informed, consumer-centered, and cost-effective manner is the most effective approach

Dane County Department of Human Services Data

In conclusion, the Madison Model continues to provide "state of the art services" for people with the most severe psychiatric disorders in keeping with evidence based and best practices. The continuous service evolution is in keeping with consumer direction and meeting all of the health and related needs through greater service integration. As a result of consumers being able to spend most of their time in the community, they are increasingly becoming better integrated experiencing a higher quality of life, including being a significant part of the service delivery system.

Porto Alegre: Brazil, 1998 to 2016 (Dinarte Ballester, Loiva dos Santos Leite, Sara Jane Escouto dos Santos, Ana Cristina Tietzmann, Ellis Busnello)

Situation in 1996 (This paragraph is written by D. Goldberg and G. Thornicroft)

This had been the site for an earlier WHO study to improve services in primary care and had developed community participation in the health system. A mental health committee had been established in 1992 in response to the Psychiatric Reform Law, and Local Health Councils were established at District level, responsible for both users and providers of medical services. This ensures that local needs are identified, and either solutions to problems are initiated. Centers for Integrated Mental Health Care provide care by mental health teams, including psychologists, psychiatrists, nurses, and social workers. There are sheltered workshops and sheltered housing to facilitate care of severely ill people. The emphasis had been on de-institutionalization, sectorization, and preventive work.

National Changes

In Brazil, mortality rates of noncommunicable diseases declined 20% between 1996 and 2007, mainly due to reductions in cardiovascular and chronic respiratory diseases, in parallel with falling smoking rates. By contrast, neuropsychiatric disorders are currently the major contributor to disease burden, and homicide and traffic-related death are epidemics (Victora et al. 2011).

The policies for mental health care in the last two decades have been concentrated on the way to change from a hospital-based to a community-based model. In 2002, hospitals spent 75.2% of the Federal budget for mental health care, but this had dropped to 20.6% in 2013. On the other hand, community-based services moved from 24.8% to 79.4% of this budget in this decade (Brasil. Ministério da Saúde 2015).

One important development has been a Federal program for deinstitutionalization, created in 2003, the "Volta para casa" (Return to home), that gives a benefit in cash for users' long time living in mental health hospitals, when they moved to their home families or sheltered housing. In 2005, with the establishment of the Social Care National System (Sistema Único de Assistência Social), other social protections became more accessible for citizens with mental health problems.

During these years, there was also an important change in the formation of health professionals. The Ministry of Education, with an intensive participation of universities and health services, created new guidelines to orient the curricula of all courses related to the health system, in order to approximate with cultural and epidemiological characteristics of Brazilian population and its needs. In Medicine, the curricula of the general practitioner now include mental health practices, and in Psychiatry, the national guidelines for these specialists propose the training in a community-based model. It is a slow but progressive movement, and new mental health professionals have shifted to practice in general hospitals, community centeres and primary care.

If it is true that there is no health without mental health, in the same way there is no mental health without citizenship and democracy. There are social determinants strictly related with mental health: violence, new family configurations, women as home providers, child and adolescent abuse, and premature sexual behaviors. It is not only a job for health providers, but mental health policies need to contribute in any way. If the economic model, based on consumption of goods more than values, impact directly in the business of drugs and violence, there is a need to act also by health promotion.

Considering the broad scenario, the last years proved that policies to improve economy of the poorest stratum of society – that means the majority of Brazilians – by the way of cash transference programs as "Bolsa Família" (Brasil. Ministério do Desenvolvimento Social (2016)), with the resources managed mainly by women, had a huge impact, moving millions of people to the above social class. Conversely, the political controversies in one of the most unequal countries in a globalized world tend to maintain the concentration of resources in fewer hands and increase social conflicts.

The expansion of primary care services through the Family Health Program is one of the most important achievements. In the beginning of 1998, the coverage of primary care teams for all Brazil was 4.40% and in June of 2016 it was 64.28% and in Porto Alegre the numbers changed from 7.40% to 50.65% in the same period (Brasil. Ministério da Saúde. Departamento de Atenção Básica 2016).

The City of Porto Alegre

In the city of Porto Alegre, there have been the same developments as in the whole country (Leite et al. 2013). In these last 18 years, one emblematic event was the closure of one large out-patient department that centralized the assistance of users from all the municipality and the progressive opening of 10 new mental health centers and other 8 mental health teams in different points of the territory.

The mental health center is the leading service in the strategy to move from the old mental hospital to the community. Nowadays, the 12 centers running are divided in 3 for child and adolescents, 4 for users with severe mental health problems, and 5 for people with alcohol and drug problems. Considering the population of Porto Alegre (1.4 M inhabitants), the coverage from these services is still clearly insufficient. In the last 10 years, especially for a massive media campaign, advising an "epidemic" of crack, the health policies moved the focus from the institutionalization of mental hospital users (or better, residents) to drug users and the violence derived from traffic. As in Brazil the "war on drugs model" remains, this issue needs urgent revision, because the results are disappointing. Violence and massive imprisonment related to drug-traffic are rising without coherent and effective measures for needs of people involved with drugs.

In this period also two 24-h mental health emergency clinics were created, to improve access and also regulate admissions in hospitals or following on in community services. The decrease in the number of beds in mental hospitals, 50% in one of these, was not attended by the expected improvement of small units in general hospitals. There are different reasons for that, but social stigma and the economic model of the health system are probably involved.

While the assistance for users with severe mental disorders and drug users is still challenging the local health system, more recently the policies of mental health and primary care started to interact. The strategy of collaborative care, named "matrix support" in Brazil, is being the way to integrate mental health approaches in primary care services, building networks into the health system and with other institutional and community resources. If in the beginning of the mental health reform the emphasis was mostly in specialized care, this new approach includes all the network of services. The lack of technically capable professionals to work at the interface between mental health and primary care services and the high turnover of the teams are problems that affect the quality of care in these settings.

Common mental disorders, specially depression and anxiety, were mostly recognized by the population and professionals as a responsibility of specialists (mainly psychiatrists and psychologists), but the mental health centers are unable to shoulder the whole burden. In recent years this vision is slowly changing, as other physicians are training and the health system incorporate the collaborative care model. There are teams specialized in this function, the "Family Health Support Nucleus or Teams" to support primary care services.

Conclusion

The mental health reform in the City of Porto Alegre is a "work in progress." Some important steps have been done: deinstitutionalization of people living in mental hospitals, a model to integrate mental health in primary care, and changes in the training of health professionals.

Other old and new challenges remain but, as in all developments in the field of health care, the reform is an inevitable movement towards civilization: more health, more citizenship, more citizenship, and more health. In this equation there are thousands of professionals engaged in the construction of the National Health System, other thousands of community representatives, actively working in the Health Councils, NGOs, and users' associations.

Sydney, Australia (Gavin Andrews and Mathew James)

Situation in 1996 (This paragraph is written by D. Goldberg and G. Thornicroft)

The National Mental Health Strategy (1992) was aimed at reducing the reliance on mental hospitals and increase the focus on community based care. New South Wales (NSW, containing Sydney) had the highest excellence score of any State in the country, yet the lowest expenditure per capita. A total of 41% of psychiatric beds are in general hospitals in NSW. The inner city mental health services of St Vincent's Hospital, consisting of 100 staff: 10% psychiatrists and trainees, 55% psychiatric nurses and 18% occupational therapists, psychologists, or social workers. There are three teams, the Crisis Team, the Case Management Team, and the Rehabilitation team. There are also three Liaison Teams dealing with the needs of the general hospital, of local general practitioners, and a community liaison program. This is a self-confident community service in a city famous for innovative community programs.

National Developments

Australia's healthcare funding system has not changed over the past 20 years, with the public hospital and fee-for-service sectors being funded by general taxation revenue. The proportion of population with private health insurance has increased from about one third to over half and is used to offset the cost of private hospital care. While health expenditure has increased from 8.5% to 9.4% of GDP (1994/5–2009/10), this appears to be insufficient with our ever aging population and the rising cost of healthcare. The last 20 years has seen patient out of pocket contributions double. These two factors, increasing private health cover and increased gap fees, will lead to a two tiered system of public and private services that is particularly obvious in mental health care.

From a mental health perspective, the National Mental Strategy, initially conceived in 1992 as an attempt to decrease reliance on psychiatric hospital care and assist in the transition to expanded community based care, continues to guide mental health reform. This transition has seen a switch of expenditure from inpatient to community services (from 29% to 51%) and a 50% increase in professionals and peer workers who work in specialist mental health settings. Parallel to this diversion towards community-based settings is improved funding and access to specialist treatment in primary care. Since its introduction in 2006 more than 1.3 million Mental Health Treatment Plans were developed by general practitioners and 4.95 million services provided by psychologists and other allied health professionals. There are a number of concerns: the annual ration of 10 sessions are inadequate for the severely mentally ill; the practice of charging a gap fee puts such services out of the reach of many with severe mental illness; and that there is no system to ensure quality or effectiveness means that inefficiencies flourish.

Medications have been the mainstay of treatment for the high prevalence disorders of anxiety and depression for some decades. They are relatively cheap, quality is assured, and they are considered acceptable by consumers. Traditional face-to-face psychotherapy appears equally effective to medications, but because of cost (including out of pocket expenses), being less widely available outside of metropolitan areas and with no mechanism of ensuring quality, their use will stall. A significant new contribution to our therapeutic arsenal has been the development of eHealth programs in which five different Australian Universities are now pioneering. In line with recent technological advances, Internet delivered CBT has a cost comparable to 2 months of medications, has equal efficacy, is widely available, and is acceptable to consumers, but most importantly, quality can be assured.

It is disappointing and unfortunate that at present there are no evidence-based computer programs for people with psychoses, given their traditional reliance on the public service and its ever scarce funding. If we do not reconsider our approach, we run the risk of an expanding two tier system, with rising inequality disadvantaging those with psychosis.

The City of Sydney

Since its founding as an English penal colony in 1788, Sydney has undergone a massive multicultural transformation into the thriving modern metropolis that it is today. This socio-cultural shift has continued since the original publication of "Mental Health in Our Future Cities," some 20 years ago. It remains a great place to live.

The population has increased to over five million people, with a combination of urban sprawl and inner city high rise. The population continues to age and family size is decreasing, but continued immigration, from Europe and Asia, has blunted these trends and driven population and economic growth. Australia largely escaped the global financial crisis that crippled much of the western world some 5 years ago, and unemployment is lower than it was. The gap between the rich and poor is widening and the number living below the poverty line has increased.

For the disadvantaged, Sydney continues to be a difficult place to live. Property values have increased relative to average income and home ownership has decreased significantly in the last 20 years. While this is multifactorial, it appears largely driven by international investment, generous tax incentives for local investors, and the ever increasing divide between rich and poor. The idea that "no-one would choose to rent on a long term basis" may be a 20-year-old ethos as the "great Australian dream of home ownership" becomes impossible.

Tobacco and alcohol remain the most significant drugs of abuse in terms of health outcomes, but the gradual effects of increased taxation, public health campaigns, and intervention strategies have seen their use at the lowest rate in decades and users are increasingly ostracised. The most common illicit substance, cannabis, has decreased in use by 25% and the use of heroin and MDMA appears also on the decline. Despite media reports of an "ice epidemic," drug use surveys suggest that while overall use is static, substance strength and increasing intravenous use accounts for increasing presentations of drug related psychosis to hospital and mental health services. Somewhat concerning, and counter to other substance use trends, is the increased abuse of prescription medications. Harm minimization, not prohibition, was the key focus of the inaugural NSW Drug Summit in 1999 and continues to this day.

St Vincent's Hospital Services for the Inner City Sector

At a local level St Vincent's Hospital continues to provide mental health care to a relatively small, yet heterogeneous local catchment area in the inner city. As some 20 years ago, this covers a much higher proportion of individuals with drug and alcohol misuse, serious mental disorders, and homelessness than the national average. However, a gradual gentrification of the area continues to take place, with constriction of NGO services for the disadvantaged.

Like most services that cater for an ever-changing public demand, the Inner City Mental Health Service has transformed over the last 20 years. The inpatient unit has moved within the main hospital, but bed numbers have remained static at 27. The development of a 6 bed Psychiatric Emergency Care Centre, a short stay unit for those in acute crisis, has catered for admission demand without an increase in longer stay bed numbers but constant bed pressures and wait times in the Emergency Department continue. Whether this will be eased by the opening of a highly specialised Psychiatric and Drug and Alcohol Short stay unit attached to the Emergency Department, and staffed by emergency physicians, addiction specialists, and mental health clinicians, remains to be seen.

There have been two losses of community services in the past 20 years: the liaison service for primary care physicians and the 24 h Acute Care Service. Psychogeriatrics now runs a successful and award-winning, yet very specialized, liaison service for local GPs among its core business of both inpatient and community care.

The adult community mental health structure has largely remained the same, with a crisis service responding to emergencies and high acuity cases, a case management team for those with chronic mental illnesses residing in the local area, and a rehabilitation team that focuses on a recovery model including living skills and vocational intervention.

Further specialized services have begun. The Program for Early Intervention and Prevention of Disability provides a range of interventions for young people (ages 16–25) with recent onset or risk of psychosis. The Homeless Health Service provides a no-wrong-door approach to improving both physical and mental health of the area's most disenfranchised. Recent census figures showed approximately 500 rough sleepers in the area, and 7 homeless clients attend the Emergency Department each day. A combination of assertive outreach to local NGOs and street corners, along with assertive case management, has been successful in linking clients with stable accommodation and mainstream services. Other service additions cater to highly distinct patient groups such as the Dialectical Behaviour Therapy group program and the service for HIV or Hepatitis C positive clients.

The Clinical Research Unit for Anxiety and Depression continues. The "partially computerised" outcome assessment of 20 years ago has expanded to a dozen different online CBT programs catering for a raft of anxiety and affective disorders. It also now focusses on the top causes of disability, for example, chronic pain, arthritis, COPD/asthma, diabetes, and cancer, in which anxiety and depression play such a significant part. Currently 500 patients per month, from all over the globe, access high quality, evidence-based Internet and phone-based interventions at a fraction of the cost of face-to-face therapy.

Reflections on the Past and Future

With the original publication of Mental Health in Our Future Cities, Australia had no access to nationwide mental health survey data to guide the planning of service provision on a needs basis. A National Survey of Mental Health and Wellbeing was undertaken in 1997 and 2007 and has provided a greater understanding of prevalence, sex and age distributions, severity, co-morbidity, and the extent of disability and health service use associated with each mental disorder. Half the people with high prevalence mental disorders (Anxiety, Depression and Drug and Alcohol) access some form of intervention, but only one third of these receive minimally adequate treatment. The survey data reveal that we could provide optimal treatment (for those who are already accessing it), at no more cost than current treatment, provided we only funded evidence-based interventions.

The question remains whether our newly attained knowledge of mental health needs on a national level has changed practice, and unfortunately the lack of any significant shift over the 10 years separating both surveys leaves this in doubt. In a largely fee for service model, we have limited ability to monitor the quality and consistency of therapeutic interventions.

Mukherjee's work on Cancer has shown that a strong alliance between patients, researchers, and clinicians can lead to significant improvements in cure rates. It stands to reason that a comparable alliance between patients, researchers, and clinicians in the psychotic disorders could produce similar advances. Without a move in this direction, focussing on research and recovery, we expose ourselves and those at most need within our mental health community.

Mental Health in Tehran (Ahmad Mohit, Ahmad Hajebi, Maryam Abbasinejad)

Situation in 1996 (This paragraph is written by D. Goldberg and G. Thornicroft)

The most revolutionary community mental health system was that operating in Iran, where there were Health Houses that serve approximately 1500 people, each containing one male and one female lay Health Worker. The workers have received a limited training course and are trained to recognize psychological symptoms and complaints and to use a framework of minor mental illness, major mental illness, mental retardation, and epilepsy. They will recognize stress-related conditions and learn simple stress-reduction methods. Those needing special treatments can be referred by a family health technician who provides additional training for the Health Workers. They also deal with referrals to District Health Centers, with school mental health and simple preventive measures and problems of mothers and children. Infant mortality rate had dropped from 145 to 28/1000 live births by 1997 and life expectancy increased from 57 to 69. Doctors in the local health centers will be trained to recognize psychosis, epilepsy, depression and severe anxiety states, and their management by drugs and their side effects and are able to refer people to psychiatrists. The city of Tehran's mental health service serves a population of seven million, with 2600 beds divided between 4 government hospitals and 3 wards in general hospitals.

National Data

The population of Iran was 75 million in 2011, being estimated to have reached 80.4 million by 2015. More than 50% of the population of Iran is under the age 35. The infant mortality rate has significantly decreased throughout the last decades, reaching approximately 15 per thousand live births. The total literacy rate of the Iranian population is approximately 90%.

Mental Health Services in Iran

Iran has approximately 1600 psychiatrists, of which 321 work in university-affiliated medical centers and hospitals in Tehran province. A proportion of these psychiatrists also work at private office, and so the total number of psychiatrists working in Tehran either in public or private sectors reaches a figure of approximately 500 altogether. There are presently 240 psychiatric residents in medical universities of Tehran working in the six previously mentioned hospitals, studying in a 4-year residency program. About 4000 postgraduate psychologists and counselors have license to work countrywide. About 1300 clinical psychologists are now working in private and public sectors in Tehran province. Approximately 400 of the clinical psychologists have started to work in the integrated urban health centers of the three universities of medical sciences located in Tehran Province.

Mental health services in Iran have significantly evolved throughout the last century, which can be divided into four periods:

First Period: The period of old mental asylums; Primitive old mental asylums with the main role of caring for severely mentally disturbed patients who were either homeless or whose families were incapable of caring for them. Such institutions existed in cities like Tehran, Isfahan, Hamadan, and Shiraz from the early years of the twentieth century.

Second Period: Psychiatry was recognized as an independent specialty in medicine in the 1940s. Psychiatric hospitals started to be established under the supervision of the faculties of medicine and training of residents was first pioneered in Roozbeh Hospital under the supervision of Tehran University in the 1960s, soon followed by Shiraz, Isfahan, Tabriz, and Mashhad Universities.

Third Period: A modern mental health approach started to flourish in the beginning of the 1970s. It consisted of providing new outpatient mental health services and establishing modern psychiatric hospitals and clinics in some areas of Tehran

Fourth Period: The integration of mental health care into primary health care services started in the beginning of the 1980s. The National Mental Health Program was performed as a pilot program in 1986 and developed as a nationwide program throughout the years afterward.

The goals of this program were:

- Making the basic mental health services available to all, particularly the most vulnerable people in rural and urban areas.
- The provision and expansion of a type of mental health services compatible with Iranian socio-cultural structure and encouraging people to participate in the expansion of mental health
- The program also aimed at increasing the people's awareness of the principles of the provision of mental health, encouraging them to use the services in order to promote health and socio-economic development and improvement of quality of life.

- Further development of proper programs for the provision of necessary services to the people affected by war and long term planning for the mental health needs of the people after the war.

In describing the progress and evolution of integration of mental health in Primary Health Care system, it is necessary to mention the important development of the introduction of family physician program in 1995. This program soon covered the whole rural areas of the country and the cities with a population below 20,000. The family physician program provided an opportunity for the mental health department to enter the family physician system through the provision of its services in co-ordinations with the units of population, family, and school health in form of "service packages."

The National Mental Health Programs

The national suicide prevention plan started with a systematic and uniform portal data collection system countrywide. Supporting the establishment and extension of community mental health centers for the aftercare of severe and chronic hospitalized patients started from 2011 and research showed its efficacy for reducing both the need for rehospitalization and the severity of psychopathology in patients under coverage. Development and standardization of in-patient mental health services, the child and adolescent mental health promotion plan, training health staff members for providing psychosocial interventions in disasters and trauma, and capacity building for parenting skills and life skills training are among the most successful national programs.

Evolution of Mental Health Strategies in the Health Transformation Plan

The shift in the distribution of the population from rural to urban areas from 33% to more than 72% throughout the last five decades has led to rapid development of suburban areas of large cities like Tehran. Meanwhile, the rate of the improvement of health infrastructure for these newly urbanized areas has not been proportionate to the needs of these populations. Regarding the proven success in terms of health indicators in rural areas due to governmental policies put into action for active screening and disease control in those areas, new policies were made at the national level with the aim of promoting health and medical services with focus on urban and suburban populations, to address these needs. New health strategies and approaches were adopted as a part of the "Health Transformation Plan" put in action since 2013. With the contribution of the Department of Mental Health and Substance abuse of the Ministry of Health and Medical Education, mental health programs were also considered as one of the main areas in which a major evolution took place.

Promotion of urban health centers and establishment of new suburban health centers for the coverage of the populations residing in suburban underdeveloped areas is a main change that has taken place as part of the Health Transformation Plan. Establishment of integrated urban health centers for every 20,000 residents of urban and suburban areas is another important action performed in the past 2 years. In these integrated health centers, multipurpose health workers have been placed as first-line health staff. These health workers, which are chosen among graduates of public health, family, nursing, and midwifery, are counterpart to those previously called "Behvarz," who generally had utmost a high school diploma, and worked in health-houses in rural areas. They actively screen for health problems among the populations in their catchment areas and refer those in need of medical services to the general practitioner in their centers. Also, postgraduate clinical psychologists have been added to the health staff working in urban and suburban integrated health centers. Their role is mainly providing primary prevention services like basic psychoeducation, parenting, and life-skills training, and at times brief psychological interventions. After diagnosis and initial management of mental disorders or detection of psychosocial issues, the general practitioner refers cases in need of specialized psychiatric care to public-funded clinics or hospitals to be visited by a psychiatrist and refers those in need of basic psychoeducation and psychological interventions to the clinical psychologists in their centers.

All of the primary health and medical services provided in rural and urban and suburban health centers by the health staff and general practitioners are free of charge. Governmental insurance covers the services provided in the outpatient and inpatient facilities.

The City of Tehran

Tehran is Iran's capital city, located in the Tehran Province, north of the central plateau of Iran. Tehran province is the most populated province with 14 counties being estimated to have reached nearly 13 million by 2015. The city is facing a number of difficulties like traffic congestion, air pollution, general and particularly youth under employment, and continuing internal migration from other areas of the country and refugees from neighboring countries are among them. All of these difficulties are much more prominent in suburban and slum areas dealing with newly urbanized populations. Rapid and to a large extent unplanned urbanization, due to both internal and external migration, has been the most important demographic change in Iran; and the Tehran has been most affected. In 2015, 73% of Iranian population live in urban areas, and this is about 20% more than 1985. With the increase in the speed of urbanization, the need to adjust the program to the new condition became more evident and this need was much more urgent for Tehran.

Mental Health Facilities

There are 2549 psychiatric beds in psychiatric and general hospitals of the metropolitan area of Tehran, of which 634 belong to six hospitals affiliated with the three main universities of medical sciences located in Tehran Province. The University of Social Welfare and Rehabilitation has one large residential mental institute outside of the metropolitan area of Tehran with 1375 beds, including wards for residential care. Tehran also has 340 private psychiatric beds scattered in seven general and psychiatric private hospitals and around 200 beds in three hospitals affiliated with the

Foundation of Martyrs and Veterans Affairs, mainly for psychiatric services given to veterans of war.

In relation to mental health services in Tehran, the more important of these programs are the improvement of in-patient services and the formation of a number of Community Mental Health Centers. These were opened in 2011 to offer different services through more participation of psychologists and general practitioners in the cities and particularly, the newly populated suburban areas of Tehran. At present five of such centers are active and work under three main universities of medical sciences in Tehran. These centers provide services like active follow up of chronic and severe mental patients. Also the provision of collaborative services for neurotic patients and the ones endangered by social problems is also done through such centers for the people in their catchment areas.

Mental Health Services in the Community

In 2013, quantitative and qualitative promotion of urban health centers and what was called "Health Bases" started. With increasing number of such centers, particularly in suburban slum areas, a new type of human resources called "Health Protection Expert" was created, working alongside a general practitioner in the first level of health provision services. These experts have a bachelor degree and their duties are described in a flow chart called a "Service Package." They are multipurpose health workers equal to the basic personnel in a "Health House" in a village (called "Behvarz" in Iran). A defined level of mental health services was also integrated to their work.

A new group of "mental health experts" was also added to the second level of health care in the cities. They hold a Master's degree in clinical psychology and along with the physician provide the second level services. At present around 400 of these personnel work in the comprehensive health centers in Tehran. What needs to be mentioned here is that in general, the role of clinical psychologists has greatly increased in the provision of mental health services throughout the country and in Tehran.

The City of Verona, Italy (Francesco Amaddeo)

Situation in 1996 (This paragraph is written by D. Goldberg and G. Thornicroft)

The new mental health act of 1978 made it impossible to admit any more people to mental hospitals, and it became necessary to design and implement community mental health across the whole country. The most thoroughly planned, and carefully assessed community service for the mentally ill in the world is that in Verona, which owes its existence to Michele Tansella. The population of South Verona is 250 K, and 16 in-patient beds in a ward in the general teaching hospital, which has a large mental health center open on weekdays and Saturdays adjacent to the hospital. There is close contact with social services and a liaison service with the general hospital. There is a User's organization and residential hostels in the community.

National Developments

Thirty-eight years have elapsed since the passing of the Italian Law Number 180, the reform law that marked the transition from a hospital-based system of care to a model of community psychiatry that was designed to be alternative to, rather than to complement, the old hospital-centered services. The new system was intended to provide care and support to all types of patients, without back up from the public mental hospital. After more than 30 years of implementation, Law 180 is still unique in the international scenario, considering that Italy remains the only country in the world where traditional mental hospitals are outside the system of mental health care (Amaddeo et al. 2012).

The main principle of Law 180 is that psychiatric patients have the right to be treated the same way as patients with other diseases and only voluntary treatments are allowed, with a few exceptions that are strictly regulated. The Law states that compulsory admissions need to be formally authorised by the Mayor and can be undertaken in psychiatric wards located in public general hospitals only. Acute patients can be admitted to the General Hospital Psychiatric Units (GHPUs), which may have a maximum of 15 beds. Outside the general hospital a network of Community Mental Health Centres (CMHCs), each serving a population of 25,000–150,000 people, provides community care in close conjunction with other health and social local services. CMHCs provide in-house and domiciliary services, and collaborate with welfare, housing and educational services.

The City of Verona

The South-Verona Community Psychiatric Service

The South-Verona Community Psychiatric Service (CPS) was developed in 1978, according to Law 180, run by the Section of Psychiatry, Department of Neurosciences, Biomedicine and Movement at the University of Verona. This institution was designated in 1987 by WHO as "Collaborating Centre for Research and Training in Mental Health and Service Evaluation" and is running since a structured program of evaluative research linked with the implementation of the new system of community-based care.

Components of the Verona Community Psychiatric Service

The South-Verona CPS has different facilities and components. The Community Mental Health Centre is the hub of the CPS: the Centre, located in a recently completely renovated historical three storey building with garden, not far away from the Academic general hospital, is open on weekdays from 8 a.m. to 8 p.m. and on Saturday from 8 a.m. to 5 p.m. Therapeutic programs include crisis intervention, day care for acute and chronic patients, rehabilitation and socialization groups. The Psychiatric General Hospital Unit is a psychiatric ward of 15 beds located in the Academic general hospital. The Outpatient Department provides psychiatric ambulatory visits.

The Emergency Service is a consultation-liaison psychiatric service for the Emergency and Accident room of the general hospital, open 24 h a day, 7 days a week. It is run by a psychiatrist from our team, who is on call. There is also an emergency night and week-end service, run by two psychiatric nurses from our team, who are on call and may provide care in our residential facilities (flats and hostel), as well as at our patients' homes, co-ordinated by the psychiatrist on call.

Home Visits can be made to provide crisis intervention in response to emergency calls, but for chronic patients these are usually planned in advance and offer regular, long-term support and care to patients and their families with the goal of minimizing relapses and hospital admissions.

Residential facilities are also part of the South-Verona CPS. Some of them are staffed 24-h, while others are partially supervised. Group homes have increased in number over the latest years especially for the work of a mutual-help group and social enterprises.

A consumers Self-Help association is also active in the catchment area (Burti et al. 2005).

South-Verona CPS clinical facilities also include: the Service of Psychosomatics and Medical Psychology and the Service of Psychotherapy. The first one provides specialized outpatient care for patients with psychiatric co morbidities and physical diseases, as well as consultations for patients and doctors of other departments of the University General Hospital (800 beds). It maintains psychiatric integration with other hospital-based medical activities and ensures continuing contact with psychiatric patients when hospitalized for medical reasons.

In South-Verona all community staff members are divided into two multidisciplinary teams, each serving a subsector of the South-Verona catchment area. They are organized as "single staff modules" to ensure continuity of care, both longitudinal continuity (through the different phases of treatment) and cross-sectional continuity (through the different components of the CPS). Within each team, each patient is assigned to one particular member of the staff (case manager). The style of intervention is psychosocial: prompt, adequate, and coherent answers to patients' needs are provided. The multidisciplinary teams ensure that all psychiatric, psychological and social needs have been together considered.

From Service Monitoring to Service Evaluation

The main research activities of different research units focus on epidemiological and social psychiatry (including economics and geography of mental health studies), translation and adaptation of diagnostic instruments, genetics and neuroimaging of schizophrenic and bipolar disorders, clinical psychopharmacology, communication in medicine and evaluation of mental health services. The research staff (more than 50 persons) includes now psychiatrists, clinical psychologists, computer scientists, physicists, geographers and social scientists, statisticians, IT-managers, technicians and secretaries.

Several studies have been conducted in Verona during the last 20 years to explore the organisation and the performances of psychiatric services using different approaches and several dimensions.

In-Patient Care

A recent study analysed all discharges (8789) from GHPUs in the Veneto Region (about 4.8 million inhabitants) during 2007 (Pertile et al. 2011) and found that Length of Stay (LoS) ranged from 4 days for drug and alcohol abuse to 24 days for psychoses. Using a hierarchical multiple regression model, we found that diagnosis alone was able to predict about 6% of the variability in LoS while a more complex model that included also gender, age, severity of hospitalisation, patient's local health district, number of psychiatrists, psychologists, hospital attendants/nurses, social workers and educators in the general hospital psychiatric units, number of outpatient contacts in each local health district and percentages of divorced and single people allowed to increase the predicted variance to 8% only.

Psychiatric Service Utilisation

In South-Verona a series of studies were conducted to assess the effect of the socioeconomic status (SES) on psychiatric service utilisation (Tello et al. 2005). An index of SES was calculated from nine census variables and grouped into four categories, ranging from SES-I-affluent to SES-IV-deprived, for each of 328 census blocks (CB) of the catchment area. Indicators of in-patient, day-patient, out-patient and community service use showed an inverse association with SES. Only first-ever and long-term psychotic patients were equally distributed in the four SES groups. The inverse association between SES and most indicators of psychiatric service use suggests that the planning of community-based services and resource allocation should take into account the SES of residents.

The Costs of Care

A following study (Donisi et al. 2011) showed as a comprehensive model, including both patients' individual characteristics and SES status was able to predict 43% of variance in costs of care.

In the last years, moving from hospital-based to community-based mental health care reduced the mortality rate of our patients. Nevertheless mortality is still higher for psychiatric patients as compared to the general population. A study conducted using the South-Verona PCR showed that the Standardised Mortality Ratio (SMR) was higher for deaths preventable with adequate health promotion policies than for those preventable with appropriate health care (Amaddeo et al. 2007). Males, alcohol/drug addicted and younger patients had the highest avoidable SMRs. From these results it seems important to implement, by specialist psychiatric services, health promotion and preventive programs targeted to psychiatric patients. Our data (Perini et al. 2014) show that, also in a community-based mental health service, psychiatric patients are at almost twofold higher risk for death than the general population and this becomes extremely higher for those diagnosed with drug addiction or personality disorders.

Service Satisfaction

In South-Verona, Service satisfaction was measured with the Verona Service Satisfaction Scale (VSSS). The organization of service and the behaviour and manners of the professionals were the main service strengths. Weaknesses were identified in the physical layout of facilities, in the lack of involvement of relatives and in the information provided. Overall satisfaction was medium-high, while subjects with longer duration of service contact and higher disability were the most dissatisfied (Ruggeri et al. 2007).

In recent years, more emphasis was given to First-episode psychosis (FEP) and Early Interventions, a large multi-site population-based survey called Psychosis Incident Cohort Outcome Study (PICOS) with a naturalistic longitudinal design and three different modules addressing, respectively, clinical and social variables, genetics and brain imaging (Lasalvia et al. 2012).

References

Amaddeo F, Barbui C, Perini G, Biggeri A, Tansella M (2007) Avoidable mortality of psychiatric patients in an area with a community-based system of mental health care. Acta Psychiatr Scand 115:320–325

Amaddeo F, Barbui C, Tansella M (2012) State of psychiatry in Italy 35 years after psychiatric reform. A critical appraisal of national and local data. Int Rev Psychiatry 24:314–320

Bharath S, Kumar KV (2010) Empowering adolescents with life skills education in schools-school mental health program-does it work? Indian J Psychiatry 52:344–349

Brasil. Ministério da Saúde. Coordenação Geral de Saúde Mental, Álcool e Outras Drogas (2015) Saúde mental em dados. n.13. Ministério da Saúde, Brasília

Brasil. Ministério da Saúde. Departamento de Atenção Básica. Histórico de cobertura de Saúde da Família (2016) Available: http://dab.saude.gov.br/portaldab/historico_cobertura_sf.php. Accessed 30 July 2016

Brasil. Ministério do Desenvolvimento Social (2016) 10 anos do Bolsa Família. Available: http://bolsafamilia10anos.mds.gov.br/themes/bolsa/bf10anos-ingles.htm. Accessed 30 July 2016

Burns T, Rugkasa J, Molodynski A, Dawson J, Yeeles K, Vazquez-Montes M, Voysey M, Sinclair J, Priebe S (2013) Community treatment orders for patients with psychosis (OCTET): a randomised controlled trial. Lancet 381:1627–1633

Burti L, Amaddeo F, Ambrosi M, Bonetto C, Cristofalo D, Ruggeri M, Tansella M (2005) Does additional care provided by a consumer self-help group improve psychiatric outcome? A study in an Italian community-based psychiatric service, South-Verona. Community Ment Health J 41:705–720

Centre for Economic Performance's Mental Health Policy Group (2012) How mental illness loses out in the NHS. Centre for Economic Performance, London School of Economics and Political Science, London

Dekker J (1996) Het Amsterdamse Sociaal Psychiatrisch Diensten Centrum. Een vergelijkend onderzoek naar een multi-functionele eenheid en de reguliere 24-uurs zorg. Proefschrift, Van Gorcum, Assen

Docherty M, Thornicroft G (2015) Specialist mental health services in England in 2014: overview of funding, access and levels of care. Int J Ment Heal Syst 9:34

Donisi V, Jones J, Pertile R, Salazzari D, Grigoletti L, Tansella M, Amaddeo F (2011) The difficult task of predicting the costs of community-based mental health care. A comprehensive case register study. Epidemiol Psychiatr Sci 20:245–256

Duurkoop P (1995) Terug naar Amsterdam. Longitudinaal onderzoek naar het functioneren van chronische patiënten in nieuwe woonsituaties. Proefschrift, Frederik van Eeden Stichting, Amsterdam

Green H, Mcginnity A, Meltzer H, Ford T, Goodman R (2005) Mental health of children and young people in Great Britain 2004. Palgrave McMillan, London

Kalyanasundaram S, Chaturvedi SK, Jagadish A, Prabhu V, Narasimha V (2007) Detection of stress, anxiety and depression in IT/ITES professionals in the Silicon Valley of India: a preliminary study. Prim Care Community Psychiatry 12:75–80
Kastrup M (1998) Mental health in the city of Copenhagen. In: Goldberg D, Thornicroft G (eds) Mental health in our future cities. Psychology Press, New York, pp 101–123
Kato TA, Tateno M, Shinfuku N et al (2012) Does the "hikikomori" syndrome of social withdrawal exist outside Japan?: a preliminary international investigation. Soc Psychiatry Psychiatr Epidemiol 47(7):1061–1075. doi:10.1007/s00127-011-0411-7
Kitchener BA, Jorm AF, Kelly CM (2015) Mental Health First Aid International manual. MHFAI, Melbourne
Kmietowicz Z (2003) London is more likely to detain people with mental problems. BMJ 327:1184
Københavns Kommune (2015a) Videnspapir om bedre psykisk sundhed 2015. https://www.kk.dk/sites/default/files/uploaded-files/psykisk_sundhed_videnspapir_pdfa.pdf. Accessed 18 June 2016
Københavns Kommune (2015b) Sundhedspolitik (2015–2025). https://www.kk.dk/artikel/sundhedspolitik-2015–2025. Accessed 18 June 2016
Kroon H, Theunissen J, van Busschbach J, Raven E, Wiersma D (1998) Epidemiologisch onderzoek naar chronisch psychiatrische patiënten in Nederland: conclusies uit regionale prevalentiestudies. Tijdschr Psychiatr 4:199–212
Lasalvia A, Tosato S, Brambilla P, Bertani M, Bonetto C, Cristofalo D, Bissoli S, De Santi K, Lazzarotto L, Zanatta G, Marrella G, Mazzoncini R, Zanoni M, Garzotto N, Dolce C, Nicolau S, Ramon L, Perlini C, Rambaldelli G, Bellani M, Tansella M, Ruggeri M, the PICOS-Veneto Group (2012) Psychosis Incident Cohort Outcome Study (PICOS). A multisite study of clinical, social and biological characteristics, patterns of care and predictors of outcome in first-episode psychosis. Background, methodology and overview of the patient sample. Epidemiol Psychiatr Sci 21:281–303
Lau M (2016) Vi vil måle på kvalitet i behandlingen. Region Hovedstadens Psykiatri. https://www.psykiatri-regionh.dk/presse-og-nyt/pressemeddelelser-og-nyheder/Nyheder-og-pressemeddelelser/Sider/Vi-vil-måle-på-kvalitet-i-behandlingen.aspx. Accessed 15 June 2016
Leite LS, Scarparo H, Dias M, Santos SJ (eds) (2013) Saúde mental ConVida – registros da trajetória da saúde mental na cidade de Porto Alegre. Secretaria Municipal da Saúde, Porto Alegre
Loganathan S, Srinivasa Murthy R (2011) Living with schizophrenia in India: gender perspectives. Transcult Psychiatry 48:569–584
Mayor of London (2014) London mental health: the invisible costs of mental ill health. Greater London Authority, London
Mcmanus S, Meltzer H, Brugha T, Bebbington P, Jenkins R (2009) Adult psychiatric morbidity in England, 2007 – results of a household survey. The NHS Information Centre for Health and Social Care, Leeds
Ministeriet for sundhed og forebyggelse, Ministeriet for børn, ligestilling, integration og sociale forhold og uddannelses- og Forskningsministeriet (2015) Forskning til gavn for mennesker med psykiske lidelser – en national strategi. http://www.sum.dk/~/media/Filer%20%20Publikationer_i_pdf/2015/Forskningsstrategi-for-psykiatrien/Forskning-til-gavn-for-mennesker-med-psykiske-lidelser-maj-2015.ashx. Accessed 16 June 2016
NIMHANS (2016) National Mental Health Survey in India, 2015–16. Ministry of Health and Family Welfare, Government of India. NIMHANS, New Delhi. NIMHANS publication 128. Bangalore
Perini G, Grigoletti L, Hanife B, Biggeri A, Tansella M, Amaddeo F (2014) Cancer mortality among psychiatric patients treated in a community-based system of care: a 25-year case register study. Soc Psychiatry Psychiatr Epidemiol 49:693–701
Pertile R, Donisi V, Grigoletti L, Angelozzi A, Zamengo G, Zulian G, Amaddeo F (2011) DRGs and other patient-, service- and area-level factors influencing length of stay in acute psychiatric wards: the Veneto Region experience. Soc Psychiatry Psychiatr Epidemiol 46:651–660
Pradeep J, Issacs A, Shanbag D, Selvan S, Srinivasan K (2014) Enhanced care by community health workers in improving treatment adherence to antidepressant medicines in rural women with major depression. Indian J Med Res 139:239–245

Region Hovedstaden (2011) Visioner for fremtidens psykiatri. https://www.psykiatri-regionh.dk/Vi-bygger-Ny-Psykiatri/Derfor-bygger-vi/Documents/Visioner_Katalog_lilla_version.pdf. Accessed 1 Juli 2016
Region Hovedstadens Psykiatri (2016) Brugen af tvang i psykiatrien falder fortsat. https://www.psykiatri-regionh.dk/presse-og-nyt/pressemeddelelser-og-nyheder/Nyheder-og-pressemeddelelser/Sider/Brugen-af-tvang-i-psykiatrien-falder-fortsat.aspx. Accessed 17 June 2016
Ruggeri M, Lasalvia A, Salvi G, Cristofalo D, Bonetto C, Tansella M (2007) Applications and usefulness of routine measurement of patients' satisfaction with community-based mental health care. Acta Psychiatr Scand Suppl 437:53–65
Shepherd-Smith A (2012) Battling drug addiction in Bangalore. Lancet 379:2040
Shinfuku N (1997) Chapter 7. Mental health in the city of Kobe, Japan. In: Goldberg D, Thornicroft G (eds) Mental health in our future cities, Maudsley monograph, vol 42. Psychology Press, London
Shinfuku N (1999) To be a victim and a survivor of the great Hanshin-Awaji earthquake. J Psychosom Res 46:541–548
Shinfuku N (2002) Disaster mental health: lessons learned from the Hanshin Awaji earthquake. World Psychiatry 1:158–159
Shinfuku N (2012) What is happening in the mental health system in Japan: some observations. Taiwan J Psychiatry (Taipei) 26:70–76
Srinivasa Murthy R (2003) Family interventions and empowerment as an approach to enhance mental health resources in developing countries. World Psychiatry 2:35–37
Tello J, Bonizzato P, Mazzi M, Tansella M, Amaddeo F (2005) Does socioeconomic status (SES) affect the use of community-based psychiatric care? Acta Psychiatr Scand 112:215–223
"The Five Year Forward View for Mental Health" – a report from the independent mental health task force for the NHS in England. February 2016
Theunissen JR (2006) Zorgvernieuwing in de chronische psychiatrie. Een vergelijkend onderzoek onder vier projecten voor langdurig zorgafhankelijke patiënten. Proefschrift, PrintPartners Ipskamp, Enschede
Theunissen JR, Duurkoop P, Kikkert M, Peen J, Dekker J (2013) Na de extramuralisering II; gezondheid, zorgvraag en kwaliteit van leven van de groep chronische patiënten in de Amsterdamse GGZ, omvang en kenmerken van het sociale netwerk en maatschappelijke integratie. Een crossectionele studie. Tijdschr Psychiatr 55:153–163
Thornicroft G (2000) National service framework for mental health. Psychiatr Bull 24:203–206
U.S. Department of Health & Human Services data Dec. 13, 2016
van der Post L (2012) IBS admission as an outcome. Factors predicting the probability of patients qualifying for compulsory emergence admission. Proefschrift, GVO drukkers, Amsterdam
Verambally S, Gangadhar BN (2016) Yoga-based interventions for the management of psychiatric disorders. In: Lam LCW, Riba M (eds) Physical exercise interventions for mental health. Cambridge University Press, Cambridge, pp 124–146
Victora CG, Barreto ML, Leal MC et al (2011) Health conditions and health-policy innovations in Brazil: the way forward. Lancet 377:2042–2053. doi:10.1016/S0140-6736(11)60055-X
While D, Bickley H, Roscoe A, Windfuhr K, Rahman S, Shaw J, Appleby L, Kapur N (2012) Implementation of mental health service recommendations in England and Wales and suicide rates, 1997–2006: a cross-sectional and before-and-after observational study. Lancet 379:1005–1012

Migration and Urban Mental Health in China

10

Juan Chen

Abstract

In the past 30 years, China has transformed from a rural to an urban society. The combination of massive rural-to-urban migration and rapid in situ urbanization has led to significant changes in cities' neighborhood composition and characteristics, as well as residents' lifestyles. Such changes have serious implications for individuals' mental health. The reported prevalence of mental disorders has increased steadily during the past two decades. Urbanization improves access to health care, which leads to health advantages. In the meanwhile, the dynamic urbanization process has also adversely affected the mental health status of both long-term urban residents and new urbanites. As China's urbanization continues, the negative mental health effects are expected to increase further. Despite the increasing public health concerns about the population's mental well-being, resources for mental health services remain scarce and unevenly distributed. Mental health service provision is primarily hospital-based and concentrated in urban centers. People encounter both structural and cultural barriers when seeking professional help for emotional or mental problems. The government recently initiated a community-based mental health service system nationwide that incorporates monitoring, intervention, prevention, and rehabilitation. To be effective, this system should integrate professional and nonprofessional approaches, mental health treatments and family interventions, and online and offline services.

Keywords

China • Migration • Urbanization • Mental health • Help seeking • Community-based service system • Service development • Service integration

---

J. Chen (✉)
Department of Applied Social Sciences, The Hong Kong Polytechnic University, Hung Hom, Kowloon, Hong Kong SAR, China
e-mail: juan.chen@polyu.edu.hk

N. Okkels et al. (eds.), *Mental Health and Illness in the City*, Mental Health and Illness Worldwide, https://doi.org/10.1007/978-981-10-2327-9_15

Contents

Introduction

In the past 30 years, China has witnessed the world's largest ever peacetime rural-to-urban migration of over 200 million people and an unprecedented acceleration of city expansion. In 1978, only 18% of China's population lived in urban areas; by 2014, the percentage had risen to 55 (National Bureau of Statistics of China 2015). Between 1978 and 2010, the total number of Chinese cities increased from 193 to 658. Of these, the number of megacity behemoths (with a population above 10 million) increased from 0 to 6; the number of megacities (with a population between five million to 10 million) from 2 to 10; large cities (with a population between one million and five million) from 27 to 124; medium-sized cities (with a population between 500,000 and one million) from 35 to 138; and small cities (with a population below 500,000) from 129 to 380. During the same period, the number of established townships leapt from 2,173 to 19,410 (Central Committee of the Communist Party and State Council of China 2014).

In recent years, the issue of mental health has received growing attention. The reported prevalence of mental illness particularly depression and anxiety disorders based on community psychiatric epidemiological surveys has continued to increase during the past two decades (Guo et al. 2011). For instance, according to the World Health Organization (WHO) World Mental Health Survey conducted in 2001–2002, the estimated prevalence of having any WHO Composite International Diagnostic Interview or Diagnostic and Statistical Manual of Mental Disorders disorder was 9.1% in Beijing and 4.3% in Shanghai, which was low compared to that of other developed countries (e.g., 18.4% in France and 26.4% in the United States) (WHO

World Mental Health Survey Consortium 2004). In a 2004–2005 study of Chinese mental health using a Chinese version of the Structured Clinical Interview for Diagnostic and Statistical Manual, the adjusted 1-month prevalence of any mental disorder was 17.5%, of mood disorders 6.1%, of anxiety disorders 5.6%, of substance abuse disorders 5.9%, and of psychotic disorders 1.0% (Phillips et al. 2009). More recently, the statistics released by the National Center for Mental Health in 2009 reveal that more than 100 million Chinese (i.e., one in every 13) suffer from various mental illnesses (Yuen 2013).

The prominence of mental disorders has become a major public health burden and concern in China. According to the 2010 Global Burden of Diseases, Injuries, and Risk Factors Study (Yang et al. 2013), mental and behavioral disorders accounted for 9.5% of all disability-adjusted life-years and 23.6% of all years lived with disability (YLD). Seven of the top 20 causes of YLD were mental disorders and major depressive disorder is the highest on the list, which cause serious health-care burdens for patients, families, society, and the nation as a whole (Gu et al. 2013; Yang et al. 2013).

This chapter focuses on issues of mental health, help-seeking, and service provision in the context of urbanizing China. It begins with a description of the general background of the Chinese urbanization process and the recent government policies that promote further urbanization. The chapter then describes the effects of urbanization on residents' mental health, particularly depressive symptoms, and their help-seeking behaviors. The chapter outlines the structural and cultural barriers that prevent people from seeking the necessary help in times of need. Both central and local governments have tried to develop new models of in-patient and community-based psychiatric services in recent years, but it is not clear if these services are meeting the actual needs of people with emotional and mental problems. The chapter further discusses what types of services the most psychologically distressed subgroup would prefer or consider appropriate. The chapter concludes by providing suggestions for the future development of mental health service systems in China.

Urbanization in China

Over the past 30 years, China has undergone a transformation from a rural to an urban society where over 50 percent of its residents now live in cities and work outside of agriculture (National Bureau of Statistics of China 2015). Urbanization in China is the result of two distinct but interrelated phenomena. The first is the process of rural-to-urban migration – the influx of over 200 million rural migrant laborers who left their homes in the countryside to start new lives in cities (Chan 2013). The second, but an equally important process, is 'in situ' urbanization, whereby 200 million people who never left their home village were reclassified as residents of newly established urban districts (Friedmann 2005). Between 1981 and 1999, the annual expansion of urban areas averaged 800 km2 per annum. After 2000, the growth rate more than doubled to 1,700 km2 per annum. By 2014, China's urban space was almost seven times that of 1981 (Ministry of Housing and Urban–Rural Development of China 2015; Yeh et al. 2011). During this process, the urban space rapidly expanded into the countryside that surrounded villages. Rural villagers became urban residents not because they decided to try their luck in the city but because their land was reclassified as urban. They did not go to the city; instead, the city came to them (Chen et al. 2015). The current leadership in China continues to support policies that promote further urbanization in order to drive further economic and social development. According to the National New-Type Urbanization Plan (2014–2020), the central government aims to raise the percentage of the urban population to 60% by 2020, an increase that will involve the relocation or reclassification of an additional 100 million rural villagers.

Living in a Chinese city, however, is not the same as being an "officially registered" urban resident. *Hukou*, the Chinese household registration system that was introduced in the 1950s, categorizes each Chinese citizen as either rural or urban to balance the agricultural surplus with urban industrialization (Chen et al. 2015). The *hukou* system has been criticized for denying equal access to employment, education, housing, health care, and social services to those who do not have urban *hukou*, particularly rural-to-urban migrants (Wang 2005). In 2012, 53% of the Chinese population lived in cities, but only 35% were registered as official urban *hukou* holders, a disparity that reflects the approximately 250 million people who are either rural-to-urban migrants or in situ urbanized rural residents whose villages have been incorporated into cities but whose *hukou* status has not be converted accordingly, leaving them unable to access the social benefits and services associated with urban *hukou*. The National New-Type Urbanization Plan hopes to address this issue by gradually granting permanent urban status to 100 million rural-to-urban migrant workers and in situ urbanized rural residents, so that by 2020, 60% of Chinese citizens will be living in cities and 45% will enjoy full urban status (Central Committee of the Communist Party and State Council of China 2014).

In July 2014, the State Council of China introduced Suggestions for Advancing the Reform of the Household Registration System (State Council [2014] No. 25) to promote the orderly re-registration of long-term rural-to-urban migrants and to expand coverage of basic urban public services. The new policy allows rural-to-urban migrants residing in small cities and towns to apply for official urban status with almost no restrictions. Those who live in medium-sized cities will experience some restrictions, but these will gradually be phased out for those with legitimate and stable occupations and housing. For those who wish to settle in large cities, however, the restrictions remain stringent, and the population of several key mega-cities will continue to be strictly controlled. As a result of these newly released government policies, it is probable that rapid urbanization will continue in China over the next decade. The treatment of new arrivals is likely to improve in smaller cities but will still be discriminatory particularly in areas of receiving social benefits and services in larger cities.

Urbanization and Mental Health

The combination of migration and urbanization has led to significant changes in cities' neighborhood composition and characteristics, as well as residents' lifestyles. Such changes have serious implications for individuals' mental health status (Chen and Chen 2015). Urbanization has certain health advantages, such as access to better care, but it also poses substantial health risks, including ambient air pollution, occupational injuries, traffic hazards, poor diet, and reduced physical activity (Gong et al. 2012). In terms of mental health, too, urban living has pros and cons: cities usually have more, and superior, facilities for treating mental illnesses, but unhealthy lifestyles jeopardize individuals' mental health status (Macintyre et al. 2002; WHO and UN-Habitat 2010). Negative effects are experienced both directly (e.g., through exposure to polluted air and water) and indirectly (e.g., through the perception of risk and attendant chronic stress) (Chen et al. 2013; Peek et al. 2009).

As China continues its process of urbanization, negative mental health effects are likely to increase. Using population density as a measure of urbanization, Chen and colleagues (2015) estimate the actual effect of urbanization on residents' mental health. They find that county-level population density appears to be a consistent, strong, and significant predictor of individual depressive symptoms, as measured by the Center of Epidemiological Studies Depression Scale (CES-D). A one percent increase in county-level population density will lead to a 1.4 increase in the mean CES-D score, which ranges from 0 to 24. As the population density increases (from 61 to 4169 people per square kilometre) across the 31 counties and city districts in the study sample, more depressive symptoms are observed with the average predicted individual CES-D score increasing from 2.6 to 8.4, an almost 6-point difference on a 24-point scale (see Fig. 1, adopted from Chen et al. (2015), Fig. 1a).

The dynamic urbanization process in China has adversely affected the mental health status of not only long-term urban residents but also new urbanites (Chen et al., 2014). Long-term urban residents are likely to demonstrate negative mental health effects because of their extended exposure to a crowded and polluted environment, as well as other risks associated with urban life (Gong et al. 2012). For new urbanites – both rural-to-urban migrants and in situ urbanized rural residents – mental health problems may be caused, or aggravated, by the stress of adapting to an unfamiliar society and a different lifestyle. Studies on the effects of migration on health consistently show that the "healthy migrant phenomenon" – migrants report better health outcomes than the native-born – does not apply to mental health. The mental health status of rural-to-urban migrants is the same or poorer than that of long-term urban residents (Chen 2011; Li et al. 2009), and there is no significant difference in the prevalence of chronic conditions or mental illnesses between in situ urbanized rural residents and long-term urban residents (Chen 2013; Chen and Chen 2015).

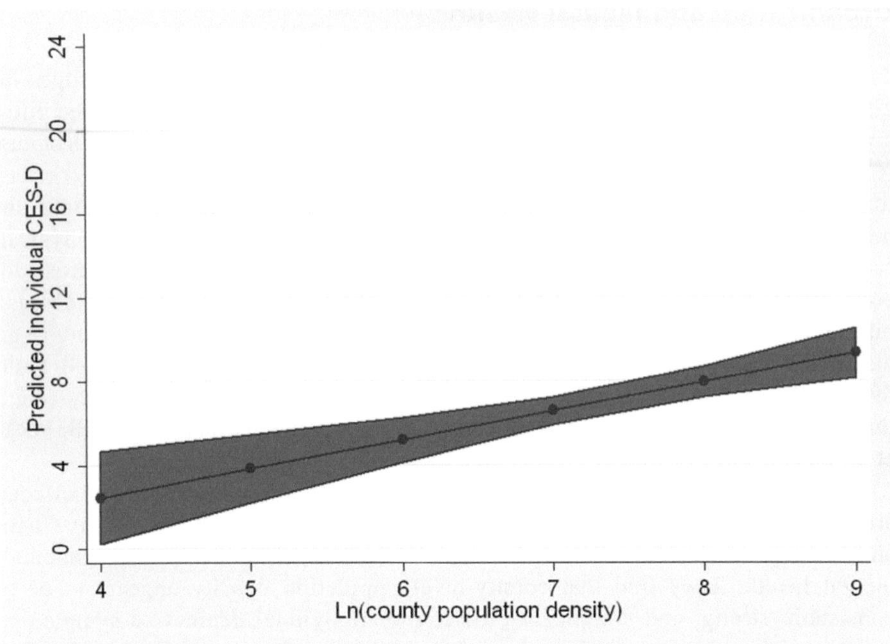

Fig. 1 Estimated effects of county population density on depressive symptoms (Note: N = 1,283. Ordinary least square (OLS) regression is estimated on individual Center of Epidemiological Studies Depression Scale (CES-D) scores. Individual socio-demographic characteristics are controlled in the model estimation. Survey design effects (strata, cluster, and individual weights) are adjusted. Data source: Migration and Quality of Life Survey (2011); figure adopted from Chen et al. (2015)

Help Seeking and Barriers to Service Use

Despite the increasing prevalence of mental disorders, very few Chinese seek professional help for emotional or mental problems. According to the 2001–2002 WHO World Mental Health Survey, 97% of Chinese people with any form of mental illness and 80% of those suffering from moderate to severe disorders received no treatment in the previous 12 months (Shen et al. 2006). Recently, Wong and Li (2014) report that about 173 million adults in China have a mental disorder, but 91.3% have never received any type of professional help. Most Chinese urban residents rely on informal sources of help, particularly from family members and friends, in times of psychological distress, leaving professional mental health and medical services underused (Chen 2012). Those who do seek professional help often experience long delays between the onset of mental illness and psychiatric treatment (Boey 1998), and about three-quarters consult an average of 3.4 caregivers before finally seeing a mental health professional (Zhang et al. 2013).

Cultural Barriers

A great deal of scholarly attention has been paid to the cultural barriers that deter the Chinese population in Hong Kong and overseas from seeking professional help for mental health problems (Ho et al. 2008; Kung 2004; Mo and Mak 2009; Spencer et al. 2010). A few studies have also examined the impact of culture on the help-seeking behaviors of those living in mainland China. For instance, the stigma associated with psychiatric illness and distrust of psychiatric services are the major deterrents to treatment (Boey 1998). Scholars also argue that stigma, poor mental health literacy, and low estimations of need all contribute to the underuse of mental health services (Shen et al. 2006). Regarding the impact on help-seeking behavior of individuals' knowledge of mental illness and their informal networks, people who endorse the psycho-social view of mental illness are more inclined to seek professional help; however, the correct self-diagnosis of depression and the perceived helpfulness of close friends have a significant negative effect on predicted professional help-seeking (Wong and Li 2014). In addition, people are reluctant to seek professional help not only because of their limited knowledge of mental illness but also because of their erroneous assumption that they are possessed of sufficient knowledge to self-diagnose; such overconfidence may be even more harmful than ignorance (Gong and Furnham 2014).

Structural Barriers

Again more studies have been conducted among Chinese living in Hong Kong or abroad (e.g., Ho et al. 2008; Kung 2004; Mo and Mak 2009) than among Chinese living in mainland China to determine the structural barriers that also contribute to the underuse of professional mental health services. Yet scholars have noted that cultural barriers are not the sole reason for the under-treatment of mental illness in metropolitan China; limited health-care resources and structural barriers also play a part (Shen et al. 2006). It is not just cultural attitudes (such as a reluctance to admit need, a lack of trust in professional competence, and embarrassment) but also concerns about accessibility and affordability that deter people from seeking professional help. The general population's tendency to seek help informally for emotional or mental problems stems primarily from a refusal to recognize the need for professional help; however, the subgroup representing the most psychologically distressed is usually deterred by concerns about the affordability of professional mental health services (Chen 2012).

In a recent study, Chen (2017) analyzes qualitative data from in-depth interviews with respondents who belong to the most psychologically distressed subgroup (with the Kessler Psychological Distress Scale (K10) score ≥ 25) identified in a household survey conducted in Beijing. The findings reinforce those of previous studies on the cultural attitudes that deter Chinese from seeking help for psychological distress, such as a psycho-social interpretation of mental illness, a reluctance to admit need, a lack of trust in mental health professional competence, and embarrassment or stigma associated with mental illness (Chen 2012; Kleinman 1986; Wong and Li 2014). The study also draws attention to the current situation of mental health service provision in Chinese cities – the lack of community-based mental health services, the unequal distribution of mental health services, the inadequate insurance coverage for such services, and the scarcity of well-trained mental health professionals (Wong et al. 2014). These factors obstruct easy access to professional mental health services for ordinary Chinese people. In addition to the cultural reluctance, the structural and practical barriers further deter people from seeking help from the appropriate professional mental health service in times of need.

Government Initiatives for Community Mental Health Services

Resources for mental health services have always been scarce and unevenly distributed in China (Gao et al. 2010). The country's mental health policy was not formulated until 1987. Because the Chinese medical system does not deal with mental health and physical health separately, psychological problems are usually treated in general health-care clinics. During the health-care reform of the 1990s, the health-care system was decentralized and privatized (Phillips 1998). The reduction of government funding to hospitals also contributed to the commercialization of mental health services. As a result, the community-based mental health system was largely eliminated. Apart from specialist consultations, mental health service provision is primarily hospital-based. Given the disproportionate concentration of hospitals and other related mental health resources in large cities and megacities, people in smaller cities, towns, and rural areas are at an enormous disadvantage (Liu et al. 2011; Yip 2005a).

In October 2012, the Standing Committee of the National People's Congress of China endorsed the first National Mental Health Law, which came into effect on May 1, 2013. One of its primary objectives is to expand access to mental health services by shifting the focus of services from specialized psychiatric hospitals in urban centers to general hospitals and community health clinics in both urban and rural communities (Phillips 2013). Various levels of government have developed new models of in-patient and community-based psychiatric services. Foremost among these is the 686 Program (named after a government grant of 6.86 million Chinese yuan in 2004), which aims to develop a service system that incorporates monitoring, intervention, prevention, and rehabilitation management of individuals with a psychosis diagnosis. Regional governments have also introduced other initiatives to develop patient-centered community-based psychiatric services (Liu et al. 2011; Wong et al. 2014; Xiang et al. 2012).

Despite these recent developments, mental health services are still unevenly distributed. Mental health prevention programs targeting the general public or potential users are rarely provided. According to the National Mental Health Working Plan (2015–2020), at the end of 2014, there were a total of 1,650 mental health institutes, 228,000 psychiatric beds, and about 20,000 psychiatrists – most located in large cities. Nonetheless, further expansion of community-based mental health services for small cities and towns is in progress. High goals have been set for the year 2020: the number of well-trained psychiatrists will double, reaching 40,000; mental health institutes shall be established at the level of counties and urban districts; and 70% of China's nearly 3,000 counties and urban districts will establish local community-based mental health rehabilitation services.

Service Preferences of Potential Users

The greater availability and higher quality of mental health services, however, will not necessarily result in substantial improvements if those suffering from mental illnesses do not use these services. What are the service needs of people with emotional or mental problems? What kind of services do they prefer or consider appropriate? In Chen's (2017) study based on in-depth interviews with the subgroup identified as the most psychologically distressed, most of the respondents indicate that they never had any direct contact or actual experience with mental health professionals. Their image of these doctors is quite stereotypical – a portrait derived mainly from television shows and movies produced in Hong Kong and other developed countries. Moreover, professional mental health services are viewed as something reserved for the upper-middle and upper classes, not for the ordinary people. The respondents assume that such services would demand a great deal of time and money, which they can hardly afford. As a result, they do not welcome the idea of having a mental health clinic or treatment center in their neighborhood. They do, however, endorse more mental health reference services and promotion programs at the community level.

Online information dissemination and service provision show promise in targeting certain user groups. Among the most psychologically distressed subgroup interviewed by Chen and Zhu (2015), 57% have access to the Internet, 25% have searched for mental-health-related information on the Internet during the preceding year, and 10% have participated in online forums or chatrooms devoted to mental health during the preceding year. The typical profile of an Internet information searcher and help seeker is an educated, young, urban-to-urban migrant. People with high levels of psychological distress are also more likely to participate in forums or chatrooms on the Internet. The number of those turning to the Internet for mental-health support is likely to increase as the Internet, particularly the mobile Internet, continues to expand in China (China Internet Network Information Center 2016). Meanwhile, given China's huge population (nearly 1.4 billion), even relatively low levels of use will result in a large number of Internet help seekers. Mental health policy-makers and practitioners should, therefore, be aware of the vast resource the Internet represents.

The preference for online service provision reflects the embarrassment and stigma associated with seeking professional help for mental health problems, as well as the concern about safeguards to preserve confidentiality (Chen 2012). The endorsement of the first National Mental Health Law is a major step in the protection of the human rights of people with mental disorders in China, but its effects have yet to be measured (Wong et al. 2014). The reasons given for participation in online forums or chatrooms also point to structural barriers to mental health help seeking, including the high expense, and time and location constraints (Chen and Zhu 2015), which need to be addressed in the process of further developing the Chinese mental health service system.

Future Development of Mental Health Service Systems

Clearly, there is an urgent need for mental health professional training and service provision in urban China (Gao et al. 2010). The development of a community-based mental health service system is a step in the right direction, but this approach to the issue must be integrated with other means (Tse et al. 2013). Effective service development must take into account not only the cultural constraints that discourage Chinese people from seeking professional help but also the structural inadequacies that make it difficult to access this help. Only social services that potential users believe will meet their actual needs are likely to succeed. In this section, we discuss directions for the future development of mental health service systems in urban China. Of critical importance is that the government-initiated community-based mental health service system should integrate professional and nonprofessional approaches, mental health treatments and family interventions, as well as online and offline services.

Integrate Professional and Nonprofessional Approaches

In the development of mental health service systems in China, the focus should not be confined to professional interventions. Nonprofessional or informal approaches, such as family networks and the Internet, are significant sources of help for Chinese in times of psychological distress and should be incorporated into the overall design of the service systems. Generous and consistent government investment over time is required to improve mental health professional training and service provision (Park et al. 2005). Increased expenditure alone, however, will not resolve problems such as the lack of universal coverage and the absence of services in many areas of the country. Public education programs that draw attention to informal helping approaches and better coordination between professional and nonprofessional service providers are urgently needed as well.

Integrate Mental Health Treatment and Family Interventions

The family is still the primary caregiving unit in China, and family services and interventions for those with mental illness are crucial (Yip 2005b). Most people with emotional or mental problems consider family members to be their preferred source of help (Chen 2012). Professional mental health treatment should therefore integrate family support and interventions with individual counseling and treatment.

Reciprocally, family services and family support systems should offer mental health counseling services. Studies have shown that the family members are oftentimes the first responders rather than mental health professionals and participation of family members in the mental health treatment process greatly increases the likelihood of positive results (Miklowitz 2008; Rea et al. 2003).

Integrate Online and Offline Services

The Internet has emerged as a widely used source of help in times of psychological distress, particularly among educated young people and recent migrants in urban China (Chen 2012; Chen and Zhu 2015). Embarrassment and the perceived stigma associated with seeking professional help have fueled Internet use (Chen 2012). The Internet has the advantages of being easily accessible, inexpensive, and, most importantly, anonymous. Unfortunately, the mental health information now available on the Internet is not always accurate and should be viewed with caution (Powell and Clarke 2006). Still, it serves a purpose, particularly for those whose fear of being stigmatized inhibits help seeking through professional or informal means (Berger et al. 2005).

Clients often lead the way forward. Online forums and chatrooms are a source of support for young people, offering interactions with empathetic peers (Horgan and Sweeney 2010). Online mental-health-related information searches are expected to grow as the Internet becomes an essential component of people's daily lives. Mental health clinicians have been challenged by clients with information gleaned from the Internet (Bell 2007). It is imperative that mental health professionals be aware of the sources of information available online, develop tools to locate reliable advice, disseminate accurate information, and recommend dependable sites to potential users (Bell 2007; Morahan-Martin and Anderson 2000).

China does not yet offer online mental health self-help treatments that are guided by professionals. Studies based on randomized controlled trials in other countries have demonstrated that online-guided self-help treatments are effective in reducing symptoms of depression and anxiety (Andersson et al. 2009). Given the current shortage of mental health professionals in China, direct or guided self-help therapy online should be encouraged as a means of providing help for a large number of people in a cost-effective manner. Without the embarrassment and stigma associated with seeking professional help, participants would likely be more willing to reveal information, and their difficulties could be addressed more directly and effectively (Bell 2007). There are, however, ethical considerations associated with providing mental health treatment over the Internet. Patient privacy and confidentiality, in particular, must be preserved, and professional liability must be reconceived (Bell 2007; Wells et al. 2007). If properly designed, online treatment programs aimed at educated young people and recent migrants in urban China, such as college students or recent graduates, would perform an important role. They may also serve as the first step in a process that leads to offline face-to-face counseling or treatment if necessary.

Broader Issues to Address

As discussed above, the potential effects on mental health of the unprecedented urbanization of the world's most populous nation require a great deal of research and policy attention. Properly designed and managed urbanization can lead to improvements in the environment and a population's health, but rapid, unplanned, and unregulated urbanization creates environmental pollution, community upheaval, and mental health problems (Zhu and Jones 2010). In order to reduce the prevalence of mental illnesses and provide effective mental health services, broader issues pertaining to urban inequalities must also be addressed.

Balance the Resources of Large and Small Cities

The Chinese government's promotion of urbanization has concentrated on large cities, where the greatest increase in mental health problems is observed. Measures to promote mental health should therefore first target cities with higher population densities (Chen et al. 2015). In the meanwhile, to relieve the pressures on large cities, the government must also create a more equal distribution of resources and opportunities between large and small cities so as to improve the living conditions and opportunities of residents in small cities and towns and reduce the flow of migrants to large and megacities.

Reduce Inequality of Access Due to Socio-Economic Disparity

Given the growing economic disparity in China, the stratification in professional mental health service utilization according to socio-economic status is likely to persist if there is no policy intervention to address it. In a commercial health-care system, the poor avoid seeking professional help even in cases of urgent need while the well-off receive treatment for much less serious problems (Shen et al. 2006). It is therefore necessary not only to provide more mental health services but also to address the inequality of access to services due to socio-economic disparity. Basic mental health services and facilities targeting citizens who are not economically self-sufficient should be the primary focus of government investment.

Narrow the Digital Divide

In order to promote access to reliable mental health information and promote online self-help and treatment, the digital divide also needs to be effectively bridged. The digital divide, which refers to the gap between people who have access to, and the skills to make use of, digital information and communication technologies and those who do not (Dewan and Riggins 2005), represents a new form of economic and social inequality. The widening income disparity has led to separate social spheres and a significant digital divide. In addition to mirroring the socio-economic divide, the digital divide is also closely tied to *hukou* status (Zhu and Chen 2012). As of June 2015, China's Internet penetration rate was 64% in urban areas and 30% in rural areas – a gap of 34% (China Internet Network Information Center 2016). Rural-to-urban migrants and in situ urbanized rural residents are also less likely to have access to the Internet, search for mental health related information, or seek help online (Chen and Zhu 2015). In addition to address the digital divide directly, the underlying socio-economic disparity and the differences between urban and rural *hukou* must also be eliminated or drastically reduced.

Conclusion

As China's urbanization continues, its negative mental health effects are likely to increase. This chapter has addressed issues of mental health, help-seeking, and service development in the context of the ongoing urbanization of China. Given the increasing concerns about urbanites' mental health status and the structural and cultural barriers that people encounter when seeking professional help, we strongly recommend that the government-initiated community-based mental health system should integrate professional and nonprofessional approaches, mental health treatments and family interventions, and online and offline services. Long-standing inequalities must also be addressed in order to reduce the disparity in mental health service use in urban China.

Acknowledgments The research undertaken for this chapter received funding from the General Research Fund of the Research Grants Council of Hong Kong (PolyU5411/12H) and the Li & Fung China Social Policy Research Fund. The author is grateful for the thoughtful comments from the editors and reviewers.

Cross-References

- ▶ Discrimination in Urban Settings
- ▶ Loneliness and Single-Person Households: Issues of *Kodokushi* and *Hikikomori* in Japan
- ▶ Mental Health Care in 11 Major Cities Over Two Decades
- ▶ Urban Design and Mental Health
- ▶ Urban-Rural Differences in Major Mental Health Conditions

References

Andersson G, Carlbring P, Berger T (2009) What makes internet therapy work? Cogn Behav Ther 38(1):55–60

Bell V (2007) Online information, extreme communities and internet therapy: is the internet good for our mental health? J Ment Health 16(4):445–457

Berger M, Wagner TH, Baker LC (2005) Internet use and stigmatized illness. Soc Sci Med 61 (8):1821–1827

Boey KW (1998) Help-seeking pattern of psychiatric outpatients in urban China. Int J Psychiatr Nurs Res 4(1):433–443

Chan KW (2013) China, internal migration. In: Ness I, Bellwood P (eds) The encyclopedia of global human migration. Wiley-Blackwell Oxford, Hoboken

Chen J (2011) Internal migration and health: re-examining the healthy migrant phenomenon in China. Soc Sci Med 72(8):1294–1301

Chen J (2012) Seeking help for psychological distress in urban China. J Community Psychol 40(3):319–341

Chen J (2013) Chronic conditions and receipt of treatment among urbanized rural residents in China. Biomed Res Int 2013:568959

Chen J (2017) Some people may need it, but not me, not now: seeking professional help for mental health problems in urban China. Transcult Psychiatry (conditionally accepted)

Chen J, Chen S (2015) Mental health effects of perceived living environment and neighborhood safety in urbanizing China. Habitat Int 46:101–110

Chen J, Chen S, Landry PF (2013) Migration, environmental hazards, and health outcomes in China. Soc Sci Med 80:85–95

Chen J, Chen S, Landry PF, Davis DS (2014) How dynamics of urbanization affect physical and mental health in urban China. The China Quart 220:988–1011

Chen J, Chen S, Landry PF (2015a) Urbanization and mental health in China: linking the 2010 population census with a cross-sectional survey. Int J Environ Res Public Health 12(8):9012–9024

Chen J, Davis DS, Wu K, Dai H (2015b) Life satisfaction in urbanizing China: the effect of city size and pathways to urban residency. Cities 49:88–97

Chen J, Zhu S (2015) Online information searches and help seeking for mental health problems in urban China. Adm Policy Ment Health Ment Health Serv Res. doi:10.1007/s10488-10015-10657-10486

China, Central Committee of the Communist Party and State Council of (2014) National New-type Urbanization Plan (国家新型城镇化规划) (2014–2020). Retrieved on 8 Oct 2015, from: http://www.gov.cn/zhengce/2014-03/16/content_2640075.htm

China, Internet Network Information Center (CINIC) (2016) The 36th Statistical Report on Internet Development in China. Retrieved on 16 Feb 2016, from: http://www1.cnnic.cn/IDR/ReportDownloads/201601/P020160106496544403584.pdf

China, Ministry of Housing and Urban-Rural Development of (2015) China's Urban Construction Statistical Yearbook 2014. China Planning Press, Beijing

China, National Bureau of Statistics of (2015) China Statistical Yearbook 2015. China Statistics Press, Beijing

China, State Council of (2014) Suggestions for Advancing the Reform of the Household Registration System (国务院关于进一步推进户籍制度改革的意见). State Council [2014] No. 25. Retrieved on 30 Oct 2015, from: http://www.gov.cn/zhengce/content/2014-07/30/content_8944.htm

China, State Council of (2015) National Mental Health Working Plan (全国精神卫生工作规划) (2015-2020). State Council [2015] No. 44. Retrieved on 16 Feb 2016, from http://www.nhfpc.gov.cn/jkj/s5888/201506/1e7c77dcfeb4440892b7dfd19fa82bdd.shtml

Dewan S, Riggins FJ (2005) The digital divide: current and future research directions. J Assoc Inf Syst 6(12):298–336

Friedmann J (2005) China's urban transition. University of Minnesota Press, Minneapolis

Gao X, Jackson T, Chen H, Liu Y, Wang R, Qian M, Huang X (2010) There is a long way to go: a nationwide survey of professional training for mental health practitioners in China. Health Policy 95(1):74–81

Gong AT, Furnham A (2014) Mental health literacy: public knowledge and beliefs about mental disorders in mainland China. PsyCh J 2014(3):144–158

Gong P, Liang S, Carlton EJ, Jiang Q, Wu J, Wang L, Remais JV (2012) Urbanisation and health in China. Lancet 379(9818):843–852

Gu L, Xie J, Long J, Chen Q, Chen Q, Pan R, Yan Y, Wu G, Liang B, Tan J, Xie X, Wei B, Su L (2013) Epidemiology of major depressive disorder in mainland China: a systematic review. PLoS One 8(6):e65356

Guo W-J, Tsang A, Li T, Lee S (2011) Psychiatric epidemiological surveys in China 1960–2010: how real is the increase of mental disorders? Curr Opin Psychiatry 24(4):324–330

Ho KP, Hunt C, Li SL (2008) Patterns of help-seeking behavior for anxiety disorders among the Chinese speaking Australian community. Soc Psychiatry Psychiatr Epidemiol 43(11):872–877

Horgan A, Sweeney J (2010) Young students' use of the internet for mental health information and support. J Psychiatr Ment Health Nurs 17(2):117–123

Kleinman A (1986) Social origins of distress and disease: depression, Neurasthenia, and pain in modern China. Yale University Press, New Haven

Kung WW (2004) Cultural and practical barriers to seeking mental health treatment for Chinese Americans. J Community Psychol 32(1):27–43

Li X, Stanton B, Fang X, Xiong Q, Yu S, Lin D, Hong Y, Zhang L, Chen X, Wang B (2009) Mental health symptoms among rural-to-urban migrants in China: a comparison with their urban and rural counterparts. World Health Popul 11(1):24–38

Liu J, Ma H, He Y-L, Xie B, Xu Y-F, Tang H-Y, Li M, Hao W, Wang X-D, Zhang M-Y, Ng C, Goding M, Fraser J, Herrman H, HFK C, Chan S, Chiu E, Yu X (2011) Mental health system in China: history, recent service reform and future challenges. World Psychiatry 10(3):210–216

Macintyre S, Ellaway A, Cummins S (2002) Place effects on health: how can we conceptualise, operationalise and measure them? Soc Sci Med 55(1):125–139

Miklowitz DJ (2008) Adjunctive psychotherapy for bipolar disorder: state of the evidence. Am J Psychiatr 165(11):1408–1419

Mo PKH, Mak WWS (2009) Help-seeking for mental health problems among Chinese: the application and extension of the theory of planned behavior. Soc Psychiatry Psychiatr Epidemiol 44(8):675–684

Morahan-Martin J, Anderson CD (2000) Information and misinformation online: recommendations for facilitating accurate mental health information retrieval and evaluation. CyberPsychology Behav 3(5):731–746

Park L, Xiao Z, Worth J, Park JM (2005) Mental health care in China: recent changes and future challenges. Harvard Health Policy Rev 6(2):35–45

Peek MK, Cutchin MP, Freeman D, Stowe RP, Goodwin JS (2009) Environmental hazards and stress: evidence from the Texas City Stress and Health Study. J Epidemiol Community Health 63(10):792–798

Phillips MR (1998) The transformation of China's mental health services. The China J 39:1–36

Phillips MR (2013) Can China's new mental health law substantially reduce the burden of illness attributable to mental disorders? Lancet 381(9882):1964–1966

Phillips MR, Zhang J, Shi Q, Song Z, Ding Z, Pang S, Li X, Zhang Y, Wang Z (2009) Prevalence, treatment, and associated disability of mental disorders in four provinces in China during 2001–05: an epidemiological survey. Lancet 373(9680):2041–2053

Powell J, Clarke A (2006) Internet information-seeking in mental health: population survey. Br J Psychiatry 189(3):273–277

Rea MM, Tompson MC, Miklowitz DJ, Goldstein MJ, Hwang S, Mintz J (2003) Family-focused treatment versus individual treatment for bipolar disorder: results of a randomized clinical trial. J Consult Clin Psychol 71(3):482–492

Shen Y, Zhang M, Huang Y, He Y, Liu Z, Cheng H, Tsang A, Lee S, Kessler RC (2006) Twelve-month prevalence, severity, and unmet need for treatment of mental disorders in metropolitan China. Psychol Med 36(2):257–267

Spencer MS, Chen J, Gee GC, Fabian CG, Takeuchi DT (2010) Discrimination and mental health-related service use in a national study of Asian Americans. Am J Public Health 100 (12):2410–2417

Tse S, Ran M-S, Huang Y, Zhu S (2013) The urgency of now: building a recovery-oriented, community mental health service in China. Psychiatr Serv 64(7):613–616

Wang F-L (2005) Organizing through division and exclusion: China's Hukou system. Stanford University Press, Palo Alto

Wells M, Mitchell KJ, Finkelhor D, Becker-Blease KA (2007) Online mental health treatment: concerns and considerations. CyberPsychol Behav 10(3):453–459

Wong DFK, Li JCM (2014) Cultural influence on Shanghai Chinese people's help-seeking for mental health problems: implications for social work practice. Br J Soc Work 44(4):868–885

Wong DFK, Zhuang XY, Pan J-Y, He XS (2014) A critical review of mental health and mental health-related policies in China: more actions required. Int J Soc Welf 23(2):195–204

World Health Organization (WHO) and United Nations Human Settlements Programme (UN Habitat) (2010) Hidden cities: unmasking and overcoming health inequities in urban settings. World Health Organization Centre for Health Development (WHO Kobe Centre) and Nairobi, Kenya: United Nations Human Settlements Programme (UN Habitat), Kobe, Japan

World Health Organization (WHO) World Mental Health Survey Consortium (2004) Prevalence, severity, and unmet need for treatment of mental disorders in the world health organization world mental health surveys. J Am Med Assoc 291(21):2581–2590

Xiang YT, Xu X, Sartorius N, Ungvari GS, Chiu HFK (2012) Mental health in China: challenges and progress. Lancet 380(9855):1715–1716

Yang G, Wang Y, Zeng Y, Gao GF, Liang X, Zhou M, Wan X, Yu S, Jiang Y, Naghavi M, Vos T, Wang H, Lopez AD, Murray CJL (2013) Rapid health transition in China, 1990–2010: findings from the global burden of disease study 2010. Lancet 381(9882):1987–2015

Yeh AGO, Xu J, Liu K (2011) China's post-reform urbanization: retrospect, policies and trends. International Institute for Environment and Development (IIED) and United Nations Population Fund (UNFPA), London

Yip K-S (2005a) A historical review of the mental health services in the people's republic of China. Int J Soc Psychiatry 51(2):106–118

Yip K-S (2005b) Chinese concepts of mental health: cultural implications for social work practice. Int Soc Work 48(4):391–407

Yuen L (2013) Unable to cope: China's inadequate care of the mentally Ill. The Atlantic 29 July 2013. Retrieved on 14 May 2015, from: http://www.theatlantic.com/china/archive/2013/07/unable-to-cope-chinas-inadequate-care-of-the-mentally-ill/278170/

Zhang W, Li X, Lin Y, Zhang X, Qu Z, Wang X, Xu H, Jiao A, Guo M, Zhang Y, Li Y, Tian D (2013) Pathways to psychiatric care in urban North China: a general hospital based study. Int J Ment Heal Syst 7:22

Zhu Y-G, Jones KC (2010) Urbanisation and health in China. Lancet 376(9737):232–233

11 Religious Perspectives in Urban and Rural Mental Health in the Muslim World

Tarek A. Okasha

Abstract

Psychiatry and religion both draw upon rich traditions of human thought and practice. In fact, psychiatry is the branch of medicine that most prominently incorporates the humanities and social sciences in its scientific base and in its treatment of illness. It also attempts to explain the full range of human behavior, including behavior associated with mental illness. Psychiatry has often needed to go well beyond the world of natural science into the philosophical realm.

Mental health providers in developed countries are increasingly treating patients whose backgrounds are much different from their own, so it is important for them to understand cultural belief systems including religious thought and practice that relate to mental health and illness.

Historically, the separate functions of religious practice and healing were performed by a single individual in most world cultures. Only with the explosive growth of scientific knowledge in the twentieth century have the roles of religious and medical healers become separate.

Psychiatry and religion are parallel and complementary frames of reference for understanding and describing the human experience and human behavior.

The understanding and interpretation of many psychiatric symptomatology by both patients and their relatives can depend greatly on their rural versus urban upbringing and education.

Keywords

Traditional and nontraditional societies • Culture and mental health • Role of religion in psychiatry • Influence of religion on psychopathology – Expressed emotions in Arab Culture • Egyptian perspective • Urban and rural backgrounds

T.A. Okasha (✉)
Institute of Psychiatry, Faculty of Medicine, Ain Shams University, Cairo, Egypt
e-mail: tokasha@internetegypt.com; tarek.okasha@gmail.com

N. Okkels et al. (eds.), *Mental Health and Illness in the City*, Mental Health and Illness Worldwide, https://doi.org/10.1007/978-981-10-2327-9_13

Contents

Introduction

An increased awareness of religion in modern urban societies has both positive and negative aspects. From a positive point of view, religious belief systems may provide understandable explanations for traumatic life events or provide meaning for individual or groups. From a negative point of view, any religious fundamentalism, regardless of belief system, can be damaging not only to individual mental health and social adjustment but also to peaceful coexistence among cultures. Unfortunately, one can look across the globe and see how the politicization of religious beliefs can destroy lives and cultures.

In fact, in the future psychiatrists will increasingly be required to confront numerous ethical and social policy issues with religious components, in addition to meeting patient demands for more comprehensive psychiatric approaches that incorporate spiritual perspectives as well as scientific ones. That is partly due to patients having higher demands of a holistic perspective from doctors, however, also the globalization and urbanization of the world is important here. Psychiatrists are increasingly treating patients with a cultural and religious background other than their own.

In many non-Western cultures, native practitioners, to whom modern psychiatry is completely unknown, treat emotionally disturbed persons. Examination of the emotional attitude and interpersonal elements in these various forms of psychological treatments gives the psychiatrist a broad perspective from which to study the basic components of present-day psychiatry and the ethics that guide it (Okasha 2000).

Traditional treatments are characterized as culturally compatible (healers are familiar with the cultural value systems of the patients) and holistic (physical, psychological, social, and spiritual aspects of healing are integrated) and are visually carried out by charismatic healers, individuals who promise to be in charge and indeed are, almost to the point of bearing the responsibility for the results of their decisions. The therapeutic process also frequently incorporates the family, tribe, or group and involves the social manipulation of the patient's immediate environment (Okasha 2000).

Traditional and religious healers in primary psychiatric care deal with minor neurotic, psychosomatic, and transient psychotic states using religious and group therapies (e.g., the Zar ceremony), suggestion, devices, amulets, and incantations (Okasha 1966). In most developing countries national health priorities and health care services are not geared toward mental health and mentally ill patients. Furthermore, 75–80% of Egyptian psychiatric patients, for example, present with somatic symptoms, and 60–70% of these patients with somatic symptoms present to traditional healers and general practitioners before seeking psychiatric help (Okasha and Karam 1998). These somatic symptoms also depend on whether the patient is coming from an urban or a rural background. Usually patients from rural backgrounds present in outpatient psychiatric clinics in urban with more somatic complaints than patients from urban backgrounds.

Thus, as mental illness and behavior associated with mental illness is highly reflecting ones cultural background, understanding of different cultures and religions is important for psychiatrist and other health persons in the globalized world. In this chapter, we will present specific issues regarding different religions and cultural backgrounds which are important to acknowledge for all people who are engaged in treatment of mental illness.

Culture and Mental Health

Culture is a matrix that constitutes the background against which we should understand the biological, psychological, and social dimensions of mental disorder. An individual's cultural background colors every facet of illness, from linguistic or emotional expression (Helman 2007) to the content of somatic complaints (Goldberg and Bridges 1988) and delusional (Yip 2003) or hallucinatory experiences (Cowen 2011). Cause, course, and outcome of major psychiatric disorders are influenced by cultural factors (Kirmayer 2001; Littlewood and Lipsedge 1997). In mental health, dysfunctional behavior is a key issue in diagnosis versus distinction from normal to disordered behavior. The social and cultural context here is important because identification of abnormal dysfunctional behavior is basically a social judgment (Kirmayer and Young 1999). Social and cultural factors are major determinants of the use of health care services and alternative sources of help. Recent changing global demography demands the recognition and response to cultural diversity in psychiatric practice (intercultural clinical work).

This understanding of social and cultural factors influencing the approach of health care services is affected by the psychiatric service being provided in a rural and/or urban setting.

Ethnicity, ethno-cultural identity, social class, cultural dimension of gender, cultural explanation and meaning of sufferings or illness, cultural codes of expression of distress, cultural value system and support network, cultural belief about religion and spirituality, cultural specificity in coping mechanism, and ways of intercultural assimilation are the few broad issues in cultural psychiatry that help psychiatrists to understand the clinical manifestation of psychopathology.

Traditional and Nontraditional Societies

One of the challenges for mental health practitioners in the city is the differences between patients from traditional versus nontraditional societies. In the Islamic culture, the humanitarian interaction with a doctor is valued as much, if not more, than his or her technical ability or scientific knowledge. The humanitarian nature of this interaction depends on the way the doctor deals with the patient and his or her family and the extent to which the doctor expresses respect for and acceptance of local cultural and spiritual norms (Okasha 2000). Besides from the perception of the doctor's role also the perception of the individual and the relatives is different.

In traditional Eastern cultures, social integration is emphasized more than autonomy; that is, the family, not the individual, is the unit of society. Dependence is more natural and infirmity is less alien in these cultures than modern societies. In traditional societies affiliation to a family or tribe is more important than a person's achievement. How one appears in front of others, especially the family is vital and the driving force for an individual to achieve is fear of shaming his/her family rather than personal guilt. In the same manner, physical illness and somatic manifestations of psychological distress become more understood and acceptable and evoke a caring response; in contrast, a vague complaint of psychological symptoms may be disregarded or be considered to indicate that the patient is "soft, weak" or lacks faith, or worse, "insane" (Okasha 2004).

In some cultures, and we argue that Arab and Islamic culture is one of them; the collectivity of the community is valued rather than the individuality of its members. Decisions are made not at an individual level but at a familial, tribal, or communal level, in the best perceived collective interest.

These differences between traditional and nontraditional societies are common in all families and are considered as the mainstream norm of thinking and behavior and not considered as a stereotyped behavior. Cultural diversity may influence the implementation of ethics in different societies. In traditional societies, the family is an extended one, decision making is group and family oriented, and Western attitude regarding individual autonomy does not exist. In traditional societies the concept of external control, dependence on God with regard to health and disease, and attribution of illness and recovery to God's will all maintain a healthy doctor-patient relationship, which makes trust, confidence, and compliance characteristic in traditional societies as shown in Table 1.

The mental health practitioner from a western urban culture must therefore have these differences in mind when treating the patient from a more traditional society.

Table 1 Comparison between traditional and nontraditional societies (Okasha 2000)

Traditional societies	Nontraditional societies
Family and group oriented	Individually oriented
Extended family	Nuclear family
Status determined by age, position in family, and care of elderly	Status achieved by one's own efforts
Relationship between kin is obligatory	Relationship between kin is a matter of individual choice
Extensive knowledge of distant relations	Knowledge of close relatives only
Family decision making	Individual autonomy
External locus of control	Internal locus control
Physician's decision making is respected	Doubt in doctor-patient relationship
Rare suing for malpractice	Common suing for malpractice
Deference to God's will	Self-determination
Trust in doctor-patient relationship	Mistrust in doctor-patient relationship
Individual can be replaced, pride is in the family	Individual is irreplaceable, pride is in self
Pride in family care of mentally ill patients	Community care of mentally ill patients
Illness and recovery attributed to God	Self-determined recovery

He must acknowledge the engagement of not only the nearest family but also the extended and try to involve them as a resource of history taking and information giving. Showing respect to the relatives as well as their cultural norms and beliefs helps the family in accepting the physician. Family and relative involvement is of great value in the management of the patient in order to ensure patient compliance on medication as well as patients' regular follow-up.

Islamic culture includes traditional beliefs in devils, jinn, the evil eye, and so on (delusional cultural beliefs). The family structure is characterized by affiliated behavior at the expense of differentiating behavior. Also, rearing is oriented toward accommodation, conformity, cooperation, affection, and interdependence as opposed to individuation, intellectualization, independence, and compartmentalization. The extended family helps in managing intergenerational conflicts. Families in some traditional societies take pride in looking after their mentally ill relatives. In these societies, it is shameful to the family if it is discovered that a mentally ill family member is homeless.

A very important issue which must be remembered by all practitioners of mental health in urban societies are the young people. Many young individuals vacillate between two worlds, one following the values of Western societies and the other following the values and beliefs of traditional societies (Okasha 2004). Treating these patients can therefore be especially difficult as the doctor must be aware that cultural beliefs of the relatives and of the patient must be respected, even if medically wrong, and the physician should not make fun or ridicule them for these beliefs, instead gradually introduce the meaning of the psychological symptoms, the diagnosis, and the management. This has to be done delicately in order to ensure that the family and the patient accept the treatment and more importantly accept the physician himself/herself. This will be helpful in reaching a better outcome for the patient as well as relative satisfaction.

The Role of Religion

Religion plays an important role in symptom phenomenology, attribution (God's will), and management of mental illness. This can be also influenced by rural and urban upbringing and education. Patients brought up in rural areas are usually more influenced with cultural norms, and societal and religious beliefs, while those from urban areas tend to have a slightly modified traditional approach influenced by the diversity in urban areas and big cities.

In the future, psychiatrists will increasingly be required to confront numerous ethical and social policy issues with religious components, in addition to meeting patient demands for more comprehensive psychiatric approaches that incorporate spiritual perspectives.

In order to understand that we need to highlight some of the concepts and philosophies of both heavenly and nonheavenly religions.

Hinduism

According to orthodox Hinduism, although humans may desire the acquisition of wealth or sensual pleasures, these cannot really satisfy them; instead, they need to tap the infinite power. This power may be actualized by yoga, a method of training to help unify internal power. The Yogi's pathway may be reflection. Love, work, or psychophysical exercises known in the West as meditation. Hinduism describes various stages of human life, such as student years marriage, retirement, and eventually renunciation of the world – a time for self-discovery. Hindus further trace an individual's journey through the migration of the body (reincarnation).

The Hindu view of mental health and illness includes magical, religious, and naturalistic elements. Unlike the West, spiritual influences on health are strongly recognized in Hinduism. While Western thought appears to be preoccupied with thinking in terms of dichotomies, be it body or mind, good or evil, or nature or nurture. In contrast, Hinduism reflects a holistic system of beliefs which views the aspects of human nature as being independent and integrated. Hindus also believe in Karma. Karma refers to the law of moral causality that is necessary in human affairs in order to control imperfection.

A person's present life is a product of what he or she has done and wanted in the past – that is, Karma. Because Karma dictates that every action has inexorable consequences, Hindus are committed to moral responsibility. As such, mental illness is often understood as resulting from the Karma of a past life. Therefore, Hindu concepts of mental health cannot be separated from beliefs about physical and spiritual health (Okasha 1999).

Buddhism

Buddha's first teachings, the Four Noble Truths, can be briefly summarized as follows: (1) life is suffering, (2) the cause of suffering is desire for private fulfillment,

(3) suffering can be relieved by release from narrow self-interest, and (4) the method of this release is through the Eightfold Path. The Eightfold Path involves the right knowledge, speech, behavior, effort, and mindfulness toward others. A basic concept of Buddhism denies a personal God, but Nirvana is described as permanent, ageless, deathless, bliss, unborn, a real truth, supreme reality, eternal, and seems to describe the theological term Godhead. The Buddhist system does not incorporate a belief in spirits. However, animism usually sits comfortably alongside Buddhism and the belief in spirits and ghosts arises out of this older tradition. These spirits and ghosts can cause mental illness (Okasha 1999).

Confucianism and Taoism

Confucius known as the First Teacher, brought some basic values of Chinese culture to a focus, and his influence has lasted to the current time. Chinese do not regard Confucianism as a religion. It is considered a religion by others and has had a major influence on Chinese thought and ethics.

Confucianism places great emphasis on five key relationships: parent and child, husband and wife, older siblings and junior siblings, older friends and junior friends, and ruler and subject. Confucianism has played the most important role in determining rules for the appropriateness of interpersonal relationships between self and others, how to classify social relationships, and how to behave toward others. Shame in Confucian cultures refers in particular to a failure to fulfill positive duties and obligations. In addition, failing to maintain one's identity in the social hierarchy can contribute to the feeling of shame. It has been found that shame can provoke damage to the individual and to social relationships.

Philosophical Taoism is not so much a movement as an attitude toward life and search for knowledge. Religious Taoism took from in the second century and promoted a multitude of deities and sacred texts.

Taoism in all its forms stands on several values important in Chinese life: humility, the rejection of self-assertiveness and competition, attunement toward nature, simplicity in thought and action, and, most notably, the tolerance of opposite as pictured in the classic Yin-Yang symbol. Yin-Yang summarizes life's opposites: male-female, hot-cold, light-dark.

Chinese medicine never really defined a dichotomy between mind and body. Emotions were known to affect body organs and vice versa. Medicines and acupuncture could reduce the Yin-Yang imbalance (Okasha 1999).

Judaism

Judaism has had a profound impact on Western culture and, specifically Western medicine. Jews found meaning in their history, with its relationship to God's intervention in their lives, and a strong moral code as illustrated by the Ten Commandments. Justice was also a basic Jewish value. The basic manual for respecting and honoring life is the Torah, the first five books in the Bible. The Mosaic Law and its oral traditions were brought up to date by Jewish scholars in a series of treatises called the Talmud. The Rabbis thought that God did not want people to suffer and that the human body was to be honored and cared for. Judaism has had a long relationship with medicine. One of the earliest statements was by Maimonides the twelfth- century rabbi-physician who promoted good physical health as a means of serving God (Okasha 1999).

Christianity

Christianity began with Jesus, born in Palestine. Jesus was a charismatic worker of wonders, his history evolved through his followers, to the Christ who had become God in human form, and thus an elaborate theology was produced. Healing and exorcism played a central role in Jesus' ministry, according to the Gospel writers. These miracles often were portrayed not as ends in themselves but as a sign of Jesus' messianic credential.

During the European Middle Ages, the understanding of mental health was dominated by witchcraft, demonology, and possession. Sin was considered the cause of illness, and especially mental illness, and psychiatry consisted of predominantly prescientific demonology.

The age of enlightenment and the moral treatment of the mentally ill continued in an uneven manner in the eighteenth and nineteenth centuries in Europe and America (Okasha 1999).

Islam

Islam means submitting to God. This submission entails that at the end it is God who decides everything. It follows that everything that happens carries with it a certain wisdom or rationale. Even if the individual fails to grasp that wisdom, Islam demands that a Moslem believe in their presence and in God's final judgment.

The Quran is divided into 114 chapters or *surahs*, whose main purpose is to proclaim God's omnipotence and mercy and man's total dependent on Him. Islam brought a vast change to the moral and social order of Arabia and established a specific social order in which faith, politics, and society were joined.

Psychological symptoms are attributed to weakness of personality, lack of faith, lack of conformity, laziness, or other factors, hardly factors that entitle an individual to a right of choice. Statements such as "if God is willing," "I seek refuge in God from the accursed Satan," and "God is the healer" are widespread in the Moslem world, indicating a belief that the final decision is made where no human has control and, therefore, that human choice is a marginal variable in the determination of the final outcome (Okasha 2000).

The Arabic word majnoun ("insane") appears five times in the Koran; however, there it alludes not to insanity but rather to how people perceived prophets when they first attempted to guide the people to enlightenment with beliefs not conforming with the traditions of their societies.

During the dark ages of Islam comes the first concept that majnoun referred to possession by jinn, but the proper Arabic meaning of the word is that of a shield or barrier on the mind. In addition to the concept that the mentally ill individual is possessed, there is the second concept that a person with mental illness is one who dares to be innovative, original, or creative or attempts to find alternatives to a static and stagnant mode of living. This concept is evident in attitudes toward certain mystics. The third concept is that mental illness is the consequence of disharmony or constriction of consciousness, to which nonbelievers are susceptible. It is related to denaturing of one's basic structure (Al Fitrah) and disruption of harmonious existence by egoism, detachment, or alienation, partly due to loss of integrative insight (Okasha 2000, 2004). In Islam, the unity of body and psyche is also recognized. The psyche (Elnafs) is mentioned 185 times in the Koran, with the word referring variously to the body, behavior, affect, and conduct (Okasha 1999).

The WHO Atlas on Mental Health (2005) compared suicide in different regions in the world and found that the Eastern Mediterranean Region, mostly a Moslem population, had the lowest incidence of suicide in all WHO regions, and suggested that adherence to religious beliefs about suicide in Islam may have an effect. Similarly, lower suicide rates were also found in catholic countries most probably for the same reasons (Okasha and Okasha 2009).

Influence of Religion on Psychopathology in Some Psychiatric Disorders

As explained above different religions have different perceptions of mental illness and its causes. However, the symptomatology of mental illness also to some extend varies with cultural and religious beliefs. These differences are important to know for the mental health practitioner since the clinical presentation or symptomatology of the patient like in obsessive compulsive disorder (OCD) and schizophrenia may have a religious or cultural coloring. Delusions and hallucinations usually have a religious content and meaning, where both the patient and their relatives explain them usually by pseudo-philosophical and superstitious religious beliefs which are not part of the core religion.

Below we present some examples on how different mental illnesses are differently presented in patients with different cultural backgrounds.

Schizophrenia

Schizophrenia appears relatively similar across a range of cultures; however, variability has been noted in symptom presentation and development (Swartz 1998). The WHO international study of schizophrenia (1973) found that schizophrenia was a fairly ubiquitous disorder with an almost similar picture over many cultures.

However, the way in which the particular types of symptoms appear may vary from individual to individual and culture to culture (WHO 1973). Furthermore, owing to differences in social customs and expectations, cultures differ in their assessment of the importance of different symptoms (Edgerton and Cohen 1994; Lucas and Barret 1995).

Historically, there are a number of studies comparing the manifestations of schizophrenia across cultures (Varma 2000). Differences in the symptomatology of schizophrenia in Arab and Islamic cultures have been a subject of many investigations with interesting results. While Taleb et al. (1996) found few clinical differences in his comparative study between patients with schizophrenia from Morocco and France, Gawad et al. (1981) showed a number of important differences in the diagnosis of schizophrenia in Egypt compared to the USA and the United Kingdom. Gawad and his colleagues studied the cross-national differences in symptom importance in the diagnosis of schizophrenia among the three countries. They found that restriction and incongruity of affect ranked first in the Egyptian study compared to the British and American ones. Their results are in agreement with other studies who stated that what is normal emotional expression in an Anglo Saxon culture may suggest a schizoid reduction of emotional response in a Mediterranean culture (Lehman 1967). The top ten symptoms in the Egyptian hierarchy for diagnosis of schizophrenia were incongruity and restricted affect, formal thought disorder, thought block, thought withdrawal, incoherence, passivity feeling, neologism, hallucination, delusions, and ideas of reference. While in the British study, formal thought disorder ranked first followed by incongruity of affect, neologism, thought block, passivity of feeling, paranoid delusions, stereotype of other delusions, thought withdrawal, and ideas of references (Willis and Bannister 1965). The Americans ranked symptoms of importance for diagnosis of schizophrenia as follows: formal thought disorder, delusions, paranoid delusions, incongruity of affect, hallucinations, ideas of reference, neologism, depersonalization, and thought block (Edwards 1972).

There was a striking difference in ranking neologism, while it is the third in the Anglo American study, it was the seventh in Egyptian study because neologism is perhaps among the commonest symptom in hysterical dissociation particularly occurring in religious pseudomystic ceremonies (Okasha 1966). The same finding was reported in Libya (Khalid 1977). The cultural and religious heritage absorbs many features, which would otherwise be considered symptomatic of a psychiatric disorder (El-Islam 1980). It was noticed that the difference between Egyptian hierarchy and British hierarchy are less than those between the Egyptian and American ones. This observation might be explained by the fact that higher qualification and training of Egyptian psychiatrists is predominantly British (Gawad et al. 1978, 1981). Moreover, at that time, the most common popular textbooks of psychiatry in Egypt were British. Nowadays Egyptian psychiatrists are more acquainted with the American classification and American textbooks, in spite of the fact that DSM-IV includes a number of features designed to enhance its cross-cultural applicability, yet it seems that it is not sensitive to cross-cultural differences (Thakker and Ward 1998).

Clinical analyses have shown differences in the clinical presentations of schizophrenia as a result of cultural pathoplastic influences (Rakhawy et al. 1987; Okasha 1988).

Findings highlighting these influences are summarized in the following sections.

Delusions

The themes of delusions are affected by the individual characteristics of the patients in relation to their culture. The most frequent themes of delusions in Egypt are religious, political, and social delusions related to health, while autistic delusions are less common (Rakhawy et al. 1987). Religious delusions are most common among Muslims and Christians. Religious delusions are frequent due to highly religious standards in Egypt. Sexual delusions are most common in groups in whom sexual behavior is severely suppressed, for example, delusions of sin are frequent in relation to masturbation, in the younger, single, and the students group (Okasha 1988; Ashour et al. 1986). Political delusions are positively correlated to the level of political sanctions and pressure. Fear of political persecution is a reality of life for people living under totalitarian regimens. Such fears may contribute to a higher prevalence of paranoid delusions (Westermeyer 1988).

The content of the patient's delusions varies directly in relation to his/her social class and education. For most of the lower class men and women, the delusional symptoms are about the cultural religious institutions. Middle and upper class patients, however, far more frequently "secularized" their restitutive narcissistic and self-esteem delusions in terms of science and class conception of power (El-Sendiony 1976).

The fate of delusions in schizophrenia in an Arabian population was studied by El-Islam (1980), who found that the involution of a delusion may be through their disappearance into a system of socially shared delusional such as cultural beliefs or their metamorphosis into less sinister symptoms (e.g., dreams and fantasies). The culturally shared belief system (magic, jinnis, evil eye, etc.) provides a container for delusions that lose their pathological quality and content.

Hallucinations

There are cultural variations in the frequency of different kinds of hallucinations within and between cultures (Sartorius et al. 1986). Many investigators in the Arab and Moslem countries studied the phenomenology and frequency of visual, kinesthetic, and haptic hallucinations. It is concluded that these types of hallucinations in schizophrenia are of common occurrence in these countries. However, the part played by cultural and other factors needs further investigations (Khalil 1990).

The cross-cultural study of the content of auditory hallucinations in patients with schizophrenia living in Saudi Arabia compared to those living in the United Kingdom showed striking differences. Much of the content of the hallucinations of Saudi Arabian patients with schizophrenia were religious and superstitious in nature, whereas instructional themes and running commentary were common in the UK schizophrenic patients (Kent and Wahass 1996). Patients from both cultures had several coping mechanisms with auditory hallucinations, but these varied between cultures. The majority of Saudi patients used strategies associated with their religion whereas UK patients were more likely to use distraction or psychologically based approaches. This study suggests that clinicians, when they attempt to facilitate the use of coping strategies, may find greater patient acceptance and efficacy if they are familiar with cultural specific factors (Wahass and Kent 1997a, b).

Behavior

Behavioral problems in schizophrenia are traditionally attributed to acts of possession by spirits (jinn), sorcery, or envy by the evil eye in many Arab and traditional societies. Families who maintain these beliefs take their patients to native healers who are endowed with powers of exorcising evil spirits, undoing sorcery, or ending the harm of envy. When efforts along these lines are judged to have failed, and this may take months or years, the patient is finally brought under medical care (El-Islam 1980).

Positive and Negative Symptoms

Negative symptoms in many Moslem and Arab countries may be attributed to religious asceticism, so patients can be accepted and assimilated in the society and is considered as a social trait. Families usually do not go for medical or psychiatric help until the appearance of positive symptoms which leads to the delay in seeking proper therapy for many years, thus increasing the duration of untreated psychosis (Okasha and Maj 2001). This delay in seeking proper therapy can also be seen in Egypt and many other Arab countries. When comparing patients coming from rural and urban areas, with the delay being more in rural patients. Many families in rural areas protect the patients at home especially females to avoid the stigma of mental disorder and the possibility of the patient not getting married. So, psychotic patients with negative symptoms can remain for years under the umbrella and care of their immediate families at home as long as they are quite and obedient. However, positive symptoms will force the family to seek treatment as soon as possible to avoid social humiliation and the disgrace of mental illness.

Obsessive Compulsive Disorder

The impact of Egyptian culture is evident in many aspects of transcultural studies in Egypt.

The religious nature of upbringing and education in Egypt, the emphasis on religious rituals, especially related to ablution remaining for hours washing, and the warding-off of blasphemous thoughts through repeated religious phrases such as "I seek refuge with the Lord from the accursed Satan" can explain the high prevalence of religious obsessions and repeating compulsions among Egyptians, even if the subjects are not practicing their religious duties (Okasha et al. 1991).

To elaborate further, Moslems, who constitute almost 90% of the Egyptian population, are required to pray five times a day. Each prayer is preceded with a ritualistic cleansing process (*El Woodoo* or ablution), which involves the washing of several parts of the body in specific order, each three times. Muslims are required to pray five times/day: dawn prayer with (2) genu flexion followed by prosternation; noon prayer with (4) genu flexion followed by prosternation; afternoon prayer with (4) genu flexion followed by prosternation; sunset prayer with (3) genu flexion followed by prosternation; evening prayer with (4) genu flexion followed by prosternation. It is required that the prayer should be clean, this requires the ritual of ablution (Woodoo) which is ritualistic washing of mouth, nose, face, ears, hands to elbow, and feet each three times, provided he/she urinated or defecated between two prays (he should wash the orifices after these physiological functions). In rare radical Muslims, ablution is necessary after touching or saluting the opposite sex or being touched by a dog as it produces annulment of the ablution. The emphasis on cleanliness' (In Arabic, Tahara; in Hebrew, Taharat) or ritual purity is the cornerstone of most of the compulsive rituals. The number of prayers, the verbal content can be a subject of scrupulous, checking and repetition. A bath is necessary after ejaculation, and washing orifices after excretion whatever its nature (Okasha et al. 1991).

Women are not allowed to pray or touch the Koran during their menstruation, after which they should clean their bodies through a ritualistic bath. The prayers themselves are different in length and consist of certain phrases and *souras* from the Holy Koran that have to be read in certain sequence (Okasha et al. 1994a).

The ritualistic cleansing procedures also can be a source of obsessions and compulsions about religious purity, for example, in some compulsives, the color red (reminder of menstruation) may trigger a compulsive washing. Another evidence of the religious connotation inherent in OCD in Moslem culture lies in the term *El Weswas (in Arabic)*. This term is used in reference to the devil, and at the same time is used as a name for obsessions.

It is also characteristic of a conservative society like that of Egypt to expect sexual obsessions to be among the most frequent in female patients. Although it is accepted socially (but prohibited religiously) for Egyptian males to have a wide range of sexual freedom in all stages of their lives, sexual matters remain an issue of prohibition, sin, impurity, and shame for Egyptian women. The female gender is surrounded by so many religious and sexual taboos that the issue becomes a rich pool for worries, ruminations, and cleansing compulsions in women susceptible to developing OCD (Okasha el al. 1994a).

Christians represent approximately one tenth of the population in Egypt. In this group, symptoms are similar to the Moslems in terms of obsessions, where religious and sexual thoughts are predominant. However, there is a marked difference in rituals, which are more frequent in Moslems, emphasizing the role of ritualistic Islamic upbringing as compared with Christian upbringing in our community (Okasha et al. 1994a).

A comparison in this context could also be drawn between the most prevalent symptoms between Egyptians and those of other studies performed in India, England, and Jerusalem (Table 1). Contamination obsessions were the most frequently occurring in all studies. However, the similarities of the contents of religious obsessions, for example, blasphemous between Moslems and Jews as compared with Hindus and Christians signify the role played by cultural and religious factors in the

Table 2 Common themes of obsessions in Egypt, India, England, and Jerusalem

Country	No.	Contamination (%)	Aggressive (%)	Ordering (%)	Sexual (%)	Religious (%)
Egypt (Okasha et al. 1994)	90	60	41	53	47	60
India (Akhtar et al. 1975)	82	46	29	27	10	11
England (Stern and Cobbs 1978)	45	38	23	11	9	0
Jerusalem (Greenberg 1984)	10	40	20	10	10	50

presentation of OCD. The obsessional contents of the samples from Egypt and Jerusalem were similar, dealing mainly with religious matters and matters related to cleanliness and dirt. Common themes between the Indian and British samples, on the other hand, were mostly related to orderliness and aggressive issues. It is interesting to note that English sample had no religious obsession (Okasha et al. 1994a) as seen in Table 2.

Emotional Expression (EE)

Emotions can no longer be considered private intrapsychic or psychobiological phenomena. Emotions, no less than other attitudes, beliefs, or behaviors, are mediated by culture and cannot be examined except in their context. Normal emotional expression in Anglo-Saxon culture may suggest a schizoid reduction of emotional response in a Mediterranean culture (Okasha et al. 1994b). Emotional expression is explained by overprotection, critical comments, inner hostility, positive regards, and warmth.

Critical comments in different cultures showed more relapses in Schizophrenia and depression: English studies: Two critical comments (Vaughn and Leff 1976), USA Studies: Three critical comments (Hooley et al. 1986), Los Angeles: Six critical comments (Montero et al. 1992), Egyptian studies: Seven critical comments which explain the high tolerance of Egyptian families (Okasha et al. 1994b).

Level of family criticism in traditional cultures is higher than in the West. High EE can be viewed as a desirable social trait; criticism can be a sign of care and concern in traditional families. Little is known about the role of warmth and positive remarks in family life and patient outcome and clinically significant levels of warmth could protect the patient against relapse.

The cultural and religious heritage absorbs many features, which would otherwise be considered symptomatic of a psychiatric disorder. Judging disability in a cultural context is a more sensitive and valid indicator of the need for treatment than diagnosis (El-Islam 1980).

Conclusion

In this chapter we have presented some of the major issues for the mental health professional when treating patients from different cultures and religions. This is an important field, as globalization and urbanization is making cities more diverse in regards to these matters. Psychiatrists and mental health professionals are regularly seeing and dealing with patients coming from diverse ethnic and religious backgrounds whose interpretation of mental disorder and mental symptoms differ. Understanding and respecting such diversities is key to successful treatment and a good doctor-patient relationship. These diversities can also be seen in patients from the same country coming from urban and rural backgrounds which can have an influence on setting policies in dealing with patients either as inpatient or outpatients as well as the need to have psychotherapeutic methods which take into account religious, cultural, and social beliefs and norms.

There are numerous areas, directions, and goals that cultural psychiatry with the influence of religion can aim for in the future. This includes focusing on cultural psychiatry issues broadly for all people in their own society whether from urban or rural areas in order to strengthen the training of cultural psychiatry and theology at different levels for medical students, residents, and future psychiatrists. Also, to work on theoretical issues to expand our knowledge and theoretical understanding for universal and culture-specific applications, and consolidate the organization of knowledge in the field of cultural psychiatry, in order to establish a solid subfield of applied science. It is equally important to promote culture-suitable positive mental health rather than treating only mental illness paying attention to religious and philosophical aspects of human life and its implication on therapy. We should also be minded by the importance of helping people to adjust to the sociocultural change that is taking place in almost every contemporary society around the world. Minimizing negative interethnic, interracial, interfaith, and intercultural conflict frequently happening in the world resulting in terrorism, genocide, or war. Finally, focusing on culture, religion, and mental health issues around the world is essential in the coming years in order to improve the services given to our patients in order to try and improve their quality of life.

Cross-References

- ▶ Discrimination in Urban Settings
- ▶ Urban-Rural Differences in Major Mental Health Conditions

References

Akhtar S, Wig N et al (1975) A phenomenological analysis of symptoms of obsessive compulsive disorders. Br J Psychiatry 127:342–348

Ashour A, Okasha A, Hussein M, Sobhi M, Khalil A (1986) The P.S.E. and social skills of chronic schizophrenics: descriptions and correlation. Egypt J Psychiatry 9:18–30

Cowen M (2011) Cultural factors influence hallucination types in schizophrenia. Compr Psychiatry 52:319–325
Edgerton RB, Cohen A (1994) Culture and schizophrenia: the DOSMD challenge. Br J Psychiatry 164(2):222–231
Edwards G (1972) Diagnosis of schizophrenia: an Anglo–American comparison. Br J Psychiatry 120:385–390
El-Islam MF (1980) Arabic cultural psychiatry. Transcult Psychiatr Res Rev 19:5–24
El-Sendiony MF (1976) Cultural aspects of delusion: a psychiatric study in Egypt. Aust N Z J Psychiatry 10(2):201–207
Gawad M, Mahfouz R, Rakhawy Y (1978) Diagnosis of schizophrenia in Egypt: a within-national comparison. Egypt J Psychiatry 1(1):100–113
Gawad M, Rakhawy Y, Mahfouz R, Howaidy M (1981) Relative symptom importance in the diagnosis of schizophrenia. Egypt J Psychiatry 4(1):39–56
Goldberg DP, Bridges K (1988) Somatic presentation of psychiatric illness in primary care setting. J Psychosom Res 32(2):137–144
Greenberg D (1984) Are religious compulsions religious or compulsion. A phenomenological study. Am J Psychother 38(4):524–532
Helman CG (2007) Culture, health, and illness. 5th edn. Hodder Arnold, London. ISBN 978-0-340-914- 502
Hooley JM, Orley J, Teasdale JD (1986) Levels of expressed emotion and relapse in depressed patients. Br J Psychol 148:642–647
Kent G, Wahass S (1996) The content and characteristics of auditory hallucinations in Saudi Arabia and the UK: a cross cultural comparison. Acta Psychiatr Scand 94(6):433–437
Khalid MS (1977) Symptomatology of schizophrenia in Libya. Psychiatr Neurol Med Psychol 29(1):46–48
Khalil AH (1990) Psychological assessment of delusional parasitosis in Egypt. A paper presented in the international congress of parasitology, ICOPA, Paris, 20–24 August
Kirmayer LJ (2001) Cultural variations in the clinical presentation of depression and anxiety. Implications for diagnosis and treatment. J Clin Psychiatry 62(Suppl 13):22–28
Kirmayer LJ, Young A (1999) Culture and context in the evolutionary concept of mental disorder. J Abnorm Psychol 108:446–452
Lehman HE (1967) Clinical features of schizophrenia. In: Freedman AM, Kaplan HI (eds) Comprehensive text book of psychiatry. The Williams and Wilkins Company, Baltimore
Littlewood R, Lipsedge M (1997) Aliens and alienists: ethnic minorities and psychiatry, 3rd edn. Taylor & Francis, London. ISBN 0-415-15725-0
Lucas RH, Barret RJ (1995) Interpreting culture and psychopathology primitivist themes in cross-cultural debate. Cult Med Psychiatry 19(3):287–326
Montero I, Gomez-Beneyto M, Ruiz I, Puche E, Adam A (1992) The influence of family expressed emotion in the course of schizophreniain a sample of Spanish patients. Br J Psychiatry 161:217–222
Okasha A (1966) A cultural study of El Zar cult in Egypt. Br J Psychiatry 1(12):1217–1221
Okasha A (1988) Clinical psychiatry. The Anglo Egyptian Book shops, Cairo, Egypt, pp 1131–1165
Okasha A (1999) Religion and mental health at the turn of the century. Plenary Lecture delivered at the XI world congress of psychiatrty, Hamburg
Okasha A (2000) The impact of Arab culture on psychiatric ethics. In: Sartorius N, Okasha A, Arboleda Florez J (eds) Ethics, culture and psychiatry: international perspectives, vol 2. American Psychiatric Association, Washington, DC, pp 15–28
Okasha A (2004) Focus on psychiatry in Egypt. Br J Psychiatry 185:266–272
Okasha A, Karam E (1998) Mental health services and research in the Arab world. Acta Psychiatr Scand 98:406–413
Okasha A, Maj M (eds) (2001) Images in psychiatry; an Arab perspective. Scientific Book House, Cairo

Okasha A, Okasha T (2009) Suicide in Islam. In: Wasserman D, Wasserman C (eds) Oxford textbook of suicidology and suicide prevention; a global perspective. Oxford University Press, New York, pp 49–55
Okasha A, Raafat M, Seif El Dawla A, Effat S (1991) Obsessive compulsive disorder in different cultures : an Egyptian perspective. Egypt J Psychiatry 14:15–30
Okasha A, Saad A, Khalil AH et al (1994a) Phenomenology of obsessive-compulsive disorder: a transcultural study. Compr Psychiatry 35(3):191–197
Okasha A, El Akabawi A, Wilson A, Youssef I, Seif El Dawla A (1994b) Expressed emotion, perceived criticism, and relapse in depression: a replication in an Egyptian community. Am J Psychiatr 151:1001–1005
Rakhawy Y, Amin Y, Hamdi E, Mahfouz R, Howaidi M, Arafa M (1987) Characteristics of thought disorder in schizophrenia. Egypt J Psychiatry 10(1 and 2):131–150
Sartorius N, Jablensky A, Korten A (1986) Early manifestation and first contact incidence of schizophrenia in different cultures. Psychol Med 16:909–928
Stern RS, Cobbs JP (1978) Phenomenology of obsessive compulsive neurosis. Br J Psychiatry 132:233–239
Swartz L (1998) Culture and mental health: a southern African view. Oxford University Press, Oxford, p 204
Taleb M, Rouillon F, Petitjean F, Gorwood P (1996) Cross-cultural study of schizophrenia. Psychopathology 29(2):85–94
Thakker J, Ward T (1998) Culture and classification: the cross – cultural application of the DSM-IV. Clin Psychol Rev 18(5):501–529
Varma VK (2000) Transcultural perspective on schizophrenia, epidemiology, manifestation and outcome, Report of section of transcultural psychiatry, WPA Newsletter 2000; XVIII(1):17–24
Vaughn GE, Leff JP (1976) The influence of family and social factors on the course of psychiatric illness: a comparison of schizophrenia and depressed neurotic patients. Br J Psychiatry 129:124–137
Wahass S, Kent G (1997a) A cross-cultural study of the attitudes of mental health professionals towards auditory hallucinations. Int J Soc Psychiatry 43(3):184–192
Wahass S, Kent G (1997b) Coping with auditory hallucinations: a cross – cultural comparison between western and non-western patients. J Nerv Ment Dis 185(11):664–668
Westermeyer J (1988) In: Othmanns TF, Maher BA (eds) Some cross-cultural aspects of delusional beliefs. Wiley, New York
Willis JH, Bannister D (1965) The diagnosis and treatment of schizophrenia: a questionnaire study of psychiatric opinion. Br J Psychiatry 111:1165–1171
World Health Organization (WHO) (1973) Report of the international pilot study of schizophrenia. WHO, Geneva
World Health Organization (2005) Mental health atlas. World Health Organization, Geneva
Yip K (2003) Traditional Chinese religious beliefs and superstitions in delusions and hallucinations of Chinese schizophrenic patients. Int J Soc Psychiatry 49(2):97–111

Mental Health and Urban Life: Experiences from Colombia

12

Carlos López-Jaramillo, Jorge Mauricio Cuartas Arias, and Ana M. Díaz Zuluaga

Abstract

This chapter looks into the main aspects impacting on mental health and daily life in Colombia, taking Bogotá as central reference. Bogotá, the capital city of Colombia, is where different problems affecting the entire national territory converge and where new strategies are devised for the improvement of the mental health and the lifestyle of its population. The lack of implementation of public policies for social inclusion, life quality, and mental health in the city of Bogotá is a reflection of the difficulties faced across the national territory. Colombia was the epicenter of drug trafficking in the 1960s leading to an upsurge of violence countrywide. Although the drug business has been gradually declining in Colombia, corruption, outlawed armed groups, and social inequality contribute to phenomena such as displacement, addictions, and lack of mental health-care programs. This situation has undermined the development of a health-centered culture centered on the quality of life of its inhabitants. In this chapter, we describe in detail the most relevant social problems that Colombia faces and how they affect the population's life expectancy and mental health. In addition, we advocate for the development of a state policy that designs and creates mechanisms of action for the management of social and scientific mental health platforms that provide healthy lifestyles in Colombia.

C. López-Jaramillo (✉) • A.M. Díaz Zuluaga
Research Group in Psychiatry, Department of Psychiatry, Faculty of Medicine, Universidad de Antioquia, Medellin, Colombia
e-mail: calberto.lopez@udea.edu.co; carloslopezjaramillo@gmail.com; ana.diazz@udea.edu.co

J.M. Cuartas Arias
Research Group in Psychiatry, Department of Psychiatry, Faculty of Medicine, Universidad de Antioquia, Medellin, Colombia

Faculty of Psychology, Universidad de San Buenaventura, Medellin, Colombia
e-mail: mauricio.cuartas@usbmed.edu.co

N. Okkels et al. (eds.), *Mental Health and Illness in the City*, Mental Health and Illness Worldwide, https://doi.org/10.1007/978-981-10-2327-9_5

Keywords
Mental health • Colombian population • Colombia • Poverty • Bogotá • Social inequality • National Mental Health Survey

Contents

Introduction

Mental health contexts in the daily life of Colombia, and more specifically of Bogotá, are linked to the political and social circumstances common to other Latin American capital cities. Undoubtedly, healthy lifestyles associated with good mental hygiene require to build and maintain a social justice where the aspects that prevail are social inclusion and balanced access to basic services and to opportunities by the government to develop people's talent, that is, access to education and to the free development of personality, in a social niche where tolerance to differences is fostered.

In countries exposed to violence and armed conflict, the mental health of their people poses a major challenge in the public policies of mental health, where prevention, diagnosis, and treatment of psychiatric diseases are a pillar of health care.

In this chapter, we develop some aspects relevant to the promotion of mental health in the daily life of Colombians, paying special attention to the capital city, Santa Fe de Bogotá. The aspects discussed highlight the relevance of social justice based on equality and social inclusion, condemn moral and administrative corruption, mitigate poverty, and offer scenarios which guarantee proper and timely mental health care as the cornerstone of the quality of life of the Colombian population.

Colombian Population

In 2014, Colombia had a population of approximately 47,000,000 people, with growth trends indicating a population of close to 51,000,000 by 2020. The ratio of male to female population is very similar: 49.4% males vs. 50.6% females. Taking into account its geography, the country shows different patterns of human settlements, low population density is found toward the peripheries (the Amazon, jungle plains, and areas of the low Magdalena), and the greatest concentration of people is found in the Andean region, the eastern area, and the *Altiplano Cundiboyacense* (Cundinamarca-Boyaca high plateau). The last one includes Bogotá, the capital city of Colombia and the political, economic, and administrative center of the country, with a population approximately of eight million people.

As with other countries in the region, Colombia has entered into a phase of demographic transition, with falling birthrates and stable moderate to low rates of mortality. It is projected that by 2020 the population under the age of 15 (31% in 2005) will have decreased 25.4%. The over 65-year-old population, representing 7% in 2005, is expected to rise to 8.5% by 2020.

The structure of the Colombian population presents challenges for public and mental health, with the aging index (calculated as the number of persons 60 years old or over per 100 persons under age 15), rising to close to 33.48%, when in 2005 it stood at 20.17%. Currently the most vulnerable populations in mental health are the child and aging adult populations. More emotional suffering, failures in access to services and medication, and limitations in social inclusion programs for older adults can be seen in both populations, as well as insufficient strategies for developing attitudes and aptitudes in the child population.

Colombian Definition of Mental Health

In 2013, Colombian law defined mental health as a condition inherent to the human being that exhibits itself within the environment and within a social context through interaction with others. This relationship constitutes a "cognitive niche," a concept developed by Steven Pinker (Pinker 2004), related to the basis of the stability of social interactions in culture, these being: technological understanding and the development of knowledge and tools for the social group, the language that transmits perceptions about oneself and others in temporal space, human features and characteristics that are shaped in large part by history and diverse cultures and traditions, and, perhaps most relevant when tackling the concept of mental health, cooperation. This feature underlies all associations, alliances, and emotional relationships and together with observable reciprocity, even with nonfamily members, supports the notion of mental health as a collective feature. Despite having connotations of the individual, its regulation implies human interactions in favor of collective goals that promote well-being and adaptation to the complexities implicit in social, political, and technological changes within societies.

In this scenario, people deploy their emotional and cognitive resources in order to combat individual and collective stress brought about by the need for varied adjustments and which impact the "allostatic load" which refers to "the wear and tear on the body." Through allostasis, the central nervous system activates the autonomic nervous system, especially the hypothalamic pituitary-adrenal system together with the metabolic and immunological systems. These are made ready in order to respond to different physiological and psychological situations experienced in the environment so as to maintain self-regulation (Turner, et al. 2016). The concept of allostatic load refers to an adaptive response that involves a momentary and particular way of acting so as to be effective. However, when there are occurrences of allostatic overload, the repeated and prolonged presence of multiple stress factors could generate responses that are nonadaptive or insufficient, impacting regulatory systems, such as changes in the bioavailability of glucocorticoids, a rise in cytokines, or metabolic changes which affect levels of leptin and melatonin, among others. Therefore, when stress surpasses the adaptive capacity, a disease may evolve, which in the context of mental health can involve psychological and behavioral alterations that affect individual and social functioning.

The effects of stress are expressed in differential allostatic loads, and they vary from individual to individual. Within this variability are genetic, organic, nutritional, and psychological factors that, in addition to the subjective perception of the stressor, give rise to a response which is different, dimensional, and temporally distinct. Social and contextual dynamics play a large role in restructuring the perceptions of the stressor, and from this, individual and collective well-being can be derived.

Mental Health in Colombia

To date, Colombia has carried out four mental health surveys in 1993, 1997, 2003, and 2015 which form part of the national system of studies and surveys for understanding public mental health in Colombia. These surveys have helped guide public health policies that are included in the Ten-Year Public Health Plan 2012–2021.

Deterioration in the mental health of Colombians and an increase in mental disorders were confirmed in 2010 by the research carried out by the Ministry of Health and Social Protection and the Center for Development Projects. In this research, depression is the second most prevalent cause of burden of disease, surpassed only by arterial hypertension, a variable which may also suggest that current strategies are inefficient in creating healthy lifestyles which reduce negative stress and which exhibit inadequate physiological responses.

The National Mental Health Survey (NMHS) carried out in Colombia, which guides the scope of this chapter, had regional and national representation and included 15,351 persons from a sample population ranging from 7 years to 96 years and allowed the researchers to assess mental health by taking into account different factors, dimensions, and access to mental health services (Ministerio de Salud y Protección Social 2015).

The NMHS survey analyzed a sample of 2727 children between the ages of 7 and 11 years and found that one of the most prevalent diseases was attention deficit hyperactive disorder (ADHD) at 3%, followed by separation anxiety disorder at 1.8%, and challenging negative thoughts and general anxiety disorder, each with 0.4%. These diseases were more prevalent within cities with greater levels of poverty.

For adolescents, analyzing a sample of 1754 young people between the ages of 12 and 17, the NMHS found anxiety to have a prevalence of 5.0%. It should be highlighted that the most common pathology was primarily social phobia and that again poverty was most often associated with this indicator. Also, suicide is more common in adolescents than in adults in Colombia. With regard to suicide, the survey found a prevalence of 1.8% and 37.6% of adolescents that have thought about suicide or have attempted it.

For mental disease in adults (45 years and above), a sample was obtained of 4981 individuals. The most prevalent diseases were affective diseases at 6.7% and 7.4% of the subjects that had attempted suicide. Similarly, to the infant and adolescent samples, conditions of poverty were related to the prevalence of mental disease.

Mental Health Policies in Colombia

In 1998, under the current General Health Social Security System (SGSSS: *Sistema General de Seguridad Social en Salud*), accomplished via Law 100 of 1993, the first National Policy of Mental Health was adopted. It established parameters which foster the psychological welfare of Colombians. However, this policy could not be fully implemented as its execution coincided with the change of government, which resulted in constant modifications to SGSSS regulations, and to a lack of such factors as political decision, funding, and dissemination of the new standard.

The Colombian Ministry of Social Protection published the Guidelines of the Policy of Mental Health in Colombia and the National Policy of the Field of Mental Health, in 2005 and 2007, respectively, which served as provisional Policies of Mental Health. It wasn't until 2013, however, that the Law of Mental Health was passed. This law modified the National Policy of Mental Health of 1998 and considered the epidemiological and legislative changes that the Colombian population faced in the last 15 years. As a result, the National Mental Health Plan was created (Congress of the Republic of Colombia 2013).

The Mental Health Act of 2013 adapts, in Section 6, to the rights of the citizens in the area of mental health as stated in the Constitution of the Republic of Colombia and in the legal provisions established by WHO (Freeman and Pathare 2005), so that there can be policies which consider the needs arising in each population subgroup. As an example, in Bogotá, several public health entities promote sexual and reproductive rights considering the diverse sexual orientations and gender identities, and they have even participated in the making of specific policies.

The current National Mental Health Plan, integrated to the Ten-Year Public Health Plan 2012–2021, defined three primary areas to meet the current challenges in mental health: mental illness prevention, mental health promotion, and integral mental health care. These three primary areas must, in turn, focus on the three main problems Colombia is facing: violence, common mental disorders, and psychoactive substance abuse (Ministerio de Salud y Protección Social y la Organización Panamericana de la Salud (OPS/OMS) 2014).

Colombian Health-Care System

The current health-care system in Colombia is regulated by Law 100 of 1993. Through this law, a mandatory basic health-care plan known as "mandatory health-care plan" (POS, *Plan Obligatorio de Salud*) must be created, which all insurance companies providing health insurance plans must cover in their totality. The plan consists of a package of basic services of promotion, prevention, and care, addressed to all citizens, regardless of their socioeconomic regime or their monetary contributions (or lack thereof) to the system (Congress of the Republic of Colombia 1993).

However, the POS (mandatory health-care plan), established until right before year 2013, did not guarantee full coverage of health services, as some technologies in mental health, medication, hospitalization services, and psychotherapy were restricted to the acute stage of mental disorders. In view of the above, the government passed Law 1616 in 2013 (Congress of the Republic of Colombia 2013), which in Section 4 warrants mental health care to all citizens and highlights the need to guarantee promotion, prevention, integral and integrated care, diagnosis, treatment and rehabilitation for all mental disorders, and prioritizing children and adolescents. The main aspects of the abovementioned law are outlined below:

- Campaigns to promote cessation of discrimination against patients with mental illness, elimination of stigma and harassment, prevention of suicide, and the use of psychoactive substances.
- Mental health promotion and prevention in the work environment, regulated by labor risks entities. Impact on psychosocial risk factors to "protect, improve, and recover the mental health of workers."
- Complementary actions, including social, educational, work, and family integration.
- Every insurance company must have an integral network to provide mental health services and guarantee service provision at all levels of complexity. They must ensure warm, timely, continuous, and readily available care and have a system of reference and counter reference which enables patients to receive hospital care appropriate to the complexity of their pathology.
- Service modalities must include outpatient care, home care, prehospital care, drug addiction care centers and drug dependency services, mental health community centers, patients and family support groups, day hospital care for the elderly, day hospital care for children and adolescents, community-based rehab, metal health units, and psychiatric ER.
- The team in charge must be interdisciplinary, including professionals of psychiatry, psychology, nursing, social work, occupational theory, psychosocial therapy, and a general practitioner.

- The National Mental Health Council is created for the assessment and regulation of the abovementioned national policies of mental health.

Both in rural and urban areas, the outpatient care process starts with initial assessment by a general practitioner who can correctly diagnose and manage the most common psychiatric disorders. The General Medicine Service decides on the referral to a specialized psychiatry service to continue with the care of the patients and provides pathology-specific treatment. Similarly, acute care services, provided by the emergency services of the clinics and hospitals across the country, include patient stabilization, initial management of the patient's crisis, and determination of treatment and follow-up strategies, whether through general medicine or psychiatry services.

In rural areas, where access to health-care services is difficult, there are "health promoters": regularly auxiliary nurses trained in health prevention and promotion who are able to identify alarm signs or symptoms of medical and psychiatric conditions.

Actual Access to Mental Care Services

The National Survey of Mental Health carried out in 2015 looked into the actual access to mental health-care services. The survey covered separate age groups, from which the following results were obtained:

- Among the population between 7 and 11 years old, 49% of the legal caregivers of the children considered that mental health care was excellent, 30.8% rate it as very good, 17.8% as good, 1.7 as mediocre, and the rest considered it was bad. In the previous 12 months, 48.1% of this population had requested health services, and 92.7% of them received the requested care. However, these services were paid by health-promoting organizations EPS (*Entidades Promotoras de Salud*) to hospitals in only 69.4% of the cases. Of this population, 53.8% had a psychiatric condition which needed referral to psychotherapy, but only 83.1% of them received it.
- For the population between 12 and 17 years old, access to health care is even more reduced, as 35.2% had requested health services in the previous 12 months, 3.9 times (IC:0.85–6.9) on average, and only 88.5% received the requested care. The EPS covered the expenses 70.9% of the times. Additionally, 0.8% of adolescents were under some psychiatric medication. Only 86.5% of the psychiatric drugs had been prescribed by a doctor, and 80% were paid for by the patients' EPS.
- Among the population between 18 and 44 years old, 38.5% had requested access to mental health services in the previous 12 months, an average of 23.5 times, and access was granted 94% of the times. For this population group, the EPS paid for the services only 66.1% of the times. Prescribed medication was given to 83.1% of the patients.
- Among adults over 45 years, 34.3% requested access to health services, with an average of 2.4 times, and only 79.4% of them received health care. Prescribed medication was obtained by 97.9% of them.

The main difficulties in accessing mental health services are the social stigma of psychiatric illnesses along with geographical barriers. The latter affects mostly the population of rural areas, where long distances, high travel costs, the lack of training in primary health care, and the lack of articulation within hospital networks hinder access to psychiatry services.

Psychosocial Context of Colombia

The psychosocial reality which involves mental health in Colombia is not all that different when compared to the rest of the world. The WHO recently showed that mental and neurological disorders will explain 15% of years lost to disability in 2020 across the globe, a proportional increase which is greater than cardiovascular disease (Min Protección Social and Universidad de Antioquia 2010). The historic and social context in Colombia requires us to revisit those psychosocial components that put the healthy lives of Colombians at risk. Undoubtedly a preliminary indicator of mental health is a rise in clinical consultations, independent of age, to receive guidance and/or treatment for situations that affect behavior or indeed indicate the presence of a mental disease. It is furthermore appropriate to point out that complaints related to mental health are appearing much earlier, and therefore the burden of disease is considerably greater.

In Colombia, and through the Ten-Year Public Health Plan (TYPHP), there has been an attempt at tackling mental health as a fundamental right of top national priority by proposing and analyzing three components:

- Rights
  - Consistent with UNESCO's Bioethics and Human Rights Declaration, in favor of supervising and directing public health policies that respect dignity, human rights, and diversity and reject discrimination, while championing plurality and promoting values and policies which respect the conditions of life
- Social determinants in health
  - In line with WHO guidelines and related to the distribution of wealth, socio-political scenarios, and access to basic needs within a context of social equity
- Population approach
  - Safeguarding inclusion and social equality, in a multicultural and multiethnic context

The second National Health Survey was carried out taking into account these guidelines (Ministerio de Salud y Colciencias 2015) and the following dimensions included in the TYPHP:

- Social cognition: allows us to analyze individuals based on their cognitive and behavioral repertoires, such as recognizing and understanding others and promoting inference, decisions, and recognition of individual and social autonomy
- Social perception: perceptions concerning the levels of trust, security, citizen participation, and inclusion

- Assessment of family functioning: family communication, tolerance and flexibility, as well as emotional bonds
- Stressful life events: coping strategies when faced with negative situations, both personal and on others, which can alter subjective well-being and affect social bonds
- Perception of security: the influence of violence, crime, armed conflict, and displacement

These dimensions contribute to examining the subjective-relational and socio-collective relationships in mental health. Accordingly, the proper balance and perception of emotional self-regulation as the constituent axis of mental health is loaded with imaginary social meanings around specific expectations and contexts. For the social cognition dimension, previous studies have shown how cognitive functions, the genetic background and the social environment correlate closely with the remodeling of neural processes underlying the human condition. The skills and abilities of inference, attribution, empathy, and social perception are linked to the processes of inhibition and cognitive self-regulation, flexibility, decision-making, working memory, and planning, all of which are part of the executive function, and that is fundamental for the human adaptation as fundamental key in the biological concept of species.

In addition, the differences in age and level of education were permeated by the indicator for poverty, given that lack of access to educational opportunities and to a healthy environment which provokes cognitive reserve underlies the poverty indicator. Similarly, that which is related to the perception of public trust and family functioning is permeated by level of schooling and a state of poverty.

On the other hand, and in recognition of the idiosyncrasies of American cities and the social dynamics which interact in people's daily lives and their social fabric, emotional suffering which is not necessarily present in a psychological or psychiatric disease would indicate the loss or absence of support networks and the development of unstable social bonds which entail problems with adaptation and survival when faced with the demands imposed by culture and society.

Furthermore, stressful life events and the lack of cognitive devices which enable people to confront adversity in the face of pain or critical social situations, such as crime, forced displacement, natural disasters, and a lack of employment and opportunities for developing talent, are generators of negative stress loads and affect daily life and social insertion. Moreover, ruptures and family dysfunction, mistreatment, sexual exploitation, and child abuse evolve as structural dominions in young people and adults, increasing the risk of suffering from personality problems and psychiatric diseases.

Social Factors and Mental Health

It is important to note that mental health in Colombia is permeated by diverse factors which generate problems that affect the healthy lifestyles and lifetimes of the Colombian population. Below are detailed some of the social and contextual factors which most impact on mental health.

Inequality and Poverty

In Colombia, approaches which include plans and programs for promoting healthy lifestyles and habits require substantial changes to public policy to promote an overall decrease in poverty.

The global level of extreme poverty has been gradually decreasing since 1990, moving from 1.9 billion to 836 million in 2015. Recently, the National Planning Department released the Multidimensional Poverty Index (Rosenkranz et al. 2016), composed of parameters such as level of income and deficiencies which affect peoples' lives and which is articulated based on three principles for determining the level of individual poverty: education, access to health services, and standard of living.

In 2014 was published a report stating that 49% of households suffered from multidimensional poverty, with outlying cities showing the greatest Multidimensional Poverty Index and the lowest found in the city of Bogotá.

However, poverty does not mean inequality; poverty relates to the average income of a society, while inequality refers to wealth distribution. This does not indicate a linear relationship; while poverty has been in decline across the world, inequality has remained relatively stable.

Inequality is also a huge problem in Colombia. A short while ago, the World Bank published the list of countries with the highest rates of economic inequality, the first five places of which were occupied by African countries, with Colombia taking second position after Honduras with regard to Latin American countries.

Violence

In 2014, the National Institute of Legal Medicine and Forensic Science recorded 25,225 cases of violent death, giving a rate of 52.92 cases per 100,000 inhabitants, placing Colombia in first place for homicide. Domestic violence as a structural axis for health care occupies a relevant position for analyzing the country's mental health, being a measure of social protection or vulnerability.

Based on legal medical examination, the NMHS reported a total of 284,718 cases of violence in Colombia in 2014, where domestic violence came in second place with 75,939 cases at 159.33 per 100,000, finding that 13.7% of the children and adolescent experienced domestic violence related to parent in 65% and other relative in 20.11%.

Suicide and Suicide Behavior

In Colombia in the last 20 years, the suicide rates have fluctuated between 3 and 5 cases per 100,000 (Ceballos Ospino 2004). In 2013, legal medicine reported 1810 suicides, of which 1461 were men. The greatest number of cases was recorded in Antioquia (n = 329), Bogotá (n = 236) and Valle del Cauca (n = 158), which are the main urban areas of Colombia. However, Antioquia has a rural area with the highest rate of suicide behavior and suicide of the country due to the high prevalence of major mental disorders, such as bipolar disorder, schizophrenia, and depression. This has been mainly explained by the special genetics features of this population (endogamy and inbreeding) which makes it a genetically isolated worldwide known as the "Paisa population."

Historically, in Colombia, the suicide rates have not been as high as in developed countries; however, the trend has been toward a rise, with the rate passing from 3.84 in 2013 to 3.94 in 2014.

Current Situation of Suicide in Bogotá

According to the Legal Medicine Institute, the suicide rate in Colombia started decreasing gradually in 2002 but rose again between 2008 and 2010, when 4.1–4.52 suicides were reported per 100,000 inhabitants.

Up to 2013, 1498 suicides were reported, of which 516 had been committed by 20–24 year olds, 388 by 15–19 year olds, and 415 by 25–29 year olds.

According to the Department of Health's System of Epidemiological Surveillance of Suicidal Behavior, in Bogotá there were 24 cases of suicide in 2012 and 220 in 2013, corresponding to a rate of 2.9 per 100,000 inhabitants. During that period, it was confirmed that suicide frequency was higher in two specific towns: *Kennedy* (166 cases) and *Ciudad Bolivar* (165 cases).

According to the study, the number of suicides was higher among men than among women, with a ratio of 5:1. By 2014, according to the FORENSIS report and the Colombian legal medical system, 1878 suicide cases were reported in Colombia. In the decade previous to that report, 18,336 suicides were reported, which amounts to an average of 1833 cases per year in the country (Ministry of Health and Social Protection 2015).

Fourteen more cases were reported in 2015, adding up to a total of 195 suicides in Bogotá, where a higher number of cases were reported compared to Antioquia: 24 and 22, respectively. In 2016, between January and March alone, 83 underage suicide deaths were reported, of which 14 were in Bogotá and 12 in Antioquia (Revista Semana 2016).

The Issue of Drug Addiction

In Colombia, the problematic consumption of psychoactive drugs brings as a consequence a deterioration in cognitive performance and social functioning in people, having a negative impact on health systems and generating social dynamics which put social well-being at risk and promote aggression and exclusion (OEA 2012; Min Salud y Protección Social 2013a, b).

The national government, through the Colombian Drugs Observatory from the Ministry of Justice and Law in its national study on the consumption of psychoactive substances, which evaluated a sample of 32,605 from all socioeconomic strata of Colombian society between the ages of 12 and 65, determined that there had been an increase in the overall consumption of psychoactive drugs (cannabis, cocaine, base or *basuco*, ecstasy, and heroin), finding a prevalence of 12%, with cannabis being the most consumed substance across the country. It likewise indicated that the age group of 18–24 year olds consumed the most illicit substances, and in general 57% of consumers for this year showed signs of abuse or dependency, exhibiting itself in significant physical, mental, familial, and social difficulties.

The Use of Psychoactive Substances in Bogotá

Colombia is ranked in a middle position when considering drug use in relation to other South American countries. However, drug use has been increasing throughout the years not only because of cultural changes (wider dissemination of allusive materials; increasingly accepted practice by the society, especially among the new generations) but also due to the diversification of psychoactive substances in the market, which has resulted in an exponential growth of drug supply (Colombian Drug Observatory 2017).

In the second national study of psychoactive substances use in Colombia, it is reported that Bogotá is in a middle position in relation to other parts of the country, with a substance-use prevalence rate of 3.5% (use of any psychoactive substance in the last year). According to the survey, Medellín and its metropolitan area are the regions with the highest prevalence, 8.2% (Ministry of Justice and Ministry of Health et al. 2014).

Drug use is more related to urban areas, probably because of their social dynamics and greater supply. It is estimated that in Bogotá 197,938 individuals have used drugs recently (in the last year). Men use more drugs than women, with a ratio of 4:1. Age of first use is around 17 years for both men and women, and the drug-use rate is higher in socioeconomic level 3, yet not significantly higher than in other socioeconomic levels. Drug abuse and drug dependence in Bogotá was reported for 177,692 individuals, which represent 3.12% of this specific population. Additionally, the lower the socioeconomic level, the higher the prevalence of drug abuse or dependence (Ministry of Defense and Ministry of Health et al. 2014).

The authorities are currently alarmed, as a recent study in Bogotá schools has demonstrated that drugs are being used by even younger age groups, and that between 2014 and 2015, the number of drug users in schools has doubled (Guevara 2016).

The most widely used substances in the capital district are marijuana (180,667 users) and cocaine (27,307 users), followed by *basuco* (cocaine paste), ecstasy, and inhalants. In Bogotá schools, the substance most frequently used is also marijuana. The second most commonly used substances are processed substances known as "handmade acids," in low socioeconomic level districts and ecstasy in schools of higher socioeconomic level districts. Handmade acids are manufactured in Colombia and sold for USD 2–7, whereas ecstasy is of higher quality and imported from countries such as Italy, Germany, or the Netherlands and much more expensive (Guevara 2016).

These studies show that one of the major problems in a city such as Bogotá is that the easy access of young population to a large number of substances promotes drug use and increases the possibility of dependence or abuse in the long term.

Armed Conflict and Mental Health

In 2016, Colombia was the country with the largest displaced population (6.9 million), outranking countries in conflict such as Syria (6.6 million) and Iraq (4.4 million) (Agencia de la ONU para los Refugiados ACNUR 2016). This is not a new phenomenon, since as early as the mid-twentieth century, there were massive displacement movements of civilians of some population sectors, first because of bipartisan violence and later due to the emergence of guerrillas. In the 1990s, the paramilitaries appeared and have played an important role in the escalation of the problem, as they aimed not only at stopping the expansion of the guerrillas but also at expelling all the inhabitants from the lands rich in energy and mining resources or where big infrastructure projects could take place.

Kidnapping, murder, rape of women and girls, forcible recruitment of child and adolescent soldiers, and bombs and massive attacks to towns and municipalities have been the cause for different psychiatric pathologies and comorbidities among the rural and urban population of Colombia. They have constituted a risk factor for anxiety disorders, posttraumatic stress disorder, depressive disorders, substance use and abuse disorders, suicide, and suicidal behavior, among others.

Conflict victims experience bigger social and family problems in the future, especially to build interpersonal relationships, to consolidate and maintain their families and to establish new emotional ties, which results in a more complex social and work reinsertion process.

Additionally, civilians who are not direct victims of the armed conflict display feelings of fear, concern, anxiety, insecurity, and hopelessness, which constitute a risk factor for the development of new psychiatric disorders.

Violence and Forced Displacement in Bogotá

Bogotá has become the main city to host the victims of forced displacement. According to the national Unique Victims Registry, the number of displaced inhabitants arriving to the city has increased considerably since 2001, reaching its peak in 2014 and decreasing in 2015 and 2016 (Registro único de víctimas RUV 2016). Approximately 23% of the victims of forced displacement have arrived in Bogotá looking for new opportunities for themselves and for their families (Administrative Department of District Planning 2004). Most of the victims of displacement relocate in the poorest areas, thus contributing to the widening of the misery belts and the escalation of other social problems, such as homelessness, overcrowding, and violence. This is corroborated by data from the Colombian Planning Department that shows that socioeconomic level 1 had the highest growth between 2011 and 2014. They face the hostility of a large city in which they arrive under economic vulnerability, with little or no support from the government, and seeing displacement as their only way to safeguard their lives and their personal integrity. According to the first survey by the Colombian General Comptroller about the Victims' Law of 2015, 35% of the displacement victims surveyed live in extreme poverty.

While this increase in displaced population may be one of several violence factors in the city, it is by no means the only one. As Olga Cubides points out in her work, violence in Bogotá is considered a "multicausal phenomenon." In her research, the sale of drugs was found to be a determining factor, for the existence of high homicidal violence in the city, along with other aspects such as sex work, micro-extortion, and the presence of criminal bands. The expansion of the drug business goes hand in hand with the increase in the recruitment of young people in order to strengthen these structures, since they are the ones in charge of the control and regulation of the micro-trafficking business through intimidation and violence. In these cases, displaced youths and children are easily recruited due to the lack of opportunities they have and to the economic needs of their families. Many of them are recruited forcefully to do tasks that range from "campaneros" (watchers, literally "bell ringers") and "carritos" (messengers, literally "little cars") to extortionists and murderers.

While the rise in insecurity in Bogotá has several components, we can't ignore that the areas of Bogotá that have received a high number of displacement victims are the ones which have presented an increase in all their violence indexes. This is why many of the inhabitants of these sectors have the perception that the arrival of displacement victims translates into an increase of insecurity levels. An example of this situation is Ciudad Bolívar, the area of the city that receives the highest number of migrants and that, in turn, has experienced the highest increase in the number of homicides over the last years.

Although the homicide rate in Bogotá in 2016 was 14.4 homicides per 100,000 inhabitants (the lowest in the last years), in 2005, it was 24.6 per 100,000 inhabitants. However, citizens still have very high perception of insecurity (above 80%) and little confidence in the police and justice administration entities. This is probably due to the increase in other crimes such as pickpocketing, mugging, or car theft and to the existence of areas where the homicide rate has increased because of micro-trafficking and fights resulting from alcohol use.

By other hand, Bogotá is ranked first in the country for domestic violence, with women and children being the primary victims. In the first term of 2016, 11,687 complaints were lodged, which included 6343 cases of women attached by their partners or husbands (El Tiempo 2016).

However, in spite of overall insecurity- and forced displacement-related problems, Bogotá is not ranked among the most dangerous cities in the world. According to a report presented in 2015 at the Conference on Homicide Data Quality in Latin America and the Caribbean, held in Bogotá by the international organization Open Society Foundations, the Ministry of Justice of Colombia, the District Attorney's Office, and the Chamber of Commerce of Bogotá, this is because considerable effort has been made to improve the social and economic conditions of the population as a mechanism for crime prevention (El Espectador 2015).

Conclusions

Colombia is a country with important prevalence and incidence of different mental disorders in all age groups. Since the country has been affected along its history by violence, poverty, and corruption, the well-being and mental health of its citizens has been clearly impacted by these phenomena in both rural and urban areas.

Although in the last decade the health-care system has been experienced multiples changes in order to ensure and increase coverage health issues of the population, there still exists a long pathway of political, educational, and economical facts which allow the improvement of the human development, the social inclusion, and the access to health care and education, among others, that finally leads to a better social growth.

Cross-References

- ▶ Crime, Illegal Drugs, and Mental Health
- ▶ Development of Mental Health Services in a Major City in Latin America: Santiago
- ▶ Homicide and Suicide in Megacities
- ▶ Prospects and Challenges of Mental Health Services in a Major City in Latin America: Rio de Janeiro
- ▶ Urban-Rural Differences in Major Mental Health Conditions

References

Agencia de la ONU para los Refugiados ACNUR (2016) Tendencias Globales sobre refugiados y otras personas de interés del ACNUR. Informe de la Agencia de las Naciones Unidas para los Refugiados. Retrieved from http://www.acnur.org/recursos/estadisticas/

Ceballos Ospino, Guillermo Augusto (2004) Características de las personas que consumaron suicidio en la ciudad de Santa Marta (Colombia) durante el año 2002: un informe de casos 1: 24–28. http://revistas.unimagdalena.edu.co/index.php/duazary/article/view/264. doi:http://dx.doi.org/10.21676/2389783X.264

Congreso de la República de Colombia (1993) Ley 100 de 1993. Bogotá D. C. Retrieved from http://www.alcaldiabogota.gov.co/sisjur/normas/Norma1.jsp?i=5248

Congreso de la República de Colombia (2013) Ley 1616 de 2013 "Por medio de la cual se expide la ley de salud mental y se dictan otras disposiciones". Retrieved from http://wsp.presidencia.gov.co/Normativa/Leyes/Documents/2013/LEY%201616%20DEL%2021%20DE%20ENERO%20DE%202013.pdf

Departamento Administrativo de Planeación Distrital (2004) El desplazamiento en Bogotá: una realidad que clama atención. Retrieved from http://www.acnur.org/t3/uploads/pics/2756.pdf?view

Departamento Nacional de Planeación (2014). Bases del Plan Nacional de Desarrollo 2014–2018. Retrieved from https://colaboracion.dnp.gov.co/cdt/prensa/bases%20plan%20nacional%20de%20desarrollo%202014-2018.pdf

El Espectador (2015) Bogotá no figura entre las 50 ciudades más inseguras del planeta. Retrieved from http://www.elespectador.com/noticias/bogota/bogota-no-figura-entre-50-ciudades-mas-inseguras-del-pl-articulo-586000

El Tiempo (2016) Bogotá, la ciudad con más casos de violencia intrafamiliar en el país, El Tiempo. Retrieved from http://www.eltiempo.com/bogota/violencia-intrafamiliar-en-bogota/16688792

Freeman M, Pathare S (2005) WHO resource book on mental health, human rights and legislation. World Health Organization. Retrieved from http://ec.europa.eu/health//sites/health/files/mental_health/docs/who_resource_book_en.pdf

Guevara C (2016) Cuatro de cada diez escolares han probado drogas. In: El Tiempo. Retrieved from http://www.eltiempo.com/bogota/consumo-de-drogas-en-los-colegios/16624127

Ministerio de Justicia y del Derecho, Observatorio de Drogras de Colombia, Ministerio de Salud, Protección Social, Oficina de las Naciones Unidas contra la Droga y el Delito UNODC, Comisión Interamericana para el Control del Abuso de Drogas CICAD, Embajada de los Estados Unidos en Colombia (2014) Estudio Nacional de Consumo de Sustancias Psicoactivas en Colombia-2013. Retrieved from https://www.unodc.org/documents/colombia/2014/Julio/Estudio_de_Consumo_UNODC.pdf

Ministerio de la Protección Social (2005) Estudio Nacional de Salud Mental Colombia 2003. Retrieved from https://www.minsalud.gov.co/Documentos%20y%20Publicaciones/ESTUDIO%20NACIONAL%20DE%20SALUD%20MENTAL%20EN%20COLOMBIA.pdf

Ministerio de la Protección Social (2007) Política Nacional del Campo de la Salud Mental. Documento – Propuesta para Discusión y Acuerdos. Dirección General de Salud Pública & Asociación Colombiana de Psiquiatría. Retrieved from http://psiquiatria.org.co/web/wp-content/uploads/2013/05/Pol%C3%ADtica-Nacional-del-Campo-de-la-Salud-Mental-versi%C3%B3n-final-2.pdf

Ministerio de la Protección Social (2010) Análisis de la situación de Salud en Colombia 2002–2007. Retrieved from https://www.minsalud.gov.co/Documentos%20y%20Publicaciones/ASIS-Tomo%20VII–Situacion%20de%20Salud%20en%20Colombia%202002-2007-%20Resumen.pdf

Ministerio de Salud (1998) Resolución número 2358 de 1998. Bogotá D. C. Retrieved from http://antigua.mamacoca.org/docs_de_base/Legislacion_tematica/RESOLUCIoN_2358_DE_1998.pdf

Ministerio de Salud y Colciencias (2015) Encuesta Nacional de Salud Mental 2015 Tomo 1. Retrieved from http://www.visiondiweb.com/insight/lecturas/Encuesta_Nacional_de_Salud_Mental_Tomo_I.pdf

Ministerio de Salud y Protección Social (2013a) Dimensión convivencia social y salud mental. Retrieved from https://www.minsalud.gov.co/plandecenal/Documents/dimensiones/convivenciasocial-ysalud-mental.pdf

Ministerio de Salud y Protección Social (2013b) Plan Decenal de Salud Pública 2012–2021. Retrieved from https://www.minsalud.gov.co/Documentos%20y%20Publicaciones/Plan%20Decenal%20-%20Documento%20en%20consulta%20para%20aprobaci%C3%B3n.pdf

Ministerio de Salud y Protección Social (2015) Encuenta Nacional de Salud Mental 2015 Tomo 2: Anexos.Retrieved from http://www.odc.gov.co/Portals/1/publicaciones/pdf/consumo/estudios/nacionales/CO031092015-documento_final_ensm_tomoII.pdf

Ministerio de Salud y Protección Social y la Organización Panamericana de la Salud (OPS/OMS) (2014) Plan Nacional de Salud Mental. Documento propuesta de ajuste de la política nacional de salud menta para Colombia 2014. Retrieved from http://www.consejonacionaldetrabajosocial.org.co/cnts/images/POLITICA-NACIONAL-DE-SALUD-MENTAL-2014.pdf

Observatorio de Drogras de Colombia O. D. C (2017) Situación del consumo de drogas Retrieved from http://www.odc.gov.co/problematica-drogas/consumo-drogas/situacion-consumo

Organización de los Estados Americanos OEA (2012) Escenarios para el problema de drogas en Las Américas Retrieved from http://www.odc.gov.co/Portals/1/publicaciones/pdf/destacados/CO031032013-informe-escenarios-.pdf

Pinker S (2004) Current biology. Cell Press 14(21):R909. doi:10.1016/j.cub.2004.10.009

Registro único de víctimas RUV (2016) Retrieved from http://rni.unidadvictimas.gov.co/RUV

Rosenkranz M, Kerimi N, Takenova M, Impinen A, Mamyrov M, Degkwitz P, Martens MS (2016) Assessment of health services for people who use drugs in Central Asia: findings of a quantitative survey in Kazakhstan and Kyrgyzstan. Harm Reduct J 13(3). doi:10.1186/s12954-016-0093-2

Semana (2016) Las alarmantes cifras de menores de edad que se suicidan en Colombia, Semana Retrieved from http://www.semana.com/nacion/articulo/suicidios-en-menores-aumentan-en-colombia-segun-medicina-legal/485131

Turner RJ et al (2016) Childhood adversity and adult health: evaluating intervening mechanisms. Soc Sci Med 156:114–124. doi:10.1016/j.socscimed.2016.02.026

Implementing Community Care in Large Cities and Informal Settlements: An African Perspective

13

Lesley Jane Robertson and Christopher Paul Szabo

Abstract

Africa, the poorest World Bank region, is a continent with high levels of migration and rapid urbanization. In a transitional state, a growing middle class and changes in traditional culture are taking place alongside the persistence of severe poverty and underdeveloped rural areas. Cities absorb migrants into sprawling informal settlements. In the southern sub-Saharan region, the escalating prevalence of interpersonal violence and substance use compounds other public health concerns. With pressing public health issues, mental health care is largely neglected.

Community mental health care, together with promotive and preventative mental health care, is central to the objectives of the World Health Organization's Mental Health Action Plan for low- and middle-income countries. Although African countries have adopted the principles of community-based care in mental health policy and legislation, implementation is poor. There are multiple competing priorities, particularly in the cities with their rapidly expanding informal settlements and lack of basic infrastructure. Multiple urgent health conditions take priority, together with an increasingly fragmented society.

L.J. Robertson (✉)
Department of Psychiatry, School of Clinical Medicine, Faculty of Health Sciences, University of the Witwatersrand, Johannesburg, Gauteng, South Africa

Sedibeng District Health Services, Johannesburg, Gauteng, South Africa
e-mail: Lesley.Robertson@wits.ac.za

C.P. Szabo
Department of Psychiatry, School of Clinical Medicine, Faculty of Health Sciences, University of the Witwatersrand, Johannesburg, Gauteng, South Africa

Charlotte Maxeke Johannesburg Academic Hospital, Johannesburg, Gauteng, South Africa
e-mail: Christopher.Szabo@wits.ac.za

N. Okkels et al. (eds.), *Mental Health and Illness in the City*, Mental Health and Illness Worldwide, https://doi.org/10.1007/978-981-10-2327-9_16

This chapter explores the provision of mental health service in Africa, with a focus on urbanization and community mental health care. Finally, the mental health services of three African cities, Lagos, Kinshasa, and Johannesburg, are discussed.

Keywords
Mental health services • Africa • Urban • Community care

Contents

Introduction

This chapter explores mental health services in sub-Saharan Africa, with particular reference to community-based care and urbanization. Although many African countries still practice under colonial era custodial mental health laws, a growing number have promulgated new legislation and policy advocating community mental health care (CMHC), consistent with international human rights and World Health Organization (WHO) recommendations (Flisher and Lund 2011; WHO 2015; Gureje 2011). Community-based care is particularly applicable to African cities, which are currently affected by rapid urbanization without corresponding infrastructure development (Oni et al. 2016; United Nations Economic Commission for Africa 2016; WHO 2014; UrbanAfrica.Net 2017). Previously almost completely rural, at least 40% of the sub-Saharan African population now live in urban areas, a figure set to increase to 50% by the year 2030. In the 1990s and early 2000s, the rate of urbanization outstripped economic growth, resulting in severe urban poverty. Approximately two-thirds of urban dwellers in Africa now live in informal settlements, with high levels of youth unemployment and severe psychosocial strain.

There is a marked paucity of research in Africa regarding the effects of urbanization on general and mental health and an absence of health planning. A call for "health in all policies" has been made which advocates a collaborative intersectoral approach to urban health (Oni et al. 2016). However, there appears to be little, if any, emphasis given to urban psychiatry in this call. There is an urgent need for greater understanding of the biopsychosocial effects of urbanization in Africa, together with an integrated, responsive approach to mental illness in such a rapidly changing environment. Such an approach should be decentralized and locally based, as part of broader community development. Much more than hospital-based psychiatric care, CMHC is poised to meet this challenge, as inherent in its underlying principles is the need to understand the cultural dynamics of the community served, to engage with both health and non-health sectors, and to foster wellness as well as treat illness (Thornicroft et al. 2016).

Mental Health in Africa: The Urban/Rural Setting

One cannot speak of "Africa" as a monolithic entity. Very different ethnic, linguistic, and cultural roots characterize the different countries and regions (Flisher and Lund 2011; WHO 2014, 2015; Gureje 2011; Oni et al. 2016; United Nations Economic Commission for Africa 2016; UrbanAfrica.Net 2017). This diversity is further influenced by the historical effect of Arabic and European colonizers on cultural and religious traditions as well as on economic and health-care systems. The WHO African region is made up of 47 member states. Apart from Algeria, all are in sub-Saharan Africa, as the other North African WHO member states are grouped with Middle Eastern countries as the Eastern Mediterranean region (WHO 2015). Notwithstanding the diversity, there are many shared features in the WHO African region, the most prominent of which is poverty. According to the World Bank income groups of 2013, as defined on the 1 July 2014 (Table 1), 53% of the 47 countries were categorized as low income and 28% as lower-middle income. Only Equatorial Guinea, a small Central African state with a population of less than a million, was classified as a high-income country. As a whole, the African region accounts for over 70% of all the WHO regions' low-income countries.

Other factors common to African countries include a massive population growth, rapid urbanization, continued colonial style governance whereby poor

Table 1 World Bank income groups for 2013

Income group	Gross national income per capita in US dollars
Low income	$1,045 or less
Lower-middle income	$1,046–4,125
Upper-middle income	$4,126–12,745
High income	$12,746 or more

Source: World Health Organization Mental Health Atlas 2014 (WHO 2015)

majorities are neglected, weak administrative systems, variable expressions of traditional versus modern culture, and an 80% plus treatment gap for mental illness. Africa is still the least urbanized continent globally. However, the rate of urban growth is 1.7 times more than the global average, with an increase in urban inhabitants from 33 million in 1950 to over 350 million in 2005. The WHO African region is now home to two megacities (Lagos and Kinshasa) and three emerging megacities (Johannesburg, Dar es Salaam and Luanda) (UrbanAfrica.Net 2017). Although urbanization is usually accompanied by improved access to health care, this has not always been the case in Africa, due largely to the rapidity of growth and a general lack of government responsiveness. Rather, high degrees of social inequity with increasing numbers of marginalized poor and distressed youth predominate. Chaotic city expansion has resulted in huge informal settlements with closely packed shack dwellings and little or no amenities. As these settlements spread out over vast tracts of land, they are remote from health services, which are usually based in colonial era hospital infrastructure located in the more developed city centers.

In terms of closing the mental health treatment gap, CMHC is widely seen as the most appropriate answer (WHO 2013; Patel et al. 2016; Petersen et al. 2011). However, there is a paucity of government programs for CMHC in sub-Saharan Africa, even in countries with modern human rights-based mental health legislation or policy. Instead CMHC in Africa is mainly reliant on research-driven projects. The Mental Health and Poverty Project (Petersen et al. 2011; Nsereko et al. 2011; Kapungwe et al. 2010), PRIME (the Programme for Improving Mental health carE) (Hanlon et al. 2016), and the BasicNeeds Mental Health and Development Model (de Menil et al. 2015) have all yielded valuable information from various African countries and, through their projects, assisted in service provision. Their extensive scientific publications, policy briefs, and lay literature are available via their respective websites (Mental Health and Poverty Project 2017; PRIME 2017; BasicNeeds 2017).

An important limitation with regard to understanding urban CMHC is that the research and project sites are all rural. The rationale for this is that African rural areas are more impoverished, have less infrastructure, and have far fewer healthcare resources than urban areas. In addition, despite the rapid pace of urbanization, 60% of sub-Saharan African people still live in rural areas. As psychiatrists in Africa are generally located in city-based university hospitals, there is at least some mental health care for urban inhabitants, whereas there is often none at all in rural areas. However, this does not take into account the limited capacity of academic hospitals to serve the burgeoning township and informal settlement populations. Nor does it address the complex needs of people living in a rapidly changing and transitional urban society. Hence, there are important unknowns: firstly, the extent to which existing city-based psychiatric services meet the needs of African urban communities; secondly, whether or not the findings of rural based research projects may be generalized to African cities; and thirdly, how mental health expenditure and resource distribution should be prioritized across the urban/rural divide.

Funding for Mental Health Care: A Global Perspective

A comparison of mental health expenditure by the WHO member states according to World Bank income groups for the year 2013 is provided by the WHO Mental Health Atlas 2014 (Table 2). The results reveal an extremely low expenditure in low- and middle-income countries (LMICs). In most of these countries, less than 1% of total health expenditure is spent on mental health. Furthermore, those resources that

Table 2 Mental health expenditure and services by World Bank income group for the year 2013

		Low	Lower-middle	Upper-middle	High
Median mental health expenditure per capita (US $, 2013)		Insufficient data	$1.53	$1.96	$58.73
Mental health workforce per 100,000 population		0.9	3.2	15.9	52.3
Distribution of expenditure by care setting	Mental hospital	–	$1.20 (78%)	$1.35 (52%)	$20.77 (34%)
	Hospital inpatient and day care	–	$0.16 (11%)	$0.23 (9%)	$18.52 (31%)
	Outpatient and primary care (CMHC)	–	$0.16 (11%)	$1.01 (39%)	$21.42 (35%)
Number of beds per 100,000 population		1.6	4.0	14.4	41.8
Distribution of beds by care setting	Mental hospital	1.56 (84%)	4.0 (87%)	8.89 (80%)	30.9 (59%)
	Acute and day care psychiatric unit	0.15 (8%)	0.58 (13%)	2.25 (20%)	11.47 (22%)
	Residential care (CMHC)	0.15 (8%)	–	–	9.95 (19%)
Number of admissions per 100,000 population		7.6	14.4	41.9	142.3
Treated cases per 100,000 population	Total	32	69	131	555
	Psychosis	20 (63%)	31 (45%)	66 (50%)	324 (58%)
	Bipolar disorder	2 (6%)	8 (12%)	22 (17%)	42 (8%)
	Moderate-severe depression	10 (31%)	30 (43%)	43 (33%)	189 (34%)

Source of data: World Health Organization Mental Health Atlas 2014 (WHO 2015)
Mental hospital = stand-alone psychiatric hospitals and institutions, *CMHC* community mental health care

are available in LMICs are concentrated in mental hospitals, and CMHC, whether urban or rural, is almost nonexistent.

It is important to note that the data in the WHO Mental Health Atlas only reflects the information provided by those member states who completed the Atlas questionnaire. Although 88% of member states at least partially completed the checklist, only one low-income country provided information regarding per capita mental health expenditure. Regarding the predominant source of funding for mental health care in the WHO African region, 67% of respondent countries reported government as the primary source, 22% reported private household expenditure, and the remaining 11% depended on employers and nongovernmental organizations for health funding. In contrast, government was the primary source of funding in all the Southeast Asia countries (all of which are also LMIC) and 86% of European countries (of which 63% are high income). A situational analysis of mental health resources in Africa by Flisher and Lund revealed an even greater lack of government funding than that described in the WHO Mental Health Atlas (Flisher and Lund 2011). They found that out of pocket payment by the family or patient was the primary source of funding in 18 out of 48 (38%) African countries, and taxation, social insurance, or government grants were predominant in only 13 (27%).

While mental health care in the lowest-income countries may be provided almost entirely by nongovernmental and donor organizations, that in high-income countries is supplemented by private sector health care. However, private health care may have an even greater influence in LMICs (Dixon et al. 2006), despite the fact that only 3% of African countries included a private sector in their submissions to the Mental Health Atlas. In Kenya, three quarters of doctors reportedly work in the private sector (de Menil et al. 2014). In Ethiopia, Alem et al. (2008) lament that "many psychiatric hours are devoted to private practice, because of the poor remuneration within the government health system." In South Africa, approximately two-thirds of psychiatrists work in the private sector (South African Society of Psychiatrists 2015), and those in the state sector tend to be concentrated in the university-linked tertiary academic hospitals, leaving CMHC severely under-resourced (Szabo 2010).

Of concern is that, although private sector mental health care may be viewed as a form of CMHC, its tendency is to serve only those with illness of mild to moderate severity or good family support. As it depends on the ability to pay for services, either up front or through employment-linked insurance, it excludes those who are unemployed or destitute due to the severity of their mental illness and poor family support (Dixon et al. 2006). A further cause of out of pocket private expenditure in many African countries is a user co-payment charged for government clinic services. This impacts negatively on the usage of district health services and may affect the distribution of mental health-care setting, regarding both community versus hospital-based care and the urban/rural divide (WHO 2014; Dixon et al. 2006).

In comparing services across World Bank income groups, it needs to be remembered that the rates of care in high-income countries do not necessarily denote a suitable goal for LMICs. Firstly, prevalence rates of common mental illness differ across countries, both between and within regions (Gureje et al. 2006; Williams et al. 2008). Secondly, clinical outcomes are not represented, and therefore it is not known if the greater resource allocation and higher admission rates in high-income countries actually reflect optimal care. Lastly, the degree of unmet need in LMICs is not known. Although the high mental health coverage in high-income countries is likely to be due to better recognition of mental illness by both the general population and clinicians in wealthier societies, a different cultural context in LMICs may contribute to their lower coverage. In LMICs, the need for mental health care could be ameliorated by different expectations of personal well-being and quality of life stemming from a collective rather than individualist value system, a higher tolerability of psychological distress, and a greater capacity of traditional communities to manage behavioral disturbance (Patel et al. 2016; Alem et al. 2008).

Nevertheless, high levels of stigma, discrimination, and human rights abuses of the severely mentally ill have been documented in sub-Saharan Africa (Kapungwe et al. 2010; Mental Health and Poverty Project 2017). Hence, it cannot be assumed that there is a lower need for mental health care in LMICs in comparison to high-income countries, although the interventions required may be different. It is essential to understand local needs and perceptions in order to draw up culturally appropriate, cost-effective goals for mental health coverage and resource allocation. However, whether or not the current trend toward national health schemes in some African states will result in an increase in resource allocation to CMHC remains to be seen and will depend upon the health priorities for each country.

Mental Health, General Health, and Competing Priorities

Possible reasons for the low expenditure on mental health in sub-Saharan Africa include competing health priorities and the low demand for care by the African public (Patel et al. 2016; Gureje et al. 2006; Williams et al. 2008). Why many people with mental illness in Africa do not demand care requires exploration; it does not necessarily mean that they do not want care, including westernized mental health care.

Mental Health and General Health

General Considerations

In the 2014 WHO African regional report, mental health care is mentioned only as one of the noncommunicable diseases (WHO 2014). The need to treat communicable diseases and improve maternal and child health, with their high mortality rates, overshadows that of people living with mental illness, which is responsible for very few deaths. In addition, as the Global Burden of Disease studies code suicide as death by injury due to self-harm and not to an underlying psychiatric disorder, most deaths are attributable to substance use and almost none to mental illness (Patel et al. 2016). When using disability adjusted life years, which combine years lost to life with years lived with disability, communicable diseases and birth complications still predominate in Africa (Table 3) (GBD 2015 DALYs and Hale Collaborators 2016; Murray et al. 2012). The high negative impact of these conditions is due to the very high number of years lost when death occurs at an early age as well as to a high prevalence in Africa. Overall, in sub-Saharan Africa, communicable, neonatal, and

Table 3 Comparison of the top ten causes of disease burden in 2015 in high-income countries and sub-Saharan Africa (SSA) using disability-adjusted life years

Rank	Global	High-income countries	Southern SSA	Eastern SSA	Central SSA	Western SSA
1	Ischemic heart disease	Ischemic heart disease	HIV/AIDS	HIV/AIDS	Malaria	Malaria
2	Stroke	Back and neck pain	LRTIs	LRTIs	LRTIs	Diarrheal diseases
3	LRTIs	Sense organ disease	Tuberculosis	Diarrheal diseases	Diarrheal diseases	LRTIs
4	Back and neck pain	Lung cancers	Diarrheal diseases	Malaria	HIV/AIDS	HIV/AIDS
5	Preterm birth complications	Stroke	Diabetes mellitus	Congenital anomalies	PEM	Neonatal sepsis
6	Diarrheal diseases	Diabetes mellitus	Road injuries	Neonatal encephalopathy[a]	Preterm birth complications	Neonatal encephalopathy
7	Sense organ diseases	Alzheimer's and other dementias	Interpersonal violence	Preterm birth complications	Congenital anomalies	Preterm birth complications
8	Neonatal encephalopathy[a]	Depression	Ischemic heart disease	Neonatal sepsis	Neonatal sepsis	Meningitis
9	Road injuries	COPD	Stroke	PEM	Neonatal encephalopathy[a]	PEM
10	HIV/AIDS	Skin diseases	Back and neck pain	Meningitis	Other neglected tropical diseases	Congenital anomalies

Source: Global Burden of Disease Study 2015 (GBD 2015 DALYs and Hale Collaborators 2016)
COPD chronic obstructive pulmonary disease, *LRTIs* lower respiratory tract infections, *PEM* protein-energy malnutrition
[a]Due to birth asphyxia/trauma

nutritional diseases account for 60–70% of the total burden of disease, as opposed to less than 10% by neuropsychiatric and substance use disorders.

The negative impact of mental illness is better revealed in the years lived with disability analysis for the Global Burden of Disease Study 2015, as it often has its onset in youth and persists for many years into later life. In this analysis, depression is the third leading cause of disease burden both globally and in sub-Saharan Africa. Anxiety disorders, the ninth leading cause globally, rank tenth in Southern, ninth in Western, and eight in Eastern Sub-Saharan Africa. Of note though is that communicable, maternal and neonatal diseases, and trauma-related conditions do not feature in the ten leading causes when years lived with disability are used, due to the high mortality and shortened life span they cause. The implications for prioritization of health funding are significant; where resources are scarce, mortality takes precedence, particularly when children and young adults are most affected.

Notwithstanding the very comprehensive nature of the Global Burden of Disease Study, some circumspection is required when interpreting the findings. Firstly, the results are dependent on country-level data, which are lacking in many African countries. Secondly, complex multi-morbidity and the role mental illness may play in precipitating or perpetuating other causes of disease burden is not identified. Thirdly, the impact on the individual and caregivers of disorders such as schizophrenia, which cause severe long-term disability but have a low prevalence, is not reflected. Fourthly, the direct and indirect economic and societal costs are not included in the assessment of burden. Lastly, even where data is available, there may still be underreporting of mental illness in Africa due to stigma and various cultural interpretations of psychiatric symptoms.

Considerations with Respect to Urbanization

Although urban/rural differences are not reflected in the overall Global Burden of Disease analyses, country-level analyses could reveal important information where there is sufficient data. This is particularly pertinent to Africa, where disease surveillance is generally poor. Only two African cities have urban health monitoring systems, Nairobi in Kenya and Ouagadougou in Burkina Faso, and these focus primarily on general health concerns (Oni et al. 2016). The marked paucity of urban specific mental health data not only affects disease priorities, it also perpetuates the lack of government responsiveness to urbanization and its biopsychosocial consequences.

Demand for Care

General Considerations

A commonly cited explanation for low demand for mental health care in Africa is a preference to seek help from traditional healers and faith healers rather than from westernized medicine, with rates of about 60–70% (Nsereko et al. 2011; Mental Health and Poverty Project 2017; Adeosun et al. 2013). Traditional healers tend to live in the same communities as their clients, with whom they have shared cultural beliefs. In Africa, this often means spiritually divining the cause of disease, particularly with respect to mental illness. This may alleviate psychological distress even if it does not eliminate the disorder. By contrast, psychiatrists, being few in number and located in the more affluent areas of cities, are less accessible. In addition, they may be culturally distant from their patients with a subsequent poor understanding of their psychological needs and psychosocial stressors. Furthermore, not only might the medical care they offer not alleviate distress, it usually does not eliminate the disorder and may induce adverse effects which could be experienced as worse than the illness itself.

Nevertheless, the perception of greater help seeking from traditional healers rather than mental health professionals was not supported by the two World Mental Health Surveys performed in Africa (Gureje et al. 2006; Williams et al. 2008). In both Nigeria and South Africa, the majority of people who received treatment for common mental illness during a 12-month period had done so from general medical or mental health service providers. Indeed, what was striking in both studies was the very high proportion of people with mental illness who had not received any treatment, from either traditional or westernized medicine. However, psychotic disorders were not included in these surveys, and it is possible that that there may be different cultural interpretations of psychotic symptoms which could result in preferential help seeking from alternative healers.

Supportive extended families may be another cause for a low demand for care in Africa. As Alem et al. (2008) explain, "the mainstay of care for mentally ill persons in traditional African societies . . .. [are the] families and folk. There is almost always an extended family to rely upon, and even severely ill persons are usually living with their family. . . . Only in more extreme cases of violence or extremely deviant behaviour, or when the family's resources are stretched to breaking point, will mentally ill persons be excluded from their families, although they may be chained up or neglected."

As suggested by the tail end of Alem et al.'s comment, less positive reasons also exist for the low demand for care. Stigma and discrimination against mental illness is common in Africa, as are human rights abuses of the psychiatrically disturbed. Low-income population groups may be particularly affected by stigma. Interestingly, not only did the Nigerian World Mental Health survey find less mental illness among low-income population groups, it also did not find an increase in depression among women, despite documented evidence of high levels of maternal depression in LMICs (WHO 2013; PRIME 2017). Furhtermore, it found a zero prevalence of bipolar disorder, in contrast to other African studies which reveal a similar prevalence to European countries (Esan and Esan 2016). The authors suggest that low-income groups may have been less willing to report symptoms and that the screening questions may not have been applicable to local interpretations of symptoms. It may be that somatic complaints are culturally more acceptable than those of mental illness.

The scarcity of mental health services and trained community mental health workers in Africa may itself perpetuate a low demand for care (Patel et al. 2016; Nsereko et al. 2011; Williams et al. 2008). As well as contributing to a lack of awareness of psychiatric disorders as treatable illness, inadequate services may result in inappropriate and unsuccessful treatment. This in turn may confirm traditional beliefs regarding mental illness or entrench stigma.

Considerations with Respect to Urbanization

The cited reasons for the low demand for care may not be applicable to people living in cities and informal settlements. Not only is the population demographic characterized by a preponderance of youth, but African cities become a melting pot of different nationalities and cultures. While traditional and spiritual healers are still sought out in African cities, it is not known if this is to the same extent as in rural areas, especially among young people who may be more influenced through technology by western ideas. The containing effect of a supportive extended family is also less likely with the fragmentation of families across vast urban/rural distances and national boundaries. The escalating levels of interpersonal violence in southern sub-Saharan Africa betray a breakdown in traditional community living. In South Africa, 20% more deaths due to injury occur in urban rather than rural areas (Oni et al. 2016).

Factors such as lack of awareness, stigma, and poor help-seeking behavior secondary to the illness itself may play a greater part than in rural areas, especially given the relative absence of family support and village elders. As with rural areas, poor access to hospital psychiatric care and community-based services may entrench the low demand. In Johannesburg, those elements of CMHC which have been provided in urban and peri-urban townships have been utilized, with increasing demand over the years (Robertson and Szabo 2017).

That there is a need for mental health care in African urban areas, there can be no doubt. All the predisposing factors for severe mental illness, and many of the consequences, are present in African cities (WHO 2013; National Department of Health 2012). However, even if some demand better mental health services, local governments are faced with broader population demands for water, sanitation, electricity, roads, schools, and other basic infrastructure, all of which are urgent. Finally, demanding care requires some sense of value as a person and a citizen, a sense which is undermined by mental illness itself as well as by overcrowded yet socially isolating settlements and the ongoing political conflicts of many African countries.

Increasing Prioritization of Mental Health Care: A Strategic Approach

Given the very low level of available resources, the vast array of urgent health and social concerns, and the low demand for care, questions arise as to the extent to which mental illness is a priority condition in Africa and how it is best addressed.

Generating the slogan "no health without mental health," the Lancet global mental health series argued that addressing public health priorities requires the management of underlying, predisposing mental health problems (Prince et al. 2007). Although Prince et al. cited extensive evidence of the association between mental illness and a range of health priorities, including HIV/AIDS, tuberculosis, childhood illness, and violence, evidence that mental health interventions would achieve the desired outcomes was lacking. In addition, the basic human right to care and the need to treat mental illness for the sake of the individual themselves still had to be addressed. Thus, the Lancet series was accompanied by the launch of the Movement for Global Mental Health which is underpinned by two principles: human rights and scientific evidence (Movement for Global Mental Health 2017). This movement continues to stimulate evidence-based research aimed at improving health policies and plans as well as access to mental health services and reducing stigma.

At the same time, the Mental Health and Poverty Project (2017) worked on identifying and developing means to break the negative cycle between mental illness and poverty in four African countries (Ghana, South Africa, Uganda, and Zambia). This project strengthened the argument that treating mental illness is essential to poverty alleviation and highlighted the economic benefit of providing mental health care. In addition, it advanced the human right to care and protection through examining legislation and policy. Detailed situational analyses were conducted in the four countries, as well as the scaling up of national mental health policies and plans. Not only did the project yield a wealth of information on mental illness in Africa, but it generated a way forward for greater discussion with African governments on improving mental health service delivery.

In terms of cost-effective, sustainable evidence-based interventions for LMICs, the recent synthesis by the third edition of Disease Control Priorities (Patel et al. 2016; Disease Control Priorities 2015) emphasizes the need to supply mental health care at the community level. However, doing this in practical terms requires the strengthening of health systems. The Emerald Project, which stands for emerging mental health systems in low- and middle-income Countries, aims to do exactly this (EMERALD 2017). Emerald, an ambitious project launched in November 2012, partnered institutions in five European Union member states with researchers in Ethiopia, Nigeria, South Africa, Uganda, India, and Nepal. Its activities include building capacity in mental health system research, addressing questions on health system financing and resources, and identifying best practices for the implementation of integrated mental health care.

For these projects to have any impact on the prioritization of mental health care in African, it is essential that the findings are disseminated in a manner that is accessible and meaningful to government health authorities. In addition, advocacy groups and activists are needed to raise awareness of mental health needs and to ensure that policy briefs generated by the research projects are acted upon by the relevant authorities. To this end, both the Movement for Global Mental Health and Emerald launched user-friendly websites and social media platforms, through which latest news items, media activities, and scientific literature may be accessed. A further website, the Mental Health Innovation Network (2017), provides a forum for the entire global mental health community to share knowledge, resources, ideas, and innovations. The Africa page provides links to almost 60 mental health innovations, enabling discussion with practitioners, researchers, policy makers, and user advocates across the continent. The aim of the network is to improve mental health care and coverage by facilitating learning, building partnerships, and enabling the leverage of resources.

Implementation of Community Mental Health Care in Africa

CMHC in Africa

General Considerations

As explained by Thornicroft et al. (2016) CMHC encompasses much more than the simple provision of psychiatric treatment from a community health facility. In balance with general hospital acute psychiatric services, it implies a comprehensive service package comprising primary mental health care, multidisciplinary general specialist care, residential care, community outreach, and engagement with non-health sectors. Integration of mental health care with general health services is vital, as are case management and access to specialist care for complex patients. The actual package varies according to the resource constraints of the country. The lower the income group, the greater the dependence on integrated primary mental health care and "task sharing," whereby more specialized functions are performed by general and community health-care personnel. Where greater resources are available, more specialized community mental health teams as well as residential homes and day care centers are included in the package. Ultimately, CMHC aims to promote the mental health of the local population, meeting their needs in an accessible, culturally acceptable manner.

However, integrated care requires the full cooperation of general and primary health-care personnel, and the degree to which this is feasible is not known. Another unknown is the degree to which task sharing is acceptable both to the community and to the involved health-care workers. How much specialist support is required and whether the country can afford it, or not afford it, needs to be established. Finally, adapting to cultural interpretations of mental illness and traditional approaches to its management is new to those trained in the westernized model of mental health care, and there is little evidence available for practical guidance.

These issues form part of PRIME's research agenda (Hanlon et al. 2016; PRIME 2017). Working closely with Emerald, PRIME aims to evaluate the implementation of integrated primary mental health care at pilot studies in Ethiopia, South Africa and Uganda, India, and Nepal. It is hoped that the findings may be generalized more broadly and incorporated into government programs.

The PRIME pilot sites were all established within the existing district health services. At the outset, all sites were poorly resourced and there was no practice of integrated care by the primary care nurses. Prior to implementation, extensive community consultation took place with local traditional and religious healers as well as health and non-health sectors. This contributed to the sociocultural appropriateness of the planned interventions, a willingness among primary care practitioners to participate, and local political support of the project. To ensure applicability of the mental health-care package, the project included analyses of local priority mental health conditions. Integrated care within the medical model was facilitated by using the WHO mhGAP manual in all countries except South Africa and complemented by community-level psychosocial interventions. Awareness raising and stigma reduction took place via community outreach and at the primary health facility.

So far, three very important concerns have been highlighted by the project. Firstly, although the utilization of community health workers may be successful in awareness raising and outreach, they may quickly become overloaded with the huge degree of unmet need in the community. Secondly, provision of psychosocial interventions appropriate to the level of need required a clinical psychologist to lead the counselling teams and therefore could not be done in all the sites. Thirdly, integrated primary care could not be carried without specialist support and it became necessary for PRIME to employ psychiatrists at those sites where there were none available. As the project progresses, the research findings, news items, and policy briefs are all made accessible to the public via the PRIME website.

Instead of the WHO mhGAP manual, the PRIME site in South Africa used Primary Care 101+ (information available via the Mental Health Innovation Network) for the facilitation of integrated primary care, to be consistent with the country's integrated chronic disease program. Primary Care 101 is a desktop manual which provides a symptom-based guideline to the management of comorbid conditions. An advantage of Primary Care 101+ is that it is a general health guide which incorporates mental health and so offers a truly integrated approach to the patient. In addition to the biomedical guideline, counselling guidelines for depression appropriate for lay counsellors are provided.

A very interesting and unique approach to CMHC is offered by the BasicNeeds Mental Health and Development Model, which has now been established in six African countries (de Menil et al. 2015; BasicNeeds 2017). BasicNeeds work across the health system, with individuals and communities, local governments, ministries of health, and international organizations. A key aim is to build capacity at each level of the health system. In terms of patient care, the model is recovery orientated and focusses on community support, poverty alleviation, and work in order to enable people living with mental illness. In addition to a strong psychosocial and collaborative approach to developing CMHC, generating evidence through information systems and research is emphasized. However, the Mental Health and Development Model differs from a research project in that it aims for long-term sustainability in partnership with government health services.

When setting up a program site, local community leaders are identified and consulted. They are then trained as carers and given skills to facilitate user self-help groups, increase mental health awareness, and refer people identified as mentally ill to the primary health clinic and specialist services where necessary. In addition to recruiting volunteer facilitators, a "field consultation" approach is used to detect people with mental illness. In this way, hitherto unrecognized or hidden illness is identified and stigma reduced. As well as being a vehicle for psychosocial support, the self-help groups take part in livelihood programs and income-generating activities. Challenges to the program include retaining volunteers, difficult attitudes toward mental illness by local health-care practitioners, inadequate supplies of medication, and inconsistent funding and availability of required resources from government.

One further consideration with regard to the provision of CMHC in Africa is collaboration with local traditional and spiritual healers. A working relationship with mutual referral between traditional healers and westernized health-care practitioners has been suggested in South Africa (Campbell-Hall et al. 2010). Such a relationship could be very beneficial, especially given the paucity of psychosocial services. However, there is very little evidence of such relationships occurring in practice.

Considerations Related to Urbanization

In the same manner as in rural areas, the development of CMHC in urban areas should contribute economic benefits through early identification of illness due to greater accessibility with the prevention of disability and downward social drift. In addition, the proximity to the community allows sensitive collaboration with traditional healers and development of support systems. If coupled with poverty alleviation and income-generating projects, as with the Basic Needs Model, CMHC should have high applicability to urban areas, with their high youth demographic and need for self-empowerment and work.

Evidence-based, effective, community-level interventions identified by third edition of the Disease Control Priorities (Patel et al. 2016) include parenting programs during infancy, life skill training in schools, self-management, and primary mental health-care outreach by community health workers (Patel et al. 2016). Such interventions may have particular value in urban areas where society is unstable, there is little extended family support, and economic and social stress on young parents is severe.

Central, practical themes for effective CMHC emerging from these projects are as follows:

- Consultation with the community and their inclusion in planning and service support is essential for successful CMHC. It involves meeting with community leaders, forming collaborative relationships with traditional and religious healers and engaging with non-health sectors.
- The training of lay people from the community as health workers and support group facilitators is effective in increasing awareness, reducing stigma, encouraging self-help, and providing adherence support. The possibility exists that it could also contribute to the general development of a cohesive sense of community in urban areas.
- "Task sharing" by nursing staff and primary care doctors to provide psychosocial interventions and medication appears to be a cost-saving viable option to deliver care that would traditionally require mental health specialists.

- Although task sharing appears to be acceptable, it is not a substitute for specialist supervision of primary care workers and care of complex patients. A minimum number of mental health professionals is needed at community level.

Illustrative Examples: Lagos, Kinshasa, and Johannesburg

Lagos, Nigeria

National Context

Nigeria is a lower-middle-income West African country with a population of almost 160 million and an area of 924,000 km^2 (WHO 2011). Almost half of the population (48%) is under the age of 18 years and only 3% over 60 years. Data collected by the WHO Mental Health Atlas for 2011 revealed a total expenditure on health of 5.8% of GDP, with no dedicated mental health budget. According to Flisher and Lund (2011), the primary funding mechanism is a mixture of private out of pocket payments, government grants, and social insurance. Eight stand-alone mental hospitals and 44 outpatient facilities serve the entire population across the country. Per 100,000 population, there are 0.06 psychiatrists, 0.09 nonspecialized mental health doctors, and 0.19 mental health nurses. Although both primary care nurses and doctors are authorized and expected to provide mental health care, there is no official training and no officially approved manuals on the primary care management of mental illness.

A recent situational analysis conducted as part of the Emerald Project (Abdulmalik et al. 2016) found that dedicated Mental Health Policy and Plans passed in 1991 were poorly implemented. A major reason for the lack of implementation identified by the authors was that the National Health Policy and National Strategic Health and Development Plan 2010–2015 completely excluded the Mental Health Policy and Plan. This, together with mental health legislation based on the Lunacy laws of 1948 meant that mental illness was not included in routine national health data collection systems or in the policies of non-health sectors such as disability, education, and criminal justice. Positive findings were that the Mental Health Policy was revised in 2013 with stronger recommendations for promotive and preventative mental health care and intersectoral collaboration. However, key stakeholders still expressed a view of mental illness being less urgent than other life-threatening health conditions.

Lagos

Dubbed the "megacity of slums," Lagos, the largest city in Africa, has a population of at least 15 million on approximately 3600 km^2 of mostly coastal land (UrbanAfrica.Net 2017; Adeosun et al. 2013; Okunlola 2017; Adewuya et al. 2016). Heavy migration from other parts of Nigeria and neighboring countries has resulted in a diverse population with over 250 different ethnic groups. Two-thirds of the city live in large informal settlements with population densities allegedly running up to 20,000 inhabitants per square kilometer. There are major deficits in urban planning, with sanitation coverage for only 40% of the population, heavy traffic congestion, and overcrowded schools with an average teacher: pupil ratio of 1:150. Writing for UN-Habitat, Paul Okunlola identifies malaria, diarrhea, and HIV/AIDs as health threats and cites high maternal and infant mortality rates of 8/1,000 and 84.6/1,000, respectively, as health priorities (Okunlola 2017).

Regarding mental health care, the city is served by a university-linked psychiatric hospital, the Federal Neuro-Psychiatric Hospital, and one or two private psychiatric clinics. The Federal Neuro-Psychiatric Hospital, established in Yaba, Lagos in 1907 as a lunatic asylum by British colonial rule, has 535 beds and about 1,000 outpatients per week. Ethnic interpretations and cultural expressions of mental illness have been found to be important factors in both identifying psychiatric disorders and accessing care. Adeosun et al. found that 69% of 139 patients presenting to the hospital with schizophrenia had consulted traditional or spiritual healers prior to seeking psychiatric help (Adeosun et al. 2013). Psychiatrists were the first contact for 17.4% and other medical practitioners for 13.8%. Of note, although those who consulted traditional and spiritual healers did so sooner after the onset of symptoms than those who consulted medical practitioners, their psychiatric treatment was delayed by an average of 9 months, resulting in a significantly longer duration of untreated psychosis. The study highlights the fact that living in a city does not necessarily change traditional beliefs and that collaborative relationships between psychiatrists, general practitioners, and traditional and spiritual healers are important in this environment.

A cultural expression of illness which may delay appropriate care among city inhabitants is that of somatic symptoms in depression. Somatization had been questioned as a possible reason for the low prevalence of depression nationally in the Nigerian Survey of Mental Health and Well-being (Gureje et al. 2006). In the Lagos State Mental Health Survey, a recent city-based survey conducted for health planning purposes, Adewuya et al. found somatic symptoms to be significantly associated with suicidal ideation, which had a prevalence rate of 7.28% among a sample of over 11,000 inhabitants (Adewuya et al. 2016). Suicidal ideation was also significantly associated with depression and anxiety. This suggests that somatization is common to both rural and urban areas of Nigeria. The health-care implication is that greater awareness and relevant screening for mental illness are needed among primary health practitioners when managing somatic complaints.

Kinshasa, Democratic Republic of Congo

National Context

The Democratic Republic of Congo, a low-income country in Central Africa of 2,300,000 km^2, has a population of almost 68 million, 54% of whom are under the age of 18 years (WHO 2011). In 2011, the total expenditure on health was 2.04%, with no dedicated mental health budget. In their analysis, Flisher and Lund found private household expenditure to be the primary method of mental health-care funding (Flisher and Lund 2011). According to the WHO Mental Health Atlas, there is no dedicated mental health legislation and no provision for mental illness in non-health sectors. However, a mental health policy and a national plan were both approved in 1999, with implementation time lines and provision made for a shift from hospital-based care to the community. There has been no implementation of the plan; primary health-care nurses are not permitted to diagnose mental illness or prescribe psychotropic medication, there has been no mental health training of primary care practitioners, and there are no clear referral pathways between primary, secondary, and tertiary facilities. The country is served by six mental hospitals, with a total of 500 beds, and one outpatient facility. It is unknown to the WHO if there are any day care or community-based facilities. Per 100,000 population, there are 0.066 psychiatrists, 0.044 nonspecialized mental health doctors, and 0.236 mental health nurses.

There are no health information data systems and no routinely collected mental health data. There is a dearth of information, and understanding the mental health services of the country is difficult. No formal situation analyses could be found on searching PubMed or the Internet in general. What could be ascertained is that the six mental hospitals are donor run, funded by Fracarita International, a Belgium-based Roman Catholic organization (www.fracarita-international.org). Additional mental health support is provided by Médecins sans Frontières in the conflict ridden northeastern region of the country (www.msf.org).

Kinshasa

The only mental hospital in Kinshasa is the Centre Neuro Psycho Pathologique (Université de Kinshasa 2017). Linked to the University of Kinshasa, the center was established in 1973 as an inpatient hospital with 420 beds, of which over 90% are unusable according to a 2010 report on the hospital website. During that year, the hospital was staffed with 30 psychiatrists and 26 nonspecialist mental health-care doctors. It serves a population of approximately 12 million, with no general hospital psychiatric beds and no formal community-based mental health services.

As with mental health services, there is a paucity of information regarding the city itself or its associated mental health problems. An online Mail and Guardian Africa article (Iazzolino 2017) paints a desperate picture of a city with no sewer or waste disposal system, a high prevalence of HIV/AIDs, and repeated cholera epidemics. The city is reportedly expanding at a rate of 390,000 people a year, as they pour in from war-torn, poverty stricken areas in the Democratic Republic of Congo and across its borders. Attempts to upgrade urban infrastructure appear to be poorly organized, dependent on international groups, with no cohesive planning.

In terms of mental health problems, a search of PubMed using the search terms "Kinshasa psychiatry" and "Kinshasa mental health" yielded a total of 25 articles going back to 1981, in English and French. These were mainly disorder-orientated small studies focussing on psychiatric manifestations of physical illness and, more recently, the validation of westernized screening tools and prevalence figures in localized samples for conditions such as attention deficit disorder and postpartum depression. A search of the Mental Health Innovation Network revealed no Kinshasa-related activities and only three links for the Democratic Republic of Congo, one of which was notable as being to a blog on maternal mental health.

Even UrbanAfrica.Net, a communication network for researchers, advocacy groups, and policy makers regarding African cities, provides very few links to articles specific to Kinshasa. Abstracts and commentary available refer to a city of gang violence, worsening poverty, declining access to westernized health care, and increasing belief systems around witchcraft. These negative images are however counterbalanced by evidence of resilience and an indomitable will to live. A theme that runs through the few available articles is the need for, and possibility of, a collective stand against corruption and political authoritarianism, seen by some as a root cause of the city's chaos.

However, while political instability is an obvious cause for the dearth of mental health services in Kinshasa, a lack of collaborative international partnerships with respect to research and development of CMHC may contribute. As a Francophone country, the Democratic Republic of Congo has partnerships with Médecins sans Frontières and Fracarita, organizations founded in France and Belgium which provide desperately needed charitable mental health care. However, the pioneer projects in mental health systems and CMHC and evidence-based research that may be used for lobbying funds, Emerald, PRIME, and BasicNeeds were all founded in England. Accordingly, the LMIC partners in these projects are predominately Anglophone countries.

Johannesburg, South Africa

National Context

South Africa, an upper-middle-income country, is said to have the highest disease burden of all middle-income countries, with quadruple epidemics of HIV/AIDS, poor maternal and child health, noncommunicable diseases, and interpersonal violence (Oni et al. 2016; DOH 2015). Reasons for the high disease burden are complex and include the racially bound severe socioeconomic inequity of 40 years of apartheid, rapid urbanization, and increased availability of illicit substances since the country entered the global economy in 1994.

With an approximate area of 1,200,000 km^2, South Africa is home to over 50 million people, of whom 36% are under the age of 18 years (WHO 2011). Total health expenditure is approximately 8.5% of GDP, with no dedicated mental health budget. However, the health expenditure is split almost in half between the public health sector, which serves around 80% of the population, and the private health sector which caters for the remaining 20% (DOH 2015). While the public sector is funded by government via taxation, the private sector is funded through private health insurance and out of pocket payments. As a whole, there are 3,500 mental health outpatient facilities, 63 community residential facilities, and 62 mental hospitals. These are staffed by 0.27 psychiatrists, 0.43 nonspecialized mental health doctors, and 9.72 mental health nurses per 100,000 population. However, the majority of facilities and personnel operate within the private health sector.

With the end of apartheid and a new human rights-based constitution in 1996, existing mental health and national health legislation were both revised, with new Acts promulgated in 2002 and 2003, respectively. The Mental Health Care Act of 2002, which promoted deinstitutionalization and the rights of the mentally ill to access care close to their homes, was followed 10 years later by the National Mental Health Policy Framework and Strategic Plan 2013–2020, which set out a plan for upscaling CMHC. However, similarly to the situation in Nigeria, the precepts of the mental health legislation and policy were completely excluded from the National Health Care Act of 2003 and the National Health Strategic Plan 2014/15–2018/19. This situation has been perpetuated in the white paper for National health insurance, a document which marks a drastic overhaul of the country's health system toward achieving universal health coverage.

Johannesburg

An emerging megacity with a population of 9,3 million, Greater Johannesburg, is characterized by marked social and health inequity, with the vast majority of people living in sprawling townships and informal settlements (UrbanAfrica.Net 2017). The youth of the city have been identified to be of particular concern, with 34% of 15–34-year-olds being unemployed and evidence of high rates of substance use, interpersonal violence, and high-risk sex behavior among high school learners (Statistics South Africa 2011; Reddy et al. 2010). In addition, a pervasive low mood was identified in the 2011 Quality of Life Survey, despite an overall increase in quality of life (Gurney 2017). Racism, homophobia, gender-based violence, and xenophobia as well as a growing sense of political powerlessness and alienation and of a breakdown of traditional standards and values are all suggested as possible causative factors. However, despite surveys such as these, there are no prevalence figures of mental illness for Johannesburg, and the level of unmet need is not known.

The public sector mental health-care services in Johannesburg are mainly provided by psychiatrists and allied mental health professionals linked to the University of the Witwatersrand (Robertson and Szabo 2017). These services are distributed between two specialized psychiatric hospitals, three academic general hospitals, and district psychiatric clinics. Apart from the hospital providing involuntary and forensic services, all the academic units are located in the most developed area of the city region, with an immediate catchment area serving approximately 4,4 million people. The other five million inhabitants rely on the district clinics and under-resourced nonacademic regional and district hospitals for care.

The district psychiatric service is the most poorly organized and poorly staffed of the university-linked mental health services. This is related to their beginning in the mid-1990s as a university-based outreach to recently deinstitutionalized patient with severe mental illness. The services gradually spread in an ad hoc manner dependent on individual arrangements with the different district health directors and eventually became established as a secondary level community-based psychiatric service, with data regarding numbers of patients seen collected routinely by the District Health Information Systems.

Using this routinely collected data, an audit of the service for the 2014/2015 year was conducted (Robertson and Szabo 2017). The district mental health staffing and the coverage of the population was found to be much lower than the minimal coverage recommended by the National Mental Health Policy Framework (National Department of Health 2012). Approximately 0.3% of the general population for adults and 0.02% for children and adolescents had received CMHC over the 12-month period, as opposed to the minimal coverage of 2.7% for adults and 1.5% for children and adolescents advocated by the policy. With regard to primary mental health care, less than 1% of clinic visits were for mental illness, with little indication in the data of continued follow-up of psychiatric patients at primary care level. Interestingly, the district with the highest staff: 100,000 population ratio also had the largest population mental health coverage. This could suggest that the demand for mental health care is present if the service is provided.

In terms of specialist district: hospital staffing ratio, the ratio was far below the 1:2 ratio of district: hospital psychiatrists recommended by the policy. Overall, the ratio was 1:9. However, there was considerable variation across Johannesburg, with the majority of psychiatrists located in the academic hospitals of the central, most developed district. Finally, the district psychiatric services focussed mainly on providing medication, with very little emphasis on psychosocial care or intersectoral engagement. In conclusion, after 10 years of deinstitutionalization, CMHC had not been fully developed and did not match the norms and standards of the mental health policy with respect to human resources, mental health coverage, or the ratio of district to hospital psychiatrists.

Reasons for the lack of implementation of CMHC in Johannesburg require exploration. The divide between mental health and general health policy and legislation may play a role as it affects the distribution of staffing posts and resources by the provincial health authorities. At clinic level, a lack of confidence on the part of general health practitioners toward integrated mental health care could also be a factor. Finally, poor implementation of the mental health policy could be perpetuated by mental health professionals, who may have their own reservations regarding CMHC.

Recommendations

As previously recommended by Gureje (2011), partnership building and collaborative processes are key to improving mental health services in Africa. This review has highlighted the advantages of international collaborative relationships with regard to the provision of mental health care where there is none (as with Fracarita), strengthening existing health services, and most of all building an evidence base from which to leverage resources and establish effective, national government funded programs.

However, these partnerships are mainly among members of the mental health fraternity.

From the examples of the three cities, Lagos, Kinshasa, and Johannesburg, if integrated care is to be implemented, then partnerships need to stretch beyond those already interested in mental health. Two key recommendations may thus be made:

- Partnership building and liaison with the general health sector to take place, with an integrated approach to reducing priority health conditions. Collaboration may be a means to closing the gap between mental and general health reports, policy, and legislation.
- Partnership building with the non-health sector. This is particularly vital to the urban environment in Africa, which if developed sensitively could contribute to the prevention and management of mental illness.

Conclusion

With the rapid pace of urbanization, fragmentation of family structure, high levels of economic stress, and increasing interpersonal violence and substance use, the socio-political conditions in Africa seem to demand improved CMHC in its cities. In addition, CMHC offers important benefits to African cities, with their high youth demographic and mobile society. As well as addressing the human rights needs of the mentally ill, CMHC offers a flexibility in approach to the management of mental illness, such that local variations in illness presentation and traditional belief systems may be accommodated.

The provision of CMHC in Africa however depends upon the success of integrated care by general health practitioners and supportive government health administration. This in turn is reliant on the outputs of international partnerships, intersectoral collaboration, and user advocacy.

Cross-References

▶ Crime, Illegal Drugs, and Mental Health
▶ Homelessness and Challenges of Community Care
▶ Homicide and Suicide in Megacities

References

Abdulmalik J, Kola L, Gureje O (2016) Mental health system governance in Nigeria: challenges, opportunities and strategies for improvement. Glob Ment Health 3:e9. doi:10.1017/gmh.2016.2

Adeosun II, Adegbohun AA, Adewumi TA, Jeje OO (2013) The pathways to the first contact with mental health services among patients with schizophrenia in Lagos. Niger Schizophr Res Treat 2013:769161

Adewuya AO, Ola BA, Coker OA, Atilola O, Zachariah MP, Olugbile O et al (2016) Prevalence and associated factors for suicidal ideation in the Lagos State Mental Health Survey. Niger BJPsych Open 2(6):385–389

Alem A, Jacobsson L, Hanlon C (2008) Community-based mental health care in Africa: mental health workers' views. World Psychiatry 7(1):54–57

BasicNeeds Mental health and development model (2017) (cited Jan 2017). Available from: http://www.basicneeds.org

Campbell-Hall V, Petersen I, Bhana A, Mjadu S, Hosegood V, Flisher AJ et al (2010) Collaboration between traditional practitioners and primary health care staff in South Africa: developing a workable partnership for community mental health services. Transl Psychiatry 47(4):610–628

de Menil V, Ndetei D, Waruguru M, Knapp M, McDaid D (2014) A hidden face of community mental health care in Africa: specialist care from private providers in Kenya. World Psychiatry 13(1):100

de Menil V, Knapp M, McDaid D, Raja S, Kingori J, Waruguru M et al (2015) Cost-effectiveness of the mental health and development model for schizophrenia-spectrum and bipolar disorders in rural Kenya. Psychol Med 45(13):2747–2756

Disease Control Priorities 3rd Edition (2015) Mental, neurological and substance use disorders. Available from: http://dcp-3.org/mentalhealth

Dixon A, McDaid D, Knapp M, Curran C (2006) Financing mental health services in low- and middle-income countries. Health Policy Plan 21(3):171–182

DOH (2015) National Health Insurance for South Africa, White Paper. Department of Health, South Africa, Pretoria

EMERALD (2017) Emerging mental health systems in low- and middle-income countries (cited Jan 2017). Available from: https://www.emerald-project.eu

Esan O, Esan A (2016) Epidemiology and burden of bipolar disorder in Africa: a systematic review of data from Africa. Soc Psychiatry Psychiatr Epidemiol 51(1):93–100

Flisher A, Lund C (2011) Resources for mental health services in Africa: situation analysis and way forward. In: Ndetei DM, Szabo CP (eds) Contemporary psychiatry in Africa: a review of theory, practice and research. Acrodile Publishing Limited, Kenya

GBD 2015 DALYs and Hale Collaborators (2016) Global, regional, and national disability-adjusted life-years (DALYs) for 315 diseases and injuries and healthy life expectancy (HALE), 1990–2015: a systematic analysis for the Global Burden of Disease Study 2015. Lancet 388(10053):1603–1658

Gureje O (2011) The challenge of mental health service in Africa: the role of partnerships and networks. In: Ndetei DM, Szabo CP (eds) Contemporary psychiatry in Africa: a review of theory, practice and research. Acrodile Publishing Limited, Kenya

Gureje O, Lasebikan VO, Kola L, Makanjuola VA (2006) Lifetime and 12-month prevalence of mental disorders in the Nigerian survey of mental health and well-being. Br J Psychiatry 188:465–471

Gurney K (2017) The real issue is not poverty, it is inequality 2012 (cited Jan 2017). Available from: https://www.urbanafrica.net/news/real-issue-not-poverty-it-inequality/

Hanlon C, Fekadu A, Jordans M, Kigozi F, Petersen I, Shidhaye R et al (2016) District mental healthcare plans for five low- and middle-income countries: commonalities, variations and evidence gaps. Br J Psychiatry 208(Suppl 56):s47–s54

Iazzolino G (2017) Kinshasa, megapolis of 12 million souls, expanding furiously on super-charged growth 2016 (cited Jan 2017). Available from: http://mgafrica.com

Kapungwe A, Cooper S, Mwanza J, Mwape L, Sikwese A, Kakuma R et al (2010) Mental illness – stigma and discrimination in Zambia. Afr J Psychiatry (Johannesburg) 13(3):192–203

Mental Health and Poverty Project: Outputs and Activities (cited Jan 2017). Available from: http://www.who.int/mental_health/policy/development/mhapp/en/

Mental Health Innovation Network (2017) A global community of mental health innovators (cited Jan 2017). Available from: http://www.mhinnovation.net

Movement for Global Mental Health (cited Jan 2017). Available from: http://www.globalmentalhealth.org/

Murray CJ, Vos T, Lozano R, Naghavi M, Flaxman AD, Michaud C et al (2012) Disability-adjusted life years (DALYs) for 291 diseases and injuries in 21 regions, 1990–2010: a systematic analysis for the Global Burden of Disease Study 2010. Lancet 380(9859):2197–2223

National Department of Health (2012) National mental health policy framework and strategic plan 2013–2020. Government of South Africa, Pretoria

Nsereko JR, Kizza D, Kigozi F, Ssebunnya J, Ndyanabangi S, Flisher AJ et al (2011) Stakeholder's perceptions of help-seeking behaviour among people with mental health problems in Uganda. Int J Ment Heal Syst 5:5

Okunlola P (2017) The power and the heartbeat of Africa's biggest urban jungle: UN-Habitat; Undated (cited Jan 2017). Available from: http://unhabitat.org/

Oni T, Smit W, Matzopoulos R, Hunter Adams J, Pentecost M, Rother HA et al (2016) Urban health research in Africa: themes and priority research questions. J Urban Health 93:722–730

Patel V, Chisholm D, Parikh R, Charlson FJ, Degenhardt L, Dua T et al (2016) Addressing the burden of mental, neurological, and substance use disorders: key messages from disease control priorities, 3rd edition. Lancet 387(10028):1672–1685

Petersen I, Ssebunnya J, Bhana A, Baillie K, Mha PPRPC (2011) Lessons from case studies of integrating mental health into primary health care in South Africa and Uganda. Int J Ment Heal Syst 5:8

PRIME, the Programme for Improving Mental health carE: evidence on scaling up mental health services for development (cited Jan 2017). Available from: http://www.prime.uct.ac.za

Prince M, Patel V, Saxena S, Maj M, Maselko J, Phillips MR et al (2007) No health without mental health. Lancet 370(9590):859–877

Reddy SP, James S, Sewpaul R, Koopman F, Funani NI, Sifunda S et al (2010) Umthente Uhlaba Usamila – the South African national youth risk behaviour survey 2008. Medical Research Council of South Africa, Cape Town

Robertson LJ, Szabo CP (2017) Community mental health services in Southern Gauteng, South Africa; an audit using data routinely collected by the District Health Information Systems. Submitted for publication to the South African Journal of Psychiatry

South African Society of Psychiatrists (2015) Membership database. Johannesburg

Statistics South Africa (2011) Gauteng Provincial profile. Report No. 03–01-76. Available from: http://www.statssa.gov.za/publications/Report-03-01-76/Report-03-01-762011.pdf

Szabo CP (2010) Should state sector community psychiatry be hospital based? A local, and personal, perspective. Afr J Psychiatry (Johannesburg) 13(1):1

Thornicroft G, Deb T, Henderson C (2016) Community mental health care worldwide: current status and further developments. World Psychiatry 15(3):276–286

United Nations Economic Commission for Africa (2016) Contribution to the 2014 United Nations Economic and Social Council (ECOSOC) Integration Segment (cited July 2016). Available from: http://www.un.org/en/ecosoc/integration/pdf/economiccommissionforafrica.pdf

Université de Kinshasa: Centre neuro psycho pathologique (cited Jan 2017). Available from: http://cnpp.e-monsite.com/

UrbanAfrica.Net: Communicating about African cities (cited Jan 2017). Available from: https://www.urbanafrica.net

WHO (2011) Mental health atlas-2011 country profiles. World Health Organisation, Geneva. Available from: http://www.who.int/mental_health/evidence/atlas/profiles/en/

WHO (2013) Mental health action plan 2013–2020. World Health Organisation, Geneva

WHO (2014) The health of the people: the African regional health report. Regional Office for Africa: World Health Organisation, Geneva

WHO (2015) Mental health atlas, 2014. World Health Organisation, Geneva

Williams DR, Herman A, Stein DJ, Heeringa SG, Jackson PB, Moomal H et al (2008) Twelve-month mental disorders in South Africa: prevalence, service use and demographic correlates in the population-based South African Stress and Health Study. Psychol Med 38(2):211–220

14 Prospects and Challenges of Mental Health Services in a Major City in Latin America: Rio de Janeiro

Giovanni Marcos Lovisi, Lúcia Abelha, Sara Schilling, and Rubén Alvarado Muñoz

Abstract

The Organic Health Law, which established the Unified Health System *(Sistema Único de Saúde - SUS)* in 1990, declared the universality of health care, the equity of access, and the regionalization of service. However, even though SUS is a national universal public health program, in practice, Brazil has a dual system, which combines public health care with private health plans. The discussion about the need for mental health care reform in Brazil began in the late 1970s and was characterized by a range of complaints and critiques of the hospital-centered model, the privatization of health, and the deplorable treatment of hospitalized patients. The new mental health law in Brazil defined a drastic reduction in the number of beds in psychiatric hospitals and the installation of a community treatment network in 2001. The city of Rio de Janeiro, the capital of the state of Rio de Janeiro, is the country's second largest city, with 6,476,631 inhabitants. Given the current financial crisis of Rio de Janeiro, and Brazil in general, decision criteria for health care are often based on the benefit-cost ratio. In the context of a health system with scarce funds, it is important to consider how the reforms under way will affect care of individuals with mental disorders, especially those living in deprived and violent areas like slums in Rio de Janeiro. The fragmentation of the mental health service network is seen as one of the main obstacles facing mental health care in Rio de Janeiro.

G.M. Lovisi (✉) • L. Abelha
Institute of Studies in Collective Health, Federal University of Rio de Janeiro, Rio de Janeiro, Brazil
e-mail: giovannimlovisi@gmail.com; lucia.abelha@gmail.com

S. Schilling • R.A. Muñoz
Mental Health Unit, School of Public Health, Faculty of Medicine, University of Chile, Santiago, Chile
e-mail: shschilling@gmail.com; ralvarado1104@gmail.com

N. Okkels et al. (eds.), *Mental Health and Illness in the City*, Mental Health and Illness Worldwide, https://doi.org/10.1007/978-981-10-2327-9_11

Keywords
Health policy • Health services • Community mental health services • Financial management • Mentally ill persons

Contents

Introduction

In the years between 2003 and 2014, Brazil went through a phase of economic and social development, reducing both poverty and inequality, with the Gini Index (measure of the degree of inequality in the distribution of family income in a country) dropping by 11% over the period. Nevertheless, since 2015, this reduction of social inequalities has shown signs of stagnation (World Bank 2016).

The current population of Brazil is approximately 200 million, and around 25% of these have private health insurance, with the remaining 75% pertaining to the country's Unified Health System *(Sistema Único de Saúde - SUS)*. National health expenditure in 2012 was 8.5% of the total Gross Domestic Product (GDP), with 3.5% corresponding to the public system (Brasil 2016a, b).

The city of Rio de Janeiro, the capital of the state of Rio de Janeiro, is the country's second largest city, with approximately 6.5 million inhabitants. The proportion of the city's population with private health plans (51.6%) is greater than the national average, but it maintains the same income distribution inequality. The city has 763 *favelas* (slums), with approximately 1.4 million inhabitants, representing 22% of the city's population (IBGE 2010).

The Organic Health Law, which established the Unified Health System in 1990, declared the universality of health care, the equity of access, and the regionalization of services in Brazil. However, even though the Unified Health System is a national universal public health system, in practice, Brazil has a dual system, which combines public health care with private health plans. Recently, the management of some Unified Health System health units were privatized, and transferred to private Social Organizations, as recommended by the World Bank, which has supported the mixing of private and public health systems. It is important to note that while the Unified Health System permits a private complementary sector, the functions of the public system cannot be completely replaced by the private sector (Mendes 2015).

The discussion about the need for mental health care reform in Brazil began in the late 1970s and was characterized by a range of complaints and critiques of the hospital-centered model, the privatization of health, and the deplorable treatment of hospitalized patients. While the mobilization of various social movements, institutions, and civil society organization played a decisive role in the psychiatric deinstitutionalization process and the formulation of mental health policies, truly major changes in the legal and political spheres began to take place with the adoption of Mental Health Law in 2001. This law protects the rights of patients with mental disorders and regulates psychiatric hospitals and the implementation of mental health polices in Brazil (Amarante 1998; Paulin and Turato 2004).

The Mental Health Law defined a drastic reduction in the number of beds in psychiatric hospitals and the installation of a community treatment network, with Psychosocial Care Centers (CAPS). According to the new policy, psychiatric inpatient beds should be located in general hospitals, and the associated closure of long-stay psychiatric hospitals required the development of different services and programs, such as Residential Treatment Services (*Serviço Residencial Terapêutico - SRT*), return-to-home support programs, social and labor inclusion programs, and mental health programs in primary care services.

Psychosocial Care Network (*Rede de Atenção Psicossocial - RAPS*) in Rio de Janeiro

Psychosocial Care Network was created in 2011, as a part of the Unified Health System, providing services for individuals with mental illness, including those with abuse and dependence issues with psychoactive substances, alcohol, crack, and other drugs. The Psychosocial Care Network represents the public mental health policy in Brazil nowadays (Fig. 1; Brasil 2015). The mental health services in Rio de Janeiro, which are part of the Psychosocial Care Network, are shown below.

Strategic Psychosocial Care

The Strategic Psychosocial Care is composed of the Psychosocial Care Centers (CAPS). The main responsibility of Psychosocial Care Centers (CAPS) is the care for individuals of all ages who suffer from psychological distress resulting from severe and persistent mental disorders, as well as those related to substance use and other medical conditions that make it difficult to establish social relationships and lead a fulfilling life.

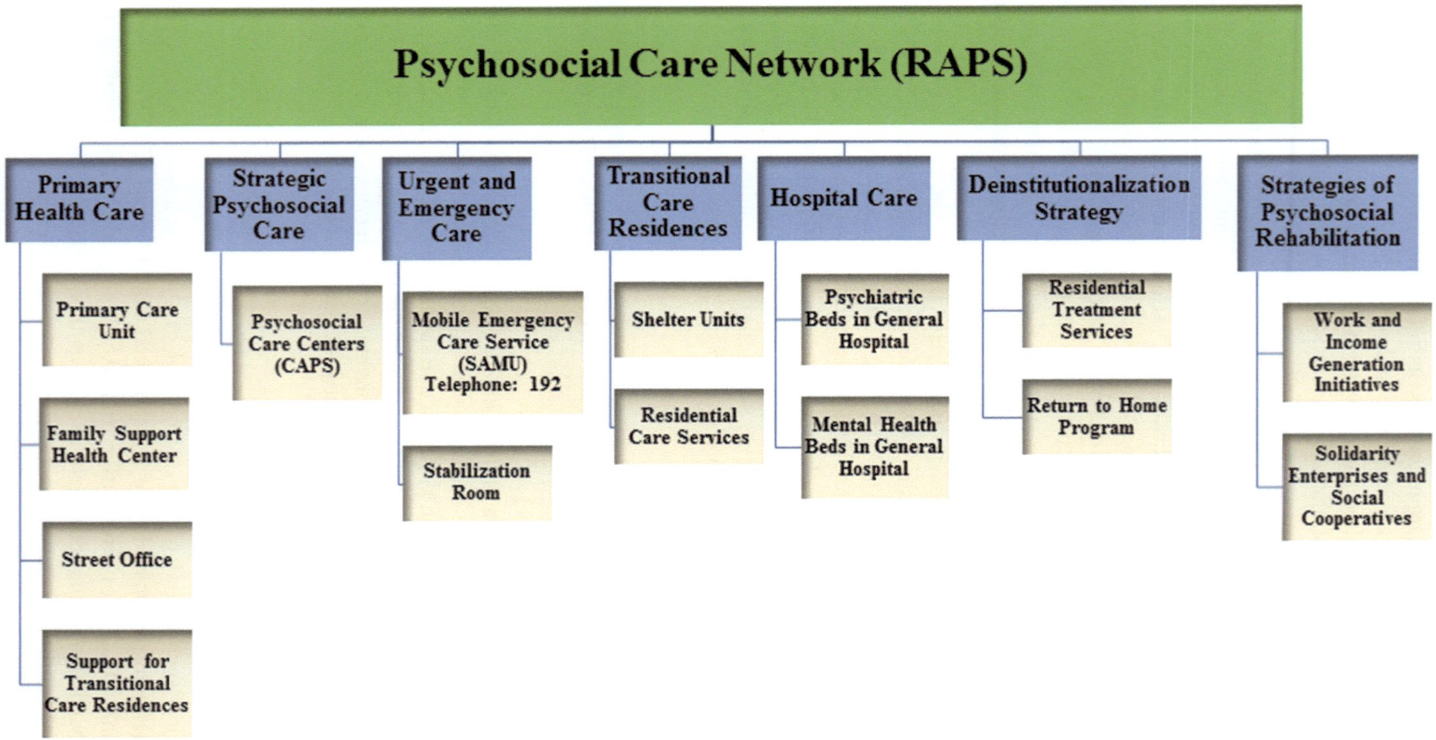

Fig. 1 Components of the Psychosocial Care Network (*Rede de Atenção Psicossocial - RAPS*) (Brasil. Biblioteca Virtual emSaúde. http://bvsms.saude.gov.br/bvs/folder/conheca_raps_rede_atencao_psicossocial.pdf. Accessed in: 12 May 2016)

Psychosocial Care Centers (CAPS) have specific centers for different target populations: for individuals with severe and persistent mental disorders (CAPS I, II, and III), for those with drug use (CAPS AD II and III), and for children and adolescents with mental disorders (CAPSi) (Brasil 2015; Couto et al. 2008). Furthermore, the Psychosocial Care Centers (CAPS) vary in their activities, coverage area, and human resources.

Rio de Janeiro is divided into five catchment areas, characterized major socio-economic contrasts, with some neighborhoods having Human Development Index (HDI) similar to those of developed countries and other neighborhoods with Human Development Index (HDI) like those of underdeveloped countries. The catchment area 1, which corresponds to downtown area, has approximately 298,000 inhabitants and the other ones (catchment areas 2, 3, 4, and 5) have a population of 900,000 to 2.4 million inhabitants.

The first Psychosocial Care Center (CAPS) was established in Rio in 1996. The CAPS coverage in Rio de Janeiro was 0.45/100,000 inhabitants in 2014, which was lower than the national average of 0.86/100,000 inhabitants. Some Psychosocial Care Centers (CAPS) are located in areas of high social risk such as the *favelas* (slums) in Rio de Janeiro, which are marked by escalating violence. The *favelas* are marked by numerous social problems, high rates of violence and violent deaths, drug trafficking, and little urban planning. The majority of *favela* inhabitants are young, non-white, and poorly educated, with a lower life expectancy compared with the rest of the city population. These problems are exacerbated even further in individuals with mental disorders who live in these areas, since they must deal with the stigma of mental illness, along with discrimination and health and social risk determinants associated with living in a *favela* (IBGE 2010; Brasil 2015; Zwarcwald et al. 1999; Rietra 1999).

There are 29 CAPS distributed among the catchment areas in Rio de Janeiro: 13 Psychosocial Care Centers II (CAPS II), 3 Psychosocial Care Centers III (CAPS III), 7 Psychosocial Care Centers for children and adolescents with mental disorders (CAPSi), and 6 Psychosocial Care Centers for those with drug use (CAPS Ad II and III). The Psychosocial Care Centers (CAPS) are mainly located in catchment areas 2 and 3. In contrast, catchment area 1 currently has no accredited Psychosocial Care Centers, and there is a shortage of Psychosocial Care Centers III (CAPS III) throughout the city (as is also the case with other Brazilian municipalities larger than 200,000 inhabitants).

The lack of Psychosocial Care Centers III in major Brazilian urban areas should not be attributed only to the failures of local governments but also as a failure of the national policy, since Psychosocial Care Centers III (CAPS III) are the most complex Psychosocial Care Centers and yet each center must offer services to a significant proportion of the population (Costa et al. 2011).

Transitional Care Residences

Shelter Units offer continuous, 24-h health care in residential communities for individuals of either sex with psychoactive substance use issues (associated with alcohol or other drugs) and social and/or family vulnerabilities, and thus who require therapeutic support and temporary protection. There are two types of Shelter Units:

residencies for individuals older than 18 years of age and others for adolescents between 12 and 18 years old (8). There are only two of these units operating in Rio de Janeiro (Fagundes et al. 2016).

Hospital Care

An important strategy used in higher-income countries during the psychiatric deinstitutionalization process was to developed Psychiatric Units in General Hospitals (PUGHs). In Psychiatric Units in General Hospitals (PUGHs), patients are less stigmatized and have easier and more expedite access to other treatment by medical specialties, if need be. These Units follow the usual dynamic of general hospitals, with early diagnosis and rapid treatment, in contrast to psychiatric hospitals, where users tend to be hospitalized for longer periods. In Brazil, however, Psychiatric Units in General Hospitals (PUGHs) have still not been widely implemented, with only a small proportion of psychiatric hospitalizations occurring in general hospitals, as opposed to psychiatric hospitals. The Ministry of Health has reduced the number of Psychiatric beds in psychiatric hospitals without a significant increase of Psychiatric beds in Psychiatric Units in General Hospitals (PUGHs).

In 2012, a new category of hospitalization, called "Mental Health beds," was created for individuals with mental disorders and issues related to alcohol, crack, and other drug use. These beds present specific parameters of coverage and the hospital guidelines operating within Psychosocial Attention Network. Additionally, the public funds used for these beds are considerably greater than those for Psychiatric beds in Psychiatric Units in General Hospitals (PUGHs), in order to incentive short term hospitalizations, as shown below (Brasil 2012a):

For Mental Health beds:

I. 85.70 US Dollars (2016 US dollars) per day, from the first to the seventh day of hospitalization.
II. 28.50 US Dollars (2016 US dollars) per day, from the eighth to the fifteenth day of hospitalization; and
III. 16.30 US Dollars (2016 US dollars) per day, after the fifteenth day of hospitalization.

Meanwhile, for psychiatric hospitalizations in Psychiatric Units in General Hospitals (UPHG), which have a capacity of 160 beds, the payment is 14.20 US Dollars (2016 US dollars) per day (Brasil 2016c).

In 2010, there were 4,105 Psychiatric beds in Psychiatric Units in General Hospitals (PUGHs) and 34,390 beds in psychiatric hospitals in Brazil; by 2015, Psychiatric beds in Psychiatric Units in General Hospitals (PUGHs) increased only 10.9% (to 4,552), while there were a 32.7% decrease in the number of psychiatric hospital beds (to 23,149). Despite attempts to prioritize the use of Mental Health beds for psychiatric and drug abuse hospitalizations in Brazil, they are still far from replacing the traditional Psychiatric beds. The 674 Mental Health beds which existed in 2013 only increased to 950 in 2015 (Brasil 2016d; Fig. 2).

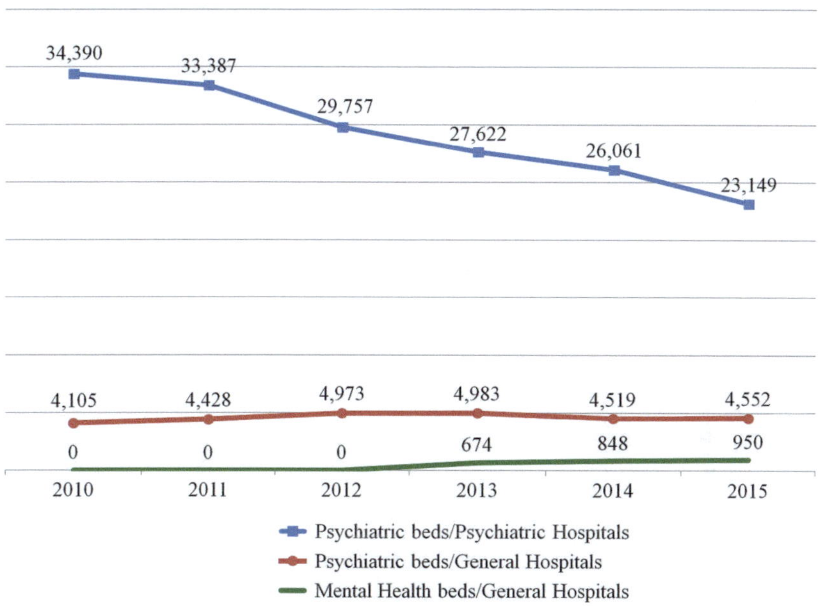

Fig. 2 Distribution of Psychiatric beds in psychiatric hospitals, Psychiatric beds in Psychiatric Units in General Hospitals (PUGHs), and Mental Health beds in general hospitals in Brazil, 2010–2015 (Ministério da Saúde. Secretaria Executiva. Datasus. Informações de Saúde. Rede Assistencial, 2015)

There were 1,474 Psychiatric beds in psychiatric hospital in the city of Rio de Janeiro in 2015, which corresponds to a ratio of 0.23 beds/1,000 inhabitants. As seen in Fig. 3, there has been a reduction of Psychiatric beds in psychiatric hospitals, a strong reduction of Psychiatric beds in Psychiatric Units in General Hospitals (PUGHs), and a constant number of Mental Health beds in general hospitals since their inception in 2012.

The last point that the number of Mental Health beds in general hospitals has never exceeded 49 beds in Rio de Janeiro, despite the existence of an economic incentive for their use, is important. In addition, Psychiatric beds in Psychiatric Units in General Hospitals (PUGHs), which were already insufficient for the existing demand, were progressively reduced, thus contributing to an overall lack of beds for patients in crisis. This is compounded by the fact that there are only a few Psychosocial Care Centers III (CAPS III) in Rio de Janeiro, since these services also offer beds for patients recovering from acute psychiatric episodes.

Deinstitutionalization Strategy

In line with the psychiatric reform process, a large number of patients have been deinstitutionalized from psychiatric hospitals and, after extended stays in hospitals,

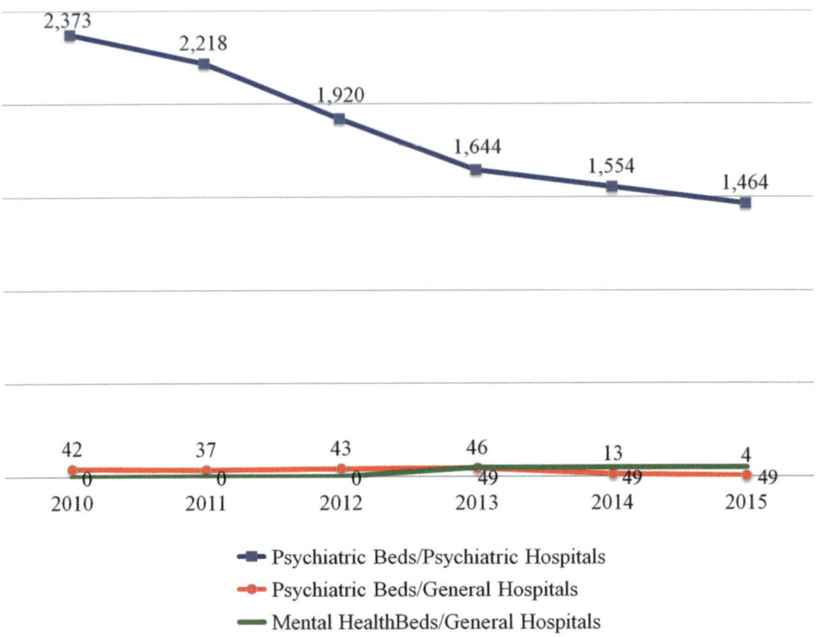

Fig. 3 Distribution of Psychiatric beds in psychiatric hospitals, Psychiatric beds in general hospitals, and Mental Health beds in general hospitals in Rio de Janeiro city, RJ, Brazil, 2010–2015 (Ministério da Saúde. Secretaria Executiva. Datasus. Informações de Saúde. Rede Assistencial, 2015)

have been sent to live in Residential Treatment Services. These are homes or apartments located in urban zones and designed to meet the housing needs of individuals with severe and persistent mental disorders who are transitioning from a psychiatric hospitalization of longer than 2 years. Residential Treatment Services are considered to be a hybrid service, given that they simultaneously offer housing and mental health care service. There are two types of Residential Treatment Services: Type I services are for users that do not require intensive care for their overall health, while Type II services are for patients that need more specific care for their physical health conditions and receive daily technical support (Brasil 2011).

At the end of 2014, there were 289 SRTs in Brazil accredited by the Ministry of Health (160 Type I and 129 Type II), which house and care for a total of 2,031 users. The number of Residential Treatment Services (SRTs) in Rio de Janeiro is still insufficient: in 2014, there were 51 Type I SRTs, with 284 patients, and 23 Type II SRTs, with 31 patients, for a total of 315 residents accompanied by 14 teams linked to Psychosocial Care Centers (CAPS) in their catchment areas (Brasil 2015; Rio de Janeiro 2014a).

The "Return to Home" Program was established in July 2003 to guarantee the long-term rehabilitation and recovery of individuals who were discharged from psychiatric hospitals. The supportive program contemplates a monthly payment to contribute to the social inclusion process and the expansion of the individuals' social support network, along with the empowerment and exercise of their civil and political rights. The Program is a yearlong, with renewal if necessary. As of 2014, 4,349 users had passed through the program in Brazil, and only 70 users had passed through Return to Home Program in Rio de Janeiro until the end of 2016 (Brasil 2015; Rio de Janeiro 2014b).

Strategies of Psychosocial Rehabilitation

The Psychosocial Care Network also maintains initiatives for income generation, inclusion in the formal job market, and cultural projects and activities to assist the patients. In 2016, 691 patients benefited from income-generating initiatives (a 15% increase over the previous year), 130 included in the formal labor market (80% more than in 2015), and 2,462 participated in cultural projects and activities in Rio de Janeiro (Rio de Janeiro 2014b).

Mental Health in Primary Care Services: An Innovative Program for Mental Health Care in Rio de Janeiro, Brazil

Mental disorders or psychopathological symptoms are seen in the general population among patients who consult physicians for other health problems, and their mental conditions tend to be underdiagnosed, undertreated, and/or underestimated by health professionals, especially when physical symptoms of other disorders are present. The World Health Organization (2001) highlights the role of primary health care centers in the promotion of mental health in communities. In the Brazilian context, the Family Health Program (*Programa de Saúde da Família - PSF*) and Social Welfare acquire a fundamental importance in this strategy (WHO 2001).

Family Health Strategy (*Estratégia de Saúde da Família - ESF*)

In an effort to reorganize health care offered through the Unified Health System, the Ministry of Health adopted the Family Health Strategy in 1994. Its aim was to replace the traditional care model for a catchment area model, based on the delimitation of territories into geographic areas (with health teams responsible for between 3,000 and 4,500 residents) and micro areas (the service area of a community health worker in the community with 400 to 750 people), as well as family records. The main innovative feature of the Family Health Strategy was transforming the primary care centers' individual healthcare focus into a family approach, which includes taking into account individuals' physical and social environments (Brasil 2012b; Florindo 2009).

Matrix Support System

In 2003, the Brazilian Ministry of Health launched the technical Matrix Support System for Family Health Strategy, based on the guidelines for the organization of mental health services in primary health care. The Matrix Support System has four basic aspects: (1) teamwork and clinical reference teams for each patient; (2) commitment to demedicalization, for example, pain cannot be solved only through medication, and sadness and fatigue are not resolved with anxiolytics; (3) health promotion for the population; and (4) empowerment of individuals, by viewing them comprehensively, in their contexts, with their families, and in their communities (Chiaverini 2011; Campos and Domitti 2007).

The role of the Matrix Support System is to provide technical and educational support for health teams and to improve operational interactions between primary care professionals and other municipal services (health, education, judicial, sports, etc.). Its objective is to form a large, non-bureaucratic network of effective services. The Matrix Support System is led by professionals who support clinical reference teams, which are composed of several professionals responsible for a specific territory, under the Family Health Strategy (Chiaverini 2011).

The Matrix Support System approach arises from realization that psychiatric reform cannot proceed without primary health care. In an effort to integrate mental health within primary care, the Ministry of Health stimulates the creation of psychosocial rehabilitation services in the community, promoting intersectoral coordination and prioritizing collective approaches to improve family bonds.

Family Support Health Centers (*Núcleo de Apoio à Saúde da Família - NASF*)

In 2008, Family Support Health Centers (NASF) were created, based on interdisciplinary and the Matrix Support System, to contribute to the work of Family Health Strategy teams. Their function is to promote respect for regional peculiarities and to train professionals in appropriate actions for difficult cases. The Family Support Health Centers addressed professionals and territories that were previously not covered under this strategy (Brasil 2008, 2009).

Family Support Health Centers aim to improve individuals' quality of life, reduce injuries and damage caused by noncommunicable diseases, and reduce drug use. They also encourage the creation of social support networks and the active participation of users in the development of diverse therapeutic projects. The Family Support Health Centers professionals are responsible for different activities, such as to work together with the Family Health Strategy and the community to identify the target population for each of the actions developed, welcome patients, humanize attention, stimulate intersectoral work, and develop individual therapeutic projects through regular discussions of clinical cases (Brasil 2008, 2009).

Family health teams and mental health teams share numerous principles, such as the importance of intervening in the family environment, establishing trusting relationships to ensure continuity of care, and organizing networks that link mental health care to other supports in the territory. Therefore, mental health services are part of the Family Health Strategy structure, and they work to support primary care teams to address patients' psychological suffering, which permeates all experiences related to the health-disease process.

Between 2008 and 2013, with the expansion of the Family Health Strategy's coverage from 3.5% to 41%, more than 2.5 million people in Rio de Janeiro gained access to health care. Under the strategy, more than 800 family health teams were deployed, with an average of 3,178 individuals registered for each team. There are currently 27 Family Support Health Centers in the city (Rio de Janeiro 2013).

In the city, and Brazil in general, the Family Health Strategy and Family Support Health Centers have fundamental importance in the mental health program. The challenge we face today in Rio de Janeiro is the low coverage of primary health care services, which have inadequate resources and are staffed by health workers who are not trained to provide adequate mental health care. Moreover, primary care services make few referrals to mental health services such as Psychosocial Care Centers (CAPS), and their connections to these secondary services are fragile. In light of this, Family Support Health Centers professionals, especially mental health and family health team professionals, are currently focused on implementing innovative strategies to intervene in the health-disease process. It is important to point out that there are large catchment areas in Rio de Janeiro where there is a great lack of mental health services.

The Street Office, which is also part of primary care component of Psychosocial Care Network (RAPS), was established in 2011 to improve access to health and social care services for homeless people, offering comprehensive health care for this population group. Street Office teams consist of health professionals who work out of a mobile office, offering activities and services to vulnerable individuals, with special attention paid to their specific health needs. The teams are responsible for offering mental health care to: (1) homeless individuals, (2) those with mental disorders, and (3) substance users. There are 111 of these units in Brazil and only two are in Rio de Janeiro (Fagundes et al. 2016; Brasil 2015).

Challenges Faced in Implementing Mental Health Care in Rio de Janeiro/Brazil

The treatment of individuals with severe and persistent mental illness is one of the biggest challenges for mental health reform, as their care requires continuous therapeutic interventions and the development of social supports, which will be long term. Especially in the context of a health system with scarce, sharply disputed funds, it is important to consider how the current reforms under way will affect these individuals. How can individuals who require continuous treatment, especially those who are historically served by the public sector, be adequately cared for in conditions of limited resources, especially nowadays with the current financial crisis in Rio de Janeiro and Brazil?

The experience of managed care organizations (such as HMO America) has clearly shown that providers have strong incentives to rationalize the services they offer, transferring the responsibility of the care of patients who require more financial resources to the public sector. Thus, individuals with severe physical and mental illnesses are at greater risk of exclusion from essential services in health care systems based on this model. Despite being a concept developed in the private sector, the public sector is increasingly interested in the managed care model. This growing interest is mainly due to increased demand for services, reduced public funding for the social sector, and the fragmentation of health services, which accompanied the transfer of hospital-centered care to the community (Hoge et al. 1994; Fogel 1993; Dauncey et al. 1992).

In times of economic restraint, when it is nearly impossible to obtain funds for new investments, it is natural that public administrators may decide to transfer financial resources from hospitals to community settings. However, once other factors are incorporated into treatment cost calculations, such as the increased demand posed by chronic patients, as they transition into the community and the greater costs associated with their care, the financial advantages of community-based care decrease (Häfner 1987).

Rio de Janeiro has the second highest internal Gross Domestic Product (GDP) in Brazil, a fact which has attracted the attention of the private sector, and the Family Health Strategy, emergency medical units, some hospitals, and other special programs have been administered by Social Organizations since 2009. This poses a problem, since when private entities assume responsibility for service management and hiring professionals, they do so with a business mindset, to increase profit, while leads to precarious contractual conditions for these professionals and consequential instability and organizational issues in health services (Lima and Bravo 2015).

The fragmentation of the service network is seen as one of the main obstacles facing mental health care in Rio de Janeiro today. Many patients are confused by the complexity of the mental health network and are left without the treatment referrals they need. The fragile relationship between service levels is explained by the high turnover of professionals due to the gradual privatization of the SUS, which makes treatment continuity nearly impossible. This is especially problematic for those patients with severe mental disorders, who live under very vulnerable social conditions, which is the case for many Psychosocial Care Centers (CAPS) users (Ribeiro and Inglez-Dias 2011).

Along with the shortage of financial resources, there are also problems in the use and management of these funds. To ensure fairness and the universality of care, inequalities in access and the utilization of mental health services among different catchment areas of the city and population groups must be reduced. Despite the increase in the number of mental health services and programs like Psychosocial Care Centers (CAPS) in recent years, they are still not enough, and the units that do exist are concentrated in certain areas of the city, leaving social groups that are most in need of services with psychological care that is neither sufficient nor adequate. Similarly, the number of mental health professionals is not enough to meet demand,

and these professionals are distributed very unevenly within the country and the city, especially psychiatrists who are concentrated in the richest regions of Brazil. There must be political will to increase the implementation of community mental health services and to incentive professionals to live and work in marginalized and underserved areas (Mateus et al. 2008; WHO 2007).

Given the person-centered nature of their work and the importance of psychosocial interventions, mental health professionals are a pillar of treatment and the "technological leaders" in the field of mental health. Psychiatric treatment plans in the community are known as Singular Therapeutic Projects, which layout the actions discussed, agreed upon, and implemented by a multidisciplinary team (Brasil 2016d).

The satisfaction of professionals working in the field mental health is one of the main indicators of the quality of services provided, and this is closely related to their relationship with colleagues, working conditions, and role in decision-making processes. A low level of satisfaction directly affects the effectiveness of services and leads to disorganized functioning, both of which are potential producers of work overload (Ishara et al. 2008).

The approach of mental health professionals in working with users and their family members necessitates specialized training and experience, acknowledging the complexity of family dynamics and the importance of knowing the patient's and family's history, their current status, and future goals. Professionals are taught to work with the entire family unit (including both sick and healthy individuals), to consider the family as a system that is affected by changes in any of its members, and to assist families in recognizing that users are dynamic and transform over time, a perspective that promotes mutual support and understanding among family members. Furthermore, mental health workers take into account a family's social and cultural context, to facilitate their relations with the community (Brasil 2013).

The process of psychiatric reform in Rio de Janeiro, with the consequent increase in community-based services and the expansion of mental health in primary care centers, has placed a high demand upon healthcare professionals, which often leads to their physical and emotional overload. The level of performance expected from professionals working under the Family Health Strategy is quite high, and their resulting burden and overload is manifested in job stress, absences from work, and elevated turnover of professionals, all of which end up hurting the quality of services. The correct training of mental health professionals working in primary care is key to improving the quality and efficiency of care (Leal et al. 2012; WHO 2007).

Regrettably, there is still a great deal of prejudice and stigma against those with mental disorders. Negative attitudes toward these individuals are largely responsible for the difficulties encountered in their recovery process. For their part, primary care professionals have a strategic position for their proximity to this population, and therefore it is essential to know their attitudes, be they negative or positive, toward users. It is quite possible that health professionals who are subject to higher demand, and are less satisfied with their work, increase their negative attitudes toward those with mental disorders. Their overload can prevent professionals from carrying out more intensive interventions in the most vulnerable and at-need individuals, such as those with severe and persistent mental illness, and especially those living in areas of great poverty and extreme violence such as the *favelas* in Rio de Janeiro (Chambers et al. 2010; Bandeira et al. 2007).

The impact of the current national economic crisis on the psychiatric reform process may prevent many advances that were the result of years of struggle and advocacy to obtain more dignified care for mental health service users. It seems that if professionals who are committed to quality care do not unite in the defense of these achievements, they will not be able to effectively fight against the ideology of market economy that has been begun to dominate the health system. In terms of the mental health care model, the reduction of Psychiatric beds meets the objectives of a policy promoting more humane, nonrestrictive treatment in the community, in contrast to service changes which are only intended to adapt to contexts of fewer resources, due to reduced funding.

All of the aforementioned are of importance for the development of effective proposals for the care of individuals with severe mental disorders living in developing countries, as has been proposed by Mari et al. (2009), especially when these proposals consider economic sociocultural characteristics and regional policies. For example, it would be nearly impossible to implement a program to treat people with severe mental disorders living in the *favelas* of Rio de Janeiro without taking into account the extreme poverty and violence that prevails in these locations. It is these considerations that determine the probability of whether or not a program will achieve its performance targets in such settings (Souza et al. 2011).

Some studies conducted in Rio de Janeiro and Brazil have found that a large proportion of the population treated in primary health care centers suffer from what are known as Common Mental Disorders (CMD), marked by sadness, anxiety, various types of body aches, sleep difficulties, and above all discouragement with life in general. At the same time, social services must deal with the consequences of CMD in families, such as abandonment, violence, suicide, and unemployment. Currently, Rio de Janeiro and Brazilian society is largely unable to cope with the populations' psychological distress, and there are increasing problems of alienation of labor, domestic violence, and drug abuse. The prevalence of Common Mental Disorders (CMD) in primary health care is high in Rio de Janeiro (51.9%), which raises concern about the level of disability they could eventually cause at work, in family relationships, and in social life in general; in fact, the resultant suffering may be incalculable (Nunes et al. 2016; Gonçalves et al. 2014; Gonçalves and Kapczinski 2008).

Unified Health System is in the process of implementing programs and policies to correct the current system distortions and achieve universal public coverage and comprehensive and responsive treatment that respond to the real needs of the population, in practice and not just on paper. The idea is that the previous model that favored the commodification of health services be replaced by another in which the public sector is a major supplier and has the ability to supervise hired and insured private providers, who would be subordinate to the public sector. Nevertheless, a significant barrier to the full application of this new model is funding – lack of investment in the public sector and questions concerning the transfer of resources to their services pose great difficulties for the Unified Health System (Campos and Domitti 2007).

In Brazil, the recent privatization of health services management involved the transfer of the administration and management of some of the Unified Health System to Social Organizations (*Organizações Sociais - OS*), which since the 1990s have permitted the combination public and private sectors, as has been seen in several Brazilian states and municipalities. Therefore, in an increasingly competitive market, a large part of health care is organized based on maximizing profitability, and every day more and more important health care decisions are being made by administrators, rather than professionals who work directly in mental health services and are in touch with the real needs of users and their families (Ribeiro and Inglez-Dias 2011; Koyanagi et al. 1993).

It should be noted that spending on mental health programs represented 2.35% of the Ministry of Health budget in 2011. The proportion of resources destined for psychiatric hospitals decreased from 93.0% to 28.9% from 1997 to 2011, while funds for extra-hospital services increased from 7.0% to 71.1% over the same period (Brasil 2012c). Despite attempts to prioritize community mental health services, they are still insufficient services to attend existent demand.

Alternatives to services centered around psychiatric hospitals involve building a network of community-based services – including primary care centers for people with Common Mental Disorders along with more specialized clinics for those with more severe condition – that are staffed with a sufficient number of human resources to effectively avoid prolonged psychiatric hospitalizations. Additionally, it is particularly important that the integrated community service networks include therapeutic residences, tailored to each user's degree of autonomy. Ideally, mental health service managers would not have to discuss the rationing of financial resources, and thus be able to fully offer the benefits of community-based mental health to all users. Nevertheless, as managed care policy has been growing globally, in both public and private sectors, such a discussion is inevitable. The main question, however, is not whether rationing or prioritizing financial resources is needed; what truly matters is whether or not this process is carried out ethically. As the prioritization and rationing of funds cannot be avoided, it is important that these issues be discussed openly and that basic mental health care is offered at all levels of care, especially for a system such as the one in Brazil.

Conclusion

Given the current economic context of Rio de Janeiro, decision criterion is often based on the benefit-cost ratio. Unfortunately, the main objective of these decisions is usually to reduce costs (and consequently the benefits conferred to patients), which undoubtedly leads to decreased quality care. In addition to their treatment considerations, individuals with severe mental disorders face numerous difficulties in their daily lives, which make them more vulnerable to adverse effects and have increased risk of suicide, drug use, job loss, and homelessness.

The social problems of Rio de Janeiro such as high rates of violence, drug trafficking, unemployment, and social inequality are exacerbated by the severe economic crisis that affects the city, making people more vulnerable to Common Mental Disorders. Those with Common Mental Disorders should be treated in the primary health care services. The low coverage of the Family Health Strategy and Family Support Health Centers, which have inadequate resources and are staffed by health workers who are not trained to provide adequate mental health care, prevents a suitable care for this population.

The positive development of psychiatric reform in recent years in Rio de Janeiro must be emphasized, especially the innovative initiative of incorporating mental health care into primary care services. Nonetheless, these newly implemented programs require further, more in-depth evaluations to ensure that users are offered better quality of services.

Cross-References

- ▶ Religious Perspectives in Urban and Rural Mental Health in the Muslim World
- ▶ Implementing Community Care in Large Cities and Informal Settlements: An African Perspective
- ▶ Mental Health Care in 11 Major Cities Over Two Decades
- ▶ Migration and Urban Mental Health in China
- ▶ Mental Health and Urban Life: Experiences from Colombia

References

Amarante P (ed) (1998) Loucos pela vida: a trajetória da Reforma Psiquiátrica no Brasil, 2nd edn. Fiocruz, Rio de Janeiro

Bandeira M, Ishara S, Zuardi AW (2007) Satisfação e sobrecarga de profissionais de saúde mental: validade de construto das escalas SATIS-BR e IMPACTO-BR. J Bras Psiquiatr 56(4):280–286. doi:10.1590/S0047-20852007000400007

Brasil (2008) Ministério da Saúde.Portaria n°. 154 de 24 DE JANEIRO DE 2008. Cria os Núcleos de Apoio à Saúde da Família – NASF. Publicada no D.O.U. n° 43, de 04/03/2008, Seção 1, pp 38–42

Brasil (2009) Ministério da Saúde. Secretaria de Atenção à Saúde. Departamento de Atenção Básica. DIRETRIZES DO NASF – Núcleo de Apoio a Saúde da Família (Série B. Textos Básicos de Saúde) (Cadernos de Atenção Básica; n. 27)

Brasil (2011) Ministério da Saúde. Portaria n° 3090 de 23 de dezembro de 2011. Altera a Portaria n° 106/GM/MS, de 11 de fevereiro de 2000, e dispõe, no âmbito da Rede de AtençãoPsicossocial, sobre o repasse de recursos de incentivo de custeio e custeiomensal para implantação e/ ouimplementação e funcionamento dos ServiçoscResidenciaisTerapêuticos (SRT). (*) Republicada por ter saído, no DOU n° 247, de 26-12-2011, Seção 1, pp 233–234

Brasil (2012a) Ministério da SaúdePortaria. n 148 de 31 DE JANEIRO DE 2012. Define as normas de funcionamento e habilitação do Serviço Hospitalar de Referência para atenção a pessoas com sofrimento ou transtorno mental e com necessidades de saúde decorrentes do uso de álcool, crack e outras drogas, do Componente Hospitalar da Rede de Atenção Psicossocial, e institui incentivos financeiros de investimento e de custei. Publicada no DOU n° 23, de 01-02-2012, Seção 1, p 33

Brasil (2012b) Ministério da Saúde. Secretaria de Atenção à Saúde. Departamento de Atenção Básica. Política Nacional de Atenção Básica/Ministério da Saúde. Secretaria de Atenção à Saúde. Departamento de Atenção Básica, Brasília

Brasil (2012c) Ministério da Saúde. SAS/DAPES. Coordenação Geral de Saúde Mental, Álcool e Outras Drogas. Saúde Mental em Dados – 11, Ano VII, n° 11

Brasil (2013) Ministério da Saúde. Secretaria de Atenção à Saúde. Departamento de Atenção Básica. Saúde mental/Ministério da Saúde, Secretaria de Atenção à Saúde, Departamento de Atenção Básica, Departamento de Ações Programáticas Estratégicas, Brasília 176p. http://bvsms.saude.gov.br/bvs/publicacoes/cadernos_atencao_basica_34_saude_mental.pdf. Accessed 4 Sept 2016

Brasil (2015) Ministério da Saúde. Saúde Mental em Dados – 12, ano 10, n° 12. Informativo eletrônico. www.saude.gov.br. Accessed 28 Mar 2016

Brasil (2016a) Ministério da Saúde. Secretaria Executiva. Datasus. Informações de Saúde.Saúde Suplementar. http://www.ans.gov.br/anstabnet/#. Accessed 12 Mar 2016

Brasil (2016b) Ministério da Saúde. Secretaria Executiva. Datasus. Informações Financeiras. http://portalsaude.saude.gov.br/index.php/o-ministerio/principal/siops. Accessed 12 Mar 2016

Brasil (2016c) Ministério da Saúde. Secretaria Executiva. Datasus. Epidemiológicas e Morbidade. http://www2.datasus.gov.br/DATASUS/index.php?area=0203. Accessed 12 Mar 2016

Brasil (2016d) Ministério da Saúde. Secretaria Executiva. Datasus. Informações de Saúde. Rede Assistencial. http://www2.datasus.gov.br/DATASUS/index.php?area=0204. Accessed 12 Mar 2016

Campos GWS, Domitti AC (2007) Matrix support and reference team: a methodology for interdisciplinary health work management. Cad Saude Publica 23(2):399–407. doi:10.1590/S0102-311X2007000200016

Chambers M, Guise V, Vlimaki M, Botelho MA, Scott A, Staniuliené V, Zanotti R (2010) Nurses attitudes to mental illness: a comparision of a sample of nurses from five European countries. Int J Nurs Stud 47:350–362. doi:10.1016/j.ijnurstu.2009.08.008

Chiaverini DH (2011) Guia Prático de Matriciamento em Saúde Mental. MS. Centro de Estudo e Pesquisa em Saúde Coletiva, Brasilia

Costa NR, Siqueira SV, Uhr D, da Silva PF et al (2011) Reforma Psiquiátrica, Federalismo e Descentralização da Saúde Pública no Brasil. Ciên Saúde Colet 16(12):4603–4614. doi:10.1590/S1413-81232011001300009

Couto MCV, Duarte CS, Delgado PGG (2008) A saúde mental infantil na Saúde Pública brasileira: situação atual e desafios. Rev Bras Psiquiatr 30(4):390–398. doi:10.1590/S1516-44462008000400015

Dauncey K, Thornicroft G, Strathdee G (1992) Setting mental health targets for England. Psychiatr Serv 43(11):1069–1072. doi:10.1176/ps.43.11.1069

Fagundes JHM, Desviat M, Silva PRF (2016) Reforma Psiquiátrica no Rio de Janeiro: situação atual e perspectivas futuras. Ciên Saúde Colet 21(5):1449–1460. doi:10.1590/1413-81232015215.00872016

Florindo AA (2009) Núcleos de Apoio á Programa de Saúde da Família e a promoção das atividades físicas no Brasil de ondeviemos, onde estamos e para aonde vamos. Rev Bras Ativ Fís Saúde 14(2):72–73

Fogel B (1993) Mental health services and outcome-driven health care. Am J Public Health 83(3):319–321. doi:10.2105/AJPH.83.3.319

Gonçalves DM, Kapczinski F (2008) Transtornos mentais em comunidade atendida pelo Programa Saúde da Família. Cad Saude Publica 24(7):1641–1650. doi:10.1590/S0102-311X2008000700019

Gonçalves DA, Mari JJ, Bower P, Gask L, Dowrick C, Tófoli LF, Campos M, Portugal FB, Ballester D, Fortes S (2014) Brazilian multicentre study of common mental disorders in primary care: rates and related social and demographic factors. Cad Saude Publica 30(3):623–632. doi:10.1590/0102-311X00158412

Häfner H (1987) Do we still need beds for psychiatric patients? Acta Psychiatr Scand 75(2):113–126. doi:10.1111/j.1600-0447.1987.tb02762.x
Hoge MA, Davidson L, Griffith EEH et al (1994) Defining managed care in public-sector psychiatry. Psychiatr Serv 45(11):1085–1089. doi:10.1176/ps.45.11.1085
IBGE-Institute Brasileiro de Geografia e Estatística (2010) Censo Demográfico 2010. http://www.ibge.gov.br/home/estatistica/populacao/censo2010/default.shtm. Accessed 12 Apr 2016
Ishara S, Bandeira M, Zuardi AW (2008) Public psychiatric services: job satisfaction evaluation. Rev Bras Psiquiatr 30(1):38–41. doi:10.1590/S1516-44462006005000064
Koyanagi C, Maries J, Surles R, Goldman HH (1993) On being very smart: the mental health community's response in the health care reform debate. Psychiatr Serv 44(6):537–539
Leal RMAC, Bandeira MB, Azevedo KRN (2012) Avaliação da qualidade de um serviço de saúde mental na perspectiva do trabalhador: satisfação, sobrecarga e condições de trabalho dos profissionais. Psicol Teor Prática 14(1):15–25
Lima JB, Bravo IMS (2015) O manejo do financiamento das Organizações sociais do Rio de janeiro em questão. In: Bravo MI, Menezes JSB, Lima JB, Souza RO (eds) A mercantilização da Saúde em debate: as Organizações sociais no Rio de Janeiro. Faperj, Rio de janeiro, pp 42–53
Mari JJ, Razzouk D, Thara R, Eaton J, Thornicroft G (2009) Packages of care for schizophrenia in low- and middle-income countries. PLoS Med 6(10):e1000165. doi:10.1371/journal.pmed.1000165
Mateus MD, Mari JJ, Delgado PGG et al (2008) The mental health system in Brazil: policies and future challenges. Int J Ment Heal Syst 2(12):1–8. doi:10.1186/1752-4458-2-12
Mendes A (2015) O subfinanciamento e a mercantilização do SUS no contexto do capitalismo. In: Bravo MI, Menezes JSB, Lima JB, Souza RO (eds) A Mercantilização da Saúdeem Debate: As OrganizaçõesSociais no Rio de Janeiro. Faperj, Rio de Janeiro, pp 11–21
Nunes MA, Pinheiro PA, Bessel M, Brunoni AR, Kemp AH, Bensenor IM, Chor D, Sandhi Barreto S, Schmidt MI (2016) Common mental disorders and sociodemographic characteristics: baseline fndings of the Brazilian Longitudinal Study of Adult Health (ELSA-Brasil). Rev Bras Psiquiatr 38:91–97. doi:10.1590/1516-4446-2015-1714
Paulin LF, Turato ER (2004) Antecedentes da Reforma Psiquiátrica no Brasil: as contradições dos anos 1970. Hist Cienc Saude-Manguinhos 11(2):241–258. doi:10.1590/S0104-59702004000200002
Ribeiro JM, Inglez-Dias A (2011) Policies and innovation in mental health care: limits to decoupling from the performance of the SUS. Ciên Saúde Colet 16(12):4623–4634. doi:10.1590/S1413-81232011001300011
Rietra RCP (1999) Inovações na gestão em saúde mental: umestudo de caso sobre o CAPS na cidade do Rio de Janeiro. Dissertation, Fundação Oswaldo Cruz, Escola Nacional de Saúde Pública
Rio de Janeiro (2013) Secretaria Municipal de Saúde. Plano Municipal de Saúde do Rio de Janeiro. http://www.rio.rj.gov.br/dlstatic/10112/3700816/4128745/PMS_20142017.pdf. Accessed 7 Nov 2015
Rio de Janeiro (2014a) Secretaria Municipal de Saúde. Subsecretaria de Atenção Hospitalar, Urgência e Emergência, Superintendência de Saúde Mental. As Residências Terapêuticas no Município do Rio de Janeiro: habitando a casa, acidade e a vida, Série E. Comunicação e educação em Saúde. Secretaria Municipal de Saúde, Rio de Janeiro, 28 p
Rio de Janeiro (2014b) Secretaria Municipal de Saúde. Notícia: Município do Rio encerra 2016 com avanços importantes na política de saúde mental. http://prefeitura.rio/web/sms/exibeconteudo?id=6631645. Accessed 28 Dec 2016
Souza FM, Valencia E, Dahl C, Cavalcanti MT (2011) A Violência urbana e suas consequências em um centro de atenção psicossocial na zona norte do município do Rio de Janeiro. Saúde Soc 20:363–376. doi:10.1590/S0104-12902011000200009
WHO (2001) World health report 2001. Mental health: new understanding, new hope. World Health Organization, Geneva

WHO (2007) WHO-AIMS report on mental health system in Brazil, WHO and Ministry of Health, Brasília, Brazil. http://www.who.int/mental_health/evidence/who_aims_report_brazil.pdf. Accessed 3 Aug 2016

World Bank (2016) Brasil: Aspectos Gerais, 2016. http://www.worldbank.org/pt/country/brazil/overview#1. Accessed 26 Apr 2016

Zwarcwald CL, Bastos FI, Viacava F, de Andrade CL (1999) Income inequality and homicide rates in Rio de Janeiro, Brazil. Am J Public Health 89(6):845–850

Development of Mental Health Services in a Major City in Latin America: Santiago

15

Rubén Alvarado Muñoz, Sara Schilling, Lúcia Abelha, and Giovanni Marcos Lovisi

Abstract

Two health systems coexist in Chile, with 80% of the population belonging to the public system, supported by general taxation and workers' regular contributions, and 18% to the private health sector, financed by copayments and prepaid insurance plans, with varied coverage and costs, depending on the characteristics of each client and plan.

In 1990, the public sector initiated the formation of mental health service networks based on the community-centered model of care. In the capital city of Santiago, there are currently 147 primary care centers, 35 secondary-level community mental health centers (COSAMs), 8 day hospitals, and 7 short-stay psychiatric services, which serve approximately six million people. Primary care centers resolve around 85% of mental health complaints, typically common mental disorders, while more complex and chronic conditions are referred to COSAMs. However, problems persist in the detection and management of certain disorders, such as psychotic episodes. The lack of effective community-based psychosocial interventions involving users' support networks, as well as prevalent stigma, has made it difficult to achieve treatment adherence and full social inclusion.

R.A. Muñoz (✉) • S. Schilling
Mental Health Unit, School of Public Health, Faculty of Medicine, University of Chile, Santiago, Chile
e-mail: ralvarado@med.uchile.cl; shschilling@gmail.com

L. Abelha • G.M. Lovisi
Institute of Studies in Collective Health, Federal University of Rio de Janeiro, Rio de Janeiro, Brazil
e-mail: lucia.abelha@gmail.com; giovannimlovisi@gmail.com; glovisi@uol.com.br

N. Okkels et al. (eds.), *Mental Health and Illness in the City*, Mental Health and Illness Worldwide, https://doi.org/10.1007/978-981-10-2327-9_20

This chapter focuses specifically on mental health service developments in Santiago, while analyzing future challenges, related to incorporating psychosocial interventions into the care model, increasing capacity to detect mental health cases early on in their evolution, and orienting policies toward the promotion of the population's mental health and subjective well-being.

Keywords
Community Mental Health Services • Mental Health in Cities • Reform of Psychiatric Services • Mental Health Services • Santiago, Chile

Contents

The present chapter will provide a detailed description of the development of mental health service networks in Chile and will conclude with a specific focus on the metropolitan area of Santiago, Chile's capital city, and the unique challenges it faces looking forward. However, to begin, we provide a general introduction to the Latin American and Caribbean (LAC) region, in term of its diverse health and social profiles and mental health services.

The General Context of Mental Health Services in Urban Areas of Latin America

The LAC region is marked by great demographic and sociocultural diversity, leading to variations in epidemiological profiles and the organization of health services, both between and within countries (PAHO 2016). The Pan American Health Organization estimates that in the LAC region in 2010, there were approximately 2.2 million avoidable deaths, 60% of which could have been addressed through prevention measures, and 40% of which were attributable to a lack of effective health services. Although the proportion of avoidable deaths decreased in 2012 in comparison to 2011, from 272 to 231 avoidable deaths per 100,000 individuals (or 10.5% of deaths), fourfold differences still exist within countries in the region (PAHO 2016).

In large part, this reality can be traced to the presence of great social inequities across the region, which has not changed significantly over the past few decades. For example, in spite of the establishment of numerous social policies, the percentage of the LAC population living in poverty (on no more than $2 USD a day) has remained around 30% for the last 20 years, a situation thus considered to be "chronic poverty" (Vakis et al. 2015). Additionally, the regional income inequality gap between men and women is second only to sub-Saharan Africa (in eight LAC countries, the nominal wages of urban women are 73% that of urban men).

Moreover, most LAC countries have a mixed health system, consisting of: (a) a public subsector, serving low- and middle-income individuals, that is administered by the Ministries of Health or Social Security Systems and financed by general taxation and/or workers' contributions and (b) a private subsector that tends to cater to the higher-income population and is financed by prepayment systems (insurance premium plans with varied coverage) and/or copayments when services are received. This latter, private health subsector has grown in recent decades: in Chile, nearly 20% of people are affiliated with private health insurance, and in Brazil, for instance, approximately 25% of the population has private health insurance, even though free coverage under the Brazilian Unified Health System (known as SUS, for its Portuguese acronym) is provided to the entire population (Becerril-Montekio et al. 2011a, b).

With respect to mental health, regional epidemiological studies have reported 12-month prevalence rates for all mental disorders if between 18.7% and 24.2%. In terms of specific disorders, the median prevalence rates in the last 12-months for adults were: 1.0% for nonaffective psychosis, 5.2% for major depressive episodes, and 4.5% for alcohol abuse/dependence (Kohn et al. 2005). Additionally, psychiatric disorders account for 21.3% of the total disease burden in LAC countries (Kohn et al. 2005). Depressive disorders, in particular, are the leading cause of disease burden in the region and are twice as common among women than among men. It is estimated that between 20% and 40% of pregnant women suffer a depressive episode during pregnancy or in the postpartum period (Kohn et al. 2005; PAHO 2014a).

Past decades have also seen an increase in alcohol and illicit drug use, as well as associated disorders, especially among the younger population. However, adequate policies and services to prevent and treat these problems have yet to be established (PAHO 2014a; WHO 2014).

Violence is a major social and health issue facing the region. Studies show that one out of every three women report that they have been subjected to gender violence, which is linked to multiple negative health consequences (depression, suicide, etc.). Similarly, studies have documented a high frequency of physical and psychological punishments of children, a situation that is perpetuated through intergenerational cycles of violence. According to national surveys carried out in several countries, more than a third of women and at least half of all men admit to having been beaten during childhood (PAHO 2014a; WHO 2014).

In LAC, the mortality rate due to suicide, adjusted by age, is 5.2 per 100,000 habitants (8.4 in men and 2.1 in women). As is commonly the case, the suicide mortality rates for men are higher than those of women (the male to female ratio is 3.8), while suicide attempts are more frequent among women. Suicide is the third leading cause of death in the 20-to-24 age group, and the fourth for the 10-to-19 and the 25-to-44 groups. The highest mortality rate for suicide is seen in individuals older than 70 years of age (12.4 per 100,000 habitants) (PAHO 2014a, b).

LAC countries have attempted to address the aforementioned mental health problems with a variety of strategies, but their responses have been insufficient; there are still significant gaps in service coverage, and quality is poor. It is estimated that nearly 75% of adults in LAC who suffer from depression, anxiety, or a substance use disorder do not receive adequate treatment (Kohn et al. 2005).

This situation is partially explained by the absence of effective legal and regulatory instruments. Only eight countries in the LAC region have specific mental health laws that are currently valid. On the other hand, even though the majority of countries have some sort of national mental health plan or policy (only six have neither of the two), mental health problems do not receive sufficient financial support: 73% of LAC countries allot only 1–5% of their health budget to mental health, in stark contrast to the fact that mental health conditions produce 29% of the regional disease burden. Furthermore, most LAC countries spend over half their mental health budget on psychiatric hospitals, with little investment in the development of community-based services (PAHO 2014b).

In this vein, another part of the problem facing mental health services in LAC lies precisely in the scarcity of community-based services and treatment options. While some countries in the region have begun to develop community services, they are still incipient and only reach a portion of the population. The majority of LAC countries have a highly centralized mental health system, with services focused in large psychiatric hospitals in major cities, which are difficult to access even for urban residents (with no access for individuals living in rural and/or distant areas). In LAC, 66% of psychiatric beds are in psychiatric hospitals, 11% are in general hospitals, and only 3% are in community residences (PAHO 2014b).

With regards to human resources, there is a great disparity between LAC countries. When mental health systems are centered upon psychiatric hospitals, most resources are also concentrated there. In general, the LAC region has a median of 2.1 psychiatrists, 6.0 nurses, and 4.2 psychologists per 100,000 habitants (PAHO 2014b).

The lack of mental health services offered within primary care centers is also an issue. Despite studies indicating that between 30% and 50% of primary care consultants have a mental disorder, these mental conditions are usually not properly identified, and therefore, not appropriately treated in primary care services. In the case that a diagnosis is made, patients are typically referred directly to specialized services (which are usually in psychiatric hospitals), as primary care centers tend to have limited resolutive. Only recently, in the past decade, has this situation started to change, through the development of innovative initiatives in primary care services (PAHO 2014b).

In summary, in the LAC region, mental disorders make up a high proportion of the disease burden, but countries' social and health responses have been inadequate, inequitable, and of low quality.

Development of Mental Health Services in Santiago, Chile

The Chilean Health System

Chile has a population of approximately 18 million inhabitants, almost 30% of who live in the capital city of Santiago. The country is in the process of an advanced demographic transition, with individuals older than 60 years of age constituting 15% of the population and a demographic dependency ratio (those not in the labor force divided by those in labor force, on a scale from 1 to 100) greater than 55, as of 2015. The total fertility rate is 1.8 and has been gradually declining for the past few decades; consequently, the population growth rate is decreasing (INE Chile 2017). The rate of migration, in contrast, has been increasing, with immigrants from nearby countries, such as Peru, Bolivia, Colombia, Ecuador, and most recently, Haiti, coming to Chile in search of a better quality life and higher wages (INE Chile 2017).

The Chilean Gross National Income (the country's total income, from domestic and foreign sources) per capita in 2014 was US$14,910, thus classifying it as a high-income country (OECD 2014). Chile, however, is also marked by significant socioeconomic inequities. As measured by the GINI coefficient, an indicator of the income distribution of a country's residents, ranging from 0 to 1, with 0 representing complete equality and 1 complete inequality, Chile's inequality index is 0.50, making it the most unequal country in the Organization for Economic Cooperation and Development (OECD), ahead of Mexico, Turkey, and the United States (OECD 2014). The percentage of the population living in poverty varies greatly within the country; for instance, while the poverty rate of the municipality of Ñuñoa, located in Santiago, is only 0.6%, in the municipality of Puerto Saavedra, in the southern Araucanía Region, 41.7% of the population is classified as impoverished.

Two health systems coexist, with relatively little intercompetition, in Chile. On the one hand, the public sector is financed by general taxation and workers' regular contributions. Over 80% of the Chilean population belongs to this system, and they are served by the network of public health services (which are typically used by persons with low income and range from primary care centers, distributed throughout the entire country, to high-complexity hospitals) and/or by private centers and practitioners that are affiliated to the public system (Becerril-Montekio 2011b; Superintendencia Chile 2016).

The private health sector, meanwhile, covers 18% of the population and is financed by copayments and prepaid insurance plans, with varied coverage and costs, depending on the characteristics of each client and plan. This sector's services are predominantly specialized and tend to be more sophisticated than those of the public health network. Primary care centers practically do not exist in the private system (Becerril-Montekio 2011b; Superintendencia Chile 2016).

Mental health services, for their part, vary greatly between the public and private sectors. The public health sector has an extensive network of mental health services, from large psychiatric hospitals and short-stay w ards in general hospitals to community mental health centers and mental health programs within primary care centers. The public sector emphasizes the importance of treatment continuity, with programs and services actively working to coordinate between distinct care levels. Psychosocial interventions are also highly valued (Alvarado et al. 2012a; Minoletti et al. 2014).

In contrast, mental health services in the private sector are fragmented, and coordination between them is minimal or nonexistent. The majority of services are based in psychiatric hospitals and the private outpatient clinics of psychiatrists and psychologists. Community-based services, such as day hospitals or rehabilitation centers, are scarce, and psychosocial interventions are few and far between. While the last decade has witnessed the incorporation of general practitioners into private sector care networks, this development is principally a cost-reduction strategy (Alvarado et al. 2012a).

History of Mental Health Care Reform in Chile

In order to situate, and better understand, the reform of mental health services in Chile, we will start with a brief history of psychiatric care in the country. Until 1852, Chile did not offer specialized care for individuals battling mental illness. In that year, the *Casa de Orates de Nuestra Señora de Los Angeles* (The Insane Asylum of Our Lady of Angels) was established in Santiago, through an initiative led by the provincial intendant, motivated by his visit to the Mental Hospital of San Andrés in Lima, Peru. The *Casa de Orates* was later renamed the National Insane Asylum, and then as the Psychiatric Hospital – the first of its kind in Chile – which is still functioning today (its current name in Spanish is *Instituto Psiquiátrico Dr. José Horwitz Barak*) (Escobar 1990).

In the first half of the twentieth century, a few psychiatric institutions were built in cities other than Santiago, though they were generally smaller and not as long lasting, as was the case with the Andrés Bello Hospice in Concepción, to the south, and the Insane Ward of San Andrés Hospital in the northern city of Iquique.

In 1927, the General Insanity Law was enacted, declaring the necessity of installing "Open Door" establishments in the capital city of each province in the country. Nonetheless, only one such center was constructed, in the rural outskirts of Santiago, which today is *El Peral* Psychiatric Hospital (Gomez 2012).

In 1952, Chile formed the National Health System, on the basis of the British National Health Service. Its creation integrated the various health care services that already existed at the time, and during the following decade, extensive treatment networks were developed, ranging from primary care centers to high-complexity hospitals. More than 90% of Chileans received medical care under this system. Psychiatric services, however, remained separate, and care was concentrated in a small number of psychiatric hospitals throughout the country.

In the course of the 1960s, three pioneering initiatives installed community mental health care in Chile. The first, led by Dr. Luis Weinstein, expanded psychiatric services in the central and western sectors of Santiago (Weinstein 1978), and in the second, a team directed by Dr. Juan Marconi enhanced treatment offerings in the southern zone of the capital city (Marconi 1973, 1976). Both of these experiences grew out of an alliance between the University of Chile and the National Health Service. For its part, the third initiative was established in the southern city of Temuco, under the direction of Dr. Martín Cordero. Unfortunately, all three of these programs were abandoned with the instauration of the military-civic dictatorship in Chile in 1973. Nevertheless, given that Dr. Marconi's community-based initiative was well integrated into the academic world and with the existing health services, it was able to maintain considerable influence in those spheres during the dictatorship, and his program was reestablished once democracy returned in 1990 (Sepúlveda 1998).

In the 1980s, the military-civic dictatorship began to alter the Chilean health system, in a process characterized by four central elements: (1) the unification of the steering and planning roles within the Ministry of Health; (2) decentralization of the administration of primary care centers, passing from Ministry of Health to the direction of each municipality; (3) decentralization of the administration of care networks, with the creation of 27 health services throughout the country; and (4) the introduction of a private insurance system (ISAPRE), which led to the installation of the private health sector and the growth of private industry related to the provision of medical services.

The predominant model of psychiatric care during the military-civic dictatorship continued to be centered around psychiatric hospitals, although two elements should be noted: (1) the creation of adolescent centers, administered by the municipalities (just like the primary care centers), which developed activities to prevent drug use among teenagers, and to treat those with dependence, and (2) since private insurance plans had minimal coverage of mental health conditions, there was little development of private psychiatric institutions (in stark contrast to the vast private development of other medical specialties).

The return of democracy to Chile in 1990 initiated a series of debates about the transformation of psychiatric care, and at the end of 1991, these conversations culminated in the "Santiago Agreement," laying the foundation for mental health reform (MinSal Chile 1991). The next crucial step was the formulation of the National Mental Health and Psychiatry Plan in 1993, which defined the mental health care paradigm and outlined a framework for the policy's development (USM MinSal Chile 1993). That same year, the community mental health model was adopted in Chile, and in 2001, with the second edition of the National Mental Health and Psychiatry Plan, its use was cemented (USM MinSal Chile 2001).

Recent Developments of Chilean Mental Health Services

The past 20 years of mental health reform have been marked by the following strategies (USM MinSal 2001; Minoletti et al. 2012):

Creation of the Mental Health Department of the Ministry of Health, which has been actively involved in the development of mental health policies and the accreditation of services to ensure quality care.

Progressive increase of the health budget dedicated to mental health, which has resulted in improved physical infrastructure, more highly trained human resources, and the establishment of detailed treatment programs for specific conditions.

Development of mental health and psychiatry service networks, based on the community-centered model of care; these are divided into catchment areas and include a range of centers, with different levels of care complexity, to provide comprehensive and continuous treatment for users.

Interdisciplinary collaboration with other social sectors, especially education, justice, labor, and community development.

Installation of programs, with evidence-based interventions, that prioritize and dedicate resources to mental disorders with the highest prevalence and greatest functional and social impact (depression, schizophrenia, and substance use disorders).

Establishment of Mental Health Units in each Health Department in the country, tasked with implementing the Ministry of Health guidelines and locally administering programs and resources.

Formation of user and family groups, which were gradually integrated into the planning and evaluation of programs and interventions.

The following figures reflect the progress of the National Mental Health and Psychiatry Plan over this period (Minoletti et al. 2012, 2014):

The process of de-institutionalization of users with severe mental illness is seen in the reduction of the number of long-stay beds, from 2,516 in 1990 to 441 in 2012.

Moreover, protected residences have been established; although there were no such facilities in 1990, as of 2012, there were 182 (with a capacity for 1,544 users). A similar phenomenon has occurred with community-based rehabilitation activities, which, while almost nonexistent in 1990, are now present throughout Chile.

Day hospitals were integrated into the mental health care network, and there are now 45, distributed across almost every region of the country. The number of outpatient facilities also increased, from 22 centers in 1990 to the current 53, in addition to 83 new Community Mental Health Centers (known by the acronym COSAM).

At the same time, short-stay beds in large psychiatric hospitals decreased from 604 in 1990 to 329 in 2012, while short-stay psychiatric beds were established in general hospitals (there were 534 such beds in 2012, located in 31 adult and adolescent wards).

Furthermore, the political decision to prioritize mental health is evident in the increase of the proportion the health budget that is dedicated to mental health –

from only 1% of the health budget in 1990 to 2.16% in 2012. This percentage, however, has not increased since 2004.

While these advances have undoubtedly been significant, more resources, more widely distributed throughout the country, are still needed, as are additional community-based services.

Mental Health in Chilean Primary Care Services

Primary care services in Chile were first instated in the beginning of the twentieth century, though it was during the 1950s that they were truly consolidated, with the creation of the Chilean National Health Service in 1952, under a model similar to the United Kingdom's National Health Service.

Currently, primary care services are characterized by community-based centers, with multidisciplinary team of health professionals that include physicians, nurses, midwives, dentists, physical therapists, nutritionists, social workers, and psychologists, as well as administrators and technicians. Each center is responsible for a defined geographic zone and treats residents who are beneficiaries of the public health system (with each center covering a population of between 10,000 and 30,000 inhabitants). Centers work with a family health orientation, which is why most of them are known as Family Health Centers (CESFAM, for its Spanish acronym). This family-centered comprehensive focus has characterized the primary care system in Chile for the last two decades, so even though older primary care centers (known as Urban Health Clinics, or CSU) differ slightly, they function under the same paradigm as CESFAMs.

The activities of CESFAMs and CSUs include health promotion; general and specific prevention initiatives (such as immunization programs); early detection of specific illnesses (e.g., cervical cancer with a papanicolaou test); routine checkups for children, pregnant women, adults, and older adults; and general (nonspecialized) care of all health problems. These primary care centers are able to resolve close to 85% of all health complaints.

The aforementioned primary care centers are associated through the public service network with secondary- and tertiary-level services. Territories have their own referral networks, and the public health system is regionalized and organized based on coverage and level of complexity (with services ranging from primary care centers that provide basic response capacity for a large number of people to extremely specialized hospitals, which are highly advanced but can only attend to a smaller population). For example, patients whose health conditions cannot be resolved in primary care but do not require hospitalization are referred to more specialized outpatient centers (secondary-level), which include Health Reference Centers (CRS), for specific health problems; Diagnostic and Treatment Centers (CDT), for general health problems; and COSAMs, for mental health issues. Hospitals, the tertiary-level of care, are also available for patients with illnesses that require more intensive treatment.

In the specific case of mental health problems, primary care teams provide mental health care using preestablished protocols (clinical practice guidelines) that are established by the Ministry of Health. There are now four mental health conditions with specific protocols for primary care: depressive disorders, first episode of psychosis, bipolar disorder, and youth drug and alcohol consumption.

Starting in 1990, with the return of democracy to Chile, numerous mental health initiatives were implemented in primary care centers. Programs developed in the southern sector of Santiago (one of the most populous areas in the country) received special recognition for their innovation and extensive community involvement, and for the formation of a mental health treatment network comprised of different care complexity levels, based on the community approach. On the basis of the aforementioned, the southern zone of Santiago was considered an example for the rest of the country's health services (Alvarado et al. 1991). Of particular note, this zone is also where Juan Marconi implemented his pioneering intra-community mental health program, 20-years prior (Marconi 1973).

In later years, Chile began to show particular concern for the development of mental health programs in primary care services after an international multicentric study found that over half (52.5%) of all individuals seeking assistance in primary care centers for a general health condition also had a mental health problem, and that the most frequent mental health conditions were depressive disorders (29.5%) (Sartorius et al. 1993).

In light of these results, a mental health program in primary care services was formed in 2001 – the National Program for the Detection, Diagnosis, and Treatment of Depression in Primary Care (Sepúlveda 1998). Evaluations of the program have demonstrated its clinical effectiveness and success, and 15 years after its initial formation, more than one million people have been treated through the program (Alvarado et al. 2012b).

With regard to first episode psychosis, primary care teams are responsible for detecting early signs and symptoms and provide general counseling, before referring patients to a more specialized center. In Chile, every year about 1,600 cases of first episode psychosis are diagnosed, though a study found that only between a third and half of first episode cases presenting in primary care centers are correctly detected and referred. There is, therefore, still much to improve in terms of detection in primary care services, in order to reduce the time of untreated psychosis among affected individuals (Alvarado et al. 2014).

The Network of Mental Health Services in Santiago

More than six million people live in the metropolitan area of Santiago, located in the middle of the country, which is made up of 37 municipalities (26 of which are completely urban in nature and 11 with both urban and rural sectors).

Few psychiatric epidemiological studies have been conducted in Chile in general, let alone Santiago specifically. The only population-based study of the country was carried out in the 1990s, using the Composite International Diagnostic Interview (CIDI) and showed a lifetime prevalence of mental disorders of 33.7% in Santiago (Vicente et al. 1994), which is very similar to the figures in Concepción (36.5%), another large city in Chile toward the south (Vicente et al. 1992), as well as the rural zones of the southern Cautín province (31.4%) (Vicente et al. 2000). Among the studied conditions, the most frequent were affective disorders, with a lifetime prevalence of 16.3%, followed by anxiety disorders (11.6%), and substance use disorders (11.0%) (Vicente et al. 2006).

Aside from the national policies, the Metropolitan Area of Santiago does not have specific mental health promotion programs or policies, nor do social sectors outside of health (education, housing, environment, etc.) have strategies that explicitly aim to improve the mental health and subjective well-being of the city's inhabitants. This is also a pending challenge.

Santiago currently has 147 primary care centers (117 CESFAM and 30 CSU), all of which provide basic care for mental health problems. When a more severe disorder is detected, or when primary care teams are not able to adequately improve a patient's condition (in cases of treatment-resistant depression, for example), these cases are referred to the corresponding COSAM in the network, to receive more specialized treatment, while maintaining a community-centered approach. If patients need to be hospitalized, there are two options: day hospitals or short-stay services in general hospitals. In Santiago, there are 35 COSAMs, 8 day-hospitals, and 7 short-stay psychiatric services.

Santiago is also home to two of the three psychiatric hospitals that still exist in Chile, one of which, the *Instituto Psiquiátrico Dr. José Horwitz Barak*, located in the northern zone of the city, was the first psychiatric hospital in the country and is over 150 years old. These hospitals are in the midst of a transformation process and are progressively reducing their number of medium- and long-stay beds, to focus on those with acute needs, better suited by short-stay beds.

Moreover, in addition to psychiatric hospitalization, *Instituto Psiquiátrico Dr. José Horwitz Barak* also offers one of the two specialized psychiatric emergency rooms located in Santiago. The other corresponds to a high complexity general hospital in the southern zone of Santiago, *Hospital Barros Luco Trudeau.* Though large, these two hospitals are not able to meet the city's current demands for urgent psychiatric care, which needs to be addressed in the near future.

In the past few years, a number of studies have evaluated the impact of the transformation of mental health services toward a community-based model. This research has been largely focused on the southern zone of Santiago, given that it is where this transformation began, and where it has been most fully implemented (Alvarado et al. 1991). While these results have yet to be published, it is important to highlight a few of the preliminary findings. There have been a reduced number of psychiatric emergency treatments, fewer psychiatric hospitalizations, and a lower rate of rehospitalizations among users who have been hospitalized. These results provide the empirical evidence needed to continue advancing the transformation of mental health services in the city of Santiago, and throughout Chile.

Challenges Facing Mental Health Services in Santiago

The development of the mental health service network in Santiago, as detailed in the present chapter, formed part of a larger countrywide process, based on the Ministry of Health's National Mental Health and Psychiatric Plan, which laid out a framework for the transition from the traditional paradigm of hospital-centered psychiatric treatment to a model focused on community-based mental health services.

In the early 1990s, a large part of mental health care was still based out of two psychiatric hospitals in Santiago. Once implemented, one of the first measures of the National Plan was to remove long-term stay beds from psychiatric hospitals, which led to the discharge of many individuals who had been hospitalized for years on end, and who had lost many of their family relationships and social supports in the process. At first, there were few outpatient centers to receive this population, and services that did exist were difficult to access and offered care with a predominantly biomedical approach.

In contrast, there is now an extensive network of community services, ranging from primary care services to psychiatric units in general hospitals, and including day hospitals and community-based rehabilitation centers. The mental health reform process has necessitated a substantial increase in resources (financial, human, material, and infrastructural) that are devoted to mental health services, and which, without a doubt, have permitted increased access to mental health care for a greater proportion of the population. Today, more individuals that require specialized mental health treatment in Chile are able to receive it.

There are, however, still challenges pending. We will explore three aspects of the reform process that are in early stages of development and could face obstacles, and we will recommend possible solutions.

The first is the strengthening of community-based strategies for mental health promotion. Currently, several initiatives are focused on promotion activities, but they are largely based in primary care centers, and there is little work within the actual community. The most promising program – "Chile Grows with You" (*Chile Crece Contigo*) – aims to support child development, starting during pregnancy until children are 5 years old. The national program includes activities such as the detection of depressive disorders in mothers during pregnancy and the postpartum period, stimulation of the mother–child bond, and growth monitoring (Bedregal et al. 2014; Jadresic et al. 1995; Alvarado et al. 2015a).

Socially, mental health issues are still very personal and stigmatized, and associated problems are viewed as something to be overcome on one's own. This outlook must be replaced by a culture that values mental health promotion, prevention, and care as a public good, where subjective well-being is acknowledged as a major social and economic development goal. This mentality change will require the involvement of many social actors, from professionals to civil institutions to general citizens.

The second challenge is to change the emphasis of the current treatment model, to achieve true integration between biomedical and psychosocial aspects. In previous studies in Chile, we have shown that psychosocial interventions are utilized for only a very small proportion of patients with first episode psychosis, despite being mandated by clinical guidelines and a national treatment policy (Markkula et al. 2011). This situation is partly due to the fact that very few professionals who work in community services have received formal training in the implementation of psychosocial interventions. Additionally, although the proposed models have proven, evidence-based effectiveness (Bustillo et al. 2011), they have not been adapted to the Chilean reality, thus reducing their local adoption and implementation in services.

Further, an assessment of mental health service users' needs, and whether or not they are fully satisfied, has allowed us to conclude that while there are still existing problems with psychopathological symptoms, in terms of their management with psychotropic drugs and psychoeducation, these needs are generally met by existing service offerings. However, users reported numerous unmet needs associated with their daily life (engaging in activities they view as important), their social life (companionship, having a romantic partner, and spending time with others), and their degree of autonomy (desiring more independence) (Alvarado et al. 2012c). These aspects are not being appropriately addressed by mental health services, in concordance with the previously mentioned deficit of psychosocial interventions (Alvarado et al. 2012c; Jorquera et al. 2015).

To advance in this regard, we are working on the local adaptation of a psychosocial intervention known as Critical Time Intervention (CTI), an evidence-based practice which was created to support homeless individuals with severe mental disorders in times of transition and has proven effective for diverse groups of individuals with mental disorder in numerous countries, to increase their long-lasting support networks and improve their quality of life (Susser et al. 1997). We have carried out a pilot application of the CTI model, adapted to the local Chilean reality, for individuals experiencing their first episode of psychosis in several community mental health centers in Chile (most of which are in Santiago), and the results are promising (Alvarado et al. 2015b). Moreover, through a collaborative regional mental health research network called RedeAmericas, we are working on perfecting this intervention, adapting its use to low-income countries, and we have tested the adaption in a pilot RCT in multiple LAC cities (Susser et al. 2011).

Thus, the next pending step in Chile is to refine the mental health care model, by putting greater emphasis on psychosocial interventions. This will require adapting interventions to the local reality of communities, testing their effectiveness, and then training professional and technical teams to properly implement this model on a larger scale.

The third and final challenge we anticipate is the continued incorporation of early detection strategies for different mental health problems within the routine activities of health services, especially in primary care centers, and in particular for severe mental disorders, where the application of early treatment could greatly improve users' quality of life, as is the case with psychotic disorders (McGorry et al. 2008; Alvarado et al. 2006a). Specialized training sessions for health teams should be established, since several studies have shown that their detection capability can be very low, as observed in a southern region of Chile, where only 1 in every 20 individuals with depression were correctly diagnosed in primary care centers (Alvarado et al. 2006b). In the case of individuals with first episode psychosis, we have found that the detection rate in primary care centers is close to a third of all cases and that this rate is significantly improved after a brief training of professional and technical teams (Alvarado et al. 2014; Valenzuela et al. 2012).

Final Reflections on Santiago's Mental Health Services

Over the past two decades, the city of Santiago has made significant progress in mental health services and infrastructure, and the developing of network of community-based services is replacing psychiatric hospital-centered care. Although this process is a part of a larger national policy, its first steps, nonetheless, were taken in Santiago (specifically in the southern zone), which served as a model for the rest of the country. Similarly, in the past few years, data collected through studies conducted in this area of Santiago have shown that the community-based care model produces better outcomes among service users, thus serving to support the continued transformation of services.

In the future, attention must be paid to transforming the care model through the incorporation of more psychosocial interventions, improving early detection of mental disorders, and advancing policies that promote mental health and subjective well-being of the entire population. Achieving these objectives could require a new way of organizing and implementing national public policies in specific cities, a topic which has yet to be addressed. Chile is currently beginning the transition from a very centralized, "one-size-fits-all" method of governing to a more decentralized approach, in which regional and municipal governments have more control over local policy development and implementation. Nevertheless, large cities made up of many municipalities, such as Santiago, do not have a mechanism to facilitate the coordination and conduction of public policies across municipalities, in order to resolve issues that affect the city as a whole (without ignoring differences present within zones of capital). This is especially important for the operation and coordination between different mental health services and between different social sectors, which is fundamental for the success of strategies that promote the mental health and well-being of the larger population.

Cross-References

▶ History of Urban Mental Illness
▶ Implementing Community Care in Large Cities and Informal Settlements: An African Perspective
▶ Mental Health Care in 11 Major Cities Over Two Decades
▶ Prospects and Challenges of Mental Health Services in a Major City in Latin America: Rio de Janeiro
▶ The Bogota Experience

References

Alvarado R, Raffo S, Gómez B, Riveros A, Sagaceta A, Salinas J, Pérez J (1991) Programa de Salud Mental para la Atención Primaria, en el Servicio de Salud Metropolitano Sur. Rev Psiquiatr 8(2):805–814

Alvarado R, Erazo C, Poblete C (2006a) ¿Es posible y útil la detcción precoz como estrategia de salud pública para el primer episodio de psicosis? Rev Chil Salud Pública 10(3):158–163

Alvarado R, Numhauser J, Galán L (2006b) Cuadros depresivos y utilización de servicios de salud en mujeres de la Isla de Chiloé. Rev Chil Salud Pública 10(1):35–46

Alvarado R, Minoletti A, Torres-Gonzalez F, Moreno-Küstner B, Madariaga C, Sepulveda R (2012a) Development of community care for people with schizophrenia in Chile. Int J Ment Health 41:48–61

Alvarado R, Rojas G, Minoletti A, Domínguez C, Alvarado F (2012b) Depression program in primary health care: the Chilean experience. Int J Ment Health 4(2):38–47

Alvarado R, Torres F, Schilling S, Alvarado F, Domínguez C, Moreno B, Aliste F (2012c) Factors associated with unmet needs in individuals with schizophrenia in Chile. Cad Saude Colect 20(4):466–472

Alvarado R, Valenzuela MT, Aliste F, Madariaga C, Minoletti A (2014) Intervención comunitaria para mejorar la detección de casos con primer episodio de psicosis en Atención Primaria en Chile. In: Fernandez AR, Enders JE (eds) Experiencia y reflexiones en Salud Mental Comunitaria, 1st edn. Kronos Editor, Córdoba, pp 169–183

Alvarado R, Jadresic E, Guajardo V, Rojas G (2015a) First validation of a Spanish-translated version of the Edinburgh postnatal depression scale (EPDS) for use in pregnant women; a Chilean study. Arch Womens Ment Health 18(4):607–612

Alvarado R, Schilling S, Jorquera MJ (2015b) Uso de la intervención Critical Time Intervention en personas con primer episodio de psicosis. Rev Fac Cienc Med Cordoba 73(3):331–338

Becerril-Montekio V, Medina G, Aquino R (2011a) The health system of Brazil. Salud Publica Mex 53(Suppl 2):S120–S131

Becerril-Montekio V, Reyes JD, Manuel A (2011b) The health system of Chile. Salud Publica Mex 53(Suppl 2):S132–S143

Bedregal P, Torres A, Carvallo C (2014) Chile Crece Contigo: el desafío de a protección social a la infancia. PNUD. Available http://www.cl.undp.org/content/dam/chile/docs/pobreza/undp_cl_pobreza_cap5_chile_crece.pdf. Accessed 27 June 2016

Bustillo JR, Lauriello J, Horan WP, Keith SJ (2011) The psychosocial treatment of schizophrenia: an update. Am J Psychiatry 158:163–175

Escobar E (1990) Historia del Hospital Psiquiátrico (1852–1952). Rev Psiquiatr 7:36 368

Gomez M (2012) El Peral Psychiatric Hospital. Int J Ment Health 41:73–86

Instituto Nacional de Estadísticas de Chile (2017) Available http://www.ine.cl/canales/chile_estadistico/familias/demograficas_vitales.php. Accessed 22 Jan 2017

Jadresic E, Araya R, Jara C (1995) Validation of the Edinburgh postnatal depression scale (EPDS) in Chilean postpartum women. J Psychosom Obstet Gynecol 16:187–191

Jorquera N, Alvarado R, Libury N, de Angel V (2015) Association between unmet needs and clinical status in patients with first episode of schizophrenia in Chile. Front Psychol 6:57. doi:10.3389/fpsyt.2015.00057

Kohn R, Levav I, Caldas de Almeida JM, Vicente B, Andrade L, Caraveo-Anduaga JJ, Saxena S, Saraceno B (2005) Los trastornos mentales en América Latina y el Caribe: asunto prioritario para la salud pública. Rev Panam Salud Publica 18(4/5):229–240

Marconi J (1973) La revolución cultural chilena en programas de salud mental. Acta Psiquiatr Psicol Am Lat 19:17–33

Marconi J (1976) Política de Salud Mental en América Latina. Acta Psiquiatr Psicol Am Lat 22:112–118

Markkula N, Alvarado R, Minoletti A (2011) Adherence to guidelines and treatment compliance in the Chilean national program for first-episode schizophrenia. Psychiatr Serv 62:1463–1469

McGorry P, Killackey E, Yung A (2008) Early intervention in psychosis: concepts, evidence and future directions. World Psychiatry 7:148–156
Ministerio de Salud (1991) Actas de la Conferencia Chilena: Desafíos en Salud Mental y Atención Psiquiátrica. Ministry of Health, Santiago. 13, 14, 15 y 16 de Noviembre
Minoletti A, Sepulveda R, Horvitz M (2012) Twenty years of mental health policies in Chile: lessons and challenges. Int J Ment Health 41:21–37
Minoletti A, Alvarado R, Rayo X, Minoletti M (2014) WHO-AIMS: Sistema de Salud Mental de Chile. Segundo Informe. Ministerio de Salud, Santiago
Organization for Economic Cooperation and Development (2014) Society at a glance 2014: OECD social indicators. OECD Publishing. Available http://dx.doi.org/10.1787/soc_glance-2014-en. Accessed 23 Jan 2017
Pan American Health Organization (2014a) Health in the Americas, 2012 Edition. Available http://www.paho.org/saludenlasamericas/index.php?option=com_content&view=article&id=62&Itemid=167&lang=en. Accessed 23 Jan 2017
Pan American Health Organization (2014b) Plan de Acción sobre Salud Mental. Pan American Health Organization, Washington, DC
Pan American Health Organization (2016) Health situation in the Americas: basic health indicators 2016. Available http://www.paho.org/hq/index.php?option=com_content&view=article&id=2470&Itemid=2003&lang=en. Accessed 23 Jan 2017
Sartorius N, Ustun TB, Costa e Silva JA, Goldberg DP, LeCrubier Y, Ormel J et al (1993) An international study of psychological problems in primary care. Arch Gen Psychiatry 50:819–824
Sepúlveda R (1998) Una reflexión sobre la organización de los servicios de salud mental y psiquiatría en Chile. Cuad Med Soc 39:16–21
Superintendencia de Salud, Gobierno de Chile (2016) Available http://www.supersalud.gob.cl/documentacion/569/w3-propertyvalue-3741.html. Accessed 23 Jan 2017
Susser E, Valencia E, Conover S, Felix A, Tsai W-Y, Wyatt RJ (1997) Preventing recurrent homelessness among mentally ill men: a "Critical Time" Intevention after discharge from a shelter. Am J Public Health 87:256–262
Susser E, Galea S, Alvarado R, Rojas G (2011) Regional network for mental health research in Latin America. Grant from National Institute of Mental Health
Unidad de Salud Mental, Ministerio de Salud (1993) Plan Nacional de Salud Mental y Psiquiatría. Primera Edición, Santiago
Unidad de Salud Mental, Ministerio de Salud (2001) Plan Nacional de Salud Mental y Psiquiatría. Segunda Edición, Santiago
Vakis R, Rigoli J, Lucchetti L (2015) Left behind: chronic poverty in Latin America and the Caribbean, overview. World Bank, Washington, DC
Valenzuela MT, Pastorino M, Alvarado R, Villalón M, Durán E, Hirmas M, Venegas J (2012) Diseño y evaluación de proceso de una intervención comunitaria para la detección precoz del primer episodio de psicosis en Chile. Rev Chil Salud Pública 16(2):123–130
Vicente B, Rioseco P, Vielma M, Uribe M, Boggiano G, Torres S (1992) Prevalencia de vida de algunos trastornos psiquiátricos en la provincia de Concepción. Rev Psiquiatr 9:1050–1060
Vicente B, Rioseco P, Vielma M, Uribe M, Boggiano G, Torres S (1994) Trastornos psiquiátricos en diez comunas de Santiago: prevalencia de seis meses. Rev Psiquiatr 11:194–202
Vicente B, Rioseco P, Vielma M, Boggiano G, Silva L, Saldivia S (2000) Prevalencia de vida y seis meses de algunos trastornos psiquiátricos en la provincia de Cautín – Chile. Rev Psiquiatr 12:75–86
Vicente B, Kohn R, Rioseco P, Saldivia S, Levav I, Torres S (2006) Lifetime and 12-month prevalence of DSM-III-R disorders in the Chile psychiatric prevalence study. Am J Psychiatry 163:1362–1370
Weinstein L (1978) De salud y autogestión. España, Madrid
World Health Organization (2014) Global status report on violence prevention 2014. World Health Organization, Washington, DC, 2016

Part V

Urban Mental Health: Solutions

Homelessness and Challenges of Community Care

16

Sam Tsemberis and Eric Macnaughton

Abstract

This chapter describes the Pathways Housing First (PHF) model. This program pioneered the practice of offering choices to individuals with co-occurring psychiatric and addiction diagnoses including the option to move directly from homelessness into a place of their own with supports. Pathways offered housing as a right, not something that must be earned through participating in psychiatric treatment or attaining a period of sobriety. Intensive treatment and support services are provided using community-based multidisciplinary teams that are mobile and make home visits. The program's philosophy is guided by principles of self-determination and empowerment, and interventions are primarily directed by the service user. Participants are not discharged from the program if they relapse or if they are evicted from their housing; the support team services are continuous and assist with rehousing. The Pathways Housing First (PHF) program represents a 180-degree turn from traditional supportive housing or other group home programs serving this population which typically use a treatment-first approach and require participation in psychiatric treatment and a period of sobriety as a prerequisites for obtaining and maintaining housing. Research studies are presented comparing Housing First to Treatment First programs on measures of housing stability, quality of life, and cost-effectiveness. Results of these studies built an evidence base that influenced policy makers to advocate for implementing the Pathways Housing First model to address homelessness for

S. Tsemberis (✉)
Pathways Housing First Institute and Department of Psychiatry, Columbia University Medical Center, New York, NY, USA
e-mail: sam@pathwaysnational.org

E. Macnaughton
Department of Psychology, Wilfrid Laurier University, Waterloo, ON, Canada
e-mail: emacnaug@telus.net

N. Okkels et al. (eds.), *Mental Health and Illness in the City*, Mental Health and Illness Worldwide, https://doi.org/10.1007/978-981-10-2327-9_6

individuals with complex needs which has led to the programs' dissemination across the United States, Canada, and Europe.

Keywords
Housing First • Homelessness • Mental illness • Client-driven services • Harm reduction • Supported housing • Recovery

Contents

Introduction

Chronic Homelessness as an Urban Problem

Over the past 30 years, homelessness has emerged as a growing and vexing public health problem. It has been linked to many causes, such as gentrification and rising rents, deinstitutionalization, and the gradual erosion of the social safety net. It remains primarily an urban phenomenon as busy downtown commercial areas of a city offer the opportunity to remain unseen in plain sight and therefore less likely to draw unwanted attention from police or others who may mean harm. Some pedestrians may simply walk by with an air of important direction and not look at the person who is homeless and sitting still. Some linger and offer a sympathetic look or coins. The very existence of so many others provides safety because as Jane Jacobs pointed out long ago, busy city streets "have eyes," and for those who are homeless this may mean a decreased likelihood to be victimized because of this informal collegial policing (Jacobs 1992). Added to the attraction of anonymity and relative safety offered by cites is the concentration of services typically offered within these urban centers.

As the influence of capitalism and its practices continues to grow increasing income disparity, poverty, and the number of people who are homeless, the resulting structural economic conditions have also permeated smaller cities, towns, suburbs, and rural areas. The disparity between earned income, housing costs, and the growing lack of affordable housing creates pressure for all, and those least able to compete in the market place end up homeless. People on the lower rungs of socioeconomic ladder, especially those relying on a fixed government benefit or entitlement, are among the first to be priced out of the real estate rental market. Public or social housing has long ago stopped keeping up with the increasing demand.

Thus for individuals with severe psychiatric disabilities who rely on government benefits and who have tenuous economic opportunities, rising rents have resulted in housing loss and a disproportionately larger increase in the prevalence of homelessness. While many individuals who become homeless because of an economic or personal crisis typically exit homelessness on average in about a month, individuals with mental illness or severe addiction are at risk for remaining homeless for months or years (Culhane et al. 2002). While this group represents a relatively small percentage of the homeless population, it is the group that is most visible and typically represents homelessness in the public's mind. And too often the public's attribution of root causes of homelessness tends to blame the victims as lazy or unmotivated to seek treatment rather than consider the economic systems' contribution to the problem (DeLio 2016).

With the increased numbers of high visibility of street-dwelling individuals, a consensus emerged that there is a need to take action. Over the years cities have adopted a variety of policy directives and strategies to address chronic homelessness, and most have met with little success. Some mayors championed a "quality of life approach" which essentially increased enforcement of laws that disproportionately target the homeless (e.g., public intoxication, vagrancy, petty theft, etc.). In many instances this approach simply moved the homeless out of public view and into jails. A more constructive approach has been the development of supportive housing specifically built to serve this population. These service-rich projects are costly to build, take years to complete, and each serves a small number of individuals relative to the magnitude of the problem. Once completed, these supportive housing programs instituted admission criteria that required participation in psychiatric treatment and sobriety, which essentially disqualifies those who have remained chronically homeless. Taken together, the results of these efforts led to a growing consensus that a new and effective intervention was needed to address chronic homelessness, especially one that can effectively meet the needs of the chronically homeless, especially those with co-occurring diagnosis.

Housing First: The Right Intervention at the Right Time

The Pathways Housing First (PHF) program, which began 25 years ago, has established a remarkable clinical and research record of successfully housing and supporting individuals experiencing long periods of homelessness who have been diagnosed with co-occurring mental health and addiction issues ((US)SAMHSA, NREPP 2007). The initial target population of PHF was the group that had been labeled "hard to house" or "treatment resistant" by traditional supportive housing providers. PHF purposefully sought out and enrolled individuals that were rejected by other providers, those who had the most severe mental health, addiction, and health problems, those with histories of difficult behavior, incarceration, and those who were isolated or ostracized. The more strikes stressors and challenges in their profile, the more likely they would go to the top of the admission list. PHF sought out the quintessential "homeless person" – the solitary figure we see on our city streets, dressed in multiple tattered layers, carrying his belongings in worn plastic bags, unwell, vulnerable, and all too visible. This group represents a small fraction, between 15% and 20% of the homeless population, but is the one that is most vividly represented in the public mind and the focus of public outcry and public policy. This cohort, referred to by providers and policy makers as the "chronically homeless," was once thought of as an intractable societal, clinical, legal, and policy problem. Today, a growing number of cities are declaring they have effectively ended chronic homelessness by using the Housing First Program (Carrier 2015). In this chapter we describe the development of the PHF model, the context from which it emerged, its philosophical and practice principles, evidence for its effectiveness, and reasons for its dissemination across the United States, Canada, Europe, Australia, and New Zealand.

Homelessness, Mental Illness, and Poverty

Over the past 35 years, homelessness has almost become a permanent and problematic fixture of city life. In the United Sates, this homelessness "crisis," the largest since the great depression of the 1930s and certainly the most prolonged, began in the early 1980s. It is a direct result of the Reagan administration's neoliberal policies that significantly reduced federal government spending on social programs and programs for the poor. Specifically, the US Department of Housing and Urban Development (HUD), the agency responsible for the development of public housing in the United States, was drastically reduced. Production of public housing was cut from 350,000 a year to below 50,000 units a year where it remains today (Dolbeare and Crowley 2002; Crowley 2016). Over the past three decades, the United States went from having a surplus of rental housing to a nationwide shortage estimated at seven million units. It is not simply that the federal government virtually eliminated subsidized housing. Reagan Era policies, still in effect today, reinforced the nation's long-standing idealization of home ownership through federal tax policies that provide an estimated annual $233 billion in tax relief to wealthy homeowners who can take advantage of itemized tax deductions (Crowley 2016). Over this same period, these economic policies and the corresponding values of individualism over collectivism have created greater income disparity and contributed to increased homelessness (Toro et al. 2007). As the influence of capitalism and neoliberalism spread north to Canada and then east to Europe, 10 years after its emergence in the United States, homelessness had emerged a growing social problem in most Western countries.

The definition of who is homeless and who is chronically homeless varies a great deal across these nations. The United States has one of the narrowest definitions for chronic homelessness: "chronically homeless" is "a unaccompanied homeless individual with a disabling condition who has either been continuously homeless for a year or more, or has had at least four episodes of homelessness in the past three years" (USICH 2013). Norway, perhaps at the other end of the continuum, has one of the broadest definitions of homelessness: a person is considered homeless if he or she does not have access to owned or rented accommodation, is left to seek casual or temporary accommodation, is staying temporarily with close relatives, friends, or acquaintances, or someone who is in the custody of the Norwegian Correctional Service or in an institution and is due to be released or discharged within 2 months and does not have a home. A person without accommodation for the coming night would also be classed as homeless" (Norwegian Institute for Urban and Regional Research). This definition allows Norway to intervene and provide services to a broad range of individuals and essentially prevent literal homelessness before it occurs.

Chronic Homelessness and Service Utilization

Individuals diagnosed with severe mental illness are disproportionately represented among the chronically homeless, and most report co-occurring addictions and acute and chronic health conditions that add considerably to their vulnerability (Bharel et al. 2013). Numerous clinical comorbidities are associated with significant utilization of emergency and acute care treatment services such as psychiatric and general emergency rooms, detox, inpatient psychiatric services, as well as overrepresentation in interactions with the criminal justice system – arrests, court appearances, and incarceration – all of which result in extremely high treatment and public service utilization costs (Hopper et al. 1997). This is costly for both the individual and the city. While those who remain chronically homeless comprise 10–20% of the total homeless population, they have been found to utilize as much as 50% of the shelter system's resources (Culhane et al. 2002).

In a recent study of service utilization among the homeless by the Health Care for the Homeless Program in Boston, Bharel et al. (2013) reported that in a sample of 6,494 patients, more than two-thirds had some form of mental illness and about 60% had substance use disorders; almost half (48%) had a co-occurring diagnosis. Chronic diseases were also prevalent including hypertension (37%), COPD or asthma (26%), diabetes mellitus (18%), and a high prevalence of infectious disease (23%) and HIV (6%). Overall, the disease burden of this sample was 3.8 times higher than the general population. The cost of providing healthcare was $2,036 (all dollar figures in US currency) per person per month compared to $568 for the general population. Almost half of the total annual program expenditures ($48M of a total

$101M) were incurred by 10% ($N = 653$) of the study population, mainly for hospitalizations (40%) and emergency department visits (11%). Healthcare costs for the highest 10% were, on average, $73,000 per person per year. It is not surprising then that the chronically homeless with high service utilization is drawing a great deal of attention. These comorbidities also tend to impede recovery and prolong time spent homeless. Health problems, mental illnesses, and substance abuses are exacerbated, and the exposure to trauma, malnutrition, and infection is greatly increased, making the causal relationship between illness and homelessness complex.

Poverty as a Disabling Condition

And then there is the disability of poverty. Those with severe clinical problems who are homeless are typically unemployed, without any source of benefit or entitlement, and rely exclusively on social services for food, clothing, and shelter. Without a source of income, obtaining and maintaining housing and not an option. In countries with a social safety net, a sizeable percentage among the homeless may be receiving a monthly government-issued benefit for those with disabling conditions that preclude the possibility of earning a living through employment. But even those who receive benefits cannot afford a place to live. Two years ago, the US national average rent for a modest one-bedroom apartment was $780, which was equal to 104% of the national average monthly income for an individual receiving benefits that same year (TAC 2014). With a monthly income insufficient for rent, let alone the requisite additional month's rent as a security or damage deposit, individuals with disabilities who are homeless, or become homeless, are essentially "priced out" of the rental housing market. In countries, such as Canada or in northern Europe, there is a recognition that the disability benefit may be insufficient and the government provides individuals receiving a disability income with an additional monthly rent subsidy around $350 per month.

Some researchers and advocates for the homeless regard poverty, not clinical problems, as the root cause of homelessness (Burt 2016; Crowley 2016). Burt argues the "*sine qua non* of homelessness, its root cause and the stage on which all else plays out, is extreme poverty" (Burt 2016, p. 49). This position is supported by the fact that there are millions of individuals diagnosed with severe mental illness, addiction, and health problems who are housed and who are never homeless. The difference between the homeless and the housed cannot be explained by their clinical or diagnostic status alone. The difference between the homeless and the housed is that those who are housed have the financial means and support they need not to be homeless.

The centrality of the role of poverty in the lives of those who are homeless may elude housing providers whose programs demand that applicants first treat their clinical problems as a precondition for housing. However, for people experiencing homelessness, clinical problems are often not their priority. Their everyday hour by hour anxiety concerns their hunger and their safety and other matters of survival and privations imposed by poverty.

Beliefs and assumptions concerning the root causes of homelessness inform program service philosophy and practice, determine the services it will provide, and influence the attitudes and practices of its staff (Henwood et al. 2013). The decision about which root causes to address, economic or clinical, has resulted in the development of two fundamentally different systems of care: the *Staircase Model*, or treatment first and then housing, and the *Housing First model*, or housing first, followed by treatment.

The Staircase Model: Treatment First and Then Housing

The Staircase Model or the "linear residential continuum of care" (Ridgway and Zipple 1990) has been and continues to be the predominant national system-wide treatment-then-housing approach to homelessness in the United States, as well as in most other Western countries. Developed in the early 1980s when the number of homeless individuals with multiple disabling conditions was increasing, this model was not based on empirical evidence but instead on an emergency response to provide care based on the existing and long-standing traditions of the mental health and addiction treatment systems. Essentially, the approach is that for a person who is both homeless and has mental health and addiction issues, these conditions must be treated first to prepare the individual for housing (Fig. 1).

In the Staircase Model, housing is offered only after he or she successfully "graduates" through a series of program steps typically beginning with low-demand outreach or drop-in, followed by compliance with treatment, sobriety, and program rules, and finally reaching shelters and transitional housing and ending in permanent housing. Program participants are routinely taught "housing readiness" skills in group settings with the expectation that this will lead to successful independent living. However, research of psychiatric rehabilitation has shown that the place-based learning is much more effective, and thus the optimal setting for learning how to live independently is to live on one's own with the right supports (Anthony and Blanch 1989). At the same time, accumulating research on addiction (Szalavitz 2016) suggests that punitive approaches to relapse involving withdrawal of housing are antithetical to emerging theories of addiction and emotion-based learning.

The length of stay in Staircase programs can range from a few months to a few years with relapses requiring a fresh start from the bottom stair. For individuals with complex problems, the Staircase Model proves to be very steep, can take years to climb, and has about a 40% chance of success (Atherton and Nicholls 2008; Goering et al. 2011). Reaching the top of the staircase means obtaining one's own housing and a having the service support needed to maintain a successful tenancy.

The treatment philosophy of the Staircase Model is based in part on the medical model and therapeutic community approach that is the basis of treatment in psychiatric inpatient services and drug rehabilitation programs. This approach operates on several assumptions including (1) clinicians are experts and guide treatment for

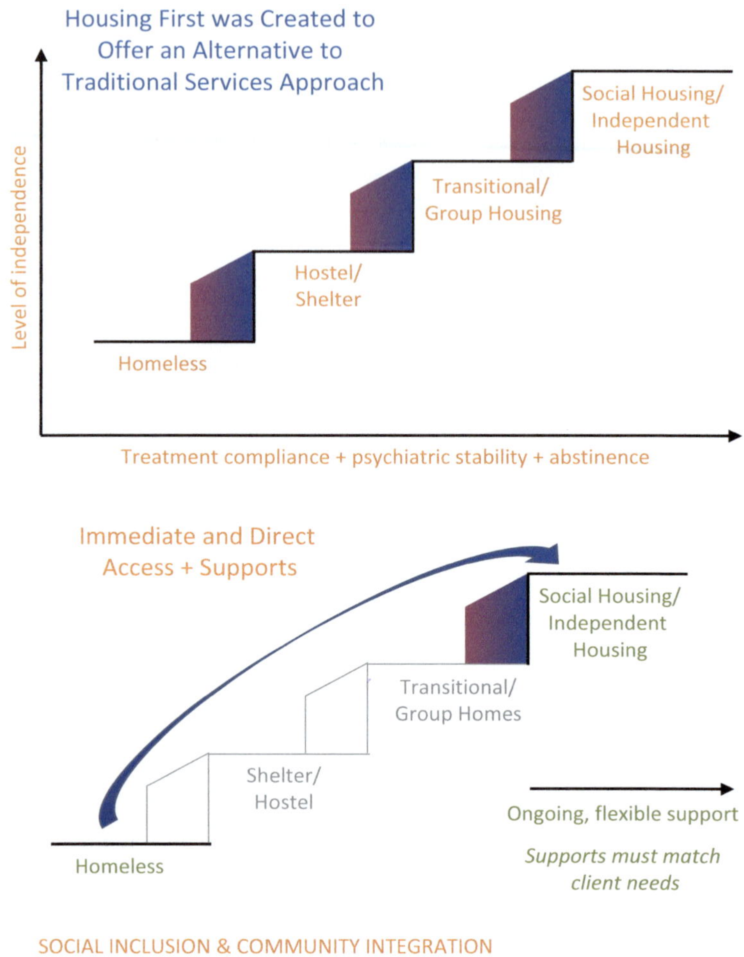

Fig. 1 Traditional staircase model vs. housing first

those in their care; the selection of the type of housing and support service; (2) individuals with severe psychiatric conditions require housing with around-the-clock supervision; (3) individuals with co-occurring mental illness and addiction diagnoses must first attain sobriety before they can make sound judgments, manage their own housing, or be trusted with other responsibilities; and (4) these individuals must achieve a period of residential stability in supervised settings and demonstrate they have learned independent living skills before they can be offered a place of their own.

In addition to these "therapeutic" assumptions of the Staircase Model, there are other implicit, unspoken but deeply held biases about those who are poor. These stem from centuries of Western or Anglo-Saxon attitudes dating back centuries to paupers' prisons and other punitive measures designed to punish the poor. These attitudes, over time, have become part of the fabric of social norms that are still very much in evidence today. DiLeo points out that in spite of what social scientists, economists, and clinicians tell us about the complex causes of homelessness, the general public regard those who are homeless as people who have somehow brought their poverty upon themselves by drinking too much, being lazy, making poor choices, or refusing societal help. DiLeo (2016) concludes, "we blame – more often than we support – people with needs" and "we seem to prefer judgment to discernment and tough love to support and meaningful institutional change" (p. 143). In fact, "tough love" is a core theme that guides the provision of most model programs.

The Origins and Mission of the Pathways Housing First Program

The Pathways Housing First (PHF) model emerged from an ongoing dialogue among service user, program staff, and researchers who had developed outreach and drop-in center programs including a National Institute of Mental Health research demonstration project called "Choices Unlimited" (Shern et al. 2000; Tsemberis et al. 2003). An ethos of respect for service user and service user's wishes was a core principle, and this was fostered by recruiting staff members that had social justice values and providing them training in service user-centric approaches such as psychiatric rehabilitation (Anthony et al. 1982), harm reduction (Marlatt 1998), and trauma-informed care (Mate 2009). A collaborative and democratic program culture was developed and nurtured by hiring several staff members that were service user themselves. Howie the Harp, an influential service user and advocate (Harp 1990), was an early advisor and brought to the program a commitment for the inclusion of service user voice ("nothing about us without us") and revolutionary fervor to changing the mental health system (Tsemberis et al. 2003).

At Choices Unlimited, neither status nor salary distinguished service user from non-service user staff; this approach reduced the power differential and artificial boundaries between staff and program participants and created a culture of collaboration focused on solving the main problem of homelessness: access to housing. In their previous efforts to secure housing, staff and service users had encountered frustration with the Staircase Model's requirements for medication, treatment compliance, and months of sobriety. They had witnessed how the need for housing was used to leverage service user acquiescence to unwanted treatment and abstinence requirements (Allen 1996).

After repeated failures to secure housing for service users through the Stairway approach, the group at the Center began a series of trial-and-error efforts to design a viable supported housing program that would be desirable to service users and worked collaboratively to develop the operational details of finding apartments, deciding who would sign the lease, policies about occupancy, the parameters of authority for the support services, and defining both program and service user fiscal responsibilities. The scattered site, independent apartment model where apartments are rented from community landlords, as described by Ridgway and Zipple (1990), was a model that met service users' requirements for normal housing, privacy, tenancy rights, and a funding mechanism that left sufficient discretionary funds after 30% of SSI was paid for rent.

The Pathways Housing First Program, or Housing First as this model is now called, originated as a service user-driven intervention and sustained its service user-centered character by attending to three lessons learned in the process of its development: (a) the importance of including service users at every level of the program and the agency, from service providers, managers, to members of the board, (b) the need for a variety of formal and informal meetings where staff and program participants have opportunities to meet informally and expand the repertoire of their dialogue, and (c) the use of accommodations to ensure full participation of service users across these contexts.

The PHF program is consistent with Rappaport's (1987) notion of empowerment. Empowerment refers to a program philosophy and practice that is intended to enhance the degree of decision-making and control that vulnerable or disenfranchised individuals exercise over their lives. Prilleltensky (1994) identifies three principal values that are empowerment's foundation, including self-determination, i.e., being able to choose a course of action; distributive justice, i.e., access to fair and equitable distribution of resources (in this case affordable housing); and democratic participation, i.e., sharing in the decision-making process (p. 360). These values are consistent with the larger aim of PHF which is not only to provide an effective approach to housing a vulnerable population but also to fundamentally change the way mental health systems and social services respond and treat those who are homeless and have co-occurring diagnoses.

Program Operations

The collaborative and democratic process used to design the PHF program may have assured it would be a desirable option and aligned with the target populations priorities and preferences. First and foremost, PHF offered immediate access to an apartment, including payment of the security deposit, furniture, and an ongoing monthly rent subsidy. By offering service users an apartment of their own at the point of admission into the program, PHF fused together several of the Staircase stairs – outreach, drop-in, shelter, transitional, and permanent housing – into a single powerful and desirable invitation (Tsemberis et al. 2003). Service users who had been through the system time and again immediately appreciated the enormity of what was being offered. They understood that the programs' financial support (rent supplement) meant they could actually rent a place of their own. In the context of addressing root causes of homelessness, PHF certainly aligned its approach with one that is consistent with the practice of distributive justice and addresses poverty first.

The program's housing component in the United States consisted of independent apartments, rented from community landlords, in regular community-based buildings, in which the majority of other tenants were members of that community. The program did not rent more than 20% of the units in any one building to ensure integrated housing in normal settings. The housing was affordable and located in a neighborhood of their choice. There were two housing requirements: (a) meet the terms and conditions of a standard lease and (b) agree to a regular home visit from the support services team.

Support services were modeled after the community-based assertive mobile treatment teams that originated in Madison, WI (Stein and Test 1980). The PHF program initially used a hybrid between Assertive Community Treatment (ACT) and Intensive Case Management (ICM) to effectively address the multiple needs of its service users. The teams made weekly visits but could be increased with necessity and were available, on call, 7 days a week, 24 h a day. People with lived experience were hired as members of the support team. The goal of the support services was to provide assistance to maintain housing and address other problems in the sequence, frequency, and intensity determined by the service user. Individuals were not evicted from apartments if they relapsed or needed hospitalization. They would be assisted to obtain treatment and, upon completion, return home. A harm-reduction approach was used to address problems of addiction, mental illness, health, troubling relationships, apartment maintenance, budgeting, and other domains in life where improvements are often seen by taking small incremental steps toward a larger goal.

PHF combined these philosophical approaches and program components into a cohesive operational program and began to offer individuals who were chronically homeless immediate access to an apartment of their own and the support services to make their housing successful (Tsemberis 2010b).

PHF Program Practice Principles

There are five major principles that govern the philosophy and practice of PHF programs: (1) service user choice (self-determination), (2) separation of housing and services (building trust and continuity of care), (3) service array (matching services to needs), (4) recovery-oriented services (hopeful and affirming service approach), and (5) program structure (assuring care and support are provided efficiently and effectively) (Stefancic et al. 2013b). These principles form the philosophical, structural, and operational framework on which the program operates, and each dimension has been defined by a small set of measurable and observable items which make up the PHF Program Fidelity Scale (Stefancic et al. 2013b).

Service User Choice

Service users actively participate in the selection of neighborhood, housing type, apartment, furniture, who they live with, and other lifestyle decisions. There are limits to service user choice; service users must adhere to terms and conditions of a standard lease and meet with the program staff for a home visit. In cases of emergency or when an individual presents a danger to himself or others, he is subject to the application of the local mental health law.

Separation of Housing and Services

The program uses a social inclusion or community integration housing approach and does not rent more than 20% of the units in any one building. Services are not located in the building; they are separate and off-site. The model is also consistent with Americans with Disabilities Act and the Supreme Court's Olmstead Decision that mandates that people with psychiatric disabilities have the right to live in the least restrictive settings, with access to the same social, cultural, educational, and employment opportunities as the rest of us. Separating housing and services allows for continuity of support and treatment through crisis, builds trust, and provides continuity of care. Some service users are evicted or leave their first apartment, but the program does not discharge them. The team continues to work with the service user through this crisis and assists the service user in securing another apartment.

Service Philosophy: A Recovery Orientation

A recovery orientation is the cornerstone of the PHF services approach. Team members convey hope, offer choice, and use shared decisions in their day to day interactions (Deegan 2007). Service plans are derived jointly and are an open, accepting, and nonjudgmental atmosphere where service users are invited to express their hopes, worries, and needs. The team uses a harm-reduction approach that creates the conditions that foster open and honest discussions about drug or alcohol issues without the fear of negative judgment or eviction.

Matching Service Array to Service User Needs

The service support team anticipates the intensity and diversity of needs of the program participants. The support services teams utilized by PHF programs are ACT (or ACT-like) for high-need service users (1:10 ratio of staff to service users on the caseload) and ICM (or ICM-like) for service users with moderate needs (1:15 or 1:20 caseload ratio). In instances where the team cannot provide the service directly, a community provider is found. The model works effectively in urban, suburban, and rural settings (Stefancic et al. 2013a). The program operates on an individualized, person-specific approach and makes rapid adjustments to the frequency and intensity of services that are needed and anticipated. Services are provided for as long as they are needed by the client.

Program Structure

This dimension refers to program organization and operation. It defines the extent to which the PHF program is operating an efficient and effective housing and services program. The program dimensions for this factor include efficiency of the team's internal operations, time taken from admission to housing, number of crises or evictions, and housing stability rates for participants. It also examines staffing patterns and administrative support for the program.

The PHF Program Fidelity Scale

In the past decade, the PHF program has spread rapidly throughout the United States, Canada, and the EU. For new programs to achieve the 75–85% housing stability rate that the program can deliver, it is essential to replicate it with a high degree of fidelity. To facilitate high-quality replication and assure consistency across PHF programs in different cities and countries, there is a manual that provides a comprehensive description of the program (Tsemberis 2010a) and a PHF Program Fidelity Scale that serves as a checklist to ensure that operations are properly aligned (Stefancic et al. 2013b).

While fidelity scales also exist for other services and housing models with which PHF has significant overlap, e.g., Dartmouth Assertive Community Treatment Scale (DACTS) (Teague et al. 1998), T-MACT (Monroe-DeVita et al. 2011), and IPS Supported Employment (Drake et al. 2013), there are important distinctions as to precisely which subsets of items are shared as well as which ones are novel to PHF and provide a unique constellation of ingredients.

The PHF fidelity scale provides specific observable and measurable items for the five key domains described above (Stefancic et al. 2013b). Since its development, the PHF fidelity scale has been used to evaluate hundreds of programs and found to be a useful quality assurance instrument, providing clear and measurable guidelines for the program operations, and it has served as a research tool to examine the relationship between program fidelity and program outcomes.

In measuring the items relating to the five core dimensions, the scale also examines the appropriateness of the target population served, staffing patterns used to provide services, admission practices, team operations, time taken to provide housing, level of support provided, the use of harm-reduction practice, integrated dual diagnosis treatment, motivational interviewing, and much more. There are 38 items in all. The main features of the model are summarized in a Summary Box (see Fig. 2).

A full fidelity assessment takes several staff members a full day to complete and includes input from multiple sources, including observing the team in action during their usual meeting, conducting individual interviews with program staff, reading a randomly selected sample of charts, conducting a focus group with program participants, and examining the program's staffing and performance data.

Research Evidence on PHF Housing Stability

From the start, the PHF program carefully studied the effectiveness of its intervention. Beginning with simply tabulating housing retention (how many people stayed housed during the first year?), the first published report on the program reported an

Housing Choice & Structure	Separation of Housing & Support	Service Philosophy	Service Array (ACT: directly ICM: through brokerage)	Program Structure
- choice - availability - permanence - affordability - integrated - privacy	- no housing readiness - no program contingencies - tenant agreement - commitment to rehouse - service continuity through housing loss - off-site mobile support	- service choice - no treatment requirements - harm reduction - goal-setting spanning wide array of life domains - motivational interviewing - non coercive engagement	- housing support - nursing - psychiatry - addictions - supported employment - peer support - social integration - 24-hour coverage	- prioritize housing instability - frequent contact with participants - low staff/participant ratio - team approach - staff meet frequently - peer representation in program

Fig. 2 Critical ingredients of housing first

84% housing retention rate after 12 months (Tsemberis 1999). Five years later, the program had grown in size, and a larger sample was available to conduct a more rigorous and robust comparative study in collaboration with the Nathan Kline Institute and New York City's Human Resources Administration. This agency had compiled residential outcome data for several thousand participants residing in the newly established NY-NY programs (part of the Staircase Model). A survival analysis was used to compare archival housing tenure data for the NY-NY's programs and the PHF program over the same 5-year period while controlling for differences in client characteristics before program entry (Tsemberis and Eisenberg 2000). Results indicated that 88% of PHF service users had remained housed as compared to 47% of program participants in the traditional Staircase programs.

In 1997, 5 years after its inception, Pathways to Housing was one of six agencies to be awarded a competitive $1.3M federal government grant from the Substance Abuse and Mental Health Services Administration (SAMHSA) to evaluate its PHF program. In collaboration with Professor Beth Shinn from New York University, Department of Community Psychology, who served as the independent evaluator for the study, Pathways began to conduct a 4-year randomized control trial known as the "New York Housing Study." The study examined the longitudinal effects of the PHF program for street-dwelling individuals with a diagnosis of severe mental illness, most with substance use (current use or history of use), on measures of service user choice, housing stability, substance use, treatment utilization, and psychiatric symptoms. Two hundred twenty-five participants were recruited and randomly assigned to receive housing contingent on treatment and sobriety (Staircase control) or to receive immediate housing without sobriety or treatment prerequisites (PHF experimental). Interviews were conducted every 6 months for 24 months. The follow-up rate for the entire sample was 94% at 12 months and 87% at study's end (Stefancic et al. 2004). The experimental group obtained housing earlier and remained stably housed 80% of the time compared to 30% for the control group. PHF participants also reported higher perceived freedom of choice. In other domains, the results were positive but on a smaller scale. Utilization of substance abuse treatment was significantly higher for the control group, but no differences were found in substance use or psychiatric symptoms. Participants in the PHF program were able to obtain and maintain independent housing even while coping with psychiatric and substance abuse symptoms (Tsemberis et al. 2004).

This study had a significant positive impact on establishing the credibility of the PHF model. The high score on residential stability served to validate the reasonableness of the PHF approach: service user choice, harm reduction, and acceptance of individuals with dual diagnoses as decision-makers capable of managing in community settings with support. Furthermore, the study addressed criticism about "creaming" (i.e., only making service available to individuals with lower support needs), because the clinical and demographic characteristics for the individuals in the two arms of the study were essentially the same.

One of the clinical studies using the New York Housing Study data also produced interesting and unexpected findings regarding the relationship between service user choice and psychiatric symptoms. There was an inverse relationship between degree of choice and psychiatric symptoms such that more choice was inversely correlated with symptoms. This relationship was partially mediated by mastery. These findings supported the practices of service user choice and self-determination, a fundamental principle of the program (Greenwood et al. 2005).

Two years later, the program had been replicated in several cities across the United States, and there were other programs based on the PHF in operation. There was some variation in the way that agencies either adopted or adapted the PHF model, some agencies essentially replicating the PHF model and others using many of the elements of the model but using modifications to fit local contexts. Most programs were essentially Housing First, if not exactly the PHF model. At this point the US Department of Housing and Urban Development (HUD), the federal agency funding most homeless programs, was interested in conducting its own evaluation of this program's effectiveness. Agencies operating Housing First programs in three different cities were selected for study (Pathways in NY was one of the sites). The researcher team was contracted directly by HUD which used a combination of data collection sources for the evaluation including interviews with key personnel and chart data to determine 12-month residential stability outcomes. These independent investigators reported 84% housing retention rate for 12 months across the three programs (Pearson et al. 2009).

Using a similar chart review methodology, another study evaluated the effectiveness of PHF program with a population of long-term shelters users. This was an extension of the program's initial engagement with a homeless population recruited primarily from the streets and drop-in centers. The results showed statistically significant rates of housing stability using the PHF approach. In that study, approximately 78% of PHF participants remained stably housed over a 4-year period (Stefancic and Tsemberis 2007). In subsequent studies, the PHF program was found to be effective for individuals released from jail and with severe addiction diagnoses (Appel et al. 2012).

In 2004, the United States Interagency Council on Homeless had generated national interest in ending "chronic homelessness" and strongly endorsed the PHF model. The Council coordinated a multi-sector $35M initiative to implement HF programs to end chronic homelessness across the United States. With the participation of four federal agencies – Housing (HUD), Health (HHS), Mental Health and Addiction (SAMHSA), and the Veterans Administration (VA) – the agency launched the national Initiative to Help End Chronic Homelessness supporting a total of 11 new HF demonstration projects. Seven of the 11 cities funded by the initiative used the PHF model and achieved 85% housing retention rates after 12 months (Mares and Rosenheck 2007).

Research Evidence on PHF Housing Stability

In 2009, the Canadian federal government approved a $110M grant for the Mental Health Commission of Canada to implement an intervention that would effectively address homelessness for individuals with mental health problems. The Commission selected PHF as the intervention and launched a 5-year longitudinal, randomized control trial to test PHF's effectiveness compared to the treatment as usual (essentially the Staircase Model) in five Canadian cities. This large-scale research demonstration project was called At Home/Chez Soi (AH/CS), in which 2,148 individuals across five cities (1,158 in the PHF group) were enrolled in programs and participated in the study. Data collection consisted of structured interviews administered every 3 months for 2 years measuring housing stability, service utilization, quality of life, psychiatric symptoms, addiction severity, and social functioning.

Except for one site, participants were randomized into one of three "arms" of the intervention across the five cities. Two "arms" were HF, one providing ACT services for those with severe needs and the other providing ICM services for those with moderate needs. At the 2 year follow-up, HF participants receiving ACT services spent more time in stable housing than participants in treatment as usual (71% vs. 29%); HF participants entered housing more quickly (74 vs. 220 days) and rated the quality of their housing more positively (Aubry et al. 2016). A similar pattern of results for residential stability was obtained for the ICM group (Stergiopoulos et al. 2015).

The implementation of the HF programs through the AH/CS demonstration project represented a paradigm shift in the Canadian context for mental health and supported housing services. Prior to AH/CS, many programs serving individuals who were homeless provided social housing or single-site supportive housing for those who were homeless and diagnosed with mental health problems. As a result of the study, the homeless service system added the option of including market rate rental housing rented from community landlords to the portfolio of options. Some social housing providers changed their waiting list practices and offered "Housing First" by allowing for a small percentage of units to be set aside for immediate occupancy by those who were homeless and engaged with the programs' support services.

Paula Goering and her colleagues examined the relationship between program fidelity and program outcomes and reported that high-fidelity programs were associated with higher social functioning, greater increases in quality of life, and spending more time in stable housing. High-fidelity programs were also associated with increased staff time spent with service users (Goering et al. 2016).

Policy makers in Canada took note of the evidence and decided to change policy and reallocate funding for homeless services in Canada. In one of the most far-reaching policy and funding shifts to date, the federal government of Canada decided that beginning in 2014, all 62 communities in Canada receiving federal funding for homeless services would now be required to spend approximately 50% (between 40% and 65% depending on population) of their federal funds on Housing First programs (Macnaughton et al. in press).

Another large-scale implementation of the PHF model involved the US Veterans Administration (USICH 2013). In 2009, the US Department of Veterans Affairs (VA) and the US Department of Housing and Urban Development (HUD) collaborated on a unique and bold initiative to address homelessness for 76,329 veterans identified by the annual HUD Point-in-Time (PIT) count. HUD supplied the necessary rent supplements using Section 8 Housing Choice Vouchers, and the VA provided case management and supportive services through its healthcare system. This joint effort, known as the HUD-VASH initiative, was soon launched, and the program ran into difficulties. The VA Medical Centers were using the Staircase Model and failing to engage and house the veterans that were homeless. In 2011, the National Center on Homelessness Among Veterans (NCHAV) began to introduce the PHF model and evaluated the effectiveness of the HF compared to the Stairway Model for a cohort of veterans ($N = 177$). The Center reported that using the HF approach reduced time to housing placement from 223 to 35 days, showed significantly higher rates of housing retention, and reduced use of emergency rooms (Montgomery et al. 2013). Based on these results, the authors recommended a national adoption of HF by the VA.

Based on these positive findings, the NCHAV selected 14 large VA Medical Centers to participate in a HF pilot implementation of HF and provided them with 50–75 additional HUD-VASH vouchers (for a total of 700 vouchers nationally). It also provided financial support for a social worker, a nurse case manager, a peer specialist, and a part-time prescriber (e.g., physician assistant, nurse practitioner, or psychiatrist). To support the adoption of the model, staff from NCHAV and Pathways to Housing provided training and technical assistance on the PHF model, including monthly conference calls, site visits, and assessment of each program's fidelity to the HF model.

Approximately 700 veterans across 14 centers were enrolled in this pilot. The overwhelming majority were single males (91%) of which 48% were divorced or separated. The majority of veterans served in the Vietnam and post-Vietnam eras. Equal proportions of the Housing First participants were black and Caucasian and more than two-thirds were between the ages of 45 and 64 years. Approximately 94% of veterans met criteria for chronic homelessness (i.e., homeless continuously for a year or more or four or more episodes of homelessness in the past 3 years). These veterans also had significant disabilities: 78% presented with an active or prior diagnosis of a mental health or addiction. Results showed that 84% of ($N = 585$) veterans were still living in permanent housing, with varying lengths of stay 1 year after the pilot began. The use of inpatient and outpatient services by this cohort had been reduced by 54.3% and 31.7%, respectively (Kane and Tsemberis 2016).

Taken collectively, these studies show a remarkable consistency of program performance across different contexts, different investigators, and a variety of populations. *Overall, there is a housing stability in the range between 75% and 85% demonstrating that the PHF program can be successfully adapted to different contexts and for different populations while still maintaining program fidelity and high performance.* The results have the effect of generating more programs, and there is evidence, at least in the United States, of a small reduction in the number of chronically homeless individuals that were counted on the streets beginning circa 2010 (Burt 2016).

These results have additional implications. These positive outcomes are achieved by engaging, housing, and keeping housed the very individuals that Staircase Model programs have labeled "not housing ready" and "treatment resistant." These results bring to light the limitations and failures of the Staircase Model. Individuals who are homeless and have co-occurring diagnoses do not need treatment first in order to be successfully housed, unless, of course, they choose it.

There are also important lessons concerning clinical practice in these findings. The correlation between functional ability and severity of diagnosis or symptoms is not as strong as clinicians are trained to believe (Anthony 1993). Individuals who may be hearing voices or who have fixed delusions, or who are struggling with addiction, are also capable of managing an apartment of their own; this entails shopping for food, cooking, and conducting routine business and personal conversations without necessarily silencing their voices or giving up their belief systems. It is a similar phenomenon for individuals with severe addiction. In other words, complete remission of symptoms or sobriety is not essential as qualifications for being able to effectively manage a place to live, especially since the program provides support services and the person does not have to manage all of this alone. These principles are also incorporated into supported employment programs and other evidence-based practices. In all, these results demonstrate that individuals with severe disabling conditions are more capable of choice and self-determination than Stairway housing programs have traditionally assumed. This is something service users have been telling us for years (Harp 1990).

PHF, Recovery, and Psychological Well-Being

In addition to residential stability, does the PHF program have other benefits for the target population? A number of qualitative and mixed methods studies examined the relationship between housing, program support, and psychological well-being and recovery. Having a place of one's own can serve as a platform for a series of life improvements. In one study of 497 AH/CS participants randomly assigned to scatter site, single site, or treatment as usual (essentially the Staircase Model), Patterson et al. (2013) reported greater overall quality of life on the safety and living situation subscales for those who were housed. A study of qualitative narrative interviews (Macnaughton et al. 2016) conducted with a cohort of 195 AH/CS participants (HF, $n = 119$; TAU, $n = 76$) across the five Canadian cities affirmed the importance of housing and supports and also elucidated that once housing stability is achieved, other services (e.g., supported employment, education, and socialization) are needed to accelerate the transitions that participants strive to make in their lives.

In another qualitative study (Padgett 2007) examining the subjective meaning of "home" among 39 persons who participated in a Housing Study, the study design sampled approximately 20 individuals from Treatment First and Housing First programs. Researchers conducted two life history interviews with each participant. Findings revealed evidence of the markers of ontological security (security, routine, privacy, and positive identity) among participants living in their own apartments. The emergent theme of "what is next" (i.e., uncertainty about the future) points to the need to address problems of stigma and social exclusion that persist even after people are housed.

In a study examining the interaction of housing and community integration, Yanos et al. (2004) interviewed tenants in a scattered site apartment PHF program and those living in a single site congregate program. Both groups reported a better sense of safety that comes from being housed (single site or scatter site) compared to be being homeless; other positive impacts included "feeling normal" (being in the mainstream of human life) and an improved self-esteem. A private apartment, however, conferred a greater sense of privacy and control to establish one's own routine (including productive and creative activities), as well as control over social life. In a significant minority of respondents, there were challenges related to feeling lonely and isolated and with fitting in.

This richly textured narrative data obtained from qualitative and mixed methods studies offers an informative and clearheaded view of life after homelessness. Overall, this narrative data shows that respondents greatly preferred being housed to being homeless, and when housed, their overall quality of life is improved. Mental health, addiction, and discretionary income are all marginally improved once individuals are off the streets and into supported and subsidized housing. Even if they do continue to struggle with recovery in a clinical sense, i.e., with mental health and addictions, as Davidson and Roe (2007) observe, participants can still achieve recovery in a social sense, i.e., improved quality of life and better housing.

Without the constant concern for daily survival, newly housed tenants turn their attention to their present surroundings and to their future. Having a place of one's own offers a safe and secure environment to take stock and consider how to construct a way forward. There is an openness and a sense of control when facing the significant challenges ahead: loneliness, unemployment, thin resume, broken family ties, cravings and urges, and others. However, being housed also opens up the possibility of achieving new aspirations.

These studies also shed some light on the important question of motivation. The Staircase Model is built upon the assumption that programs must be designed to make participants work for their rewards, and this can have negative consequences if their efforts are deemed inadequate. The aim here is to make reaching the most desired goal – housing – difficult so that when the goal is achieved, it will feel deserved or valued. This perspective can also be seen as a tacit manifestation of long-held biases and bigotries toward the poor: those who live in poverty are viewed as lacking motivation or lazy and thus need to be pushed or guided by others to achieve success; it is their fault. In this framework, poverty is considered as lacking character, not simply as lacking cash.

Contrary to these biases, the results of qualitative interviews show an abundance of character and internal motivation. The results are congruent with Sen's notion of capabilities (Sen), suggesting that people value resources (i.e., housing) that enable the "beings and doings" they aspire to (reclaiming or building a sense of belonging and a positive sense of self). Having a home of their own, people are looking ahead, examining their lives, and considering how to address the obstacles they face. When they were in the valley of homelessness, the view up the mountain, their goal, was to obtain housing. Now, newly housed, the view from their new perch is changed and unfamiliar. The way forward is not as simple and direct as getting a place to live: it consists of multiple challenges and opportunities, and it involves setting priorities, making choices, and perhaps seeking support to stay the course. Their new goals may include seeking employment, reconnecting with family, attending to their health, managing symptoms or use, finding a relationship, or wanting to join a particular community. The motivation to achieve the next goal is mostly an internal drive; it is already present, but it may need to be awakened, nurtured, or supported. There is no need to design programs that make life for these individuals any more difficult than it already is. Programs based on the values of self-determination, distributive justice, and compassion are more effective.

Cost-Effectiveness and Cost Offset Studies

Numerous cost-effectiveness and cost offset studies of HF programs have consistently found that when the program targets frequent users of acute care services, there are considerable cost offset savings to be realized. As an example of the numerous cost-effectiveness studies of PHF conducted over the years (Tsemberis 2010b), consider the following study conducted by Larimer and her colleagues with a Seattle cohort of individuals with histories of long-term homelessness and diagnosed with severe addiction and mental health problems (Larimer et al. 2009). One hundred thirty-four individuals who were chronically homeless and diagnosed with severe alcohol addiction and co-occurring mental illness and health problems were entered into a HF program. Total medical and criminal justice costs for this cohort in the year prior to entering HF were $8.2 million. Twelve months after entering the HF program, their overall service costs had been reduced by 53%.

In a literature review, looking at the potential cost offsets of HF programs, Ly and Latimer (2015) conducted a systematic review of the literature between 2007 and

2014 from published and gray literature, examining impact on service cost associated with HF and net impact on overall costs. Twelve published studies (four randomized studies and eight quasi-experimental) and 22 unpublished studies were reviewed. Shelter and emergency department costs decreased with HF, while impacts on hospitalization and justice costs were mixed. Studies using a pre-post design reported a net decrease in overall costs, while experimental studies reported a net increase in overall costs with HF. The authors conclude that HF programs cannot be expected to pay for themselves, but there will be significant cost offsets. When benefits for participants are considered, HF represents a more efficient allocation of resources than usual care.

These well-documented cost offset studies have served to strengthen the case for PHF supporters but are especially important in winning new converts who struggled to accept some of the PHF values and practices, such as giving away apartments to people who are actively using drugs. Fiscally conservative individuals and ideologically conservative political leaders often find the cost savings evidence compelling. When Ted Clugston, the mayor of Medicine Hat (the first Canadian city to end chronic homelessness), was interviewed on CBC radio, (www.cbc/asithappens) he told the interviewer that when he first heard about HF, he was a city council member and when it was explained as giving apartments to addicts without requiring sobriety, he voted against introducing the program to city repeatedly. "Why should they have granite countertops when I (work hard) and they don't" he quipped. Over time, he realized that keeping someone with severe addiction and other problems homeless costs the city $100,000 per year, whereas HF only costs $20,000 per person per year. He may not have liked how the program operated, but he was sold on the cost-effectiveness of the approach.

A recent paper by Stefan Kertesz and his colleagues (2016) makes a compelling case for adopting Housing First to end homelessness. They frame their proposal by presenting three main arguments: (a) the rigorous research and sound evidence of the effectiveness of PHF; (b) the cost offset argument, buffered with the caution that we should avoid becoming too focused on any program's "net gains" because it "implicitly devalues" the lives of people of who are homeless. They caution against an emphasis on cost and ask rhetorically: would we only deliver an intervention known to be effective only if there was a monetary gain? And (c) their moral argument is both compelling and eloquent and worth quoting directly:

> From an ethical perspective, Housing First upholds the human right to housing articulated in Franklin D. Roosevelt's "Second Bill of Rights" and solidified in the United Nations Universal Declaration of Human Rights. It also resonates with the tenets of nearly every religious tradition in prioritizing care and hospitality for poor and vulnerable people. It aims to fulfill a collective responsibility to account for and remediate the incongruity of persistent homelessness in (some of) the wealthiest countries in the world. We believe that Housing First, coupled with efforts to prevent more people from becoming homeless, represents the best possible expression of what Abraham Lincoln characterized as "the better angels of our nature." (p. 2117)

Dissemination of PHF in the EU Context

The prevalence of a nation's homelessness is in part a function of its income disparity (Toro et al. 2007). The income disparity in the United States (0.45 Gini) is higher than Canada's (0.32) and most of the EU nations (France, 0.30; Netherlands, 0.25). The EU has fewer individuals who are homeless than the United States and significantly more robust health, social housing, and social services to address the problem. One of the remarkable discoveries made possible through the work of PHF dissemination is just how ubiquitous the presence of the Staircase Model is. Perhaps the Staircase Model should be credited with the dissemination of PHF since it displays the same shortcomings everywhere it is established. Many EU nations have introduced PHF programs either as individual programs or numerous programs as part of new initiatives in national policy. In the EU, the PHF program is proving viable and effective for individuals who have remained chronically homeless and who have co-occurring diagnoses. Numerous studies of the effectiveness of the PHF model in the EU have been conducted in many member nations typically using a pre-post design except for France's national randomized control trial, after AH/CS. The results have proven to be as robust as the findings in North America (for a review of these studies see Busch-Geertsema 2011; Greenwood et al. 2013).

Conclusion

In conclusion, the PHF program, its principles and values, the adaptability of the operational design, and its demonstrated effectiveness give rise to the hope that in the near future, more places will implement the PHF program and will soon be declaring that they too have successfully ended chronic homelessness. Today, in Medicine Hat, Alberta, Canada, no one is homeless in emergency shelter for more than 10 days before they receive housing. More than 40 cities have officially ended homelessness among veterans across the United States. Norway reports it has ended homelessness in their Western region and is pushing to accomplish this at a national level. New Zealand is implementing a city-wide PHF initiative in Auckland: a growing number of points of light as an effective intervention brings people home in city after city across the globe.

We are all just walking each other home. Rumi

References

Atherton I, Nicholls CM (2008) 'Housing First' as a means of addressing multiple needs and homelessness. Eur J Homelessness 2:289–303

Anthony W (1993) Recovery from mental illness: the guiding vision of the mental health system in the 1990s. Psychosoc Rehabil J 16:11

Anthony W, Blanch A (1989) Research on community support services: what have we learned? Psychosoc Rehabil J 12(3):55–81

Anthony WA, Cohen M, Farkas M (1982) A psychiatric rehabilitation treatment program: can I recognize one if I see one? Community Ment Health J 18(2):83–96

Appel PW, Tsemberis S, Joseph H, Stefancic A, Lambert-Wacey D (2012) Housing first for mentally ill homeless methadone patients. J Addict Dis 1:270–277

Aubry T, Goering P, Veldhuizen S, Adair CE, Bourque J, Distasio J, … Tsemberis S (2016) A multiple-city RCT of housing first with assertive community treatment for homeless Canadians with serious mental illness. Psychiatr Serv 67(3):275–281. doi:10.1176/appi.ps.201400587

Author (2014) Priced Out, Technical Assistance Collaborative, Boston

Bharel M, Lin W, Zhang J, O'Connell E, Taube R, Clark R (2013) Health care utilization patterns of homeless individuals in Boston: preparing for Medicaid expansion under the affordable care act. Am J Public Health 103(Suppl 2):S311–S317

Burt RM (2016) Three decades of homelessness. In: Burnes D, DiLeo D (eds) Ending homelessness. Lynne Rienner Publishers, Boulder, pp 47–66

Busch-Geertsema V (2011) Housing first Europe: a "social experimentation project". Eur J Homelessness 5(2):209–211

Carrier S (2015) Room for improvement. Mother Jones. Online version

Crowley S (2016) Homelessness is about housing. In: Burnes D, DiLeo D (eds) Ending homelessness: why we don't and how we can. Lynne Rienner Publishers, Boulder, pp 159–175

Culhane D, Metraux S, Hadley T (2002) Public service reductions associated with placement of homeless persons with severe mental illness in supportive housing. Hous Pol Debate 13:107–163

Davidson L, Roe D (2007) Recovery from versus recovery in serious mental illness: one strategy for lessening confusion plaguing recovery. J Ment Health 16(4):459–470. doi:10.1080/09638230701482394

Deegan PE (2007) The lived experience of using psychiatric medication in the recovery process and a shared decision-making program to support it. Psychiatr Rehabil J 31(1):62–69

DiLeo D (2016) How we've learned to embrace homelessness. In: Burnes D, DiLeo D (eds) Ending homelessness: why we don't and how we can. Lynne Rienner Publishers, Boulder, pp 141–158

Dolbeare C, Crowley S (2002) Changing priorities: the federal budget and housing assistance 1976–2007. National Low Income Housing Coalition, Washington, DC

Drake R, Bond G, Becker D (2013) IPS supported employment: an evidence- based approach. Oxford Press, New York

Goering PN, Streiner DL, Adair C, Aubry T, Barker J, Distasio J et al (2011) The at Home/Chez Soi trial protocol: a pragmatic, multi-site, randomized controlled trial of a housing first intervention for homeless individuals with mental illness in five Canadian cities. BMJ Open 1(2):e000323. doi:10.1136/bmjopen-2011000323

Goering P, Veldhuizen S, Nelson G, Stefancic A, Tsemberis S, Adair C et al (2016) Further validation of the pathways housing first fidelity scale. Psychiatr Serv 67(1):111–114

Greenwood RM, Schaefer-McDaniel NJ, Winkel G, Tsemberis SJ (2005) Decreasing psychiatric symptoms by increasing choice in services for adults with histories of homelessness. Am J Community Psychol 36(3–4):223–238. doi:10.1007/s10464-005-8617-z

Greenwood RM, Stefancic A, Tsemberis S, Busch-Geertsema V (2013) Implementations of housing first in Europe: successes and challenges in maintaining model fidelity. Am J Psychiatr Rehabil 16(4):290–312

Harp HT (1990) Independent living with support services: the goal and future for mental health consumers. Psychosoc Rehabil J 13(4):85–89

Henwood BF, Shinn M, Tsemberis S, Padgett DK (2013) Examining provider perspectives within housing first and traditional programs. Am J Psychiatr Rehabil 16(4):262–274

Hopper K, Jost J, Hay T, Welber S, Haugland G (1997) Homelessness, severe mental illness, and the institutional circuit. Psychiatr Serv 48(5):659–665

Jacobs J (1992) The death and life of great American cities. Vintage Books, New York

Kane V, Tsemberis S (2016) From hospital to community: transforming the VA medical center hospital system to adopt Housing First. Paper presented at the 2nd international housing first conference, July 8, Limerick

Kertesz S, Baggett T, O'Connell J, Buck D, Kushel M (2016) Permanent supportive housing for homeless people – reframing the debate. N Engl J Med 375(22):2115–2117

Larimer ME, Malone DK, Garner MD, Atkins DC, Burlingham B et al (2009) Health care and public service use and costs before and after provision of housing for chronically homeless persons with severe alcohol problems. JAMA 301:1349–1357

Ly A, Latimer E (2015) Housing first impact on costs and associated cost offsets: a review of the literature. Can J Psychiatr 60(11):475–487

Macnaughton E, Nelson G, Piat M, Goering P (in press) Moving evidence into policy: the story of the at Home/Chez Soi initiative's impact on federal homelessness policy in Canada, and its implications for the spread of Housing First in Europe and Internationally. Eur J Homelessness

Macnaughton E, Townley G, Nelson G, Caplan R, Macleod T, Polvere L, . . . Goering P (2016) How does housing first catalyze recovery? Qualitative findings from a Canadian multi-site randomized controlled trial. Am J Psychiatr Rehabil 19(2)

Mares A, Rosenheck R (2007) HUD/HHS/VA collaborative initiative to help end chronic homelessness national performance outcomes assessment preliminary client outcomes report. VA Northeast Program Evaluation Center, Washington, DC

Marlatt GA (1998) Basic principles and strategies of harm reduction. In: Marlatt GA (ed) Harm reduction: pragmatic strategies for managing high-risk behaviors. The Guilford Press, New York

Mate G (2009) In the realm of hungry ghosts: close encounters with addiction. Knopf Canada, Toronto

Monroe-DeVita M, Teague G, Moser L (2011) The TMACT: a new tool for measuring fidelity to assertive community treatment. J Am Nurses Assoc 17(1):17–29

Montgomery AE, Hill LL, Kane V, Culhane DP (2013) Housing chronically homeless veterans: evaluating the efficacy of a housing first approach to HUD-VASH. J Community Psychol 41(4):505–514. doi:10.1002/jcop.21554

Padgett DK (2007) There's no place like (a) home: ontological security among persons with serious mental illness in the United States. Soc Sci Med 64(9):1925–1936. doi:10.1016/j.socscimed.2007.02.011

Patterson ML, Rezansoff S, Currie L, Somers JM (2013) Trajectories of recovery among homeless adults with mental illness who participated in a randomised controlled trial of housing first: a longitudinal, narrative analysis. Br Med J Open 3(9):e003442. doi:10.1136/bmjopen-2013-003442

Pearson C, Montgomery AE, Locke G (2009) Housing stability among homeless individuals with serious mental illness participating in housing first programs. J Community Psychol 37(3):404–417

Prilleltensky I (1994) Empowerment in mainstream psychology: legitimacy, obstacles, and possibilities. Can Psychol 35(4):358–375

Rappaport J (1987) Terms of empowerment/exemplars of prevention: toward a theory of community psychology. Am J Community Psychol 15:121–148

Ridgway P, Zipple A (1990) The paradigm shift in residential services: from the linear continuum to supported housing approaches. Psychosoc Rehabil J 13(4):11–31. doi:10.1037/h0099479

Shern DL, Tsemberis S, Anthony W, Lovell AM, Richmond L, Felton CJ, Winarski J, Cohen M (2000) Serving street-dwelling individuals with psychiatric disabilities: outcomes of a psychiatric rehabilitation clinical trial. Am J Public Health 90(12):1873–1878

Stefancic A, Tsemberis S (2007) Housing first for long-term shelter dwellers with psychiatric disabilities in a suburban county: a four-year study of housing access and retention. J Prim Prev 28(3–4):265–279

Stefancic A, Schaefer-McDaniel NJ, Davis AC, Tsemberis S (2004) Maximizing follow-up of adults with histories of homelessness and psychiatric disabilities. Eval Program Plann 27(4):433–442

Stefancic A, Henwood B, Melton H, Shin SM, Lawrence-Gomez R, Tsemberis S (2013a) Implementing housing first in rural areas: pathways Vermont. Am J Public Health 103:206–209

Stefancic A, Tsemberis S, Messeri P, Drake R, Goering P (2013b) The pathways housing first fidelity scale for programs serving individuals with psychiatric disabilities. Am J Psychiatr Rehabil 16:240–261

Stein LI, Test MA (1980) Alternatives to mental hospital treatment. Arch Gen Psychiatry 37(4):392–397

Stergiopoulos V, Hwang S, Godzik A, Nisenbaum R, . . . Goering P (2015) Effect of scattered site housing using rent supplements and intensive case management on housing stability among homeless adults with mental illness: a randomized trial. JAMA 313(9):905–915. doi:10.1001/jama.2015.1163

Szalavitz M (2016) The unbroken brain: a revolutionary new way of understanding addiction. St. Martin's Press, New York

Teague B, Bond G, Drake R (1998) Program fidelity in assertive community treatment: development and use of a measure. Am J Orthopsychiatry 68(2):216–232

Toro PA, Tompsett CJ, Lombardo S, Philippot P, Nachtergael H, Galand B, . . . Harvey K (2007) Homelessness in Europe and North America: a comparison of prevalence and public opinion. J Soc Issues 63(3):505–524

Tsemberis S (1999) From streets to homes: an innovative approach to supported housing for homeless adults with psychiatric disabilities. J Community Psychol 27:225–241

Tsemberis S (2010a) Housing first: the pathways model to end homelessness for people with mental illness and addiction. Hazelden, Center City

Tsemberis S (2010b) Housing first: ending homelessness, promoting recovery, and reducing costs. In: Gould EI, O'Flaherty B (eds) How to house the homeless. Russell Sage Publications, New York, pp 37–56

Tsemberis S, Eisenberg RF (2000) Pathways to housing: supported housing for street-dwelling homeless individuals with psychiatric disabilities. Psychiatr Serv 51(4):487–493

Tsemberis S, Moran LL, Shinn M, Asmussen SM, Shern DL (2003) Consumer preference programs for homeless individuals with psychiatric disabilities: a drop-in center and a supported housing program. Am J Community Psychol 32:305–317

Tsemberis S, Gulcur L, Nakae M (2004) Housing first, consumer choice, and harm reduction for homeless individuals with a dual diagnosis. Am J Public Health 94(4):651–656

U.S. Interagency Council on Homelessness (2013) Ending veterans homelessness: HUD-VASH makes housing first a priority. Retrieved from http://usich.gov/population/veterans/veterans_homelessness_in_focus/ending_veterans_homelessness_hud_vash_makes_housing_first_a_priority/

U.S. Substance Abuse and Mental Health Services Administration (2007) National registry of evidence based program and practices: pathways' housing first program. Available: http://www.nrepp.samhsa.gov/programfulldetails.asp?PROGRAM_ID=195 www.cbc.ca/radio/asithappens/as-it-happens-thursday-edition-1.3074402/medicine-hat-becomes-the-first-city-in-canada-to-eliminate-homelessness-1.3074742

Yanos PT, Barrow SM, Tsemberis S (2004) Community integration in the early phase of housing among homeless persons diagnosed with severe mental illness: successes and challenges. Community Ment Health J 40(2):133–150

Sam Tsemberis, Ph.D., is the founder of Pathways to Housing where he developed the Housing First (HF) model. The Pathways HF model is listed on the US (SAMHSA) National Registry of Evidence-based Programs and has been implemented throughout the United States, Canada, Europe, Australia, and New Zealand. Dr. Tsemberis is currently the CEO of the Pathways Housing First Institute and on the faculty at the

Department of Psychiatry, Columbia University Medical Center, New York. He has published numerous articles and book chapters on topics of homelessness, mental illness, and addiction including *Housing First, by Hazelden Publishing, and coauthored Housing First: Transforming Systems and Changing Lives by Oxford University Press.*

Eric Macnaughton, Ph.D., was on the National Qualitative Research Team of the At Home/Chez Soi project, and before that he was Policy & Research Director of Canadian Mental Health Association, BC Division in Vancouver, British Columbia, Canada. He is currently the Project Manager of the PHSI initiative, a project that is working with several communities across the country to implement Housing First and evaluate the implementation process. He is a Research Associate with Wilfrid Laurier University as well as Adjunct Faculty with Adler School of Psychology in Vancouver.

Urban Design and Mental Health

17

Layla McCay, Ingrid Bremer, Tarik Endale, Marjia Jannati, and Jihyun Yi

Abstract

Physical and social environments can affect our mental health. With increasing numbers of people living in cities, public mental health is now expanding beyond the remit of psychiatrists and psychologists. Designing to support and improve mental health has not always been an obvious priority for citymakers, even as they begin to focus on physical health – though in fact perhaps some of the greatest potential for their impact lies with mental health. Policymakers, urban planners, architects, engineers, transport specialists, developers, and others impact on mental health through their contributions to the design and delivery of the urban built environment. This chapter discusses key opportunities for smart urban design to help promote good mental health, prevent illness, and support people who have mental health challenges. We focus on green space, active space, prosocial space, and safe space across the lifespan, from ensuring safe and local play opportunities for children to empowering older people to socialize and safely navigate around their neighborhoods. Citymakers can consciously seize opportunities for sustainable improvement of population mental health, putting the research into action through policies, plans, design, development, and management. Meanwhile, more research – and its funding – is needed to realize the opportunities in this high impact field.

L. McCay (✉)
Centre for Urban Design and Mental Health, London, UK

Department of Global Health Entrepreneurship, Tokyo Medical and Dental University, Tokyo, Japan

Department of International Health, Georgetown University, Washington, DC, USA
e-mail: layla@urbandesignmentalhealth.com

I. Bremer • T. Endale • M. Jannati • J. Yi
Department of International Health, Georgetown University, Washington, DC, USA
e-mail: ib319@georgetown.edu; tte5@georgetown.edu; mj555@georgetown.edu; jy447@georgetown.edu

N. Okkels et al. (eds.), *Mental Health and Illness in the City*, Mental Health and Illness Worldwide, https://doi.org/10.1007/978-981-10-2327-9_12

Keywords
Mental health • Urban • City • Architecture • Urban planning • Design

Contents

Introduction

We are becoming a population of city dwellers. The United Nations projects that by 2050, at least two thirds of the world's population will live in cities. This concentration of people means that the question of urban public health is expanding far beyond its traditional practitioners and becoming the remit of citymakers of all disciplines. Policymakers, architects, designers, urban planners, public health officials, engineers, geographers, and others are increasingly finding that to meet population needs, they need to give serious thought to how our cities can be designed in ways that promote better public health. There are numerous, well-known examples of urban design initiatives that seek to improve physical health, such as designing activity into city life to improve cardiovascular health and reduce obesity, or improving air quality to reduce breathing problems. Citymakers of all disciplines have access to a range of helpful, evidence-informed guidelines to advise them on what they can do to actively improve population health. These are inspiring times for synergies between public health and the city. And yet, with all the discussion and research and guideline production and inspiring proposals, all too often the emphasis is largely focused on just one element of health: physical health.

Improving physical health through urban design is clearly both laudable and essential, but it is not sufficient. The World Health Organization defines health as a state of "physical, mental and social wellbeing," and the mental health aspect of that definition is essential. Most of us will experience symptoms of mental health problems in our life, and one in every four of us will have a diagnosable mental disorder. From stress and sadness to attention deficit-hyperactivity disorder (ADHD), depression, biopolar affective disorder, schizophrenia, addictions, and dementia, mental health disorders directly account for about 7·4% of disease burden worldwide and are the leading cause of years lived with disability. The global burden of disease attributable to mental and substance use disorders increased by 37·6% between 1990 and 2010, almost entirely due to population growth and aging (Whiteford et al. 2013). When it comes to living in cities, these numbers may be even greater. Compared to rural settings, city dwellers have an almost 40% increased risk of depression, over 20% more anxiety, and risk of schizophrenia doubles (Peen et al. 2010).

It would be simplistic to blame the intrinsic features of cities for the increased associations found between city living and mental health. In fact, many of the very challenges that predispose a person to develop mental health problems also predispose that person to live in the city. In particular, people with low income, unemployed people, minority groups, homeless people, those who use alcohol or illicit drugs, and those with previous mental illness often tend to gravitate towards the city in pursuit of better prospects, such as economic opportunities, housing, social support, and healthcare, including mental healthcare. This trend may lead to an elevated level of baseline risk factors for mental disorders in city populations.

Maintaining good mental health is universally important for wellbeing and success, and its impact spreads to every part of our lives. Good mental health can improve our enjoyment, coping skills, relationships, our educational achievement, employment, housing and economic potential, help reduce physical health problems, ease healthcare and social care costs, build social capital, and decrease suicides. And while mental health problems can have complex combinations of causes, urban design can play an important role in prevention, support, and health promotion. Given the elevated risk of mental disorders that many city residents have, alongside the features of cities themselves that can add to these risk factors, mental health promotion is an important consideration for thriving, resilient, sustainable cities.

Despite the clear need and the opportunity, few policies or recommendations for healthy urban environments currently address mental health in any depth. There is a common adage: "no health without mental health," and yet when it comes to city planning, this can feel like someone else's remit. This chapter will consider the range of opportunities for those working in urban design to take action to improve public and personal mental health.

How Can Urban Design Improve Mental Health?

While in many cases, there is a need for more research to help define and explore this promising field, there is already substantial evidence that identifies a wide range of opportunities with good potential to promote good mental health, help prevent mental illness, and support people who have mental health problems through good urban design. These opportunities are important to consider during any urban design project. However, many architects and urban planners report that they encounter challenges when trying to think about how to integrate mental health into urban design.

Mind the GAPS: A Thematic Framework to Conceptualize Key Opportunities to Improve Mental Health Through Urban Design

The Centre for Urban Design and Mental Health has given particular thought to an approach that helps focus policy thinking and practical planning to support policymakers, planners, designers, and developers to improve mental health in their projects. We are developing a framework through which to consider any urban project using the lenses of four key opportunity themes for good mental health, summarized by the acronym GAPS: Green places, Active places, Prosocial places, and Safe places.

While these themes are by no means fully inclusive, they aim to provide a framework that urban planning, design, and development professionals can apply to any project to help identify diverse opportunities to promote and support good mental health. The themes can be defined as follows:

Green space – There are important relationships between accessible green spaces and mental health and wellbeing. Access to natural settings in neighborhoods and in the course of people's daily routines seems to help improve and maintain mental health and wellbeing.

Active space – Exercise and mental health are closely correlated. Regular activity improves mood, wellbeing, and many mental health outcomes. There are almost limitless opportunities to design cities in ways that integrate physical activity into daily life. From facilitating accessible, convenient, safe active transport to placing outdoor gyms, action can be taken to help integrate exercise, social interactions, and a sense of agency into people's daily routines to promote good mental health.

Prosocial space – Urban design should facilitate positive, safe, and natural interactions among people and promote a sense of community, integration, and belonging. This includes potentially vulnerable groups like refugees, migrants, young, and older people, with multifaceted engagement from passive observation to active participation. Interesting, flexible, and accessible public places are important, and their creation should involve citizens at each stage of design and development.

Safe space – A sense of safety and security is integral to people's mental health and wellbeing. People benefit from feeling safe and comfortable moving around their neighborhoods, and urban design has a great deal to contribute. Urban dangers include traffic, getting lost, environmental pollutants, and risks posed by other people. Appropriate street lighting and surveillance, distinct landmarks that help people with dementia navigate their neighborhoods, and people-centric design of residential, commercial, and industry routes are good examples of important urban design interventions. However, a balanced approach is necessary: a safe environment may improve accessibility and comfort, but risk-averse city design can reduce action opportunities and people's sense of agency and choice.

These four themes are clearly interrelated and can be applied to different types of projects in the built environment, whether focused on particular settings (for example, public open spaces, parks and gardens, transportation, sidewalks, and so on) or on target populations, such as the young or the old. Specific interventions that might be applied to these projects will be considered in the following pages.

First, it can be helpful to consider specific settings where research indicates particular interventions can have positive impact on mental health.

Green Space and Mental Health

Perhaps the most frequently researched correlation between the built environment and mental health is access to green space. Green space is the vegetated land or water that exists within the urban environment, including parks, gardens, playing fields, woods, green corridors, rivers, and canals. Functionally, they can be described as breathing spaces, health spaces, play spaces, wild spaces, and growing spaces. Sometimes people refer to "green and blue spaces," which includes water features. We consider these together (Fig. 1).

What Are the Effects of Green Space on Mental Health?

Research consistently finds links between green spaces and mental health and wellbeing. A survey of 10,000 people in the UK found that communities living in greener urban areas were more likely to report lower mental distress and higher levels of wellbeing after adjusting for socioeconomic variation at the individual and regional level (White et al. 2013). Another study found a possible association between more trees per kilometer of street and a decrease of 1.18 antidepressant prescriptions per thousand population (Taylor et al. 2015). Other effects of green spaces include the positive impact on stress reduction (Grahn and Stigsdotter 2003), improved social and cognitive functioning (including for those with attention deficit disorder), reductions in depression, including mood symptoms associated with dementia, and a reduction in violence (Dannenberg et al. 2003).

Part of the positive impact of green spaces for mental health is delivered through synergies with physical health. Green spaces promote and facilitate physical activity, which can have substantial benefits for mental health. Exercise has positive effects

Fig. 1 A busy street in Tokyo, Japan prioritizes pedestrians and green space, with places to sit that encourage relaxation and social behavior (Photograph by Layla McCay)

on people's general self-esteem and sense of wellbeing. Exercise can be moderately effective for reducing the symptoms of stress and anxiety. Studies have also demonstrated regular exercise helps alleviate symptoms of depression and can be just as effective as taking antidepressants for the treatment of mild to moderate depression (Stathopoulou et al. 2006). Exercise seems to be particularly helpful for depression that is associated with chronic physical illnesses. Exercise can also help alleviate some symptoms associated with attention deficit disorder, dementia, and even schizophrenia (blunted emotions, loss of drive, and thinking difficulties). Also important is the role of physical activity in counteracting the side effect of weight gain that is associated with taking certain common antipsychotic medications – through weight management and increasing fitness, exercise can help reduce these people's risk of diabetes and cardiovascular disease (Morgan et al. 2013).

Providing spaces that promote social interaction and social cohesion for all ages is another important aspect of green spaces, and indeed may be one of the key mechanisms by which mental health gains are achieved. Again, the key impacts are on reducing symptoms of stress and depression. However, it is important to be aware that green spaces can also have negative effects on people's mental health and social cohesion. For example, if the environment feels threatening to people's safety, such as a park known to be frequented by drug dealers, others may experience anxiety and avoidance of these places. Achieving balance between nature access and safety for good mental health is an important objective in urban design.

Research is increasingly finding that the positive effects of urban green space on mental health and wellbeing extend beyond the benefits of physical exercise and social interaction (Crouche et al. 2007). After adjusting for socioeconomic variables, perceived "neighborhood greenness" has been found to be strongly associated with mental health, an effect only partially explained by physical activity and increased social interaction (Sugiyama et al. 2008; de Vries et al. 2013). Even simply seeing green space on a regular basis may benefit mental health.

How Does Green Space Exert Its Effects on Mental Health?

While there is growing evidence that the very nature of being both in and around green space is associated with better mental health and wellbeing, the mechanisms of this effect are less well understood. There are three primary theories for how this association may work, all of which evolved in the 1980s (Crouche et al. 2007). The first is Edward Wilson's biophilia theory (Sugiyama et al. 2008). Wilson's theory argues that humans have a close relationship with the natural world and as such, a predetermined biological need has evolved that drives humans to subconsciously seek contact with other species. Around the same time, Roger Ulrich proposed his stress reduction theory. Ulrich's theory maintains that natural environments promote recovery from stress by through a range of physiological and psychological responses to exposure (de Vries et al. 2013), including distance from everyday demands, possibilities for aesthetic appreciation, and activity driven by interest (Wilson 1984). Thirdly, Rachel and Stephen Kaplan proposed the attention restoration theory (Ulrich 1983). The Kaplans' theory argues that nature environments exert benefits by relieving "attention fatigue" through distancing people from performing tasks that require prolonged maintenance of attention and instead facilitating the use of attention without the need for concentration.

Is There a "Dose" of Green Space?

Given the numerous potential benefits of green space for mental health, the question of a "dose" or "prescription" for access to green space is often discussed, both as a strategy to prevent mental health problems and to address their symptoms. It is clear that quality and quantity of urban greenery are both connected to mental health (Hartig 2008) and parks and other green spaces that are walkable and have opportunities for physical activity and socializing are likely to reap greater benefits than green features in the streetscape or in buildings (Kaplan and Kaplan 1989). That said, there is a risk of limiting people's exposure to green spaces to places they need to travel to, rather than ensuring they encounter these spaces naturally in their course of their day. The more time people spend in outdoor green spaces, the less stressed they seem to feel, regardless of age, gender, and socioeconomic status (Fan et al. 2011). Both visual and physical access to green space seem to be beneficial, opening the possibility of exploring landscaping and vegetation options inside and around places of work and study, including urban parks, walkways, and incorporation into building design. Specific location is also important: distance from a person's home to green space is an independent predictor of stress (Nielson and Hanson 2007). And when people move from a green to a less green neighborhood, measures of mental health have been found to deteriorate (Alcock et al. 2014). However, the optimal location, qualities, or duration of access to parks and other green space with the specific objective of benefiting mental health are not yet fully understood. Still, green space represents an important opportunity in urban design to promote good mental health, and there are many exciting projects in this area.

Public Spaces, Streetscapes, and Mental Health

A key opportunity for urban designers to improve community mental health is through streetscapes. A streetscape refers to a street's appearance and how it is designed (particularly how the paved or pedestrian area of the street is laid out and treated). This includes faces of the buildings visible from the street, the street surface, and street-furniture and fittings from bus shelters to benches to planters. With most urban dwellers being exposed to streetscapes on a daily basis, the potential for their impact on public mental health is important (Fig. 2).

Land Use to Promote Mental Health

Land use in urban environments has become a matter of much debate, as many cities evolve from the traditional "town square" model to a more dispersed, sometimes less-cohesive layout. Mixed land use with streets that contain a diversity of lively neighborhood shops, cafes, and housing can help create a sense of community, positive perceptions of a neighborhood, and facilitate social interactions; these benefits can be lost when land use is segregated, so that driving between isolated locations becomes necessary, turning potential social trips into town into discrete, often solitary activities.

Boring Streetscapes

Another important evolution in the cityscape is the transition from smaller occupied facades that give personality and diversity to the street view to often long, monotonous walls that often lack features such as windows or entrances for whole city blocks at a time. If we are not busy processing varied information that we perceive in the external world, people are more likely to dwell and ruminate on pessimistic internal thoughts and to become bored. Boredom is associated with stress and risky behaviors, particularly addiction. Urbanists such as Jan Gehl (2011) recommend that to engage people's minds and prevent rumination and boredom, the average walker, moving at a rate of about 5 km per hour, should see an interesting new site about once every 5 s. Long, unchanging facades such as office blocks or the walls of monolithic shops that extend, unchanging, for many meters may have negative effects on mental wellbeing in that they may create boredom and support negative ruminations, limit social opportunities, and reduce people's interest and ability in interacting with their environment. While more research will improve our

Fig. 2 A public space in La Mariscal in Quito, Ecuador, welcomes people to a safe pedestrian space with seats and greenery in the city, alongside welcoming food establishments and separation from traffic, facilitating relaxation and social interactions (Photograph by Jaime Izurieta-Varea)

understanding, many urbanists consider these factors highly relevant for people's positive interactions in their environment (Fig. 3).

Nature Exposure

In terms of green space provision, urban parks seem to be one of the most effective interventions for positive impact on mental health promotion. Tree and flower planting has perhaps become the most common way to incorporate nature into urban environments. As a result, street-based trees have become an abundant and accessible form of urban nature, with the associated mental health benefits. There is much that can be achieved within the streetscape, though design can be complex, meeting the needs of both pedestrians and road users (Ichihashi et al. 2000).

Public Open Spaces

Public open spaces can provide an opportunity to facilitate nature exposure and prosocial interaction. High quality public open spaces are venues for cooperative community activities that create feelings of social integration and belonging and provide opportunities for people to volunteer for positive activities; altruistic donation of time and skills to others is associated with mental wellbeing. An Australian study showed that residents in neighborhoods with high quality public open spaces

Fig. 3 Another street in Quito, Ecuador, where an almost-featureless perimeter wall renders the street less engaging for users (Photograph by Jaime Izurieta-Varea)

had a lower risk of psychological distress – and that quality was more important than quantity in generating both positive social interactions and community activities but also a place in which to relax (Francis et al. 2012). However, care must be taken in the design and management of public open spaces to avoid them becoming settings for crime, intimidation, or abuse, which would limit their use and conversely have negative effects on people's mental wellbeing.

Transportation and Mental Health

Automobile infrastructure is increasingly taking over space in neighborhoods, marginalizing pedestrians into narrow sidewalks, reducing opportunities for social interaction, and disconnecting different parts of the neighborhood.

The realm of transportation is therefore another exciting opportunity for urban design to improve communities and positively impact mental health. The majority of people interact with transportation, and for many, transport use is an integral part of their daily routines, making transportation an important aspect of urban design to promote and improve mental health, though this is an area where further research would be helpful.

Transport plays an essential, largely positive role in urban mental health by moving people efficiently and affordably, linking communities, facilitating movement around the city, and by doing so, increasing opportunities for education, employment, housing, leisure, social interaction, and access to nature, all of which are important for mental health. Living near public transit connections is positively associated with good mental health, particularly for older people, who may be less likely to be able to drive. Bike lanes, pavements, and other infrastructure can also promote active transport options that can help improve people's mental health through physical activity, access to nature, and sometimes even social interactions.

However, transportation can also have negative impacts on people's mental health. A great deal of land is increasingly used for roadways, intersections, and car parking spaces. In many countries, roads split communities, reduce walkability, and create noise, pollution, and danger, all of which can create mental health challenges. A key challenge here is the growing global dominance of the car. The amount of time people spend in their cars is increasing across the world, with growing traffic density, and people commuting substantial distances from their home to their job every day. This trend has important consequences for mental health, particularly impacting on stress, social capital, and access to green spaces and positive streetscapes.

Commuting

The act of commuting, especially long distance commuting, can have a range of negative impacts on mental health, particularly when commutes last for over 30 min. Exposure to noise, crowds, unpredictable traffic, and traffic congestion during commutes across town can each contribute to stress, anxiety, hostility, aggression, and negative feelings (Fong et al. 2001), and this is exacerbated when people have slept poorly (Sygna et al. 2014). Commuting problems can create tardiness at work, which can contribute to work-related stress, and spending long times in transit each day means less time for social engagement with friends and family. Indeed, spending extended commuting can have substantial impact on people's social capital, with transit time coming at the expense of the very social interactions that are supportive of good mental health. A study looking at automobile duration and the quantity of time spent with friends and family found that when daily commuting surpassed 1 h per day, it decreased the amount of time men were able to spend with their spouses and children and decreased the time women spent with their friends (Christian 2012).

Long-distance commuting in particular can also impact on sleep and physical activity, essential for mental health maintenance. In one Swedish study, commuters, whether traveling by car or public transport, experienced increases in everyday levels of stress, poor sleep quality, and feelings of exhaustion (Hansson et al. 2011). On the other hand, when infrastructure is in place to ensure safety, commuting by walking or bicycle can have positive mental health benefits, particularly through exercise and often access to nature.

Interventions to improve mental health through transportation should therefore aim to improve the quality of people's commute, increase leisure and social time, increase use of active modes of transportation, reduce stress, and aim to support improved quality of sleep. Specific interventions may include protected bikeways and networks, bikeshares, bus-only routes, and pedestrian plazas to enhance active commuting and opportunities for social interaction.

Lessons from the Mental Healthcare Facility Setting

Looking beyond the cityscape to identify relevant lessons from other settings, there is a wide body of research that concentrates on the interaction between people with mental health problems and their environment in the context of mental healthcare. This research has been driven largely by architects and interior designers seeking to create high-quality mental healthcare facilities that are sensitive to the particular needs of their users and offers lessons relevant to design in the wider built environment.

Quality of Setting

In the mental healthcare context, the quality of a person's accommodation has been linked to their levels of psychological distress. Clean, comfortable, safe, and high-quality settings can help promote self-worth and reduce stress levels (Evans 2003), while residential crowding may increase psychological distress (though this has not been linked to the development of serious mental illnesses).

Balancing Privacy and Social Interaction

The need to balance privacy and social interaction is another important opportunity for design in the built environment. In particular, people appreciate the opportunity for solitary time and the ability to exert personal control to regulate their social interactions with others in their environments. In the healthcare setting, this has been achieved, for example, through the configuration of spaces, furnishing, windows, welcoming entry features on a building's exterior such as porches (Szapocznik et al. 2009), and balancing opportunities for socializing and privacy (Evans 2003).

Autonomy to Regulate One's Own Environment

Another important component of design to support mental health in the healthcare setting is the opportunity to understand one's environment and exert autonomous decision-making, as opposed to having only one option of where, when, and how to interact with a place. In the hospital, this could mean having more than one route to the common area or flexible spaces where furniture can be moved to suit different needs. Being invited to make decisions engages people and supports their cognitive functioning. The importance of autonomy, choice, and personal agency can be extrapolated to the city setting: having options supports people's development of attention, problem solving, and conscious engagement with the places they are in, engaging their brains in active decision-making.

Safety and the Urban Environment

Improving safety in the urban environment is not always thought of as a specific intervention to improve mental health, but in fact a feeling of security in one's daily life is an important contributor to mental health and wellbeing in any urban setting. People who have their property stolen or experience or witness violent crime are more likely to demonstrate poorer mental health, and these effects tend to persist for over a year after an incident (Clark et al. 2006). Crime prevention through urban design has been extensively discussed, and a full description of methods to reduce crime through urban design is beyond the scope of this chapter. However, some key principles apply.

First, taking opportunities to leverage natural surveillance is important. Crime often occurs where people feel there are no witnesses to their activity. Since the feeling of being seen is a deterrent for criminal activity, the built environment can be designed to increase public visibility of people's activity in public places. Classic examples include designing windows into the walls of buildings that overlook pedestrian areas, creating supportive infrastructure that increases pedestrian and cycling traffic, avoiding sight-limiting walls, and ensuring potentially risky areas are well-lit, with lighting particularly angled to show faces, while avoiding too-bright security lighting that cast glare and shadows.

Another opportunity is natural access control and territorial marking. This means ensuring that the difference between public and private space is well demarcated to dissuade public space users from encroaching on private space for the purpose of criminal activity. Classic examples include minimizing the number of entry points to private spaces, using impediments to access vulnerable access points such as positioning thorny bushes under windows, avoiding design features that allow access to upper levels or roofs, and using lighting, signage, and other design features to make clear what is public space and what is not.

Next is the question of maintenance. The "broken windows theory" uses a broken window as a metaphor for all types of neighborhood disorder in the built environment. This theory proposes that the presence of features such as broken windows create a neighborhood feeling that the area is not safe. Further, a dilapidated, deteriorating space implies a tolerance of disorder and less likelihood of oversight and therefore increases vulnerability to criminal behavior. On the other hand, a clean, well-maintained space implies oversight and engenders pride in an area.

Senses on the Street

Another opportunity that has recently gained much interest is how to leverage the senses for good mental health in the urban built environment. A study looking at the design of sound and visual components in urban soundscapes evaluated the effect of natural sounds in conjunction with road traffic noise and visual components in urban streets. The study found that "the perceptual dimensions of the environment were different from the noise levels (Hong and Jeon 2013).

Noise is defined as "unwanted, uncontrollable, and unpredictable" sound, and urban noise can impact people's mental wellbeing. A particularly important aspect of noise is nocturnal noise: sleep plays a key role in people's mental health and wellbeing, and disturbed sleep can exacerbate many mental health problems. Noise can also impair task performance and aggravate mood symptoms such as anxiety and anger.

The specific impact of sound on mental health has been studied, but the results are inconclusive. Loud outdoor noises, such as those associated with airports or traffic, have been reported to increase psychological distress, but links have not been found with serious mental illness (Evans 2003; Clark et al. 2006). Moderate but persistent traffic noise has been associated with symptoms of depression (Orban et al. 2015). Sudden loud sounds, such as the beeping of horns in heavy traffic, may trigger auditory reflexes that put people in a state of sound-driven high alert and potentially increase stress over time (Pheasant et al. 2010). However, a quiet street is not necessarily an optimal street for mental wellbeing; a more helpful aim is to achieve a balance of sounds. In particular, nature sounds such as birdsong have been associated with wellbeing – though urban designers need to approach this opportunity with caution: research has also found that the sound of falling water was considered to degrade the quality of the soundscape if there was concurrent traffic noise (Hong and Jeon 2013).

Smell should also be taken into account by urban designers. In the past, attempts were made to keep smells from abattoirs and tanneries and the like far from residential areas; with increasing mixed land use, smell can impact on people's experience and perceptions of their environment (Orban et al. 2015); smells can contribute to wellbeing though as yet there is no conclusive evidence about the association of smell with serious mental illness.

Target Population Groups and Urban Design

In addition to generally designing urban environments to promote good mental health, there are often occasions where planners, designers, and developers seek to understand specific opportunities for target population groups. Three of the most interesting opportunities lie with people who have addictions, the young, and the old.

Urban Design and Addiction

The built environment can have significant impact on alcohol and illicit drug consumption that interact with the other risk factors of developing addictions.

Condition of the Built Environment

People living in neighborhoods with buildings that are in a dilapidated and deteriorating condition have been found to be 150% more likely to report heavy drinking than those who live in neighborhoods in better condition (Bernstein et al. 2007). A wide range of research has also demonstrated the links between dilapidated, deteriorating built environments and urban drug use, including an association with an increased likelihood of fatal accidental drug overdose, even after taking individual factors into account. This research demonstrates the importance of good upkeep of the built environment (Jackson 2003; Hembree et al. 2005). A dilapidated environment is also conducive to the use of intravenous drug use, creating a vicious circle; certain cities have created safe injecting spaces, which may reduce fatal overdose, as well as the spread of infectious diseases (Epstein et al. 2014) as well as making the outdoor environment more welcoming to the general public.

Alcohol Outlet Density

Residential exposure to alcohol outlets is associated with harmful consumption of alcohol (Jozaghi 2012). An Australian study (Murphy et al. 2014) found that "each additional liquor store in the neighborhood was associated with an increase by 1% in the mean number of standard drinks of alcohol consumed per drinking day and by 6% in the mean number of days of harmful consumption of alcohol.". In the USA, alcohol outlet density, particularly bunching of outlets, has been found to have a negative effect on binge drinking and alcohol related injuries and violence (Pereira et al. 2013).

Locations for Drug Dealing

Research has examined the design attributes of locations that make antisocial behavior such as drug dealing more likely, particularly in urban parks. Dealers and addicts were found to favor entrances where there is disorganization, confusion, and chaos and areas where furniture is old and worn out. In terms of lighting, they favor locations where plants have created dark and shady areas, where the trees have low and vast crowns, and where the view is limited. Areas where food is served and minor routes that end at the main entrance are also vulnerable, though unsurprisingly the more a space attracts public behavior, the less likely that space will be used for purposes such as dealing drugs (Livingston et al. 2007).

Children, Urban Design, and Mental Health

The experience of childhood is becoming increasingly urban, which can have a range of important implications for their mental health. Much of the researched impact of the urban environment on children's mental health and wellbeing is characterized by limitations in the opportunity for self-discovery and play in natural environments.

Places for Play

Part of the restriction in play opportunities that many children experience in the urban environment can be attributed to the design of housing and surrounding areas. In particular, high-rise multiple dwelling units with lack of nearby play spaces result in children being confined to apartments, diminishing these children's opportunities for nature exposure, play, and socializing with other children; this effect is more marked with children who live on higher floors (Evans 2003).

Quality of Housing

However, housing can also contribute to a child's mental health and development in other ways. Children who live in a situation of residential crowding have been found to manifest more behavioral problems in day care centers compared to their peers from uncrowded homes, and girls living in crowded homes are particularly likely to experience helplessness in task persistence (Evans 2003). A further factor is toxins such as lead in building materials or house paint; in particular, lead can negatively affect self-regulatory behavior such as focused attention and frustration tolerance, which can contribute to conduct disorders and aggression (Aminzadeh and Afshar 2004). Finally, poor quality of housing, such as structural quality, maintenance, or upkeep, can help create stigma that can result in social exclusion and frequent relocations, both risk factors for socioemotional problems for children (Evans 2003).

Children, Attention, and the Built Environment

In studies in day care and residential settings, children with greater access to nature and to natural light have been found to be less impulsive and have better concentration (Evans 2003) to the extent that studies have found children with attention deficit hyperactivity disorder (ADHD) undertaking outdoor activities in nature settings have shown 30% improvement in their symptoms compared to those undertaking outdoor activities in non-nature settings, and three-fold compared to those undertaking the activities indoors (Kuo and Taylor 2004). Some physicians are now "prescribing" time playing in parks as prevention or as an adjunct to medication for children with symptoms of ADHD. While it is not known what an optimal "dose" of outdoor play in natural environments may be, it is becoming increasingly clear that this practice delivers significant benefits for children's mental health and wellbeing. Another environmental factor impacting on children's concentration and task persistence is the presence of external noise when, for instance, they are undertaking schoolwork.

Children, Stress, and the Built Environment

There is some evidence that contact with nature also helps children to deal with stress. There seems to be an association between nature exposure, higher self- worth, and lower levels of stress. This exposure has been found to be particularly effective for children who have been bullied, punished, have recently relocated, or are experiencing family strife (Evans 2003).

Adolescents, Mental Health, and the Built Environment

Studies have shown links between adolescents' perceptions of their environment and their mental health (Mair et al. 2008). Components of the built environment that may be best leveraged to support good mental health for adolescents build opportunities for physical activity and for positive social interactions. Age-appropriate public spaces where adolescents can safely spend time with their friends, including exercising, can help foster social capital, social support, and so-called "peer competence" and may have a corresponding positive mood impact and have a protective impact on adolescent mental health (Aneshensel and Sucoff 1996).

Conversely, a perceived threatening environment has been associated with symptoms of depression, anxiety, oppositional defiant disorder, and conduct disorder in adolescents, while environments that are run down and attract crime in particular tend to support antisocial behavior (Mair et al. 2008).

Older People, Urban Design, and Mental Health

Another subset of the population whose built environment often receives special focus is older people. The close relationship between older people, mental health, and urban design is well-encapsulated by Garin et al. (Garin et al. 2014): "In the elderly, the immediate urban environment becomes an important issue. People tend to spend more time in this environment because of physical decline, retirement, decreased access to transport and shrinking in social networks."

A positive built environment is linked to older people's independence, autonomy, self-esteem, and physical and mental wellbeing. Given the increased importance of the local built environment to impact older people, and its potential to both positively and negatively impact on their mental health, it is disappointing to identify comparatively few academic studies addressing this matter; of existing studies, the majority are cross-sectional, which means that it is difficult to ascribe specific cause and effect. However, they demonstrate clear links between urban design and mental health. There are two key areas of focus for older people in urban design: preventing depression and supporting good psychosocial health for people who have dementia.

Urban Design Interventions to Prevent Depression for Older People

Older people can be subject to loneliness and depression. Various urban design interventions may help. Older people having a positive impression of the internal quality of their own home have been found to be associated with fewer depressive symptoms, for instance the home being well-decorated, well-lit, clean, fully furnished, and containing all facilities that might be expected in a dwelling in a particular location (Stewart et al. 2002; Lai and Guo 2011). Having sufficient living space is also important; research suggests having more than two rooms in a home has a protective impact on depression in older people (Chan et al. 2011). Attending to the prevention of moisture build-up, lighting, and design for easy upkeep may be further important factors.

Outside the home, neighborhood walkability plays a role, though its impact is mixed. Being able to get out of the house and walk about in a local neighborhood has clear benefits through providing opportunities for nature exposure, social interaction, autonomy, and physical exercise. Good walkability seems to be associated with lower depressive symptoms for men, though the same effect has not been shown for women (Berke et al. 2007). Consideration should be given to build assisted care facilities, nursing homes, and residences for the elderly in walkable neighborhoods.

However, the picture is complex. One study showed that greater land use diversity and retail availability seems to be associated with higher depressive symptoms in older men, despite the association with increased walkability (Saarloos et al. 2011). This indicates the complexity of these urban design, depression, and older people.

Urban Design Interventions to Support the Mental Health of People with Dementia

Importantly for urban planners, increasing numbers of older people with dementia are living in urban environments (Mitchell and Burton 2006; Forbes et al. 2006). It is imperative that this rapidly growing population is considered when designing the built environment, from the design of neighborhoods and cities to the design and location of care facilities. In order to understand the opportunities for urban design to support the mental health of people with dementia, it is important to understand some of the symptoms of dementia. These symptoms can include memory loss, difficulties with thinking, orientation, problem-solving, and/or language, and mood or behavior changes. If older people tend to spend more time than other people in their immediate environment, given the challenges of these symptoms, this trend is even more marked for those with dementia. As their symptoms progress, many people with dementia find themselves unable to drive and experiencing challenges when trying to use public transport, meaning they can become increasingly limited to destinations within walking distance of their home (Mitchell and Burton 2006). Even getting around in the immediate urban environment can create particular challenges for people with dementia, such as disorientation, fear, and difficulties in getting safely between their home and their destinations, all of which can have substantial negative effects on people's health and wellbeing.

Older people who do not experience social isolation and feelings of disconnection are half as likely to experience rapid cognitive decline in multiple functional domains (Mitchell and Burton 2006). And yet the effects of dementia can contribute to both. The ability of a person to undertake basic routine tasks, take a walk, have encounters with nature, and engage socially in neighborhoods can play pivotal roles in maintaining a person's sense of self, wellbeing, and quality of life – in addition to the physical health benefits of this exercise. Being unable to go out safely and confidently withholds most of these benefits. Urban design can actively help maintain older people's independence, social engagement, stimulation, exercise, and physical wellbeing.

Two major opportunities are in design to facilitate social engagement and simple, accessible navigation. In particular, key opportunities lie with familiarity, legibility, distinctiveness, accessibility, comfort, and safety (Evans 2003), an approach currently being taken by the movement to create "dementia-friendly neighbourhoods."

Improving Street Layout to Help People with Dementia Navigate

People with dementia can have difficulties both in concentration and in ability to communicate, relying on environmental cues such as landmarks to navigate. Urban design can inadvertently make it more difficult for these people to find their locations, for instance, changes to the local area remove familiar landmarks and can cause disorientation, while loud noises can create distractions and lead to confusion.

Even when people with dementia are familiar with a neighborhood and feel confident with navigation, many still get lost, with the greatest challenge often being in spaces that lack clear, visible environmental cues, and at road junctions – particularly crossroads; forks and T-junctions tend to be easier (Mitchell and Burton 2006). People with dementia report avoiding long, wide roads due to their monotony, favoring short, narrow, gently winding streets where the ends of streets and other environmental cues can be more easily identified and help maintain interest and concentration.

Buildings as Environmental Cues for People with Dementia

Environmental cues are a key aid for everyone's location and direction in the built environment, but for people with dementia, these landmarks take on extra importance. People with dementia report that the most helpful navigational cues are distinct, easily recognizable buildings with functions that are clearly obvious from their style, image, and details, for example, church steeples or grocery shops with fruit displayed in front (Mitchell and Burton 2006). Varied shapes, features, colors, and contrasts in building form and architecture can all be helpful. As such, people with challenges navigating tend to favor routes with diverse land use, building form and architectural feature variability, particularly routes containing landmarks with distinct and obvious uses. Commonly cited types of helpful navigational landmarks include: historical or civic buildings (churches, libraries, town halls), distinct structures (clock and water towers, public art), places of activity (mixed-use squares, parks), places of personal significance, aesthetic features (gardens, trees, planters), and practical features (telephone boxes, bus shelters). In residential neighborhoods that lack more obvious navigational landmarks, distinctive rooflines, tiles, chimneys, front doors, windows, and gardens can all be employed as navigational aids. A clear hierarchy of streets including main streets, side streets, and alleyways may be helpful, with efforts made to create streets that have distinguishing features rather than blocks of almost identical streets. In all environments, whenever a distinctive building or landmark is being considered for removal, the role of that structure as a navigational cue for people with dementia should be considered.

Signage to Help People with Dementia Navigate Their Environment

While landmarks are the most effective and least complicated way for people with dementia to navigate, signs may also play a role in adding clarification at places where confusion is most likely to occur, such as direction signs placed at junctions or location signs on buildings whose function is unclear. Signs that people with dementia report being most helpful are simple, have large dark lettering on light backgrounds, convey only brief, essential information, and use highly realistic, recognizable symbols. Direction signs can be particularly helpful at junctions, while location signs should be perpendicular to walls. While signage can be helpful, they can also be confusing for people with dementia and create navigational challenges rather than solving them. People with dementia have criticized signs placed too low, too high, or parallel to walls. Signs may be too numerous, too crowded, too hard to read, too complicated, or have nothing to do with navigation, at which point they become a hindrance rather than an aid to navigating the environment.

Comfort and Safety for People with Dementia

Neighborhoods can be inviting or intimidating to their inhabitants in a range of ways, but for people with dementia, open informal spaces with a lot of activity can present a more welcoming and safe space than empty ground or imposing historical architecture that can seem forbidding. A cluttered environment with lots of street furniture such as railings, kiosks, and trashcans, or complicated patterns on the ground, can create confusion. Persistent background noise can impact people's hearing, and sudden loud noises can be startling and disconcerting, leading to confusion and anxiety. Finally, good lighting helps people with dementia to better identify both navigational cues and obstacles, avoiding both disorientation and injury.

Mental Health Is Often Overlooked by Citymakers

There are clear opportunities to integrate mental health promotion and support into urban design. But with an incoming wave of competing voices, demands, needs, funders, activists, research, and more, citymakers must prioritize. Despite the substantial benefits to health and wellbeing, mental health often features low on these priority lists. Three important barriers to prioritizing mental health in urban design are stigma, fear of complexity, and low prioritization becoming a self-fulfilling prophecy.

Stigma is often associated with mental illness and that stigma can extend beyond the individual experiences into citymaking. Policymakers and designers may lack knowledge about how common mental illness is, have negative beliefs about it, worry that people will make assumptions about their own mental health, or just never happen to think about including it as a consideration in urban planning and design for a sustainable city, even as they integrate physical health promotion. For that to change, vocal advocacy is needed to persuade citymakers of the benefits of attention, action, and investment for mental health in urban projects. Often advocates for other causes are louder, or advocates are deterred for fear of experiencing stigma, and the case for mental health goes unheard. Mental health advocates need to engage policymakers and designers with personal stories, activate champions, and demonstrate persuasive research and case studies – in other words, they need to build a good, engaging, consistent, evidence-based argument for action, just as advocates for physical health in urban design have achieved.

A second barrier seems to be the perceived complexity of mental health disorders. Mental disorders may emerge from a complex range of factors including genetics, early experiences, family relationships, and social settings that cannot be addressed through urban design – but urban design can and should play a role, just as it does for physical disorders, which have equally complex causes.

A third barrier is the self-fulfilling prophecy of low prioritization: if the health promotion aspect of urban planning, design, and development continues to focus on innovations around physical health, then it will be these initiatives that are showcased, discussed, and rewarded, while urban design for mental health will have less visibility. This exposure creates awareness, expectations, appreciation, and demand, erroneously implying that when it comes to integrating health into urban design, the key opportunities lie with physical health. Without a similar movement for mental health, citymakers are inevitably hearing much less about mental health in urban design plans, policies, and projects. This has the self-perpetuating effect of insufficient inspiration, motivation, and empowerment around mental health in urban design, despite the need and opportunity. The time to act to advocate for mental health integration in urban design is now.

Summary

With an increasing number of people growing up, living, working, and spending their retirement years in cities, the mental health component of public health is not only becoming more recognized – it is rapidly becoming the remit of those involved in designing and building cities, from policymakers and planners to architects, engineers, and developers. The links between the built environment and mental health are increasingly understood, making urban design and mental health a growing interdisciplinary field. Through considering green, active, prosocial, and safe spaces, there is a full palette of opportunities for urban designers to improve population mental health, and many of these opportunities are already being enacted, whether by creating places where people feel safe and confident, reducing noise to improve sleep, facilitating encounters with nature, promoting social interactions, or reducing the stress associated with commuting. But there is more that can be achieved. More investment in research will reap rewards, as will leveraging creative research methods, such as smartphone technology, to better understand reactions to specific places in real time. As cities strive to become "smart cities," here too is a place for mental health. Opportunities beyond this chapter will continue to emerge, and the future of the emerging urban design and mental health specialty is exciting.

Conclusion

Numerous factors contribute to people's mental health, but there is exciting untapped potential for improving public mental health by better leveraging urban design. Urban planners and designers can integrate mental health promotion into their projects to help promote mental health for diverse citydwellers across their lifespans, achieving substantial positive impact. Urban design can promote good mental health and wellbeing for the urban population, support the mental health and wellbeing of vulnerable people, and aid the recovery of those with mental health problems at all phases of their lives as residents of a city.

So the time has come to move beyond the stigma that has held back progress in this field, to clarify the opportunities, and start designing urban mental health into plans and projects. Citymakers must seek and seize these opportunities. All of us are at risk of mental health problems, so this matters personally to everyone on a personal level, to cities, and to countries.

These authors envision that as awareness and research both develop, integrating better mental health and wellbeing into urban design will become a standard component of sustainable urban policy. Funders will further invest in this exciting, high impact field, and citymakers will put research into action through policies, plans, design, development, and management. This is a moment of great opportunity, and action is needed today.

Cross-References

- ▶ Green Environment and Mental Health in the City
- ▶ Loneliness and Single-Person Households: Issues of *Kodokushi* and *Hikikomori* in Japan
- ▶ Revisiting the Virtuous City: Learning from the Past to Improve Modern Urban Mental Health
- ▶ Urban-Rural Differences in Major Mental Health Conditions

References

Alcock I, White MP, Wheeler BW, Fleming LE, Deplege MH (2014) Longitudinal effects on mental health of moving to greener and less green urban areas. Environ Sci Technol 48:1247–1255

Aminzadeh B, Afshar D (2004) Urban parks and addiction. J Urban Des 9(1):73–87

Aneshensel CS, Sucoff C (1996) The neighborhood context of adolescent mental health. J Health Soc Behav 37(4):293–310

Berke EM, Gottlieb LM, Moudon AV, Larson EB (2007) Protective association between neighborhood walkability and depression in older men. J Am Geriatr Soc 55(4):526–533

Bernstein KT, Galea S, Ahern J, Tracy M, Vlahov D (2007) The built environment and alcohol consumption in urban neighborhoods. Drug Alcohol Depend 91(2–3):244–252

Chan A, Malhotra C, Malhotra R, Østbye T (2011) Living arrangements, social networks and depressive symptoms among older men and women in Singapore. Int J Geriatr Psychiatry 26:630–639

Christian TJ (2012) Automobile commuting duration and the quantity of time spent with spouse, children, and friends. Prev Med 55(3):215–218

Clark C, Myron R, Stansfeld S, Candy B (2006) A systematic review on the effect of the built and physical environment on mental health. J Public Ment Health 6(2):14–27

Crouche K, Meyers L, Bretherton J (2007) The links between greenspace and health: a critical literature review. Available via Greenspace Scotland. http://greenspacescotland.org.uk/links-between-greenspace-and-health.aspx. Accessed 24 May 2016

Dannenberg AL et al (2003) The impact of community design and land-use choices on public health: a scientific research agenda. Am J Public Health 93(9):1500–1508

de Vries S, van Dillen SME, Groenewegen PP, Spreeuwenberg P (2013) Streetscape greenery and health: stress, social cohesion and physical activity as mediators. Soc Sci Med 94:26–33

Epstein DH, Tyburski M, Craig IM, Phillips KA, Michelle L, Vahabzadeh M, Preston KL (2014) Real-time tracking of neighborhood surroundings and mood in urban drug misusers: application of a new method to study behavior in its geographical context. Drug Alcohol Depend 134:1–16
Evans GW (2003) The built environment and mental health. J Urban Health: Bull N Y Acad Med 80:4
Fan Y, Das K, Chen Q (2011) Neighborhood green, social support, physical activity, and stress: assessing the cumulative impact. Health Place 17:1202–1211
Fong G, Frost D, Stansfeld S (2001) Road rage: a psychiatric phenomenon? Soc Psychiatry Psychiatr Epidemiol 36(6):277
Forbes DA, Morgan D, Janzen BL (2006) Rural and urban canadians with dementia: use of health care services. Can J Aging 26:321–330
Francis J, Wood LJ, Knuiman M et al (2012) Quality or quantity? Exploring the relationship between public open space attributes and mental health in Perth, Western Australia. Soc Sci Med 74:1570–1577
Garin N, Olaya B, Miret M, Ayuso-mateos JL, Power M, Bucciarelli P, Haro JM (2014) Built environment and elderly population health: a comprehensive literature review. Clin Pract Epidemiol Ment Health 10:103–115
Gehl J (2011) Cities for people. Island Press, Washington, DC
Grahn P, Stigsdotter UA (2003) Landscape planning and stress. Urban For Urban Green 2(1):1–18
Hansson E et al (2011) Relationship between commuting and health outcomes in a cross-sectional population survey in southern Sweden. BMC Public Health 11:834
Hartig T (2008) Green space, psychological restoration, and health inequality. Lancet 372:1614–1615
Hembree C, Galea S, Ahern J, Tracy M, Piper TM, Miller J, Tardiff KJ (2005) The urban built environment and overdose mortality in New York City neighborhoods. Health Place 11:147–156
Hong JY, Jeon JY (2013) Designing sound and visual components for enhancement of urban soundscapes. J Acoust Soc Am 134:2026–2036
Ichihashi H, Watanabe R, Kojima M (2000) Mental effects of street planting on pedestrians and drivers. J Jpn Inst Landscape Arch 63(5):795–798
Jackson RJ (2003) The impact of the built environment on health: an emerging field. Am J Public Health 93(9):1382–1384
Jozaghi E (2012) "a little heaven in hell": the role of a supervised injection facility in transforming place. Urban Geogr 33(6):1–19
Kaplan R, Kaplan S (1989) The experience of nature. A psychological perspective. Cambridge University Press, Cambridge
Kuo FE, Faber Taylor A (2004) A potential natural treatment for attention-deficit/hyperactivity disorder: evidence from a national study. Am J Public Health 94(9):1580–1586
Lai DWL, Guo A (2011) Gender differences in depressive symptoms of aging Chinese in urban Jiangsu in China. Clin Gerontol 34(3):190–206
Livingston M, Chikritzhs T, Room R (2007) Changing the density of alcohol outlets to reduce alcohol-related problems. Drug Alcohol Rev 26(5):557–566
Mair C, Roux AVD, Galea S (2008) Are neighbourhood characteristics associated with depressive symptoms? A review of evidence. J Epidemiol Community Health 62(11):940–946
Mitchell L, Burton E (2006) Neighbourhoods for life: designing dementia-friendly outdoor environments. Qual Ageing Older Adults 7(1):26–33
Morgan AJ, Parker AG, Alvarez M, Jimenez AF, Jorm AF (2013) Exercise and mental health: an exercise and sports science Australia commissioned review. JEP Online 16(4):64–73
Murphy A, Roberts B, Ploubidis GB, Stickley A, McKee M (2014) Using multi-level data to estimate the effect of an "alcogenic" environment on hazardous alcohol consumption in the former soviet union. Health Place 27:205–211
Nielson TS, Hanson KB (2007) Do green areas affect health? Results from a Danish survey on the use of green areas and health indicators. Health Place 13(4):839–850

Orban E, McDonald K, Sutcliffe R et al (2015) Residential road traffic noise and high depressive symptoms after five years of follow-up: results from the Heinz Nixdorf recall study. Environ Health Perspect 124:5

Peen J, Schoevers RA, Beekman AT, Dekker J (2010) The current status of urban-rural differences in psychiatric disorders. Acta Psychiatr Scand 121:84–93

Pereira G, Wood L, Foster S, Haggar F (2013) Access to alcohol outlets, alcohol consumption and mental health. PLoS One 8(1):e53461

Pheasant RJ, Fisher MN, Watts GR, Whitaker DJ, Horoshenkov KV (2010) The importance of auditory-visual interaction in the construction of 'tranquil space'. J Environ Psychol 30:501–509

Saarloos D, Alfonso H, Giles-Corti B, Middleton N, Almeida OP (2011) The built environment and depression in later life: the health in men study. Am J Geriatr Psychiatry 19(5):461–470

Sciallo WG, Alexander G, Farrell KP (1992) Lead exposure and child behavior. Am J Public Health 82(10):1356–1360

Stathopoulou G, Powers MB, Berry AC, Smits JAJ, Otto MW (2006) Exercise interventions for mental health: a quantitative and qualitative review. Clin Psychol Sci Pract 13:179–193

Stewart R, Prince M, Harwood R, Whitley R, Mann A (2002) Quality of accommodation and risk of depression in later life: an analysis of prospective data from the gospel oak project. Int J Geriatr Psychiatry 17(12):1091–1098

Sugiyama T, Leslie E, Giles-Corti B, Owen N (2008) Associations of neighbourhood greenness with physical and mental health: do walking, social coherence and local social interaction explain the relationships? J Epidemiol Community Health 62:e9

Sygna K et al (2014) Road traffic noise, sleep and mental health. Environ Res 131:17–24

Szapocznik J, Mason CA, Brown SC, Lombard JL, Martinez F, Newman FL, Spokane AR (2009) The relationship of built environment to social behaviors and mental health in Hispanic elders: the role of "eyes on the street". J Gerontol B 64(2):234–246

Taylor MS, Wheeler BW, White MP, Economou T, Osborne NJ (2015) Research note: urban street tree density and antidepressant prescription rates – a cross-sectional study in London, UK. Landsc Urban Plan 136:174–179

Ulrich RS (1983) Aesthetic and affective responses to natural environment. In: Altman I, Wohlwill J (eds) Behaviour and the natural environment. Plenum Press, New York, pp p85–125

White MP, Alcock I, Wheeler BW, Depledge MH (2013) Would you be happier living in a greener urban area? A fixed-effects analysis of panel data. Psychol Sci 24(6):920–928

Whiteford HA, Degenhardt L et al (2013) Global burden of disease attributable to mental and substance use disorders: findings from the global burden of disease study 2010. Lancet 382 (9904):1575–1586

Wilson EO (1984) Biophilia – the human bond with other species. Harvard University Press, Cambridge, MA

Green Environment and Mental Health in the City

18

Ee Heok Kua and Angelia Sia

Abstract

Green urbanism is an important component of sustainable and liveable cities in developed and developing countries. Parks and gardens in cities have aesthetic, health, and social benefits, and they improve the quality of life in an increasingly urbanized society. Densely populated megacities with inadequate provision of green spaces are associated with poorer mental health. There is growing interest and research on horticultural therapy as a strategy to promote active aging. Studies on the use of urban parks and gardens have confirmed the reduction of stress and other mental health benefits. The natural environments have restorative characteristics which could mitigate stress and aid in the recovery from fatigue. Research has shown that exposure to green spaces can be psychologically and physiologically restorative by improving mental health, reducing blood pressure and stress levels, as well as encouraging physical activity. Some cities, including Singapore, are incorporating green environments into urban design as important aspects of city planning, given their possible positive influence on mental health and even cognitive functions.

Keywords

Urban parks • Horticultural therapy • Mental health • Restorative • Exposure to green spaces

E.H. Kua (✉)
Department of Psychological Medicine, National University Hospital, National University of Singapore, Singapore, Singapore
e-mail: ee_heok_kua@nuhs.edu.sg

A. Sia
National Parks Board, Ministry of National Development Centre for Urban Greenery and Ecology Research, Singapore, Singapore
e-mail: ANGELIA_SIA@nparks.gov.sg

N. Okkels et al. (eds.), *Mental Health and Illness in the City*, Mental Health and Illness Worldwide, https://doi.org/10.1007/978-981-10-2327-9_2

Contents

Introduction

Investments in creating access to the natural environment could yield important benefits for the "mental capital" of cities and countries (Beddington et al. 2008). It is incontrovertible that the economic health of cities and countries depends on the mental health of the people. Building a healthy urban environment should be a major policy priority in all countries, because by 2050 it is estimated that 66% of humans will live in cities, and even in less developed countries, the majority of people will be in urban areas (Dye 2008; World Health Organization 2014). The United Nations General Assembly had adopted the Sustainable Development Goals as a run-up to the Paris conference in December 2015, where countries agreed to cut carbon emissions. The broad goal of sustainable development is also further advanced by the UN Global Compact. However, cities' sustainability and liveability should not just focus on carbon reduction and controlling pollution, but there should also be an effort to increase the quantity and quality of urban greenery. Some of the Sustainable Development Goals which are relevant and supportive of the green environment include:

1. Take action to combat climate change.
2. Ensure healthy lives and promote well-being for all ages.
3. Sustainably manage forests.
4. Make cities inclusive, safe, resilient, and sustainable.

Green urbanism is of vital importance for sustainable and liveable cities in all developed and developing countries. Trees improve air quality, reduce cooling and heating energy use, and make the urban environment aesthetically more preferable (Smardon 1988; Akbari et al. 2001; Nowak et al. 2014). Parks provide fascination and inspiration to enable people to gain a fresh perspective on life, and green spaces in a city play an important role in helping visitors to escape temporarily from buildings and the crowded streets. A park experience rejuvenates the city dwellers and provides a sense of peacefulness (Ulrich 1981; Kaplan 1983). To Frederick Law Olmsted and other park visionaries, parks were not "amenities" but "necessities," providing recreation, inspiration, and essential respite from the city's blare and bustle. The Central Park in New York City, Hyde Park in London, and the Singapore Botanic Gardens in Singapore are modern-day examples of parks that soften and make surrounding high-density urban environment more liveable.

The presence of natural assets (i.e., parks and forest green belts) enriches human life with meaning and the experience of nature is a source of positive feelings that contribute to a better quality of life (Chiesura 2004). Besides aesthetic, psychological, and health benefits, natural features in cities have other social benefits. Trees and grass in outdoor common spaces promote the development of social relationships through interaction among residents – it helps to create a sense of community (Kuo et al. 1998) (Fig. 1).

Green urbanism is a valuable municipal planning approach and a key factor for city sustainability. Urban nature can provide economic benefits for both municipalities and citizens. Air purification by trees, for example, can reduce costs of atmospheric pollution on people's health. Furthermore, the aesthetic, historical, and recreational values of many urban parks increase the attractiveness of the city and promote the tourism industry, generating employment and revenue. The debate on sustainable and liveable cities must also consider atmospheric pollution from burgeoning industries which can have dire consequences on the green environment.

Fig. 1 City dwellers can easily access and experience nature in the City in a Garden of Singapore, which has a provision of over 300 parks and 4 nature reserves (Photo: Francis Lee)

In Southeast Asia, the periodic haze from the burning of trees and scrublands has blanketed many cities, impacting not just the environment and health but also the economy of the region.

Impact of Green Environment on Health

In his book "Fighting for Mental Health," Sartorius (2002) wrote eruditely on the mental health needs of an urbanized planet and commented that many governments of the world, faced with rampant urbanization, have not developed an effective strategy for the provision of health care. Although urbanization has many benefits, it is also associated with increased levels of mental disorders, such as anxiety disorders and depression (Wang 2004; Lederbogen et al. 2011). City dwellers have better access to healthcare and social services, improved sanitization, and amenities; however, they are also at greater risk for mental disorders due to a more stressful environment. A follow-up study in Sweden has shown that a high level of urbanization is associated with increased risk of psychosis and depression for both women and men (Sundquist et al. 2004).

Contact with nature promotes the physical and mental health of individuals and communities. There is now more attention on the type of nature close to where people live and work, such as gardens or parks in cities and their benefits to people. In the United Kingdom, an extensive survey used data from over 10,000 individuals to explore the relation between urban green space and mental distress as assessed by the General Health Questionnaire (Goldberg et al. 1997). Controlling for individual and regional covariates, the authors found that individuals had lower mental distress when living in urban areas with more green space. This effect was traced to the living location of the same individuals as they moved closer or further from green space (White et al. 2013), indicating that being close to nature or near small-scale green areas in cities is beneficial to urban dwellers. Other correlational studies reveal that window views that include natural elements (compared with window views that do not) are associated with improved attention, memory, and impulse inhibition as well as greater feelings of subjective well-being (Kaplan 2001).

Impact of Gardening on Health

There is increasing evidence to show that gardening can contribute to well-being in many ways. Unruh (2004) suggested that gardening could be adopted as a kind of coping strategy for a person with stressful life experiences, because the natural environment supports the healing process and the recovery for a person with mental disorder(s). For many city dwellers, gardening is a meaningful and enjoyable activity, and the experience of nature is a source of positive feeling. A garden provides a temporary escape from a stressful situation and acts as a positive distraction. People performing gardening tasks also often experience a sense of

Fig. 2 The benefits of gardening have been recognized by many cities. This example of Fort Mason Community Garden (San Francisco) offers more than 100 garden plots where members may grow vegetables and flowers as they wish (Photo: Tham Xin Kai)

accomplishment. Physical activity during gardening improves mental well-being by reducing feelings of stress, anxiety, and depression (Fig. 2).

Growing house plants is a popular recreational activity in China and Japan for centuries. The art of bonsai and Zen-Buddhist rock garden are entrenched in Chinese and Japanese traditions of gardening to maintain the equilibrium of physical and mental health. Exposure to indoor plants could have a positive effect on attention, physical health, and work productivity even in windowless environment (Wiesinger et al. 2006; Lohr et al. 1996).

Attention Restoration Theory

The attention restoration theory suggests that recovery from attentional fatigue requires a restorative environment like natural settings for attentional recovery and reflection (Kaplan and Kaplan 1989; Kaplan 1995; Herzog et al. 1997). Natural environments with pleasing aesthetic qualities including open views and lack of loud, distracting noises are often chosen as preferred restorative environments (Korpela et al. 2001; Hartig et al. 2003; Herzog 2003). Studies have shown that trees near public housing improved the life satisfaction of residents and decreased levels of fear and incivilities among residents (Kuo et al. 1998; Kuo and Sullivan 2001). In a paper on the restorative effects of the natural environment, Hartig and Staats (2003) commented that participants in the nature reserve experienced an increase in positive affect and reduction in anger, compared to those in the urban environment. These findings are supported by another study by Hartig et al. (2014) who reported that urban green space can improve attention and memory, and increase positive mood. Several other studies have also demonstrated that interacting with natural environments can have beneficial effects on attention and memory for healthy individuals and for patient populations (Berto 2005; Berman et al. 2008; Kaplan and Berman 2010).

Exposure to Nature Reduces Stress

Ulrich (1984) in an innovative study found that hospital patients who could look out at trees and nature from windows recovered more quickly from surgery than those whose views were restricted to buildings – supporting the observation that the natural environment has a positive influence on health. Using quantitative and qualitative data, Schroeder (1991) concluded that natural environments with plants induced a relaxed state in people compared with urban scenes with no vegetation. The benefits were due to the regeneration of emotional equilibrium, break from the daily routine, and the stimulation of a spiritual connection with the natural world.

Other Positive Impacts of Nature Experience

Based on the attention restoration theory of Kaplan (1995), a research on views of nature and impact on the self-discipline of inner city children in Chicago was published by Taylor et al. (2002). The project examined the relationship between near-home nature and self-discipline in 169 inner city girls and boys living in identical high-rise buildings with varying levels of nearby nature. Parent ratings of the naturalness of the view from home were used to predict the children's performance on tests of concentration, impulse inhibition, and delay of gratification. The results showed that the more nature a girl could view from home, the better her test performance. However, nature views from home had no relationship with performance on the tests for boys.

Green space is thought to influence mental health through an increase in physical activity by providing a place for neighborhood residents to meet and facilitate social ties. Community studies in the Netherlands have shown that the prevalence rates of anxiety and depression are higher in areas with lower levels of local green space (de Vries et al. 2003; Maas et al. 2009; van den Berg et al. 2010). However, other studies have indicated that greater access to residential green space is associated with less depression, but there is less evidence of an association between green space and decreased anxiety (Annerstedt et al. 2012; Alcock et al. 2014). In Japan, an interesting study by Takaro et al. (2002) found that the amount of "walkable green space" around the home of elderly individuals was associated to their mortality rate 5 years later – they found lower mortality among those with higher green-space proximity.

More recently, there was research in Canada from the Ontario Health Study focusing on a large urban population. It combined high-resolution satellite imagery and tree data with questionnaire-based self-reports of general health perception, cardiometabolic conditions, and mental illnesses. The results from multiple regressions and multivariate correlation analyses showed that people who lived in neighborhoods with a higher density of trees on their streets reported significantly higher health perception and significantly less cardiometabolic conditions (controlling for socioeconomic and demographic factors (Kardan et al. 2015)). These compelling results should encourage city planners to support green urbanism to enhance liveability and promote public health.

Horticultural Therapy

There is now a growing interest and research on horticultural therapy which taps on nature's benefits for human health. In horticultural therapy, recreational activities involving plants, gardening, and closeness to nature are used as methods in conducted programs to promote health. There is a difference between horticultural therapy and horticultural activity. Horticultural therapy involves a planned program that uses plants as a therapeutic medium by a trained professional to achieve a clinically defined goal. On the other hand, horticultural activity is working with plants as a recreational activity without having a predefined therapeutic goal. Horticultural therapy is well established in many countries including Australia, New Zealand, Germany, USA, Canada, United Kingdom, Sweden, Japan, Korea, and recently in China. In some countries, it has been introduced in educational institutions like kindergartens, schools, and special-needs schools. Horticultural therapy is well accepted in medical care facilities especially in rehabilitation hospitals, psychiatric services, palliative care, and aged homes to provide patients with graded and carefully designed gardening activities to improve their quality of life. Working or walking in the healing garden promotes rehabilitation in persons with disabilities (Fling 2006). A Korean study explored its benefits in lifting the self-esteem and mood of women who were victims of domestic violence (Lee et al. 2008) (Fig. 3).

Horticultural therapy has now been widely adopted as a therapeutic activity for persons with different psychological needs (Kaplan and Kaplan 1989). It has been noted to promote the social functioning and self-esteem of patients with psychiatric illness especially those with chronic schizophrenia (Son et al. 2004). In a study on clinically depressed patients, Gonzalez et al. (2011) showed that a 12-week therapeutic horticulture program significantly lowered their depression severity during the intervention, and this beneficial change remained stable even 3 months after the program. They also found that the program fostered group cohesiveness which facilitated the recovery of the patients, and there was increased social activity even after the program in 38% of the patients. In an earlier study of clinically depressed patients, Gonzalez et al. (2010) had shown that the program helped to decrease rumination and depression.

Fig. 3 A horticultural therapy session conducted by Ms. Elizabeth Diehl, Director of Therapeutic Horticulture, College of Medicine at the University of Florida. (Photo: Jesse S. Jones)

The effects of horticultural therapy on mood and heart rate in cardiac patients were investigated by Wichrowski et al. (2005). They found that in a cardiopulmonary rehabilitation program, compared to a control group who attended education classes, patients on horticultural therapy had improved mood state and lower heart rate. Kam and Siu (2010), in a study in Hong Kong, investigated the effect of horticultural therapy on work performance and quality of life of persons with psychiatric illness. In the single-blind randomized controlled trial, 24 participants with psychiatric illness (schizophrenia, bipolar disorder, and major depression) were recruited to participate in a horticultural program and were randomly assigned to experimental and control groups. The participants were assessed on mood, personal well-being, and work behavior before and after the program. There was a significant difference in scores on the mood scale between the experimental and control groups but no difference in personal well-being or work behavior. Although the study was well designed, the sample was small and the follow-up of 2 weeks was short. With the limitations, the authors concluded that horticultural therapy was effective in decreasing the levels of anxiety and depression among participants in this pilot study but not in improving quality of life or work behavior.

In another Korean study, occupational therapists investigate the effects of horticultural therapy on the physical and psychological rehabilitation of patients with stroke (Kim et al. 2010). There were 20 stroke patients on occupational therapy plus horticultural therapy and a control group of 20 stroke patients who were on occupational therapy only. It was not a randomized controlled trial and the follow-up of 4 weeks was short. However, the authors found that those patients on occupational therapy plus horticultural therapy improved significantly on depression, communication, and self-care compared to the control group.

A study on horticultural therapy with dementia patients conducted by Jarrott and Gigliotti (2010) demonstrated that it was a desirable choice for dementia-care programs, because it engaged the patients who were often difficult to engage in activities that elicited high levels of adaptive behavior. The authors concluded that horticultural therapy represented an appropriate activity for older adults with dementia for it promoted positive engagement and could be introduced in care settings, facilitating a good person-environment fit.

However, in a systematic review by Hiroharu et al. (2014), the authors found that there were too few randomized controlled trials to draw any convincing conclusion on the efficacy of horticultural therapy. They concluded that many studies had poor methodology, small sample size, and the results were not robust. Most of these studies were naturalistic with biased samples, and the types of patients were too heterogeneous. There was also a paucity of data from Asia, Africa, or Latin America. Notwithstanding, the authors believed that horticultural therapy might still be helpful to improve the quality of life of some patients with dementia, depression, schizophrenia, and cancer. In all these disorders, horticultural therapy is not the main method of treatment and is introduced as an adjunctive therapy to improve social functions as shown in the study by Son et al. (2004) with chronic schizophrenia patients, and social connectedness as described by Gonzalez et al. (2011) in their prospect study of clinically depressed patients.

To address the gaps on horticultural therapy research, the National Parks Board of Singapore, in collaboration with the National University Health System, conducted a randomized controlled trial at the Training and Research Academy at Jurong Point (Ho et al. 2016). The study aimed to investigate the efficacy of horticultural therapy in improving the mental health and cognitive functioning of elderly people aged 65 years or older living in public housing in the community. There were 69 elderly divided into two groups – treatment (horticultural therapy) and control (wait list). Informed consent was obtained from the participants, and permission for the study was given by the university's ethics committee. The participants were assessed on neuropsychological tests using the Montreal Cognitive Assessment (Freitas et al. 2013) and questionnaires on life satisfaction (Diener et al. 1985) and social connectedness (Hawthorne 2006). To assess the psychological well-being of participants, the Zung Self-Rating Depression Scale (Zung 1965) and Zung Self-Rating Anxiety Scale (Zung 1971) were used. Blood samples were taken to check the levels of biomarkers like cytokines.

The program for the participants involved a weekly horticultural therapy session for the first 3 months, followed by monthly sessions for the next 3 months. The detailed program was designed by the National Parks Board to cultivate participants' interest in gardening and park visits. During the sessions, they planted vegetables such as lady's fingers and learnt gardening techniques like composting and transplanting. When the crops were ready for harvest, participants used the crops to prepare a soup dish together. The participants also visited the Singapore Botanic Gardens, Sungei Buloh Wetland Reserve, and Gardens by the Bay. These visits were facilitated by the respective park guides who shared their knowledge on the plants and the landscape features with the participants.

To evaluate the efficacy of the program, assessment interviews were conducted before the program, at a 3-month point during the program, and after the program ended. In the preliminary results after 3 months, the elderly who received horticultural therapy did better than the control group in terms of their scores for life satisfaction, memory, and psychological well-being, but the differences were not significant. The only significant finding ($p < 0.05$) was for positive relation, which refers to social connectedness and trusting relationships. During the assessment of the biomarkers, it was found that in the sixth month of the program, the level of cytokine interleukin-6 (IL-6), a pro-inflammatory protein, was significantly reduced, as opposed to the control group. As cytokines have been found to be involved in various negative age-related health outcomes, having high psychological well-being may have a biologically protective function. Hence, the results showing significantly lower IL-6 and higher positive well-being scores may be potential factors in reducing the risk of depression. A major beneficial effect of the green environment could thus be brought about by immunoregulation.

In attempting to understand the neuropsychological basis of horticultural therapy, Kaplan's attention restoration theory suggested that nature has specific restorative effects on the prefrontal cortex-mediated executive attentional system. High levels of engagement with technology and multitasking place demands on executive attention to switch between tasks, maintain task goals, and inhibit irrelevant actions. The theory suggests that interaction with the natural environment is particularly effective in replenishing depleted attentional resources in the brain. This theory may explain the improved mental status of busy city dwellers who spend some time at the parks or gardens. But is the same theory applicable for the patient with depression, anxiety, or cognitive impairment?

How horticultural therapy influences the brain neural systems is a new frontier of scientific research for neuroscientists. An intriguing question is: What mechanisms link decreased nature experience to the development of mental health problems? Bratman et al. (2015) have suggested that one such mechanism might be the impact of nature exposure on rumination, which is associated with increased risk for depression and anxiety. They have shown that in healthy participants, a brief nature experience in a 90-min walk in a natural setting could decrease both self-reported rumination and neural activity in the subgenual prefrontal cortex (sgPFC), whereas a 90-min walk in an urban setting had no such effects on self-reported rumination or neural activity. This suggestion draws support that at a neurobiological level, nature experience leads to decreases in sgPFC activity, a brain region that has been shown to be associated with a self-focused behavioral withdrawal linked to rumination in both depressed and healthy people. Some new studies support the finding that the sgPFC has increased activity during sadness, withdrawal, and negative self-reflective processes (Mayberg et al. 1999; Krishnan and Nestler 2008; Kross et al. 2009; Berman et al. 2011). Rumination is a maladaptive pattern of self-referential thought that is associated with risk for depression and anxiety and with activity in the subgenual prefrontal cortex (Nolen-Hoeksema 2000; Hamilton et al. 2015).

There are still many unanswered questions on the neurobiology of horticultural therapy. How does being in a park, garden, or forest alter neurotransmitters in the brain and allay anxiety and depression, or improve sociability? Future research on horticultural therapy could explore some of these questions in neurobiology using new tools of scientific studies in brain imaging, neurochemistry, neuroimmunology, or genetics.

The City in a Garden: A Case Study of Singapore

Situated at the Southeastern tip of the Asian land mass, the island of Singapore has become a busy international port along the major sea land from Europe to China. Before Singapore became a British colony in 1819, the island was covered with natural forest. During British rule, the island was designated a tropical crop-testing area, and vast forested areas were cleared for crop-testing. The deforestation also made way for the growth of rubber, pepper, and gambier plantations (Turnbull 1996). An experimental laboratory for botanical research was also established, in the form of the Singapore Botanic Gardens, which was founded in 1859. When the mental asylum, Woodbridge Hospital, was built in 1928, the Colonial Governor stipulated that the Medical Superintendent should be an officer "with some taste for gardening and farming, who will help to make the patients interested in such pursuits" (Kua 2004). The mental patients in Woodbridge Hospital were encouraged to be involved in gardening as an activity of occupational therapy. Like most mental asylums in the British Empire, Woodbridge Hospital was located away from the city and had extensive grounds around the wards – besides flowering plants, vegetables and fruits were grown for the hospital kitchen.

With a surge of immigrants from China, India, and Indonesia in the last century, Singapore's population grew. The greening of modern Singapore started in 1963 when then Prime Minister, Mr. Lee Kuan Yew launched the first Tree Planting Campaign by planting a tree and asked for 10,000 trees to be planted a year. In the early 1960s because of the rapid industrialization and urbanization of the country, there was an effort to green up areas wherever possible, to make up for trees destroyed due to economic development (Wong et al. 2014). The Garden City Campaign was later introduced by Mr. Lee Kuan Yew in 1967 to transform Singapore into a city with lush greenery and a clean environment in order to make life more pleasant for the people. In 1968, the Parks and Trees Division (which later expanded to become the Parks and Recreation Department) was set up within the former Roads Branch in the Public Works Department. This was followed by the formation of the School of Ornamental Horticulture at the Singapore Botanic Gardens in 1972. Green urbanism cannot possibly succeed without political will. There was political resolve when the Government set up a "Garden City Action Committee" tasked with overseeing policies and coordinating activities of various government agencies for building a city with abundant greenery, parks, birdlife, and lush street planting.

The land area of Singapore covers approximately 718.3 km^2. Given the small land size and a high population of 5.61 million (Department of Statistics, Singapore, June 2016), it faces multifaceted challenges in reconciling liveability, environmental sustainability, and economic competitiveness. To overcome the issue of land constraints, Singapore needs to consider industries that it needs for economic growth and competitiveness but do not require huge land areas. Working collaboratively across government departments and actively engaging both the private sector and people are also key in coming up with effective solutions (Khoo 2014).

Hence, the urban planning authority takes on a consultative approach in its master planning. To ensure a pleasant city for living in, visiting, and business, about 9% of land in Singapore has been allocated for parks and nature reserves. There are 24 designated nature areas including the four nature reserves at Bukit Timah Nature Reserve, Central Catchment Nature Reserve, Sungei Buloh Wetland Reserve, and Labrador Nature Reserve. To provide shade from the tropical heat, strong and fast-growing species like the Rain Tree and the Angsana were planted and are ubiquitous today. It is envisioned that 85% of residents will live within 400 m of a park by 2030. There will also be the proposed Round Island Route, stretching 150 km, which will provide an uninterrupted leisure route around the island, linking major cultural and historical sites, and cycling networks. These green spaces provide ample opportunities for city dwellers to exercise and interact besides enjoying the flowering plants, trees, and lakes.

Singapore's master plan for a clean, green, and blue sustainable city is making a paradigm shift from a "Garden City" to a "City in a Garden." The concept of biophilic urbanism encourages roof and wall greening as well as balcony gardening. There will be roads with complete canopy cover and heritage tree programs, and green connections across the city. To promote green urbanism, community participation is important and there are programs to encourage the public to be part of the green movement and adopt responsible practices and healthy lifestyles. There are also gardening workshops for schools and the community, and regular free concerts in the parks for the public (Figs. 4, 5, and 6).

Fig. 4 To ensure a pleasant city for living, about 9% of land in Singapore has been allocated for parks and nature reserves. Featured below is the 17.3 ha Zhenghua Nature Park, located in Bukit Panjang neighborhood (Photo: National Parks Board)

Fig. 5 Throughout the decades, Singapore has worked towards integrating greenery into its built environment so as to enhance its liveability. This example features a vertical green wall at the Ocean Financial Centre (Photo: Lok Yan Ling)

Fig. 6 Nature programs for school children conducted in the Sungei Buloh Wetland Reserve (Photo: National Parks Board)

Building technologies for a green city and capabilities to fulfill sustainable development are important. Urban agriculture includes orchards for harvesting fruits to crops growing at different scales, from large plots to small rooftop gardens and lawn spaces around residential estates. Urban agriculture is now a growing interest – it provides fresh food at a low cost at the local supermarkets and encourages the integration of composting facilities to recycle household organic waste. Many private enterprises have seized these opportunities with governmental support to establish start-ups which can potentially be expanded to other Asian cities looking for new ideas in green urbanism. A biophilic city endorses natural processes in the landscaping of buildings – the increased greenery increases energy efficiency in buildings.

The Singapore experience has shown that for green urbanism to flourish, there should be adequate water management and conservation to develop resilience against adverse climate conditions. There is a national endeavor to focus on the ecosystem services provided by the green infrastructure to ensure the sustainability of the city. Many private and public organizations work together to enhance sustainability and liveability in the dense urban environment. Singapore has now about

72 ha of rooftop greenery – the size of over 100 football fields. Contributing to this landscape are buildings, including hotels and hospitals, which have been retrofitted with greenery on their roofs and walls, and new developments that incorporate greenery in the building designs. In the area of nature conservation, legislation is in place. For instance, written approval must be obtained for the removal of trees in a designated tree conservation area, vacant land, or a heritage road green buffer. Tree planting verges are a required provision for various categories of roads.

Going beyond the provision of adequate green spaces, the National Parks Board has a keen interest to optimize recreational parkland to promote public health. Since 2015, it has started to carry out research on the health benefits of nature such as horticultural therapy and its impact on the mental health of the elderly. The outcomes of the research study on horticultural therapy mentioned earlier provide empirical evidence for the effectiveness of horticultural therapy in improving the psychosocial and cognitive well-being of the elderly and reducing healthcare costs. With its clinical implications, this initiative could potentially persuade policymakers and healthcare providers of the need to support cost-effective horticultural therapy as a strategy to engage older adults in a more active and healthier lifestyle.

Urban parks are easily accessible to all, making them an ideal and cost-effective resource to support the promotion of community health. This preventive program is intended to promote active aging and improve the mental health and cognitive function of the elderly. It offers an enjoyable experience to the elderly, and potentially they can be motivated to appreciate nature and engage in gardening even after the program ends. This change in lifestyle could enhance their physical activity levels and consequently improve their health. Community-based horticultural therapy may provide an avenue for older individuals to socialize more and improve their mental health. After retirement, many are more likely to lead a sedentary lifestyle, and for those who are living alone, participation in horticultural therapy will become a meaningful plant nurturing activity and help build social connectedness with other older individuals. Such community-based interventions supported by volunteers are important as they have the potential to be sustained (Figs. 7 and 8).

Fig. 7 Singapore implemented the Therapeutic Garden @ HortPark in May 2016 to promote mental wellbeing (Photo: Chelsea Sia)

Fig. 8 The Therapeutic Garden @ HortPark has an area specially designed for gardening, with ample shade and moveable raised beds (Photo: Chelsea Sia)

Conclusion

Green spaces are of a strategic importance for the quality of life of an increasingly urbanized society. The natural environment not only enhances the aesthetic landscape of the city but also fosters social connectedness within neighborhoods by providing settings for groups to meet formally and informally for recreational or leisure pursuits. Besides improving the environment in air purification and noise filtering, natural areas are of crucial importance for the liveability of modern cities and the well-being of city dwellers. With the fast pace in city life, urban parks offer city dwellers a respite from the daily grind.

Densely populated megacities with the inadequate provision of green spaces are associated with poorer mental health. Studies on the use of urban parks and forests have confirmed the belief about stress-reduction benefits and mental health (Hartig et al. 1991). The natural environments have restorative effects which could mitigate stress and aid in the recovery from fatigue. Several other research studies have shown that exposure to green spaces can be psychologically and physiologically restorative by promoting mental health, reducing blood pressure and stress levels, as well as promoting physical activity (Humpel et al. 2002; Pretty et al. 2005; Richardson et al. 2013). Some cities, including Singapore, are incorporating benefits of green environment into urban design as important aspects of city planning that may influence mental health and even cognitive functioning (Jackson 2003; van Dillen et al. 2012; Ward Thompson et al. 2012). The multiple benefits of greenery extend to users of different profiles. The spin-offs for "economical" health arising from natural environment include the capacity for parks and gardens to improve the productivity of workers and recovery rates of patients in hospital. Larsen et al. (1998) have suggested that contact with nature could improve overall health, as well as reduce the number of sick days and employee absences. With better understanding, the mental health benefits of nature can be incorporated into a wide array of initiatives and investments in sustainable cities and conservation.

However, as mentioned by Hiroharu et al. (2014), there should be better research in horticultural therapy in randomized controlled trials to provide cogent evidence on its health benefits. What is it in plants, forest, or gardening that is therapeutic? Should there be more neurobiological studies to understand the underlying mechanisms in horticultural therapy? At the moment horticultural therapy is still an adjunctive therapy for depression, anxiety, and early dementia. Understanding the neuroscience is important to provide a better insight into green urbanism and mental health. As a method to promote mental health, horticultural therapy is an important strategy for preventive psychiatry.

Governmental support is sine qua non in green urbanism. It is not just the responsibility of city planners and politicians but also the community including mental health professionals who can have an advocacy role. Involvement in the green movement will allow the community to articulate commonly shared values on the environment. Being volunteers in the planning of a therapeutic garden for elderly at the HortPark in Singapore has provided a platform for the psychiatrists involved to address issues concerning aging and mental health – it is also an opportunity for the destigmatization campaign to speak up for mental illness. In several Asian mental hospitals today, horticultural therapy has been introduced for patients with chronic schizophrenia. In Indonesia, Thailand, and Malaysia, a few hospitals have fruits and vegetable plantations – small enterprises which have given patients much satisfaction and better quality of life. In a recent psychiatry meeting in Singapore, some Asian leaders of psychiatry shared their experiences and favorite leisure activity, which fortuitously was gardening, a hobby decreed by His Majesty's Government in 1928 (Kua 2011).

The natural environment constitutes a health promotion platform with potential for the prevention of mental health disorders (Pryor et al. 2006). Public health needs to focus on both the social and environmental aspects of human health. A socio-ecological approach to health enhances human health, social health, and environmental health. Such a strategy with a socioecological approach to public health and the natural environment will increase sustainability for all. Collaboration among the key stakeholders including government departments, nongovernmental bodies, volunteer organizations, and the general public is crucial.

City planners will face difficult decisions about building and conserving green cities, promoting economically growing cities, and advocating social justice – these are issues of the environment, politics, and economics (Campbell 1996). There is always tense economic debate about urban redevelopment and the preservation of the city's park and green spaces. There are challenges in green urbanism that will impact the United Nations Sustainability Development Goals. It is undeniable that long-term city sustainability is dependent on having policies which can address and advocate improvement in carbon reduction, resource efficiency in energy, water conservation, and green urbanism.

Cross-References

▶ Revisiting the Virtuous City: Learning from the Past to Improve Modern Urban Mental Health
▶ Urban Design and Mental Health

References

Akbari H, Pomerantz M, Taha H (2001) Cool surfaces and shade trees to reduce energy use and improve air quality in urban areas. Sol Energy 70:295–310

Alcock I, White MP, Wheeler BW et al (2014) Longitudinal effects on mental health of moving to greener and less green urban areas. Environ Sci Technol 48:1247–1255

Annerstedt M, Ostergren PO, Bjork J et al (2012) Green qualities in the neighbourhood and mental health – results from a longitudinal cohort study in Southern Sweden. BMC Public Health 12:337

Beddington J, Cooper CL, Field J, Goswami U, Huppert F et al (2008) The mental wealth of nations. Nature 455(7216):1057–1060

Berman MG, Jonides J, Kaplan S (2008) The cognitive benefits of interacting with nature. Psychol Sci 19:1207. doi:10.1111/j.1467-9280.2008.02225.x

Berman MG et al (2011) Depression, rumination and the default network. Soc Cogn Affect Neurosci 6(5):548–555

Berto R (2005) Exposure to restorative environments helps restore attentional capacity. J Environ Psychol 25:249. doi:10.1016/j.jenvp.2005.07.001

Bratman GN, Hamilton JP, Hahn KS, Daily GC, Groso JJ (2015) Nature experience reduces rumination and subgenual prefrontal cortex activation. PNAS 112:8567–8572

Campbell S (1996) Green cities, growing cities, just cities? Urban planning and the contradictions of sustainable development. J Am Plan Assoc 62:296–312

Chiesura A (2004) The role of urban parks for the sustainable city. Landsc Urban Plan 68:129–138

De Vries S, Verheij RA, Groenewegen PP, Spreeuwenberg P (2003) Natural environments – healthy environments? An exploratory analysis of the relationship between green-space and health. Environ Plan A 35:1717–1731

Diener E, Emmons RA, Larsen RJ, Griffin S (1985) The satisfaction with life scale. J Pers Assess 49:71–75

Dye C (2008) Health and urban living. Science 319:766–769

Eling M (2006) Persons-plant interaction: the physiological, psychological and sociological effects of plants on persons. In: Hassink J, VanDijk M (eds) Farming for health. Springer, Dordrecht, pp 43–55

Freitas S, Simões MR, Alves L, Santana I (2013) Montreal cognitive assessment: validation study for mild cognitive impairment and Alzheimer disease. Alzheimer Dis Assoc Disord 27:37–43

Goldberg D, Gater G, Sartorius N, Ustun T, Piccinelli M, Gureje O, Rutter C (1997) The validity of two versions of the GHQ in the WHO study of mental illness in general health care. Psychol Med 27:191–197

Gonzalez MT, Hartig T, Patil GG, Martinsen EW, Kirkevold M (2010) Therapeutic horticulture in clinical depression: a prospective study of active components. J Adv Nurs 66(9):2002–2013
Gonzalez MT, Hartig T, Patil GG, Martinsen EW, Kirkevold M (2011) A prospective study of group cohesiveness in therapeutic horticulture for clinical depression. Int J Ment Health Nurs 20:119–129
Hamilton JP, Farmer M, Fogelman P, Gotlib IH (2015) Depressive rumination, the default-mode network, and the dark matter of clinical neuroscience. Biol Psychiatry. doi:10.1016/j.biopsych.2015.02.020
Hartig T, Staats H (2003) Guest editors' introduction: restorative environments. J Environ Psychol 23(2):103–107
Hartig T, Mang M, Evans G (1991) Restorative effects of natural environments experiences. Environ Behav 23:3–26
Hartig T, Evans GW, Jamner LD, Davis DS, Gărling T (2003) Tracking restoration in natural and urban field settings. J Environ Psychol 23:109–123
Hartig T, Mitchell R, de Vries S, Frumkin H (2014) Nature and health. Annu Rev Public Health 35:207–228
Hawthorne G (2006) Measuring social isolation in older adults: development and initial validation of the friendship scale. Soc Indic Res 77:521–548
Herzog TR (2003) Assessing the restorative components of environments. J Environ Psychol 23(2):159–170
Herzog TR, Black AM, Fountaine KA, Knotts DJ (1997) Reflection and attentional recovery as distinctive benefits of restorative environment. J Environ Psychol 17:165–170
Hiroharu K, Kiichiro T, Minoru Y, Hyuntae P, Hitoyasu O et al (2014) Effectiveness of horticultural therapy: a systemic review of randomized controlled trials. Complement Ther Med 22:930–943
Ho R, Ng KH, Chan HY, Ng M, Sia A, Mahendran R (2016) Gardening and mental health of the elderly. Presented at the annual scientific conference on ageing, Singapore, 5 May 2016
Humpel N, Owen N, Leslie E (2002) Environmental factors associated with adults' participation in physical activity: a review. Am J Prev Med 22:188–199
Jackson LE (2003) The relationship of urban design to human health and condition. Landsc Urban Plan 64(4):191–200
Jarrott SE, Gigliotti CM (2010) Comparing responses to horticultural-based and traditional activities in dementia care program. Am J Alzheimer's Dis Other Demen 25:657–665
Kam MCY, Siu AMH (2010) Evaluation of a horticultural activity programme for persons with psychiatric illness. Hong Kong J Occup Ther 20:80–86
Kaplan R (1983) The analysis of perception via preference: a strategy for studying how the environment is experienced. Landsc Urban Plan 12:161–176
Kaplan S (1995) The restorative benefits of nature: toward an integrative framework. J Environ Psychol 15:169–182
Kaplan R (2001) The nature of the view from home. Environ Behav 33(4):507–542
Kaplan S, Berman MG (2010) Directed attention as a common resource for executive functioning and self-regulation. Perspect Psychol Sci 5:43. doi:10.1177/1745691609356784
Kaplan R, Kaplan S (1989) The experience of nature: a psychological perspective. Cambridge University Press, New York, pp 177–200
Kardan O, Gozdyra P, Misic B, Moola F, Palmer LJ et al (2015) Neighbourhood greenspace and health in a large urban centre. Sci Report 5:11610. doi:10.1038/srep11610
Khoo (2014) Urbanization in Singapore. Making small more livable http://www.mayorsandcities.com/politics/world/urbanization-singapore-making-small-livable
Kim MY, Kim GS, Mattson NS, Kim WS (2010) Effects of horticultural therapy on the physical and psychological rehabilitation of patients with hemiplegia after stroke. Korean J Hortic Sci Technol 28:884–890
Korpela KM, Hartig T, Kaiser FG, Fuhrer U (2001) Restorative experience and self-regulation in favourite places. Environ Behav 33(4):572–589

Krishnan V, Nestler EJ (2008) The molecular neurobiology of depression. Nature 455(7215): 894–902

Kross E, Davidson M, Weber J, Ochsner K (2009) Coping with emotions past: the natural bases of regulating affect associated with negative autobiographical memories. Biol Psychiatry 65(5): 361–366

Kua EH (2004) Focus on psychiatry in Singapore. Br J Psychiatry 185:79–82

Kua EH (2011) Woodbridge Hospital – British psychiatry in Singapore. Br J Psychiatry 199:442

Kuo FE, Sullivan WC (2001) Environment and crime in the inner city. Does vegetation reduce crime. Environ Behav 3(33):343–367

Kuo FE, Bacaioca M, Sullivan WC (1998) Transforming inner city landscapes: trees, sense of safety, and preferences. Environ Behav 1(30):28–59

Larsen L, Adams J, Deal B, Kweon BS, Tyler E (1998) Plants in the workplace: the effects of plant density on productivity, attitudes, and perceptions. Environ Behav 30:261–282

Lederbogen F et al (2011) City living and urban upbringing affect neural social stress processing in humans. Nature 474(7352):498–501

Lee S, Kim MS, Suh JK (2008) Effects of horticultural therapy on self-esteem and depression of battered women at a shelter in Korea. Acta Hortic 790:139–142

Lohr VI, Pearson-Mims CH, Goodwin GK (1996) Interior plants may improve worker productivity and reduces stress in windowless environment. J Environ Hortic 14:97–100

Maas J, van Dillen SMJ, Verheij RA, Spreeuwenberg P, Schellevis FG, Groenewegen PP (2009) Morbidity is related to a green living environment. J Epidemiol Community Health 63:967–973

Mayberg HS et al (1999) Reciprocal limbic-cortical function and negative mood: converging PET findings in depression and normal sadness. Am J Psychiatry 156(5):675–682

Nolen-Hoeksema S (2000) The role of rumination in depressive disorders and mixed anxiety/depressive symptoms. J Abnorm Psychol 109(3):504–511

Nowak DJ, Hirabayashi S, Bodine A, Greenfield E (2014) Trees and forest effects on air quality and human health in the United States. Environ Pollut 193:119–129

Pretty J, Peacok J, Sellens M, Griffin M (2005) The mental and physical health outcomes of green exercise. Int J Environ Health Res 15:319–337

Pryor A, Townsend M, Maller C, Field K (2006) Health and well-being naturally: contact with nature in health promotion for targeted individuals, communities and populations. Health Promot J Austr 17:114–123

Richardson EA, Pearce J, Mitchell R, Kingham S (2013) Role of physical activity in the relationship between urban green space and health. Public Health 127:318–324

Sartorius N (2002) Fighting for mental health. Cambridge University Press, Cambridge

Schroeder HW (1991) Preferences and meaning of arboretum landscapes: combining quantitative and qualitative data. J Environ Psychol 11:231–248

Smardon RC (1988) Perception and aesthetics of the urban-environment – review of the role of vegetation. Landsc Urban Plan 15:85–106. doi:10.1016/0169-2046(88)90018-7

Son KC, Um SJ, Kim SY, Song JE, Kwack HR (2004) Effect of horticultural therapy on the changes of self-esteem and sociality of individuals with chronic schizophrenia. Acta Hortic 639:185–191

Sundquist K, Frank G, Sundquist J (2004) Urbanisation and incidence of psychosis and depression. Br J Psychiatry 184:293–298

Takaro T, Nakamura K, Watanabe M (2002) Urban residential environments and senior citizens' longevity in megacity areas: the importance of walkable green spaces. J Epidemiol Community Health 56:913–918

Taylor AF, Kuo FE, Sullivan WC (2002) Views of nature and self-discipline: evidence from inner city children. J Environ Psychol 22(1-2):49–63

Turnbull CM (1996) A history of Singapore: 1891–1988, 2nd edn. Oxford University Press, Oxford

Ulrich RS (1981) Natural versus urban sciences: some psycho-physiological effects. Environ Behav 13:523–556

Ulrich RS (1984) View through a window may influence recovery from surgery. Science 224:420–421

Unruh AM (2004) The meaning of gardens and gardening in daily life: a comparison between gardeners with serious health problems and healthy participants. Acta Hortic 639:67–73
van den Berg AE, Maas J, Verheij RA, Groenewegen PP (2010) Green space as a buffer between stressful life events and health. Soc Sci Med 70:1203–1210
van Dillen SME, de Vries S, Groenewegen PP, Spreeuwenberg P (2012) Greenspace in urban neighbourhoods and residents' health: adding quality to quantity. J Epidemiol Community Health 66(6):e8
Wang JL (2004) Rural-urban differences in the prevalence of major depression and associated impairment. Soc Psychiatry Psychiatr Epidemiol 39(1):19–25
Ward Thompson C et al (2012) More green space is linked to less stress in deprived communities: evidence from salivary cortisol patterns. Landsc Urban Plan 105(3):221–229
White MP, Alcock I, Wheeler BW, Depledge MH (2013) Would you be happier living in a greener urban area? A fixed-effects analysis of panel data. Psychol Sci 24(6):920–928
Wichrowski M, Whiteson J, Haas F, Mola A, Rey MJ (2005) Effects of horticultural therapy on mood and heart rate in patients participating in an inpatient cardiopulmonary rehabilitation program. J Cardpulm Rehabil 25:270–274
Wiesinger G, Neuhauser F, Putz M (2006) Farms, horticultural therapy, animal-assisted therapy. In: Hassink J, Dijk M (eds) Farming for health. Springer, Dordrecht, pp 233–248
Wong YK, Chua SE, Lee SK, Choo TS, Lee P, Lau-Tan A (2014) Garden city Singapore – the legacy of Lee Kuan Yew. Suntree Media Pte Ltd, Singapore
World Health Organization (2014) World urbanization prospects. World Health Organization, Geneva
Zung WWK (1965) A self-rating depression scale. Arch Gen Psychiatry 12:63–70
Zung WWK (1971) A rating instrument for anxiety disorders. Psychosomatics 12:371–379

Revisiting the Virtuous City: Learning from the Past to Improve Modern Urban Mental Health

19

Carolyn Stephens, Alfredo Gutierrez Carrizo, and Abbas Ostadtaghizaddeh

Abstract

This chapter looks at the evidence on current urban mental health in the context of rapid global change, both environmentally and socially. We summarize relevant epidemiological evidence on urban mental health, looking particularly at studies from Asia, Latin America, and Africa. We then look at two complementary major philosophical ideas from urban thinkers of the past – the Ideal or Utopian City and the Garden City, and discuss how these ideas are relevant for the urban mental health in the twenty-first century. We conclude with the importance of incorporating all human wisdom from the past millennia to deal with the future challenge of creating mentally healthy and sustainable cities.

Keywords

Urban equity • Urban-rural differences • Philosophy and architecture • Urban design • Climate change and mental health • Garden City • Utopian City • Sustainable cities

C. Stephens (✉)
UCL Institute of Health Equity, London, UK

London School of Hygiene and Tropical Medicine, London, UK

Universidad Nacional de Tucumán, Tucumán, Argentina
e-mail: Carolyn.stephens@lshtm.ac.uk

A. Gutierrez Carrizo
Artech, Tucumán, Argentina
e-mail: Alfredo.gutierrez@uclmail.net

A. Ostadtaghizaddeh
Department of Disaster and Public Health, School of Public Health, Tehran University of Medical Sciences, Tehran, Iran
e-mail: a-otadtaghizadeh@razi.tums.ac.ir; ostadtaghizadeh@gmail.com

N. Okkels et al. (eds.), *Mental Health and Illness in the City*, Mental Health and Illness Worldwide, https://doi.org/10.1007/978-981-10-2327-9_14

Contents

Introduction

> Damascus surpasses all other cities in beauty, and no description, however full, can do justice to its charms
>
> Ibn Battuta (Battuta and Lee 2013) (Travels of Ibn Battuta 1325–1354)

The world has had cities and towns since at least 4,000 years BC. Damascus, described by Ibn Battuta in the fourteenth century, is, in 2016, perhaps popularly best known as the embattled capital of war-torn Syria. Yet, the beautiful city of Damascus was established in the third millennium BC and is one of the few cities in the world that has been continuously inhabited since this time until 2016. Thus, for millennia there have been urban societies, and, since these early cities, diverse philosophers and intellectuals have tried to make the urban space an articulation of the aspirations of each culture and society, including aspects of moral, spiritual, and material well-being. When Al-Farabi wrote his *Treatise on the Views of the Residents of the Virtuous City* in roughly 948 AD, he was synthesizing Arabic and Greek philosophical ideas of urban human society, and, as an admirer of Plato and Aristotle, he was thinking of how humans can live together in happiness, peace, and harmony in the urban space (Tanabayeva et al. 2015).

Ibn Battuta's fourteenth-century perspective is also a fascinating reflection on the city for our understanding of today's urban world (Battuta and Lee 2013). Starting in Tangiers in 1325, at the age of 21, Ibn Battuta travelled 73,000 miles over a period of 24 years, visiting the major cities of the equivalent of 40 present day countries of the near East, Africa, and Asia. He travelled to amazing and diverse ancient cities such as Damascus, Delhi, Constantinople, Tangiers, Isfahan, Samarkand, Timbuktu, and Kabul: he describes the cities he saw as the locus of cultural advancement, architectural sophistication, and the epitome of human cultural, spiritual, and social "development."

In the globalized twenty-first century, with its everyday images of the often violent, unequal, and polluted towns and cities of the world, it is easy to forget that cities and towns may still hold cultural and spiritual appeal – and part of the puzzle of urbanization is the fostering of urban spaces that do and will nurture both material and mental well-being.

Historical and Twenty-First Century Urbanization

Why is it important to bring the wisdom of the past to debates about our current mental health situation? Perhaps because we, as a species, are at a turning point: in 2008, we reached a milestone of human demography, history – and culture – when the United Nations (UN) calculated that the world's population had become more urban than rural, and predicted that from this moment, the future home of the world's human population would be principally urban (United Nations DoEaSA, Population Division 2014).

The World Health Organization (WHO) also put a previously under-recognized health problem on the policy map in the twenty-first century – for the first time, after centuries of struggle to achieve physical health and well-being, mental health problems have emerged as a new major cause of ill health and disability (World Health Organization 2014a). In May 2013, the 66th World Health Assembly formally adopted the first-ever Mental Health Action Plan of the World Health Organization (World Health Organization 2014a). This does not mean that the only issue to be dealt with is mental health in cities – most cities and towns have severe burdens of disease and death related to the physical conditions of the urban space, but it does mean that attempts to create healthy cities must include in their definition of health the broad sense of mental and spiritual well-being.

In the twenty-first century we have both new and old urban challenges, some the same as those faced by the ancient thinkers and designers of cities, but some different. Cities and towns are now the locus of the world's most stark inequities (UN Habitat 2010). Inequities in urban living conditions have been a feature of urban life since towns and cities were developed. Now in the twenty-first century only a minority live in what could be described as "healthy" conditions, conducive to physical and mental well-being, and UN Habitat calculates that 863 million people live in "slums" (UN Habitat 2010). "Slum" is the pejorative term used to describe low-income informal settlements in the European past and is still the derogatory word used to describe settlements of millions of low-income urban citizens of Asia, Africa, and Latin America.

Cities also concentrate the world's wealth and the wealthy – we have the extremes of poverty and wealth concentrated in the city and townscapes of the world (UN Habitat 2010). Urban inequity may seem unconnected to mental well-being, but there is evidence that inequity is linked to levels of insecurity and violence in cities and towns – which have significant impacts on mental well-being (Decker et al. 2014; Dinizulu et al. 2014; Echeverria et al. 2014).

Two global challenges that past thinkers did not know, or have to face, are climate change and massive natural resource depletion. The effects of environmental changes on urban health and well-being are likely to be severe (McMichael 2005). Paradoxically, cities, or rather city dwellers, are largely responsible for these new risks they will face: the move from rural to urban corresponds with a growing problem of industrialization and material consumption, and cities and towns are a significant driver of human-induced climate change and resource depletion (UNEP 2012). Cities are the generators of environmental change, but also people in cities may be the hardest hit by these two environmental challenges: by climate change impacts as sea levels rise and extreme weather events batter towns and cities, and by resource scarcity as cities, often dependent on imported food, water, and other resources, find themselves without access to vital resources. In terms of climate change, coastal cities are particularly at risk, and large-scale disasters, displacement, and severe health impacts are predicted by the UN (UNEP 2012).

This chapter looks at urban mental health from the perspective of the past and future – considering particularly the evidence in the urban centers of Asia, Africa, and Latin America; learning from the architects, planners, and philosophers of the past; and looking at how these thinkers can help us face our newer challenges.

It is important to realize that there are widely different types of "urban" – and the complex social, material, and economic matrix of each urban setting can have a unique effect on its inhabitants' lives and mental well-being. For example, "urban" can mean anything from living in a small urban settlement of 50,000 or less to living in a megacity of over **ten million**, perhaps covering up to 100 km^2 – and the impacts of the urban experience are very different depending on the type of urban center involved. To put this in a historical context, it is notable that few cities in the past had more than 50,000 inhabitants, and, even today, many countries have smaller populations than the megacities we see of ten million and more inhabitants. The challenges and opportunities of urban life today are simply at a completely different scale.

Current urbanization patterns are also important. It is Asian, Latin American, and African cities that concentrate the majority of the world's urban population now, and will in future. Therefore, the specific conditions of the towns and cities of these regions will determine the overall urban mental health and well-being profile of the world. Particularly significant is the speed of change – African and Asian cities are the most important demographically, but they are also growing more rapidly than other cities with a subsequent massive pressure on resources.

The presence and magnitude of low-income urban settlements and the pattern of inequality between people in any city will also determine urban health patterns overall and the mental health problems within these patterns.

Summarizing the Evidence on Urban Mental Health

It is important to be clear about the term "mental health and illness." The WHO argues that mental health is an integral part of health, that mental health is more than the absence of mental disorders, and that mental well-being is determined by socioeconomic, biological, and environmental factors (World Health Organization

2014a). It then follows that the specific urban social, economic, and environmental context – the risk factors for health disorders or the protective factors for mental well-being – of each town and city will influence overall health profiles – and mental health profiles of a city, but also will influence distribution of mental health disorders and of mental well-being within each city.

To understand urban mental health we need to understand a wide range of disorders that include "depression, bipolar affective disorder, schizophrenia and other psychoses, dementia, intellectual disabilities and developmental disorders including autism"(World Health Organization 2014a). The extreme end of this spectrum also includes suicides, and for some authors it is valid to interpret violence as a mental health issue (World Health Organization 2014b).

Urban mental health disorders, unlike most other infectious or noncommunicable illnesses, do not necessarily have the same socioeconomic or environmental determinants in all cases and may not manifest in the same way. For example, "depression" for a wealthy and culturally privileged person in a city may have very different determinants to "depression" for a person living in a low-income, informal settlement or an urban minority person or an urban refugee. The depression of these different urban dwellers may also manifest itself in different ways, and its prevention or treatment may also need to be different.

The evidence base of urban mental health is extensive, but incomplete. In summary:

- While over 50% of the urban population lives in Asia, and the fastest growth is in Asia and Africa, almost all of the large-scale national studies of urban mental health, and most systematic reviews of mental health overall, are based on studies of European and North American populations (Walker et al. 2015).
- There are limited data on mental health in towns and cities of Asia, Africa, and Latin America (Cortina et al. 2012).
- While over half the world's urban population lives in towns of less than 500,000, there are few studies on mental health in smaller towns and cities.
- While some cities may have up to 60% of their population in low-income settlements, there are few studies of mental health patterns within cities, nor specific studies of mental health in low-income settlements (Fryers et al. 2003).
- Almost all studies are cross-sectional studies of mental health. There are very few data on long-term trends in mental health in towns and cities (Richter et al. 2008).

In terms of evidence, one of the main debates is about whether living in towns and cities is better or worse for mental health than living in rural areas. There is a huge split in academic opinion (Breslau et al. 2014; Reeves et al. 2013; Paykel et al. 2003; Nepomuceno et al. 2016). Perhaps the problem is that analysts ask the wrong question: when urban areas are so big and conditions within cities vary so widely, these huge aggregations of "urban-rural" make little sense. Studies of *intra-urban* health differentials expose evidence that both mortality and morbidity indicators for the urban poor are worse than those for their wealthier urban neighbors and are often worse than rural indicators (Stephens and Satterthwaite 2008; Stephens 2012; UN Habitat 2010).

Evidence from Systematic Reviews

Systematic reviews, which are scientific reviews that compile all the current evidence on any theme, are often the most helpful summary of current evidence. This is also true of evidence on urban mental health, with the caveat that most urban mental health systematic reviews accessible in English are based on studies in wealthier countries with larger databases and often miss studies published in other languages or exclude studies from poorer regions on the grounds of quality.

Recent systematic reviews suggest the following trends and patterns in common mental disorders such as depression, addictions, and anxiety.

- **Trends in urbanization and trends in mental health**: There is a strong trend towards urbanization over time, but does that translate into evidence of increasing mental disorders over time? A recent systematic review of long-term trends in mental health disorders pooled data from 44 major international studies and found that "neither general mental disorders nor specific disorders such as depression, anxiety, addictions or eating disorders showed a clearly increasing trend. Child and adolescent psychiatric disorders also failed to show an increasing trend" (Richter et al. 2008).
- **Urban-rural patterns of mental health disorder**: There is a major debate about urban-rural differences in mental health, but there is little evidence of a genuine independent effect of "urban" living on mental health disorders. A recent systematic review of urban-rural differences in common mental health disorders looked at 20 large-scale population studies, principally from mostly urban Europe and North America (with two studies from other countries: Australia and South Korea). Researchers found that, in pooled analysis, and after controlling for confounders (such as age and sex, socioeconomic conditions or relationship status), there was a slight increase in prevalence in overall mental disorders in urban areas compared to rural areas. Interestingly, the majority of the studies showed no significant urban-rural difference for mood disorders (66% of studies showed no difference), anxiety (67%), or substance abuse (69%). Intriguingly, Belgium showed an opposite trend to other countries with significantly higher prevalence of mental health disorders in rural areas (Peen et al. 2010).
- **Urban poverty, mental health, and social inequality**: A recent systematic review, again of large databases covering the urban European population, looked at the links of social inequality and mental health disorders and found that social conditions within cities were more important than the actual "urban" effect: "evidence of an association between one or more markers of less privileged social position and higher prevalence of common mental disorders . . . the more consistent associations were with unemployment, less education and low income or material standard of living" (Fryers et al. 2003).
- **Urban disasters and mental health**: The future threats to urban health from climate change are predicted to involve extreme weather events and increased risk of disaster. There are several systematic reviews looking at the impacts of disasters on mental health disorders, particularly post-traumatic stress disorders

(PTSD), depression, and anxiety. A wide-ranging systematic review looking at responses to community disasters found that "many (11%–38%) distressed individuals presenting for evaluation at shelters and family assistance centers have stress-related and adjustment disorders; bereavement, major depression, and substance use disorders were also observed, and up to 40% of distressed individuals had preexisting disorders" (North and Pfefferbaum 2013). Looking specifically at mental health responses of elderly people to disaster, a recent systematic review found that "older adults were 2.85 times less likely to experience post-traumatic stress disorder (PTSD) symptoms following a disaster (95% CI: 1.42–5.70) when compared with younger adults. There was no statistically significant difference in terms of anxiety and depressive symptoms" (Siskind et al. 2016; Parker et al. 2016).

Evidence from Asia, Africa, and Latin America

Given the focus of most systematic reviews on large-scale surveys from wealthier countries, it is important to look at data from regions that are less represented in the literature, and particularly from cities and towns in Asia, Africa, and Latin America.

Urban-Rural Differences

One recent study in Northeast Brazil looked at urban-rural differences in common mental disorders. Researchers found "significant differences between the inhabitants of the rural and urban communities regarding well-being and the prevalence of common mental disorders (CMD), with a higher average well-being score in the rural context; the urban sample had a higher average regarding the prevalence of CMD" but also found that income was an important confounder of the rural/urban difference (Nepomuceno et al. 2016). A study of urban-rural differences in depression in Nigeria found significantly higher rates of depression in rural than urban areas, again with links that suggest the importance of socioeconomic conditions in each area, which are more important than measures of "urbanicity" or "rurality" (Amoran et al. 2007). Interestingly, a study looking at urban-rural differences between a low-income rural and low-income peri-urban community in South Africa, that "a high prevalence of mental health and substance abuse problems was observed in both communities, with highest rates in the peri-urban township" (Havenaar et al. 2008). These findings, on the importance of the context of socioeconomic conditions in either rural or urban settings, are supported by studies in Europe, for example, "there are considerable British urban-rural differences in mental health, which may largely be attributable to more adverse urban social environments" (Paykel et al. 2003).

Impacts on Different Age Groups

In terms of urban-rural mental health disorders in different age groups, there is some evidence that urban adolescents and the urban elderly are particularly vulnerable to disorders such as depression, but evidence suggests that the urban-rural difference is often explained by sociodemographic factors (Mojs et al. 2012; Hesketh and Ding 2005).

For example, a study in Japan found that both rural and urban elderly experienced depression linked to living alone, being unemployed, sleep disturbance, suicidal ideation, financial strain, and poor social support (Abe et al. 2012). A recent study of mental health in the rural and urban elderly in India found that urban residence, female gender, higher age, nuclear family, poverty, and functional and cognitive impairment were significantly associated with depression (Sengupta and Benjamin 2015). A study in Korea of the rural and urban elderly found that mental social support deficits were most strongly associated with depression in people with a lifetime rural residence, followed by urban residents with a rural birthplace (Kim et al. 2004).

Urban Violence and Mental Illness

Perhaps the indicators of the most extreme mental distress can be seen from both personal and interpersonal violence (Middleton et al. 2003). Interpersonal urban violence is one of the leading causes of death in some cities, particularly in Latin America and Africa. There is also evidence that levels of interpersonal violence in cities and towns have a profound impact on mental well-being and on levels of mental disorders, particularly post-traumatic stress, not just for the victims and the witnesses but also for the perpetrators of violence (Kadra et al. 2014; Kliewer and Lepore 2015). Street children and disadvantaged young men are particularly at risk (Diez Roux et al. 2007).

Urban Suicide?

Suicides are an expression of extreme mental distress. In 2015, WHO released a major global report on suicide and reports "Risks linked to the community and relationships include war and disaster, stresses of acculturation (such as among indigenous peoples or displaced persons), discrimination, a sense of isolation, abuse, violence and conflictual relationships . . . risk factors at the individual level include previous suicide attempts, mental disorders, harmful use of alcohol, financial loss, chronic pain and a family history of suicide" (World Health Organization 2014b). This implies that urban areas might pose a particular set of suicide risks for vulnerable individuals, but studies from multiple settings internationally indicate that, at present, rural areas have significantly higher rates of suicide, particularly for young or working age men, and the elderly (Razvodovsky and Stickley 2009; Law et al. 2014; Wallis 2015).

Policy Approaches to Urban Mental Health: Learning from the Past and Looking to the Future

Looking at the diverse cities and towns of the world and their seeming chaos, it would appear that planners and architects have never thought about the mental well-being of urban citizens and that they have failed to think about inequality or environmental change. Yet, any health specialist with this assumption needs to understand that the fundamental aim of urban philosophers, planners, and architects has always been to conceptualize, plan, and manage towns and cities for the fulfillment of both mental and physical health of urban citizens. The challenge has always been to marry this aim with the reality of the urban space and the people who live in it.

As Anthony Churchill reflected (Box 1), when writing the preface to a major conference on Cities in Conflict (Lea and Courtney 1985) – in his then role as head of the World Bank's urban program in Asia – cities are created by people and each city is a creation of the interaction of social, economic, cultural, and political imperatives of each moment of city development. In this context, the role of planners and architects is profoundly linked to the historical and cultural context of the cities in which they live.

Box 1 Cities Are Built by *People*

This is an obvious but often forgotten statement. When we see the cities of the developing world, we cannot help but be impressed by the restless energy of their teeming millions. Dwellings and places of work rise in an endless and apparently unstructured fashion, almost before our very eyes. Yet what urban professionals – whether they be architects, planners, or economists – can look upon this chaotic scene without the almost compulsive urge to impose their own sense of order? Fortunately for these cities, however, the professionals usually fail.

What we have all forgotten is that the city is a creation of economic, political, and social necessity. It is the way it is for very good reasons, and unless we are prepared to address the underlying rationale for the existence of urban areas, attempts to bring order and reason into the processes of urbanization will fail – if we are lucky. If we are unlucky, we may well damage the urban fabric that we are so anxious to improve.

Anthony Churchill in (Lea and Courtney 1985) Preface speech Cities in Conflict Conference.

Learning from Architects, Philosophers, and Planners of the Past

What we have to learn, if anything, from history? Two ideas may illustrate this.

The first is the concept of the Ideal or Utopian City – combining spiritual and material well-being – a political philosophy and practice developed over centuries and put into practice in cities throughout history (Fourier 1902; Steiert 2002). The second is another utopian ideal – the Garden City – which conceives the city as a place at one with nature – first articulated centuries ago in the ancient cities such as the Persian city of Isfahan (Akhouchekian 2009).

Ideal and Utopian Cities

The Ideal or Utopian City has been an aspirational way of addressing urbanization almost since the genesis of the most ancient cities.

Essentially an Ideal or Utopian City planner looks at the architectural form of the city but most importantly explores how urban form can be designed to social and spiritual goals. In terms of urban mental health, the Utopian City philosophers and architects are important because they were perhaps the deepest thinkers about the need for city forms to generate mental and spiritual well-being, and many of these thinkers explicitly addressed issues of happiness, work, the use of color and form to create peaceful spaces, and the ideas of virtuous and congenial living (Arjmandi et al. 2011). The Virtuous City, conceived by Al-Farabi in the first century, is one of the earliest explorations of this idea, but architects and planners have drawn on the same ideas for centuries.

Philosophers of ancient western Asia have been among the most influential thinkers about ideal design of cities. Persia has centuries of historical development of urban spaces and has been one of the most important civilizations to think about specific urban design as the creation of a spiritual urban space, inspired of religious and philosophical ideologies, which either created energy or decreased environmental stressors. Their view was linked to Islamic principles: Persian philosophers such as Abu Nasr Farabi (872–950), Avi Sina (980–1037), and Nasir al-Din Tusi (1201–1274) explored the characteristics of an Ideal Spiritual City (Akhouchekian 2009). The essential elements of this design aimed to create:

- A place which ensures social security for its inhabitants
- A place where the virtuous live
- A place that is governed by wise individuals
- A place which contributes to spirituality and religion
- A place that ensures urban harmony of all environmental, social, and cultural conditions
- A place with an emphasis on public spaces and services
- A place where all citizens have and observe equal human rights

In the fifteenth century, several centuries after the Persian philosophers, Filarete, an Italian architect–philosopher was commissioned to design the ideal city of Sforzinda (Lang 1972). Filarete centered his ideas on the inhabitants' physical protection from external (human) threats: This idea of security defined the shape of the city and part of its internal structure. He developed a star-shaped, walled scheme surrounded by a moat, with all the entrance gates observed by two watchtowers. But Filarete was not only interested in the provision of security, he also created spaces where human activities (leisure, commerce, and spiritual) would take place in an organized and calm manner (Steiert 2002).

Four centuries later, the Phalanstère, designed by the nineteenth-century French architect Charles Fourier, followed these ideas of ideal spiritual design. Fourier's aim was to create an ideal of urban design and social well-being. Social happiness and success would come from cooperative work of organized communities. These communities would not exceed 1500–1600 inhabitants, occupying a one-league square piece of land, in which a multipurpose building, called the Phalanstery would sit (Fourier 1902). Utopian philosophical groups welcomed the idea of the Phalanstery, and several Phalansteries were built and inhabited in the USA in the late nineteenth century. This idea was then adapted by le Courbusier, in the mid-twentieth century, when he designed the Unité d'Habitation, a self-contained commune, at Marseilles (Coleman 2005).

Nature, Spiritual Well-Being, and Mental Health in Cities

The second key idea, again from another branch of Utopian philosophy and architecture, is the idea of the Green or Garden City. There is considerable evidence that access to urban green spaces has a positive impact on both urban mental well-being and urban violence (Toronto Public Health 2015). This idea also has a long history and again comes from some of the most ancient inhabited regions and cities of the world. This idea is perhaps not only important for mental well-being but also for the challenge of environmental sustainability – where Garden cities are seen as a potential solution to this newer problem.

One of the most ancient cities of the world, Isfahan, was already a Garden City by the early part of the first millennium. Isfahan is important as an example because it has been seen as an icon of religious and social tolerance and also was conceived and developed as a city of natural and man-made beauty. It is one of the highest-ranking examples of Islamic-Iranian architecture inspired from philosophic Utopia (Rostami et al. 2015). Other cities in Persia were modeled on the same ideas. Using natural areas such as green space, water, soil (as places of identity, memory, and belonging), the philosophers, and planners of Persian cities believed that "green" design facilitated positive psychological and social effects in the city inhabitants.

Several centuries later, these concepts reemerged with the European Garden City idealists, including the UK Garden City movement, conceived and implemented by the late nineteenth/early twentieth century architect Ebenezer Howard (Howard 1965). These cities were developed in response to the late nineteenth-century pollution and poverty within cities, which were regarded as a major concern by most urban planners (Box 2). Ebenezer Howard believed that "The garden cities were to be slum-less and smoke-less, with good-quality housing, planned development, large amounts of open space and green belts separating one settlement from another" (Howard 1965).

Box 2 The Birth of UK Garden City Movement

"We are becoming a land of great cities. Villages are stationary or receding; cities are enormously increasing. And if it be true that great cities tend more and more to become the graves of the physique of our race, can we wonder at it when we see the houses so foul, so squalid, so ill-drained, so vitiated by neglect and dirt?" Ebenezer Howard 1902 *Garden Cities of Tomorrow.*

Nine garden cities were built in the UK, and there are now over 40 garden cities in countries as diverse as Finland, Japan, Canada, the USA, Australia, India, South Africa, Malaysia, Argentina, and Venezuela (Toronto Public Health 2015).

We chose a tiny snapshot of the wealth of human thought applied over centuries to create spiritually and mentally healthy urban spaces (Rostami et al. 2015; Leach 1997; Coleman 2005). Do these effects persist today? A recent study in four of these ancient Iranian cities found that their historical Persian gardens still promote social attachment, increase public gatherings, as well as facilitating public services, employment, and local economic activities. Garden cities are regarded as having among the best quality of life indicators for residents (Toronto Public Health 2015).

Conclusions

The modern myth that "urban" life inexorably causes mental illness continues and is likely to persist (Barber 2011). But the world is still urbanizing and people are still moving to towns and cities, particularly in Asia and Africa – and many people move to improve their lives. The important thing then becomes the urban experience. Evidence suggests that complex factors such as poverty, social relations, and employment are critical determinants of mental illness or well-being, overriding the simple "Urban = Unhappiness" analysis.

With the massive challenges of towns and cities today, it is easy to see how the practical issues of urban shelter, water, sanitation, and transport may overwhelm modern urban planners. It is also easy to see how modern urban health professionals get overwhelmed with urban illness, and urban *mental* illness becomes just another thing to worry about.

But it is also the scale of today's urban challenges that demands that we look for all the ideas and wisdom, including from all periods of urban thinking. Do thinkers from the past take us forward? Of course – if we are prepared to look back, learn, and move forward. What does this mean in practice?

We avoid reinventing the wheel – Every new generation likes to think that they are inventing something new, and this is true of the modern rediscovery of green cities or the rediscovery of the links of urban design and mental health. Yet, both the Utopian and Garden City movements experimented with the same ideas, and learning what worked and why, and what did not work and why, is invaluable for the improvement of today's versions of these concepts. For example, garden cities have become a worldwide phenomenon and are very popular with those who live in them. But in many contexts they have also become home to the wealthiest urbanites, and the challenge has always been to make these as inclusive as Howard and his predecessors originally intended.

We gain the courage to dream big – In the context of today's challenges, perhaps one of the most significant things to draw from urban thinkers of the past is their vision – their ambition and courage of vision about the urban space. They were not afraid to think big. They too were living with massive urban inequalities, with "slums" and, in most towns and cities, in terms of health – shorter life expectancies, and constant threat of epidemics, illness, and violence. Yet these urban thinkers had the courage to design the Virtuous City, the Utopian City, and the Garden City. They had the sagacity to see the urban space not just as a physical planning challenge but also as a challenge of the city of spiritual, mental, and moral well-being, and it was this that guided their overarching vision of urban design.

Ancient thinking for a modern dilemma – Creating equitable and sustainable cities is cast as a uniquely twenty-first-century dilemma, and mental well-being is a key part of this. Yet over 1000 years ago, Al-Farabi was already considering this very modern predicament. In his view "authentic happiness" in the Virtuous City is found in ethical, modest living, and in congenial, honest interaction with other people. Inauthentic happiness stems from greed, arrogance, material wealth, and ostentation. This resonates with the evidence of the underlying causes of mental ill health in cities and speaks to the solution of the social inequality that we see so starkly in our cities. Al-Farabi's view on authentic human happiness is also highly relevant for our currently unsustainable cities – we seem trapped in a development model of unquenchable material consumption, which promises happiness for all, but so far has delivered happiness for the few, and planetary destruction for all.

Avoiding Hubris – Finally, learning from history involves humility – we are not the first generation to think about the city as a space for mental well-being or illness, and we will not be the last. Learning from past urban thinkers means that we take the best of the lessons from the past, learn what to adapt, and move forward. People in the future may do the same with our experiments.

In the context of our planetary crisis and our urban future, there has never been a better time to take every lesson we can from the brave experimenters of the past. We go full circle when we realize that the urban future needs to be both socially and environmentally sustainable – quite simply because our human future depends on it. We go full circle too because Al-Farabi along with many philosophers teaches us also what we have known for thousands of years: beyond a certain point of basic necessity, material "well-being" has no link to either individual or community mental health or well-being – or human happiness.

Cross References

▶ Green Environment and Mental Health in the City
▶ History of Urban Mental Illness
▶ Urban Design and Mental Health
▶ Urban-Rural Differences in Major Mental Health Conditions

References

Abe Y, Fujise N, Fukunaga R, Nakagawa Y, Ikeda M (2012) Comparisons of the prevalence of and risk factors for elderly depression between urban and rural populations in Japan. Int Psychogeriatr 24(8):1235–1241. doi:10.1017/S1041610212000099

Akhouchekian A (2009) From philosophic utopia to developmental utopia. Paper presented at the The first conference on Islamic Utopia, Isfahan

Amoran O, Lawoyin T, Lasebikan V (2007) Prevalence of depression among adults in Oyo state, Nigeria: a comparative study of rural and urban communities. Aust J Rural Health 15(3):211–215. doi:10.1111/j.1440-1584.2006.00794.x

Arjmandi M, Tahir M, Shabankareh H, Shabani MM, Mazaheri F (2011) Psychological and spiritual effects of light and color from Iranian traditional houses on dwellers. J Soc Sci Humanit 6(2):288–301

Barber N (2011) Depression stalks the city – why is urban life so depressing? Psychology Today. https://www.psychologytoday.com/blog/the-human-beast/201109/depression-stalks-the-cities

Battuta I, Lee S (2013) The travels of Ibn Battuta: in the near east, Asia and Africa, 1325–1354. Dover Publications, Mineola

Breslau J, Marshall GN, Pincus HA, Brown RA (2014) Are mental disorders more common in urban than rural areas of the United States? J Psychiatr Res 56:50–55. doi:10.1016/j.jpsychires.2014.05.004

Coleman N (ed) (2005) Utopias and architecture. Routledge, New York

Cortina MA, Sodha A, Fazel M, Ramchandani PG (2012) Prevalence of child mental health problems in sub-Saharan Africa: a systematic review. Arch Pediatr Adolesc Med 166(3): 276–281. doi:10.1001/archpediatrics.2011.592

Decker MR, Peitzmeier S, Olumide A, Acharya R, Ojengbede O, Covarrubias L, Gao E, Cheng Y, Delany-Moretlwe S, Brahmbhatt H (2014) Prevalence and health impact of intimate partner violence and non-partner sexual violence among female adolescents aged 15–19 years in vulnerable urban environments: a multi-country study. J Adolesc Health 55(6 Suppl): S58–S67. doi:10.1016/j.jadohealth.2014.08.022

Diez Roux AV, Green Franklin T, Alazraqui M, Spinelli H (2007) Intraurban variations in adult mortality in a large Latin American city. J Urban Health 84(3):319–333. doi:10.1007/s11524-007-9159-5

Dinizulu SM, Grant KE, McIntosh JM (2014) The influence of nondisclosure on the mental health of urban African-American adolescents exposed to community violence. J Prev Interv Community 42(3):208–220. doi:10.1080/10852352.2014.916583

Echeverria SE, Luan Kang A, Isasi CR, Johnson-Dias J, Pacquiao D (2014) A community survey on neighborhood violence, park use, and physical activity among urban youth. J Phys Act Health 11(1):186–194. doi:10.1123/jpah.2012-0023

Fourier C (1902) Doctrina social, el Falansterio. E. Rodríguez Serra, Madrid

Fryers T, Melzer D, Jenkins R (2003) Social inequalities and the common mental disorders: a systematic review of the evidence. Soc Psychiatry Psychiatr Epidemiol 38(5):229–237. doi:10.1007/s00127-003-0627-2

Havenaar JM, Geerlings MI, Vivian L, Collinson M, Robertson B (2008) Common mental health problems in historically disadvantaged urban and rural communities in South Africa: prevalence and risk factors. Soc Psychiatry Psychiatr Epidemiol 43(3):209–215. doi:10.1007/s00127-007-0294-9

Hesketh T, Ding QJ (2005) Anxiety and depression in adolescents in urban and rural China. Psychol Rep 96(2):435–444. doi:10.2466/pr0.96.2.435-444

Howard E (1965) Garden cities of to-morrow. MIT Press, Massachusetts

Kadra G, Dean K, Hotopf M, Hatch SL (2014) Investigating exposure to violence and mental health in a diverse urban community sample: data from the South East London Community Health (SELCoH) survey. PLoS One 9(4):e93660. doi:10.1371/journal.pone.0093660

Kim JM, Stewart R, Shin IS, Yoon JS, Lee HY (2004) Lifetime urban/rural residence, social support and late-life depression in Korea. Int J Geriatr Psychiatry 19(9):843–851

Kliewer W, Lepore SJ (2015) Exposure to violence, social cognitive processing, and sleep problems in urban adolescents. J Youth Adolesc 44(2):507–517. doi:10.1007/s10964-014-0184-x

Lang S (1972) Sforzinda, Filarete and Filelfo. J Warburg Courtauld Inst 35:391–397. doi:10.2307/750940

Law CK, Snider AM, De Leo D (2014) The influence of deprivation on suicide mortality in urban and rural Queensland: an ecological analysis. Soc Psychiatry Psychiatr Epidemiol 49(12):1919–1928. doi:10.1007/s00127-014-0905-1

Lea J, Courtney J (1985) Cities in conflict: studies in the planning and management of Asian cities. World Bank, Washington. http://www-wds.worldbank.org/external/default/WDSContentServer/WDSP/IB/1999/11/10/000178830_98101911003171/Rendered/PDF/multi_page.pdf

Leach N (ed) (1997) Rethinking architecture: a reader in cultural theory, 1st edn. Routledge, London

McMichael A (2005) Global environmental changes, climate change and human health. In: Lee K, Collin J (eds) Global change and health. Understanding public health, 1st edn. Open University Press, Maidenhead

Middleton N, Gunnell D, Frankel S, Whitley E, Dorling D (2003) Urban-rural differences in suicide trends in young adults: England and Wales, 1981–1998. Soc Sci Med 57(7):1183–1194

Mojs E, Warchol-Biedermann K, Glowacka MD, Strzelecki W, Ziemska B, Marcinkowski JT (2012) Are students prone to depression and suicidal thoughts? Assessment of the risk of depression in university students from rural and urban areas. Ann Agric Environ Med 19(4):770–774

Nepomuceno BB, Cardoso AA, Ximenes VM, Barros JP, Leite JF (2016) Mental health, well-being, and poverty: a study in urban and rural communities in northeastern Brazil. J Prev Interv Community 44(1):63–75. doi:10.1080/10852352.2016.1102590

North CS, Pfefferbaum B (2013) Mental health response to community disasters: a systematic review. JAMA 310(5):507–518. doi:10.1001/jama.2013.107799

Parker G, Lie D, Siskind DJ, Martin-Khan M, Raphael B, Crompton D, Kisely S (2016) Mental health implications for older adults after natural disasters – a systematic review and meta-analysis. Int Psychogeriatr 28(1):11–20. doi:10.1017/S1041610215001210

Paykel E, Abbott R, Jenkins R, Brugha T, Meltzer H (2003) Urban-rural mental health differences in Great Britain: findings from the National Morbidity Survey. Int Rev Psychiatry 15(1–2):97–107. doi:10.1080/0954026021000046001

Peen J, Schoevers RA, Beekman AT, Dekker J (2010) The current status of urban-rural differences in psychiatric disorders. Acta Psychiatr Scand 121(2):84–93. doi:10.1111/j.1600-0447.2009.01438.x

Razvodovsky Y, Stickley A (2009) Suicide in urban and rural regions of Belarus, 1990–2005. Public Health 123(1):27–31. doi:10.1016/j.puhe.2008.10.003

Reeves WC, Lin JM, Nater UM (2013) Mental illness in metropolitan, urban and rural Georgia populations. BMC Public Health 13:414. doi:10.1186/1471-2458-13-414

Richter D, Berger K, Reker T (2008) Are mental disorders on the increase? A systematic review. Psychiatr Prax 35(7):321–330. doi:10.1055/s-2008-1067570

Rostami R, Lamit H, Khoshnava S, Rostami R, Rosley M (2015) Sustainable cities and the contribution of historical urban green spaces: a case study of historical Persian gardens. Sustainability 7(10):13290

Sengupta P, Benjamin AI (2015) Prevalence of depression and associated risk factors among the elderly in urban and rural field practice areas of a tertiary care institution in Ludhiana. Indian J Public Health 59(1):3–8. doi:10.4103/0019-557X.152845

Siskind DJ, Sawyer E, Lee I, Lie DC, Martin-Khan M, Farrington J, Crompton D, Kisely S (2016) The mental health of older persons after human-induced disasters: a systematic review and meta-analysis of epidemiological data. Am J Geriatr Psychiatry 24(5):379–388. doi:10.1016/j.jagp.2015.12.010

Steiert R (2002) The ideal city: utopias and projects: a historical sketch. Soz Wiss Inf 3(1):38–47

Stephens C (2012) Urban inequities; urban rights: a conceptual analysis and review of impacts on children, and policies to address them. J Urban Health 89(3):464–485. doi:10.1007/s11524-011-9655-5

Stephens C, Satterthwaite D (2008) Urban health in developing countries. In: Heggenhougen K, Quah S (eds) International encyclopaedia of public health, vol 1. Elsevier, Oxford, pp 452–463

Tanabayeva A, Massalimova A, Mukhatova O, Alikbayeva M, Alikbayeva Z (2015) Al-Farabi's humanistic principles and "virtuous city". Eur Proc Soc Behav Sci. doi:http://dx.doi.org/10.15405/epsbs.2015.01.14. http://www.futureacademy.org.uk/files/menu_items/other/12j.pdf

Toronto Public Health (2015) Green city: why nature matters to health – an evidence review. Toronto Public Health, Toronto

UN Habitat (2010) State of the world's cities 2010/2011: bridging the urban divide. Earthscan, London. http://www.unhabitat.org/pmss/listItemDetails.aspx?publicationID=2917

UNEP (2012) One planet, how many people? A review of Earth's carrying capacity. A discussion paper for the year of Rio+20 (2012), UNEP, Nairobi. Accessed 12 Nov 2013

United Nations DoEaSA, Population Division (2014) World urbanization prospects: the 2014 revision, highlights (ST/ESA/SER.A/352)

Walker ER, McGee RE, Druss BG (2015) Mortality in mental disorders and global disease burden implications: a systematic review and meta-analysis. JAMA Psychiat 72(4):334–341. doi:10.1001/jamapsychiatry.2014.2502

Wallis L (2015) Rural youths commit suicide almost twice as often as urban counterparts. Am J Nurs 115(6):15. doi:10.1097/01.NAJ.0000466301.39777.65

World Health Organization (2014a) Mental health atlas 2014. http://www.who.int/mental_health/evidence/atlas/mental_health_atlas_2014/en/. Accessed 12 May 2016

World Health Organization (2014b) Preventing suicide: a global imperative. http://apps.who.int/iris/bitstream/10665/131056/1/9789241564779_eng.pdf?ua=1. Accessed 12 May 2016

Larger Cities of the World

1. Los Angeles, 2. Madison, 3. Baltimore, 4. New York, 5. Mexico City, 6. Santiago, 7. Buenos Aires, 8. Porto Alegre, 9. Sao Paulo, 10. Rio de Janeiro, 11. London, 12. Paris, 13. Amsterdam, 14. Verona, 15. Copenhagen, 16. Moscow, 17. Istanbul, 18. Cairo, 19. Lagos, 20. Kinshasa, 21. Johannesburg, 22. Tehran, 23. Delhi, 24. Mumbai, 25. Bangalore, 26. Calcutta, 27. Dhaka, 28. Beijing, 29. Shanghai, 30. Guangzihou, 31. Hong Kong, 32. Kobe, 33. Osaka, 34. Tokyo, 35. Jakarta, 36. Sydney.

Made and printed with permission from medical student Christian Kirkedal

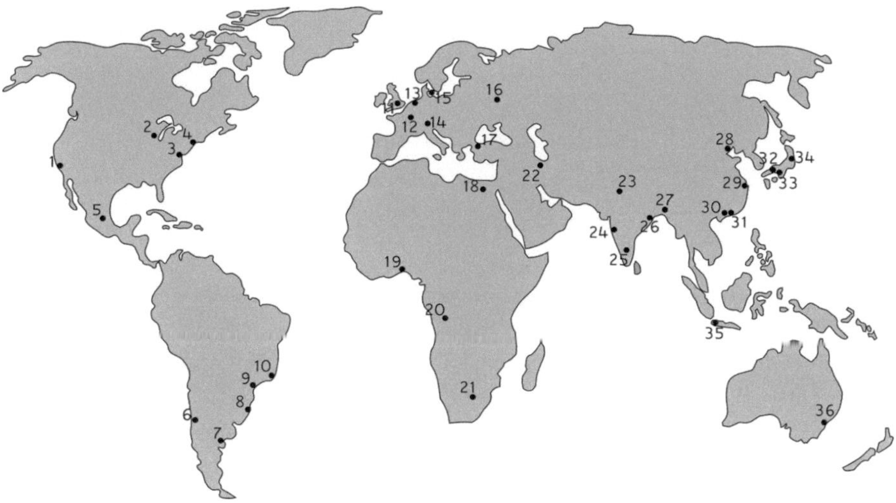

N. Okkels et al. (eds.), *Mental Health and Illness in the City*, Mental Health and Illness Worldwide, https://doi.org/10.1007/978-981-10-2327-9

Index

N. Okkels et al. (eds.), *Mental Health and Illness in the City*, Mental Health and Illness Worldwide, https://doi.org/10.1007/978-981-10-2327-9

FSC
www.fsc.org
MIX
Papier aus verantwortungsvollen Quellen
Paper from responsible sources
FSC® C105338

If you have any concerns about our products,
you can contact us on
ProductSafety@springernature.com

In case Publisher is established outside the EU,
the EU authorized representative is:
Springer Nature Customer Service Center GmbH
Europaplatz 3, 69115 Heidelberg, Germany

Printed by Libri Plureos GmbH
in Hamburg, Germany